# Engine Testing

Electrical, Hybrid, IC Engine and Power Storage Testing and Test Facilities

# Engine Testing

## Electrical, Hybrid, IC Engine and Power Storage Testing and Test Facilities

Fifth Edition

Anthony J. Martyr

David R. Rogers

ELSEVIER

Butterworth-Heinemann
An imprint of Elsevier

Butterworth-Heinemann is an imprint of Elsevier
The Boulevard, Langford Lane, Kidlington, Oxford OX5 1GB, United Kingdom
50 Hampshire Street, 5th Floor, Cambridge, MA 02139, United States



**British Library Cataloguing-in-Publication Data**
A catalogue record for this book is available from the British Library

**Library of Congress Cataloging-in-Publication Data**
A catalog record for this book is available from the Library of Congress

ISBN: 978-0-12-821226-4

For Information on all Butterworth-Heinemann publications
visit our website at https://www.elsevier.com/books-and-journals

*Cover credit:*
*Main Image* – Reproduced with permission from ZF Friedrichshafen AG
Inset image (far left) – Reproduced with permission from DRÄXLMAIER Group
Inset image (middle left) – Reproduced with permission from Kistler Instrumente AG
Inset image (middle right) – Reproduced with permission from M-Sport Ltd
Inset image (far right) – Reproduced with permission from Taylor Dynamometer

*Publisher:* Matthew Deans
*Acquisitions Editor:* Carrie Bolger
*Editorial Project Manager:* John Leonard
*Production Project Manager:* Nirmala Arumugam
*Cover Designer:* Mark Rogers

Typeset by MPS Limited, Chennai, India

# Contents

# About the authors

*Anthony J. Martyr* has been either the sole or a coauthor of all editions of "Engine Testing." For the last 50 years, he has held senior technical positions in companies, internationally involved in the design and testing of automotive and marine powertrains. His published works include a book on Project Management, and papers on subjects covering dynamometry and the international transfer of technology.

*David. R. Rogers* has been operating in technical and commercial roles, in automotive research and development for over 25 years. He has expert knowledge and experience in the areas of powertrain testing, instrumentation, test systems, tools, and workflows. Professionally he is a registered Chartered Engineer and European Engineer, and a Fellow of the Institution of Mechanical Engineers.

# Preface to the 5th edition of *Engine Testing*

One of two declared errors in the fourth edition of this book was the statement that it would be the last edition of this series of books in which I would play any creative role. I had not anticipated the rapidity of the development of electric and hybrid power trains or that I would continue in my involvement in the power train test industry. A generation of electrically qualified "incomers," sometimes employed by companies with little or no history in automotive test facility design and operation, has appeared. At the same time the established engine test industry has had to deal with ever more demanding forms of emission testing and absorb a rapidly changing form of electrical prime mover and energy storage system; all of which have demanded development in automotive test facilities and their staffs' training.

The original purpose of this book has always been, and still is, to give practical advice and relevant background information to the widest possible group of people involved in any way in the "design, building, modification, and use of power train test facilities." It is the intention of this edition to expand the breadth of the technical information and advice so that it is supportive of the world-wide test industry of the 2020s.

At the time the first edition was published in 1995, it was still just possible for a well-trained engineer with a life-time's practical experience in the testing of internal combustion engines (ICE) to have mastered all of the knowledge required to write every chapter—in 2020 such a feat is inconceivable. Therefore this edition has been co-authored with Dave Rogers, whose expertise in the modern techniques of data collection and management, in addition to experience gained over 20 plus years in power train testing and systems, will be evident to readers. Our collaboration has been supported, and the book made possible by the generous support and active assistance of many people in the industry. I want to make particular mention of the following:

Nickolas (Nick) Birger of the AVL Technical Centre in Coventry for his help and to his colleague Balaji Jagadeesan for allowing accompanied access to their very modern test facilities and answering detailed queries.

Martin Hughes and Phil Stones of Millbrook Proving Ground for their technical support before, during, and after a visit to their site which features innovative use of modular buildings to create test chambers for both e-axles and battery packs.

Mark Emery of Horiba Mira Ltd without whose expertise and generously given advice, covering the wide range of EMC legislation and testing, the relevant chapter would have been impossible to complete.

The Directors and staff of Cambustion for their help with advanced and high-speed emission testing technology. Bob McFarland of SAKOR Technologies Inc. is an industrial colleague of many years and has patiently represented the current state of the automotive test industry in the United States via emailed queries throughout the writing of this edition. Adam Muencheberg of Dyne Systems has also supplied images and information on uniquely American test products.

From India, Rujuta Jagtap, now the Executive Director of SAJ Test Plant and daughter of company's founder, my old friend Prakash Jagtap, has supplied images and information of test products peculiar to the important and developing Indian market.

Dave Rogers wishes to acknowledge the help and support of Declan Allen of Horiba MIRA and Thomas Trebitsch from ZF Test Systems. Particular thanks must go to Dr. Byron Mason and Jack Prior at Loughborough University, for their valuable assistance in the development of the Chapter covering data management.

There are many other, unnamed excolleagues and acquaintances within the automotive test industry, both suppliers and customers, who have unknowingly helped in the creation of this book, through both their questions and their answers, we acknowledge them all.

Finally, two errors were mentioned at the start of this preface as being present in the fourth edition. The second error would seem to prove that we authors, and even proofreaders can read the word we expect to find rather than the one that is written. On page 135 of the 4$^{th}$ edition, there is Fig. 6.11 the title of which should have read "Psychrometric Chart," instead it reads "Psychometric Chart." The first (intended) word means *the science of studying thermodynamic properties of moist air*. The second word (used in error throughout the text) refers to the *field of study concerned with the theory and technique of psychological testing and measurement*, not a field of testing in which I, Dave, or most of the readers, have any expertise!

Tony Martyr, Clun, England, August 2020
Dave Rogers, Winterthur, Switzerland, August 2020

# Introduction

This is the fifth edition of the series of books having the prime title of "Engine Testing."

It is acknowledged by the authors that, due to personal requirements and the availability of a downloadable, non-paper version, many users will not read the whole book. Therefore some readers will notice occasional repetition of critical advice given in more than one chapter; in order to minimize this, users will find that cross-referencing used where appropriate.

Starting in 1995, the volumes have tracked the evolution of the technologies incorporated in both automotive power trains, internal combustion engines (ICE) their test facilities, and the legislation under which they are allowed to operate around the world.

This fifth edition is written at a time of industrial and social upheaval as a pandemic threatens to change the social order while disturbing global efforts to achieve viable mass of transport systems and vehicles producing "zero" (harmful) emissions. All this encourages the rapid evolution of existing power train technologies and national infrastructure.

At the root of our problems is the need to find means of safe storage of energy at a density that is both storable, portable and usable in lighter vehicles. The almost universal concern felt about the thermal stability of high-powered electrical batteries, particularly those based on lithium, has significantly affected the design of test facilities housing them.

In the lifetime of this book the density of stored energy will increase, and the chemistry will change but, as with any energy storage system, we are wise to be cautious, even when experience and familiarity in the technology have increased our confidence, as it has with the use of volatile liquid fuels.

This edition has been written to deal with not only the continuing evolution of the ICE but also with the revolution that is represented by the development of the electric vehicle and the many variations of hybrid ICE/electric vehicles.

The implications for the automotive test industry are life-changing; some types of the automotive "engines," which were the subject of the earlier editions, will be largely phased out of production within the lifetime of this book. To date, 13 countries and about 20 cities around the world have proposed banning the future sale of passenger vehicles powered by fossil-fueled engines. In the United Kingdom, there has been a government announcement banning the sale of new ICE and hybrid cars after 2035; this has been condemned by some in the industry as "a date without a plan." Certainly, it has huge socioeconomic implications and requires colossal and urgent investment in the country's infrastructure; we all await developments.

Meanwhile, because of the increasingly complex variations of hybrid vehicles being developed, many test cells will have to continue to meet all the legislative and logistical requirements of testing ICEs not only alone but also within their hybridized power trains.

With the drive toward electrification, it may seem that combustion-related measurements could be a diminishing topic—eventually to be extinguished altogether. However, smaller and evermore efficient engines together with cleaner fuels are predicted to be needed in the next decade to supplement electric drives, and therefore this edition deals at greater depth than earlier editions with the technology and processes required to gain and use in-cylinder pressure data from the combustion chambers of the ICE during operation.

There will be a vast collection of "heritage" engines, including truck and marine designs around the world, whose lives will be greatly prolonged while the infrastructure required to support wide electrification is put in place; because of this, all those parts of the book solely concerned with their testing have been retained and where necessary, updated.

The decline in the amount of hydrocarbon fuel test engineers will have to burn in their working life will be matched by the number and type of electrical storage devices, such as batteries and "supercapacitors" that they will have to house, test, use, and emulate. Electric vehicle devolvement is currently (2020) still a "teenage" technology undergoing change at a bewildering pace, as legislation, intended to control its safe implementation, struggles to keep pace.

The unit under test (UUT) has become much more difficult to define; it will no longer be a monolithic IC prime mover supported by easily substituted vehicle-level infrastructure systems such as fuel, fluid cooling, and low voltage power supplies. Automotive power train test cells will have to be built or adapted to house a UUT that will consist of a fully integrated package of subsystems and control units—or they will house only some of those subsystems and will have to simulate or emulate the function of the missing modules. This "front-loading" concept allows more work to be done in the virtual world, hopefully reducing costs and significantly shortening development time.

A danger to be avoided is that a generation of engineering students may spend so much of their time in a virtual world of simulation that they become more and more removed from understanding some of the fundamental realities of prime movers and rotative engineering. *Power train test engineering is not solely about becoming an advanced software user. A software simulation of a dead short in a high-power electrical system is very different from the real-life experience and develops a completely different appreciation of that fault condition.*

The rapidly developing requirements and techniques of simulation are giving rise to a whole new set of acronyms that are used to define what is being simulated within the test field. In this volume the authors will use the general term and "x-in-the-loop," where "x" is all or part of the power train hardware or a software model, a processor, or a function (HiL, MiL, PiL, FiL).

In HiL testing, often used in the verification of a module containing embedded software, the UUT control module is physically present but the target hardware is not; it is represented in real time by a software model, the veracity of which is vitally important.

Some types of x-in-the-loop testing allow power train modules to be tested in buildings quite unlike those traditionally used in the industry and specified in earlier editions of this book. The testing may be carried by technicians unversed in the best practices developed by previous generations of "traditional" ICE test engineers, the safety implications of such a deficiency this edition attempts to correct.

The "full power train" test cell using hub dynamometers, rather than wheels, running on a chassis dynamometer roller set now provides a test field that allows a full range of combinations of physical and x-in-the-loop testing to be carried out with the same accuracy of measurement as the engine/dynamometer test cell. Such testing highlights the importance of having valid and proven simulation models of the absent devices or control units.

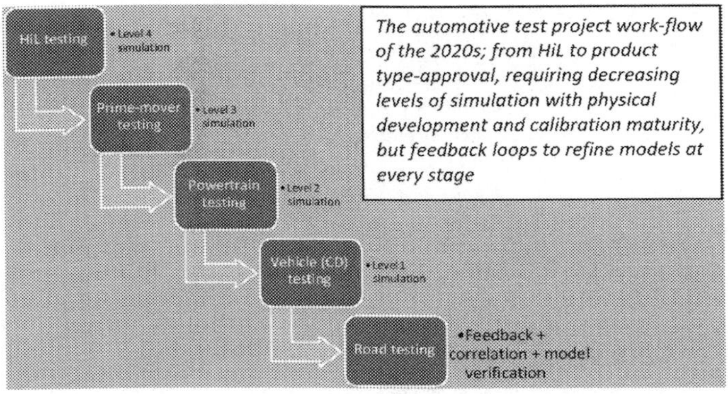
HiL testing
•Level 4 simulation
Prime-mover testing
•Level 3 simulation
Powertrain testing
•Level 2 simulation
Vehicle (CD) testing
•Level 1 simulation
Road testing
•Feedback + correlation + model verification
The automotive test project work-flow of the 2020s; from HiL to product type-approval, requiring decreasing levels of simulation with physical development and calibration maturity, but feedback loops to refine models at every stage

The technology used in the speed and load control of electrical dynamometers is very closely allied to that being used in the modern electric-powered vehicles so there is little new theory to describe but there is a considerable difference in the packaging, electrical currents and waveform, safety systems, and operational practices. What is completely new is the possibility of batteries of electric vehicles being integrated with a national power grid, not only for battery charging but also forming part of a vast dispersed power storage facility. The storage of electrical energy the testing, and emulation of batteries in both the vehicle and test facilities is covered, in addition to the storage and handling of liquid and gaseous fuels used in engines and fuel cells.

The move toward electrification and storage of electrical energy is causing some rather fundamental problems in the national electrical grid systems of many countries, the systems of which were designed for flow in one direction from large baseload power stations. These systems are having to adapt to dealing with the growing multitude of generators. Automotive test facilities pose a particular problem to the grid in that they represent both a significant load and generators and can very rapidly and unpredictably change from one to the other. Having to fit switchgear to "waste" power in load banks rated at several megawatts, which has been the case at some sites around the world, seems perverse.

In the first edition of this book the subject of computerization was restricted to test bed control, test data acquisition, and postprocessing via a host computer. The power train of today and tomorrow is and will be a digitally controlled system of such complexity that human control cannot be substituted for it, and very specialized facilities and knowledge will be required to audit it—the audit of this digital dependency starts within the modern test cell.

Within the chapters of this book, there are many references to international and national legislation covering all aspects of vehicle design and testing, much of it concerned with environmental protection. At the apex of these systems is the World Forum for Harmonization of Vehicle Regulations that is working party 29 (WP.29) of the Sustainable Transport Division of the UN Economic Commission for Europe (UNECE). The working group was established in June 1952 and currently has 62 countries in the world who are participating members.

Most countries in the world recognize the UN Regulations, either mirror their content in their own national requirements or permit the import, registration, and use of UN type-approved vehicles, or both. The United States and Canada are the two significant exceptions; the UN Regulations are not generally recognized, and UN-compliant vehicles and equipment are not authorized for import, sale, or use in the two regions, unless they are tested to be compliant with their own safety laws. However, there are many truly worldwide standards such as the ISO 8178 that is an international standard for exhaust emission measurement from a number of non-road engine applications. It is used for emission certification and/or type approval testing in most countries, including the United States, European Union, and Japan.

The legislative jungle in which we work is now so difficult to navigate that every major OEM's test facility requires access to a specialist lawyer versed in the appropriate specialisation.

Finally, the subject of human morality has never been mentioned in previous editions, but the very deliberate and sophisticated cheating, discovered by a test institution outside both an OEM or any legislative authorities' test facilities, raises a question for the test industry concerning the work we do within our national or international legislative framework. Such laws are designed to ensure that the products we produce do the least practical damage to our environment and human health, which leaves the question "in the air"—to whom do we, as test engineers, owe ultimate loyalty: our employer or the biosphere and its inhabitants?

*Note*: In common English usage, "simulate" and "emulate" tend to be somewhat interchangeable. In control engineering and in this book, they are used in their true and original meanings.

A Simulator has similar behavior to target system, but its behavior is implemented in an entirely different way, usually via a software model; a Flight Simulator cannot fly, but it simulates the experience.

An Emulator has a behavior that is identical to the target system and adheres to all the same rules; a Battery Emulator physically and functionally can "replace" the missing battery in a functional sense.

Chapter 1

# Test facility specification, system integration, and project organization

## Chapter Outline



## Introduction

Any powertrain, engine, or motor test facility will consist of a complex of machinery, instrumentation, energy storage, and support services, safely and securely housed in a building adapted or built for its purpose. For such a facility to function correctly and cost−effectively, its many parts must be matched to each other, while meeting the operational requirements of the user and being compliant with all relevant regulations.

Engine Testing. DOI: https://doi.org/10.1016/B978-0-12-821226-4.00001-2

Engine, powertrain, battery, control software, and vehicle developers may need to measure changes in performance that are so small as to be in the noise band of some available instrumentation. Such levels of measurement require that every device in the measurement chain is integrated with each other and within the total facility, such that their performance and the data they produce are not compromised by other physical or virtual parts of the unit under test (UUT), the environment in which they operate, or services to which they are connected.

Powertrain test facilities vary considerably in sources of energy they use, in physical layout, in power rating, performance, and the markets they serve. While most internal combustion engine (ICE) test cells built in the last 40 years have many common features, all of which are covered in the following chapters, there are types of cells designed for very specific and limited functions that have their own sections in this book.

In the 21st century the advent of the electric-powered vehicle in the general automotive market has led to the development of some types of powertrain test cells that are not required to or are incapable of running ICE, this obviates all of the requirements governing the reticulation of liquid fuels but requires the safe housing of battery packs or battery emulating devices. At the same time the development of hybrid powertrains requires test facilities that can house and run both automotive rated electrical motors and ICE. The task of modern test facility specification has become more complex as the variations of possible UUT have proliferated. In this volume the major cell types have been listed under the following top-level classifications:

1. Test Cells built to test parts or all of a powertrain system containing or consisting of an ICE running under power while supplied with dynamometry, fluid, and ventilation support services within a hazard containment cell, attached to a separate operator control, data acquisition, and observation space. Cells used in the testing of turbochargers that use gas generators are included in the same "hot" cell classification, all of which are considered as a "hazard containment enclosure" required to be built in compliance with fire and fuel storage regulations (see Chapter 2: Quality and Health & Safety Management and Chapter 3: Test Facility Design and Construction).
2. Test Cells or facility areas testing parts or assemblies of powertrain systems containing one or more electrical motors together with their actual or emulated energy source and their control systems. This classification of "non-ICE cells" covers a very wide range of test roles and physical plant configurations that precludes a "one-size-fits-all" building (housing) specification. Given that some of this test work may take place in facilities somewhat inexperienced in automotive powertrain testing, it is important for diligent and informed risk analysis to be carried out continuously.

3. Test cells running any type of hybrid system containing both ICE and electrical drive motors. Clearly, such "hot" cells have to meet safety standards demanded by both liquid fuel and high-powered electrical systems, which demand particular emphasis on staff training and appropriate safety systems.
4. Test cells physically, climatically, and electrically testing battery pack or fuel cell performance.
5. Whole vehicle test systems using individual wheel (hub) dynamometers, or a chassis dynamometer, are covered in Chapter 10, Chassis Dynamometers, Rolling Roads, and Hub Dynamometers.

Whatever the cell type, the maintenance of confidentiality and security of the data enforced by physical and operational restriction of access, without hindrance of work, needs to have been built into the facility design, rather than added as an afterthought.

## The product is data

Whether your facility requires hot or cold test cells, or your test work can be carried out from a computer desk, the common product is *data* that will be used to identify, modify, simulate, homologate, or develop performance criteria of all or part of the UUT.

*All post-test work will rely on the relevance and veracity of the test data; the quality audit trail starts in the test cell.*

## System integration

To build, or substantially modify, a modern powertrain test facility requires the coordination of a wide range of specialized engineering skills; many technical managers have found it to be an unexpectedly complex, challenging, and wide-ranging multidisciplinary project.

The task of putting together test cell systems from their many component parts has given rise, particularly in the United States, to a specialized industrial role known as "system integration." In this industrial model a company, more rarely a consultant, having relevant experience of one or more of the core technologies required, takes contractual responsibility for the integration of all the test facility components from various sources. Commonly the integrator role has been carried out by the supplier of test cell control systems and the contractual responsibility may, ill-advisedly, be restricted to the integration of their products and control room instrumentation.

In Europe the model was somewhat different because the long-term development of the dynamometry industry has led to a very few large test plant contracting companies making suites of matched equipment and software. Now in the 2020s, new technologies are being introduced which means that the test facility engineer has to add to their traditionally required skills in mechanical and control engineering with knowledge of high and medium power electrics, not only in the dynamometer drives but also within the powertrain under test. All this has meant that the number of individual suppliers involved in both UUT and test instrumentation development has increased, making the task of system integration ever more demanding. Thus for every facility build or modification project, it is important to nominate the role of system integrator, so that one person or company takes the contractual responsibility for the final functionality of the total test facility.

## Levels of test facility specification

*Without a clear and unambiguous specification, understood by all major stakeholders, no complex project should be allowed to proceed [1].*

This book suggests the use of three levels of specification:

1. *Operational specification*, describing "what is it for," created and agreed within the user group, prior to a request for quotation (RFQ) being issued. This may sound obvious and straightforward, but experience shows that different groups and individuals, within an industrial or academic organization, can have quite different and often mutually incompatible views as to the main purpose of a major capital expenditure. Without resolution at the outset this disunion of purpose will poison the project.
2. *Functional specification*, describing "what it consists of and where it goes," created by a user group if they have the necessary skills, if not then by buying it in via appropriate consultancy. It might also be created as part of a feasibility study by a third party, or by a nominated main contractor as part of a design study contract. This forms the document against which prospective supplier will bid.
3. *Detailed functional specification*, describing "how it all works," created by the project design authority within the supply contract.

## Note concerning quality management certification and accreditation

Most medium and large test facilities will be part of organizations certified to a quality management system equivalent to International Organization for Standardization (ISO) 9001 and an environmental management system equivalent to ISO 14000 series. Indeed, such certification is a prerequisite for any organization producing data concerning the performance of modules for use within the automotive, aeronautic, and marine industries.

It should be understood that such certification has considerable bearing on the methods of compilation and the final content of the operational and functional specifications therefore the QA function should be involved from the outset of any project.

It should also be understood from the outset that the certification or accreditation of any test laboratory by an external authority such as the UK Accreditation Service or the ISO has to be the responsibility of the operator, since it is based on approved management procedures as much as the equipment. External accreditation of a "yet-to-be built" facility cannot realistically be made a contractual condition placed upon the main contractor building or modifying a test facility.

## Why do we need an enclosed test cell?

The hot testing of automotive ICE in an open shop was a common practice 30 and more years ago but increased pressure from health and safety and fire prevention legislation and the need to attenuate noise and eliminate any form of "cross talk" has made it the rare exception to best practice. Even so some testing of automotive modules still takes place within a "semi-open" test shop environment when using whole vehicle systems to support the UUT. Some types of hydrogen fuel cell systems have been developed in large open shops with high air volumes but even in these cases casual access to energy storage systems and rotative parts must be prevented. Modern best practice dictates that

*An automotive prime-mover test cell, whether using volatile fuels or high-voltage electrical storage, is a "zone 2" hazard containment box.*

While it is possible and necessary to maintain a nonexplosive environment, it is not possible to make the Cell's interior inherently safe since the UUT, in a "rigged for test" configuration, is not inherently safe. Therefore, the cell's function is to minimize and contain the hazards by design and function and to inhibit human access when hazards may be present.

> *Note*: Zone 2 is a place in which an explosive atmosphere is not likely to occur in normal operation but, if it does occur, will persist for a short period only. These areas only become hazardous in the case of an accident or some unusual operating condition.

This concept of spatial hazard containment should also be observed in facilities testing powertrains that are purely electrical and their electrochemical power storage systems. The special requirements of 'power electric only' areas concerning hazard detection and fire suppression are dealt with in Chapter 7, Energy Storage.

The general rule to remember is that any form of energy storage system, whether electrical batteries or capacitors, volatile fluid and gaseous fuels, or kinetic energy in flywheels is potentially dangerous unless the storage and release of that energy is safely controlled. Familiarization must not breed contempt, and enclosure of hazards must be built into the facility and incorporated in its operational procedures.

A feature of modern test facilities is that they contain arrays of similar looking, large electrical cabinets having a large operational footprint and even larger when access to their internals is required. Some of these units may need to be enclosed within an "office standard" environment. It is important to label these boxes together with their isolation switches. Their containment space may or may not be within the cell housing the UUT, so the dangers posed by or to different high-powered electrical cabinets needs careful and expert consideration.

## Creation of an operational specification for a powertrain test facility

The first step is a statement giving a clear description of the tasks for which the facility is being created. Because of the range of skills required in the design and building of a "greenfield" test laboratory, it is remarkably difficult to produce a succinct specification that is entirely satisfactory to all stakeholders, or even one that is mutually comprehensible to all specialist participants.

Its creation will be an iterative task and in its first draft it need not specify the instruments required in detail, nor does it have to be based on a particular site layout. Its first role will normally be to state the business case for its creation and thus support the application for budgetary support and outline planning; subsequently it remains the core document on which all other detailed specifications and any RFQs are based.

At this early stage it is not too early, and is usually helpful, to consider the inclusion of a brief description of envisaged facility acceptance tests within the operational specification document. Consider what test could be witnessed using a UUT of what power rating; but remember that any contract acceptance tests needs to be based on one or more test objects that will be available on the project program.

It is always a sound policy to find out what instrumentation, software and service modules, appropriate to your industrial role, are available on the market and to reconsider carefully any part of the operational specification that makes demands that may unnecessarily exceed the operational range that already exists.

*A general cost consciousness at this stage can have a permanent effect on capital and subsequent running costs.*

## Key points to consider

Any operational specification should, at least, address the following questions:

- What are the primary and secondary test purposes for which the facility is intended?
- What, if any, are the implications imposed by the geographical location of the intended site(s), altitude, proximity to sensitive or hostile neighbors (industrial processes or residential), and seasonal range of climatic conditions?
- What is the realistic power range of units under test and are there more than one physical layout that will have space implications within the test cell?
- How are test data (the product of the facility) to be displayed, distributed, stored, and postprocessed?
- How many individual cells have been specified and is the number and type supported by a sensible workflow and business plan?
- What possible extension of specification or further purposes should be provided for in the initial design? *Note it is too easy at this point to suffer from cost increasing "mission creep."*
- May there be a future requirement to install additional equipment and how will this affect space requirement?
- How often will the UUT be changed in content and arrangement and what are the arrangements made for transport into and from the cells?
- Where will the UUT be prepared for test?
- How many different fuels are required to be stored and distributed and are arrangements made for quantities of special or reference fuels?
- What uprating, if any, will be required of the site electrical supply and distribution system? Be aware that modern alternating current dynamometers and battery test facilities may require a significant investment in electrical supply uprating and specialized transformers.
- To what degree must engine vibration and exhaust noise be attenuated within the building and at the property border?
- Have all local regulations (fire, safety, environment, working practices, etc.) been studied and considered within the specification?
- Have the site insurers been consulted, particularly if insured risk has changed or a change of site use is being planned?

## Audit proposed of existing site(s) early in the specification process

During the early stages of developing a facility specification, it is necessary to have an audit of the infrastructure available at the available or chosen site (s). Indeed, in the United Kingdom and Europe any civil construction specification must include preexisting site conditions or legislatively imposed restrictions that may impact on the facility layout or construction. In the United Kingdom this requirement is specifically covered by law, since all but the smallest contracts involving construction or modification of test facilities will fall under the control of a section of health and safety legislation known as *Construction Design and Management Regulations 1994* [2]. Not to list site conditions that might affect subsequent work, such as the presence of contaminated ground or flood risk, can jeopardize any building project and risk legal disputes.

Discussion on the proposal should be made with the site's electrical power provider early in the process as it is important to know what electrical power is available to the site, since any upgrade in supply will be expensive in time and money. An increasing number of automotive test facilities, particularly those operating on a 24-hour basis, are capable of exporting electrical power, a feature that may, in different parts of the world, create contractual difficulties and the need for special interface equipment (see Chapter 4: Electrical Design Requirements of Test Facilities).

## Specification of control and data acquisition systems

Specification of these systems might be considered as part of the more detailed functional specification indeed the choice of the test automation or simulation equipment and software supplier need not be part of the first draft operation specification. However, since the choice of test cell control and data acquisition software may be the singularly most important techno-commercial decision in placing a contract for a modern test facility, it would seem sensible to consider the factors that should be addressed in making that choice.

The test cell software lies at the core of the facility operation; therefore its supplier will play an important role within the final system integration and subsequent development of the facility. The choice therefore is not simply one of software suites but of a key support role.

Project designers of laboratories, when considering the competing automation suppliers, should consider detailed points covered in Chapter 4, Electrical Design Requirements of Test Facilities, and Chapter 13, Test Cell Safety, Control, and Data Acquisition, and also the following strategic points:

- Compatibility, if important, with any preinstalled system, relevant to their present requirement and targeted industrial sector.
- Does the chosen software need to be part of an integrated suite of products from the same supplier that covers a range of modules within products' development and their test tasks?
- If "x-in-the-loop" test rigs are to be used the form, availability and integration of simulation models needs to be considered.

- Does one or more of their major customers exclusively use a particular control and data storage system? (Commonality of systems may give a significant advantage in exchange of data and test sequences.)
- Level of operator training and support required. It should be noted that the software suites produced by the market leading suppliers, which particularly targeted at R&D users, are highly complex and require new operatives to attend several days of training in order to gain even the basic skills required to run live tests.
- Has the control system been proven to work with any or all of intended third-party hardware?
- How much of the core system is based on industrial standard systems and what is the viability and cost of both hardware and software upgrades? (Do not assume that a "system X lite" may be upgraded to a full "system X.")
- Requirements to use preexisting data or to export data from the new facility to existing databases.
- Ease of creating your test sequences.
- Ease and security of channel naming, calibration, and configuration.
- Flexibility of data display, postprocessing, and exporting options.

A methodical approach requires a "scoring matrix" to be drawn up whereby competing systems may be objectively judged.

Anyone charged with producing specifications is well advised to carefully consider the role of the test cell operators, since significant upgrades in test control and data handling will totally change their working environment. *There are many cases of systems being imposed on users and that never reach their full potential because of inadequacy of training or a level of system complexity that was inappropriate to the task or the grade of staff employed.*

It used to be the case that the effective time constants of many powertrain test processes were not limited by the data handling rates of the computer system, but rather by the physical process being measured and controlled. With the advent of multiple integrated control systems within electrical powertrain modules, all communicating on a common communication bus, clock speeds need to be matched to these high-speed processing devices. The skill in using such information is to identify the numbers that are relevant to the task for which the test and control system is required.

## Feasibility studies and outline planning permission

The investigatory work required to produce a site-specific operational specification may produce a number of alternative layouts, each with possible first-cost or operational impacts. Part of the investigation should be an environmental impact report, covering both the facility's impact of its surroundings and the locality's possible impact on the facility.

Complex techno-commercial investigatory work may be needed, in which case a formal "feasibility study," produced by an expert third party, might be considered.

In the United States, this type of work is often referred to as a "proof design" contract.

The secret of success of such studies is the correct definition of the required "deliverable." An answer to the technical and budgetary dilemmas is required, giving clear and costed recommendations, rather than a restatement of the alternatives; so far as is possible the study should be supplier neutral.

A feasibility study will invariably be site specific and, providing appropriate expertise is used and the proposal is indeed viable, it should prove supportive to gaining budgetary and outline planning permission. The inclusion within any feasibility study or preliminary specification of a site layout drawing and graphical representation of the final building works will be extremely useful in subsequent planning discussions. The text should be in a format that is capable of easy division and incorporation into the final functional specification documents.

Finally, it should be understood that a genuine feasibility study might find that, within the stated remit, the project is not feasible; such an outcome is not uncommon.

## Regulations, planning permits, and safety discussions covering test cells

In addition to being technically and commercial viable, it is necessary for the new or altered test laboratory to be permitted by various civil authorities. Therefore the responsible project planner should consider discussion at an early stage with the following agencies:

- local planning authority
- local electrical supply authority
- local petroleum officer and fire department
- local environmental officer
- building insurers
- other site utility providers

Note the use of the word "local." There are very few regulations specifically mentioning engine test cells; much of the European and American legislation is generic and frequently has unintended consequences for the automotive test industry.

Most legislation is interpreted locally, and the nature of that interpretation will depend on the highly variable industrial experience of the officials concerned. There is always a danger that inexperienced officials will overreact to applications for engine or battery test facilities and impose unrealistic restraints on the design or function.

It may also be useful to remind external participants in safety-related discussions that their everyday driving experiences take them far closer to a running engine or high-energy battery, in a more potentially hazardous environment, than is ever experienced by anyone sitting at a test cell control desk.

Most of the operational processes carried out within a typical engine or powertrain test cell are generally less potentially hazardous than those experienced by garage mechanics, motorsport pit staff, or marine engineers in their normal working life. The major difference is that in a cell the running automotive powertrain module is stationary in a space, connected in a way for which it was not designed and humans have, unless prevented by safety mechanisms, potentially dangerous access to it.

It is more sensible to interlock the cell doors to prevent access to an engine or motors running above "idle" state, than to attempt to make every rotating element "safe" by the use of close-fitting and complex guarding that will inhibit operations and inevitably fall into operational disuse.

The authors of the high-level operational specification would be ill-advised to concern themselves with some of these minutiae but should simply state that industrial best practice and compliance with current legislation is required.

The arbitrary imposition of existing operational practices on a new test facility should be avoided until confirmed as appropriate, since they may restrict the inherent benefits of the technological developments available.

One of the restraints commonly imposed on the facility buildings by planning authorities concerns the number and nature of chimney stacks or ventilation ducts that are often considered to produce an inappropriately "industrial" character; this is often a cause of tension between the architect, planning authority, and facility designers.

Noise breakout through exhaust ducting may, as part of the planning approval, have been reduced to the preexisting background levels at the facility border. This can be achieved in most cases, but the space required for attenuation will complicate the plant room layout (see Chapter 5: Ventilation and Air Conditioning of Automotive Test Facilities, concerning ventilation).

The use of gaseous fuels, such as liquefied petroleum gas, stored in bulk tanks, hydrogen stored in bulk cryogenic storage or as compressed gas in tanks, or natural gas supplied through an external utility company, will impose special restrictions on the design of test facilities. If they are to be included in the operational specification, the relevant authorities and specialist contractors must be involved from the planning stage. Modifications may include blast pressure relief panels in the cell structure and exhaust ducting, all of which needs to be included from design inception.

The use of bulk hydrogen, required for the testing of fuel cell–powered powertrains, requires building design features such as roof-mounted gas detectors and automatic release ventilators. Special consideration has given to the delivery, storage, and reticulation of the gas; therefore *realistic* estimates of the volume required will need to be included in the operational specification.

## Creation of functional specifications: some common difficulties

Building on the operational specification, which describes what the facility has to do, the functional specification describes how the facility is to perform its defined tasks and what it will need to contain. If the functional specification is to be used as the basis for competitive tendering, then it should avoid being unnecessarily prescriptive.

Overprescriptive specifications, or those including sections that are in some detail technically incompetent, are not rare and create a problem for specialist contractors. Overprescription may prevent a better or more cost-effective solution being quoted, while technical errors mean that a company, which, through lack of experience, claims compliance, may win the contract and then inevitably fail to meet the customer's expectations.

Examples of overprescription range from choice of ill-matching of instrumentation to an unrealistically wide range of operation of UUT subsystems within a single cell.

A classic problem in facility specification concerns the range of engines or powertrain configurations that can be tested in one test cell using common equipment and a single-shaft system. Clearly, there is an operational cost advantage for the whole production range of a manufacturer's units to be tested in any one cell. However, the detailed design problems and subsequent maintenance implications that such a specification may impose can be far greater than the cost of creating two or more cell types that are optimized for a narrower range of UUT. Not only is this a problem inherent in the "turn-down" ratio of fluid services and instruments having to measure the performance of a range of engines from say 450–60 kW, but the range of vibratory models produced may exceed the torsional capability of any one shaft system.

This issue of dealing with a problem of torsional vibration, both within the test installation and in the case of some hybrid configurations, can blight rotative test programs and can be difficult to predict and simulate. In the testing of a range of ICE it may require that cells be dedicated to particular types or that alternative shaft systems are provided for particular engine types. Errors in this part of the specification and the subsequent design strategy are often expensive to resolve postcommissioning.

At the risk of overrepetition, it must be stated that it is never too early to consider the form and content of acceptance tests, since from them the experienced test plant designer can correctly infer much of the detailed functional specification.

Failure to incorporate acceptance tests into contract specifications from the start can lead to delays and disputes at the end.

## Interpretation of quotation specifications

"Do not assume that the receiver of your documentation is as diligent in reading it as you were in writing it. Keep it appropriately simple and avoid jargon even if you believe the receiving parties are your technological peers." [1]

Employment of contractors with the relevant industrial experience is the best safeguard against misinterpretations leading to quotations containing prices containing overblown contingencies or significant omissions.

Provided with a well-written operational and adequate functional specification, any competent contractor, experienced in the relevant area of the powertrain or vehicle test industry, should be able to provide a detailed quote and specification for their module or service within the total project.

Subcontractors who do not have experience in the industry will not be able to appreciate the special, sometimes subtle, requirements imposed upon their designs by the transient conditions, operational practices, and possible system interactions inherent in our industry. In the absence of a full appreciation of the project, based on previous experience, inexperienced sales staff will search the specification for "hooks" on which to hang their standard products or designs, and quote accordingly. This is particularly true of air- or fluid-conditioning plant, where the bare parameters of temperature range and heat load can lead the inexperienced to equate test cell conditioning with that of a chilled warehouse. An escorted visit to an existing test facility should be the absolute minimum experience for subcontractors quoting for systems such as chilled water, electrical installation, and heating, ventilation and air conditioning (HVAC).

Facilities designed to test electrical powertrain components may contain components and cabling, which comply with automotive and data transmission wiring codes and practices rather than more familiar building codes and therefore selection and training of electrical installation contractors is vitally important.

### After *what* comes *how*?

From the first concept it is vital to consider not only the form and content of any new test facility but also what sort of organization is going to be responsible for its creation and operation.

In all but the smallest test facility projects, there will be three generic types of contractor with whom the customer's project manager has to deal. They are:

- civil contractor
- building services contractors
- test instrumentation contractor

How the customer decides to deal with these three industrial groups and integrate their work will depend on the availability of in-house skills and the skills and experience of any preferred contractors.

The normal variations in project organization, in ascending order of customer involvement in the process, are as follows:

- A consortium working within a design and build or "turnkey" contract based on the customer's operational specification and working to the detailed functional specification and fixed price produced by the consortium.
- Guaranteed maximum price contracts, where a complex project management system, having an "open" cost-accounting system, is set up with the mutual intent to keep the project within an agreed maximum value. This requires joint project team cohesion of a high order.
- A customer-appointed main contractor employing a supplier chain working to the customer's full functional specification.
- A customer-appointed civil contractor followed by a services and system integrator contractor each appointing specialist subcontractors, working with the customer's functional specification and under the customer's project management and budgetary control.
- A customer-controlled series of subcontract chains working to the customer's detailed functional specification, project engineering, and site and project management.

Whichever model is chosen, the two vital roles of project manager and design authority (system integrator) have to be clear to all and provided with the financial and contractual authority to carry out their allotted roles.

## Project roles and management

The key role of the client, or user, is to invest great care and effort into the creation of a good operational and functional specification. Once permission to proceed has been given, based on this specification and budget, the client has to invest the same care in choosing the main contractor.

When the main contractor has been appointed, the client's representation within the project team needs to be formalized but the day-to-day tasks of that role, and the client's user group, should, ideally, reduce to that of attendance at review meetings and being "on call."

*Nothing is more guaranteed to cause project delays and cost escalation than ill-considered or informal changes of specification details by the client's representatives.*

Whatever the project model, the project management system should have a formal system of "notification of change" and an empowered group within both the customer and contractor's organization to deal with such requests quickly. The type of form shown in Fig. 1.1 allows individual requests for project change to be recorded and the implications of the change to be discussed and quantified. Change can have either a negative or positive effect on project costs and may be requested by either the client or contractor(s).

*Note:* As originally conceived: in a *turnkey project,* once the operational specification and the site boundaries have been agreed, the main contractor takes complete charge, almost to the exclusion of the client. When the project is complete and commissioned, the client is invited to a full demonstration and acceptance test, and is then, upon acceptance, presented with the keys and takes possession. The private finance initiative contracts covering UK hospitals were modeled on turnkey concepts and have, in several cases, been practical and commercially disastrous for the customer. Shipbuilding occasionally comes closest to turnkey, but it is never realized in the building of automotive test facilities; indeed, under UK law the client cannot abrogate themselves of their project and site responsibilities.

## The project triangle

All projects have to operate within the three restraints of time, cost, and quality (content). The relative importance of these three criteria to the specific project has to be understood by the whole project management group. The model is different for each client and for each project; however, many clients may protest that all three criteria have equal weighting and are fixed, in reality though if change is introduced, one has to be a variable (Fig. 1.2).

The later in the program that change, within the civil or service systems, is required, the greater the consequential effect. The effects of late changes within the control and data acquisition systems are much more difficult to predict; they may range from trivial to those requiring a significant upgrade in hardware and software, which is why a formal "change request" process is so important.

## Project management tools: communications and responsibility matrix

Any multicontractor and multidisciplinary project creates a complex network of communications. Formal networks and informal subnetworks, between suppliers, contractors, and personnel within the customer's organization, may preexist or be created during the project; the danger is that informal communications may cause unauthorized variations in project content, cost or timing.

Customer: Variation no:

Project name: Project no:

Details of proposed change:
(include any reference to supporting documentation)

Requirement (tick and initial as required)
Design authority required design change
Customer instruction:
Customer request:
Contractor request:
Urgent quotation required:
Customer agreed to proceed at risk:
Work to cease until variation agreed:
Requested to review scope of supply:
Other (specify):

| | Name | E-mail/phone No. |
|---|---|---|
| Contractor representative | | |
| Authorized | | |
| Customer representative | Name | E-mail/phone No. |
| Authorized | | |

Actions:
Sales quote submitted: (date and initial)
Authorized for action: (date and initial)
Implemented: (date and initial)

**FIGURE 1.1** A sample contract variation request form and record sheet.

Good project management is only possible with a disciplined communication system and this should be designed into, and maintained during, the project.

The arrival of e-mail as the standard communication method has increased the need for communication discipline and introduced the need, within project teams, of creating standardized computer-based filing systems.

## Web-based control and communications

The proliferation of informal short message service (SMS), web-based communication, and social networking tools is potentially disastrous when used for project communications. Not only does the use of such systems, on

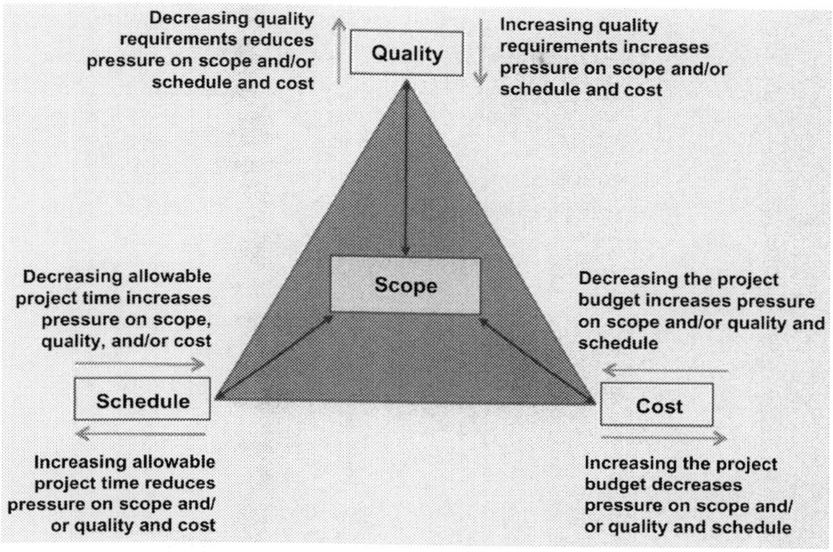


**FIGURE 1.2** One of many possible versions of the project triangle, showing the resulting effects of change in one or more of the project constraints.

company computers or mobile phones, have the potential for confusion, but also confidentially is endangered.

There are a number of powerful Document Control software packages available to use in large multidisciplinary projects, such as those developed by NextPage Inc. or BIW Technologies Ltd.

Some corporate customers prefer to create and maintain a project-specific Intranet or Internet website by which the project manager has an effective means of maintaining control over formal communications. Such a network can give access permission, such as “read only,” “submit,” and “modify,” as appropriate to individuals’ roles and the nominated staff having commercial or technical interest in the project. Note though that if such systems prove too difficult to use they will be by-passed, to the detriment of project control.

The creation of a responsibility matrix is most useful when it covers the important minutiae of project work—that is, not only who supplies a given module, but also who insures, delivers, offloads, connects, and commissions the module.

## Use of “master drawing” in project control

The use of a common facility layout or schematic drawing that can be used by all tendering contractors and is continually updated by the main contractor or design authority can be a vital tool in any multidisciplinary project.

In such projects there may be little detailed appreciation between specialized contractors of each other's spatial and temporal requirements.

Constant, vigilant site management is required during the final building "fit-out" phase of a complex test facility if clashes over space allocation are to be avoided, but good contractor briefing while using a common layout can reduce the inherent problem. If the system integrator or main contractor takes ownership of project floor layout plans and these plans are used at every subcontractor meeting and kept up to date to record the layout of all services and major modules, then most space utilization, service route, and building penetration problems will be resolved before work starts. Where possible and appropriate, contractor's method statements should use the common project general layout drawing to show the area of their own installation in relation to the building and installations of others.

## After *how* comes *when*?—project timing charts

Most staff involved with a project will recognize a classic Gantt chart, but not all will understand the relevance of their role or the interactions of their tasks within that plan. It is the task of the project manager to ensure that each contractor and all key personnel work within the project plan structure acknowledge and commit to their part of the overall program. This is not served by sending repeatedly updated, electronic versions of a large and complex Gantt chart to all participants, but by early contract briefing and strategically timed progress meetings.

There are some key events in every project that are absolutely time critical and these have to be given special attention by both client and project manager. Consider, for example, the site implications of the arrival of a chassis dynamometer for a climatic cell:

- Although the shell building must be weather-tight, access into the chassis dynamometer pit area will have to be kept clear for special heavy handling equipment, by deliberately delayed building work, until the unit is installed; the access thereafter will be closed up.
- One or more large trucks will have to arrive on the client's site, in the correct order, and require suitable site access, external to the building, for maneuvering.
- The chassis dynamometer sections will require a large crane to offload, and probably a special lifting frame to maneuver them in place. To minimize hire costs, the crane's arrival and site positioning will have to be coordinated some hours only before the trucks' arrival.
- Other contractors will have to be kept out of the affected work and access areas for the duration, as will client's and contractor's vehicles and equipment.

Preparation for such an event takes detailed planning, good communications, and authoritative site management. The nonarrival, or late arrival, of one of the key players because "they did not understand the importance" clearly causes acute problems in the abovementioned example. The same ignorance or disregard of programed roles can cause delays and overspends that are less obvious than the previous example throughout any project where detailed planning and communications are left to take care of themselves.

## The importance of final documentation

Complex fluid services and electrical systems, particularly those under the control of programmable devices, are, in the nature of things, subject to detailed modification during the building and commissioning process. The final documentation, representing the "as commissioned" state of the facility, must be of a high standard and easily accessible, post handover, to maintenance staff and subcontractors. The form and due delivery of documentation should be specified within the functional specification and form part of the acceptance criteria. Subsequent responsibility for keeping records and schematics up to date within the operator's organization must be clearly defined and controlled.

## Summary

The project management techniques required to build a modern test facility are the same as those for any multidisciplinary laboratory construction but require knowledge of the core testing process so that the many subtasks are integrated appropriately.

The statement made early in this chapter, "Without a clear and unambiguous specification no complex project should be allowed to proceed," seems self-evident, yet many companies and government organizations, within and outside our industry, continue either to allocate the task inappropriately or underestimate its importance, and consequently subject it to postorder change. The result is that project times are extended by an iterative quotation period or there develops a disputatious period of modification from the point at which the users realize, usually during commissioning, that their (unstated or misunderstood) expectations are not being met.

## References

[1] T. Martyr, Why Projects Fail, Business Expert Press, New York, 2018.

[2] The Construction (Design and Management) *Regulations* 2015. Available from: <www.hse.gov.uk/construction/cdm/2015/index.htm>.

Chapter 2

# Quality and health and safety management

## Chapter outline



## Part one: quality and test facility quality certification

The product of a test facility is data, and it is the primary task of its *senior management* to ensure the quality and security of that data, together with the safety of personnel involved in its collection. An essential role within the primary task is to ensure that all staff receive *training* that is appropriate to their roles and to the rapidly changing environments in which they work.

Since both the data that is required to allow testing to be carried out and the data the test environment produces will be held and transmitted as digital code, the role of *information technology (IT) management* has become increasingly important in the strategic planning of test facility management.

Engine Testing. DOI: https://doi.org/10.1016/B978-0-12-821226-4.00002-4

The *technical management* of any test laboratory has to ensure that the test equipment is chosen, maintained, and used safely, to its optimum efficiency in order to produce data of the quality required to fulfill its specified tasks. Those specific tasks are designated as test projects under the control of a test engineer acting in the role of *project manager*.

Automotive test facilities work under two types of legislation:

- Legislation governs how the organization should function, in terms of its processes, procedures, and disciplines in order to consistently produce products or provide services to meet customers and regulatory requirements. International Standards Organization (ISO) 9001 is the international standard for a quality management system while QS9000 and TS16949 are add-ons, specific for the automotive industry to be used in conjunction with the QMS standard. The handling of data will be defined by an organizational IT strategy that will comply with the requirements of ISO 27001.
- Legislation governing the required performance of the products they are testing and the details of the tests to which those products are subjected. While every component of a road vehicle is the subject of legislation; those setting the exhaust emission test procedures and the allowed emission limits have dominated the form and function of internal combustion engine (ICE) test facilities for the last two decades. In the third decade of the 21st century, it will be the electrical drivetrain that will increasingly take over that role.

With the possible exception of some academic organizations, all test facilities carrying out work for, or within, original equipment manufacturer (OEM) organizations or for government agencies, will need to be certified to the ISO 9001 or an equivalent national quality standard based upon it.

Independent confirmation that organizations meet the requirements of ISO 9001 is obtained from third party, national, or international certification bodies. Such certification does not impose a standard model of organization or management, but all certified test facilities will be required to create and maintain documented processes and have the organizational positions to support them.

ISO 9001 requires a quality policy and quality manual that would usually contain the following compulsory documents defining the organizations systems for:

1. control of documents;
2. control of records, including test results and calibration;
3. internal audits, including risk analysis and calibration certification;
4. control of nonconforming product/service, including customer contract and feedback;
5. corrective action; and
6. preventive action, including training.

Small uncertified test organizations should use such a framework in their development. Directors of certified organizations support the fact that the role of quality management within their organization is not that of simply feeding a bureaucratic monster but of continuous improvement of company products and services.

## Management roles

Although the organizational arrangements may differ, a medium-to-large test facility will employ staff having the following distinctly different roles:

- Executive and financial control.
- A quality group, having direct report to the executive and charged with the maintenance of ISO 9001 and related certifications, internal audit, and management of the instrument calibration system. An independent role within this Quality Assurance group should be the management of health and safety (H&S).
- IT management that controls the use of electronic computers and computer software to convert, store, protect, process, transmit, and securely retrieve information. (See Chapter 14: Data Handling and Modeling, for discussion on role of data science.)
- Facility management charged with building, maintaining, and developing the installed plant, its support services, and the building fabric.
- The project group who are the internal user group charged with designing and conducting tests, collecting data, and disseminating information.

The listed tasks and roles are not mutually exclusive, and in a small test, environment may be merged. However, in all but the smallest department management of the QA task, which includes the all-important responsibility for calibration and accuracy of instrumentation, should be an allocated to trained individual who is given the well-defined and acknowledged role.

Each group will have some responsibility for two funding streams, operational and project specific.

## Work scheduling

In many cases, facility and individual cell "efficiency" is interpreted by management as ensuring the plant achieves maximum "uptime" or "shaft rotation time." However, any test facility, like any individual, can operate as a "busy fool"; as when the tests are badly designed or undertaken in a way that is not time and energy efficient or, much worse, when the produced data is corrupted by some systematic mishandling or postprocessing. Quality of data is the paramount concern and has to be not only achieved but under ISO 9001, provided with an audit trail.

The efficiencies to be gained by complete "prerigging" of the unit under test (UUT) are covered elsewhere (Chapter 3: Test Facility Design and Construction, Chapter 11: Mounting and Rigging Internal Combustion Engines for Test, and Chapter 12: Rigging and Running of Electrical Drive Systems), while in this section, we consider the management policies that need to be considered.

Cell use and scheduling can pose complex techno-commercial problems, and the decisions taken will be substantially affected by the design of the facility (see Chapter 4: Electrical Design Requirements of Test Facilities).

A feature of 21st century test facilities is that they tend to have a larger number of test cells dedicated to testing modules or subsystems than that of the previous generations of ICE cells. This can lead to complex scheduling problems caused by the unavailability of expensive service devices, such as Battery Emulators that, through budgetary constraints, may have to be shared between small specialized cells.

Efficient test scheduling requires a test request quality gateway wherein the specifications of the test are defined, preferably using an information template. The appropriate reuse of test schedules and configurations saves time and improves the repeatable quality of the data. The management system must be capable of auditing the quality of the test request and converting it into a test schedule, including cell allocation, equipment and transducer configuration, and data reporting format, all based, but not always rigidly subject to, previously used templates.

Major OEMs will use enterprise resource planning systems, such as those developed by SAP for controlling their entire operations, but specialist tools developed specifically for large powertrain test facilities are also used within a global business management system, of which the "Test Factory Management Suite" developed by AVL List is an example.

An important line management task in ensuring cost-effective cell use is to decide how detected faults in the UUT are treated. If test work is queuing up, do you attempt to resolve the problem in the cell, turning the space into an expensive workshop, or does the UUT get removed and the next test scheduled take its place in the queue?

Ignoring the disruption to the work program, having a "plug-in" prerigged palletized UUT (Chapter 3: Test Facility Design and Construction, Chapter 11: Mounting and Rigging Internal Combustion Engines for Test, and Chapter 12: Rigging and Running of Electrical Drive Systems) means that the transition time between swapping jobs can be reduced to under one hour, providing that the control system can deal with the curtailment of the previous test and the reparametrizing of the new test without loss or corruption of data.

Key to the solution of test scheduling is the need to have the UUT available to program or to be able to simulate it, prior to its physical availability, within a wider X-in-the-Loop test setup (Fig. 2.1).

| Validation of > | xCU | ICE test cell | E-motor test cell | Transmission test cell | Battery pack test cell | Powertrain test cell | Vehicle chassis dyno | Road |
| --- | --- | --- | --- | --- | --- | --- | --- | --- |
| | | | | | | | | Maneuvering |
| | | | | | | | | RLE |
| | | | | | | | | Chassis |
| | | | | | | | Wheels | Wheels |
| | | | | | Battery | Battery | Battery | Battery |
| | | | | Transmission | | Transmission | Transmission | Transmission |
| | | | E-motor | | | E-motor | E-motor | E-motor |
| | | ICE | | | | ICE | ICE | ICE |
| | xCU | ECU | xCU | TCU | BMS | xCU + xCU+ | xCU + xCU+ | xCU + xCU++ |
| | Maneuver | Maneuver | Maneuver | Maneuver | Maneuver | Maneuver | Maneuver | Maneuver |
| Via Simulation of > | Chassis | Chassis | Chassis | Chassis | Chassis | Chassis | Chassis | |
| | Wheels | Wheels | Wheels | Wheels | Wheels | Wheels | | |
| | Transmission | Transmission | Transmission | | Transmission | | | |
| | E-motor | E-motor | | E-motor | E-motor | | | |
| | ICE | | ICE | ICE | ICE | | | |
| Emulation of | Battery | Battery | Battery | Battery | | (Optional) | | |

**FIGURE 2.1** Validation of UUT though "x-in-the-loop" testing where "x" is simulated or emulated in each type of test cell.

A particular problem experienced by all test facilities is the on schedule availability of the unit to be tested, its required rigging attachments, and its key data (Control Unit settings, performance limits, inertia data for shaft selection, etc.) that are required to set the test parameters.

When multi-departmental project management is suboptimal, test cells can lie idle while the management system tries to trace the missing link to the chain that should have been joined much earlier in the work schedule.

## The role of key life testing in powertrain quality assurance

The problem is that the real life of a powertrain system, on average, lasts for several years while the pressure on test engineers is to reduce development test times from months to weeks. Once a powertrain system has been calibrated to run within it design brief and legislative limits then released for sale, the key operational variables become the end user's "use and abuse", which may reveal component vulnerability, or design and calibration flaws undiscovered during development. Key life testing is carried out to help reveal such user-induced flaws and evaluate the design life of a vehicle component or subassembly at an accelerated pace. Key life and endurance testing of powertrains and the "fuels and lubes" test regimes also plays an important role in the determining of maintenance (service) periods. The basis of key life testing is the determination of the anticipated real-life operating conditions for the vehicle and then designs accelerated test regimes so that the system or component under test can fulfill its design life in weeks or months rather than years.

For many years, OEMs have developed and used ICE and transmission unit endurance test sequences, prior to tear down and wear measurement; similarly, mileage accumulation rigs, required by emission legislation, have been adapted to accelerate powertrain endurance testing. At component and subassembly level, key life testing is carried out at using highly accelerated life testing that involves repeated vibration at six-degree-of-freedom and rapid thermal cycling.

Battery pack testing is, almost by definition, a form of key life testing since the whole development thrust of the industry is to increase the operational life and the number of charge–discharge cycles that can be achieved, under widely variable environmental temperatures, before the pack's state of health falls below acceptable limits.

*Note*: The state of health of a battery represents a measure of the battery's ability to store and deliver electrical energy, compared with a new battery, while the state of charge parameter can be considered as a thermodynamic quantity enabling one to assess its potential energy at any point in time.

## Vehicle and vehicle systems type approval, homologation, and confirmation of production

This is a complex area of national and international legislation. In essence, in order to be sold around the world, vehicles have to meet the many and varied environmental and safety standards contained in regulations, in each country in which they are sold. These standards not only define exhaust emission limits of the ICE powertrain systems but also the performance and construction standards of every vehicle components and the whole vehicle. The 1958 UNECE (United Nations Economic Commission for Europe) agreement has largely harmonized international standards, with 50 countries and 1 region (the European Union) currently following it.

EC directives and UN regulations require third-party approval for testing, certification, and production conformity assessment by an independent organization. Each Member State is required to appoint an "Approval Authority" to issue the approvals and a Technical Service to carry out the testing in accordance with the Directives and Regulations. Note that the approval issued by one authority will be accepted in all the Member States.

The Vehicle Certification Agency (VCA) is the designated UK Approval Authority and a Technical Service for all type approvals to automotive EC directives and most UN regulations. At the time of writing, 2020 and post-Brexit, the international role of the VCA awaits redefinition.

The development of battery electric vehicle (BEV) and hybrid electric vehicle (HEV) has necessitated the drawing up of UN regulations on REESSs (Rechargeable Energy Storage Systems) to address the particular safety issues involved. Regulation No. 100 is intended to ensure the safety of drivers, passengers, and the car's mechanical systems. It includes testing mechanical integrity, resistance to mechanical shock, and resistance, during operation and when the vehicle is parked.

## Type-approval "E" and "e" marking

The e-mark is an EU mark and is part of the vehicle homologation requirements for approved vehicles and vehicle components sold into the European Union (see Fig. 2.2A). It is the type-approval mark required by devices related to all safety-relevant functionality of the vehicle. Such devices will have to be e-mark certified by a certifying authority; manufacturers cannot self-declare and affix an e-mark, unlike most CE marking, where manufacturers' self-declaration is the norm.

Since late 2002, all electronic devices intended for use in vehicles, including aftermarket electronics products, are required to obtain formal type approval for products before placing them on the market.

The E-mark is an UN mark for approved vehicles and vehicle components sold into the European Union and some other countries under UN-ECE Regulation 10.02 (see Fig. 2.2B).

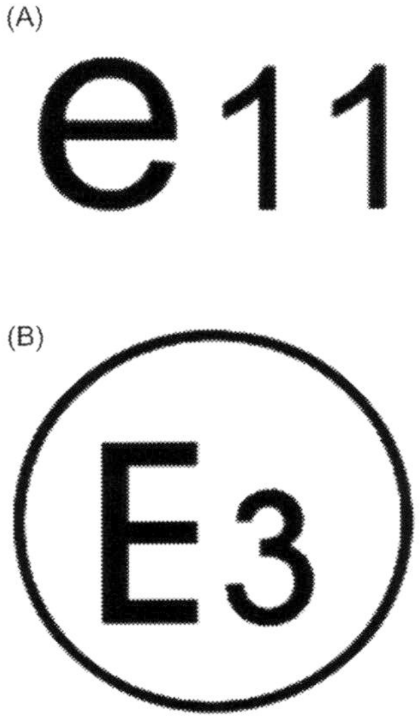


**FIGURE 2.2** (A) The "e" mark issued by the certifying authority followed by the EU country code number for the United Kingdom (as of 2019) (B) and the "E" mark followed by the EU country code, in this case for Italy.

Both marks are affixed to approved products with other numbers relating to the legislation under which the approval has been gained.

## Use and maintenance of test cell logbooks

In the experience of the author of this section, nothing is more helpful to the safe, efficient, and profitable operation of test cells than the discipline of keeping a logbook or cell diary that contains a summary of the maintenance and operational history of both the cell and the tests run therein.

The whole idea of a physical, paper based, and hand-written logbook may be considered by many to be anomalous in an age in which the computer has taken over the world. However, test cell computer software is not usually designed to report those subjective recordings of a trained technician nor are the records and data they hold always immediately available to those who require them, either because the system is shutdown or they are barred by a security system from its use, or data is hidden within the computer or network.

> *Where do the technicians, who worked until 1 a.m. fixing a problem or, what is more important, not fixing a problem record the cell's status to be read by the morning shift operator, if not in the logbook?*

The logbook is also a vital record of all sorts of peripheral information on such matters as safety (minor injury record), maintenance, and suspected faults in equipment or data recording and as an immediate record of "hunches" and intuitions arising from a consideration of perhaps trivial anomalies and unexpected features of performance.

The facility accountant will benefit from a well-kept logbook, since it should provide an audit trail of the material and time consumed by a particular contract and record the delays imposed on the work by "nonbookable" interruptions.

It must be obvious that to be of real benefit the logbook must be read and valued by the H&S and facility management team.

## Test execution, analysis, and reporting

Assuming that there is an adequate understanding of the relevant theory within the experimenting group, the basic flow of stages and prerequisites of an isolated Test Project are as follows:

- The design, execution, analysis, and reporting of a program of tests and experiments on automotive powertrain components is becoming more and more difficult with the proliferation in individual UUT modules and the need, because of development time pressures, to simulate parts of their technical environment using HiL techniques.
- First, the test engineer must understand the questions that his project is intended to answer and the requirements of the "customer" who has asked them. The form of the "deliverable" has to be agreed.
- The extreme parameters of the test sequences must be agreed, and if beyond precedent or containing novel features, a risk analysis must be carried out, acted upon and recorded.
- The necessary apparatus and instrumentation must be assembled and, if necessary, designed and constructed.
- The experimental program must itself be designed with due regard to the levels of accuracy required.
- The test program is executed, and test data is stored in uniquely labeled form.
- The test data is reduced and presented in the specified form and to the level of accuracy required to the "customer."
- The test result data is archived according to procedures compliant with ISO 27001.
- The findings are summarized, and related to the questions, the program was intended to answer. The results and the data generated form part of the intellectual property and market value of the company and should be protected as such (Fig. 2.3).

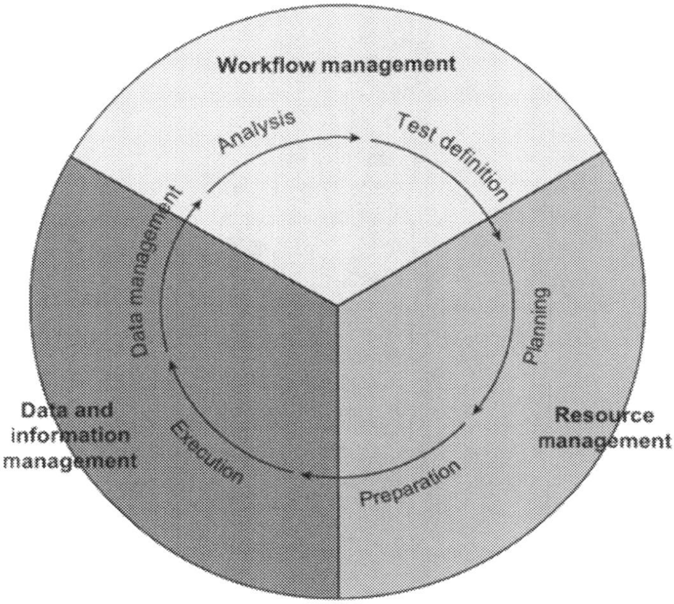


**FIGURE 2.3** Diagrammatic representation of the roles of management of a test facility: workflow, resources, and data.

## Maintaining test data quality and security

Post-test corruption of data is mentioned in Chapter 19, The Pursuit and Definition of Accuracy, and here it is examined as part of the Quality Management role. All medium and large companies running automotive test facilities will be operating under ISO 9001 and ISO 27001 and will have procedures and responsibilities for handling the test data generated. To maintain control and confidentiality of the raw data, the test cell computers are kept within a local computer network that is an isolated bubble (no Internet connection). The gateway out from the local host will be via a managed link running Secure File Transfer Protocol. Test facilities that are geographically dispersed have the challenging task of to centrally managing the access to and use of their dispersed data whether as generated or archived in a shared data warehouse. Security of access to data has to consider both online and, in the age of miniature memory sticks of gigabyte capacity, physical leakage through malicious action. It is normal for the work of test facilities to be organized into projects under the control of a designated senior test engineer acting in the role of project manager. Each project is a separate documented entity and has unique labeling of all data used and produced.

Data management has become an increasingly difficult technical and strategic management task. It is safe to assume that the volume of test data being produced will continue to increase. It has been calculated in Ref. [1] that there are now at least 60,000 parameters that require to be defined, mapped and optimized in the process of calibrating a modern vehicle. If some of this data originates from disparate sources and is delivered in a variety of file formats the efficiency and quality of the calibration task suffers. It follows that efficiency and quality can only be guaranteed if the data is common in form and format.

The commonly used data storage and retrieval system is based on the standards created and maintained by Association for Standardization of Automation and Measuring Systems (ASAM), and there are a number of commercially available analysis tools designed to process such standardized data.

The ever-increasing quantity of test data produced during the calibration of automotive powertrains is intensifying the trend toward modularization of components and software models. Using the same components in several vehicle configurations can reduce the calibration task by reusing the previous control logic with parameters based on prior knowledge and historical data. The newly calibrated model is then saved under its project identification.

Back up and archiving of test data will be the subject of company, industrial, and legal procedures and legislation. Most test cell data is automatically backed up to a local host with a time lag of, typically, 3 minutes.

Many test facilities hold their project test data on their local "data warehouse host," in searchable format for 3 years before off-loading to a remote host where it is held for typically 10 years.

## Cell-to-cell correlation of internal combustion engine performance

It is quite normal for the management of a test department to wish to be reassured that all the test stands in the test department "give the same answer." Therefore it is not unusual for an attempt to be made to answer this question of cell correlation by the apparently logical procedure of testing the same ICE on all the beds. The outcome of such, specifically ICE tests, based on detailed comparison of results, is invariably a disappointment. Over many years, such tests have led to expensive disputes between the test facility management and the suppliers of the test equipment.

The need to run conformity of production tests in test facilities distant and different from the homologation test cells has again raised the problem of variability of engine test results.

It cannot be too strongly emphasized that an ICE, or a powertrain containing one, however sophisticated its management systems, is not suitable for duty as a standard source of torque or emissions.

As discussed in Chapter 6, Cooling Water and Exhaust Systems, and Chapter 14, Data Handling and Modeling, very substantial changes in engine performance can arise from changes in atmospheric conditions. In addition, engine power output is highly sensitive to variations in fuel, lubricating oil and cooling water temperature, and it is necessary to equalize these very carefully, over the whole period of the tests, if meaningful comparisons are to be hoped for.

Finally, it is unlikely that a set of test cells will be totally identical: apparently small differences in such factors as the layout of the ventilation air louvres and in the exhaust system can have a significant effect on engine performance.

Over the last 25 years, there have been several correlation exercises carried out by different companies that involved sending a sealed "golden engine" to run identical power curve sequences in test cells around the world. Very few, if any, of the detailed results have ever been released into the public domain, simply because they were judged unacceptable.

As has been demonstrated by the test procedures of fuel and lube certification, to gain acceptable correlation results, the "round robin" experiment has to be carried out using a pallet mounted engine that is fully rigged with all its fluid, air, and immediate exhaust systems. The engines need to be fully fitted with the test transducers, and each test cell should have combustion air supplied at a standard temperature. In such cases, and only when using a reference fuel and making allowances for ambient cell air pressure, correlation on the power curve points of better than $\pm 1.5\%$ should be possible.

A fairly good indication of the impossibility of using an engine as a standard in this way is contained in the Standard BS 5514. This Standard lists the "permissible deviation" in engine torque as measured repeatedly during a single test run on a single test bed as 2%. This apparently wide tolerance is no doubt based on experience and, by implication, invalidates the use of an engine to correlate dynamometer performance.

A more recently observed reason for apparent differences in cell-to-cell data was caused by the poor management and loosely controlled use of data correction factors embedded in either the data acquisition system or, in the case of AC dynamometer systems, the drive control system.

There is no substitute for the careful and regular calibration of all the machines, and while cell-to-cell correlation testing is widely practiced, the results have to be judged on the basis of the degree of control over all critical variables external to the engine combustion chambers and by staff with practical experience in this type of exercise.

## Power test codes and correction factors

Several complex sets of rules are in general use for specifying the procedure for measuring the performance of an engine and for correcting this to standard conditions.

The most significant regulations covering powertrain test procedures in Europe are covered by ISO 3046; it is possible to purchase the full documentation from various sources (see websites section).

Quoting from the Standard:

*ISO 15550 establishes the framework for ISO engine power measurement standards. It specifies standard reference conditions and methods of declaring the power, fuel consumption, lubricating oil consumption and test methods for internal combustion engines in commercial production using liquid or gaseous fuels. It applies to reciprocating internal combustion engines (spark-ignition or compression-ignition engines) but excludes free piston engines and rotary piston engines. These engines may be naturally aspirated or pressure-charged either using a mechanical pressure-charger or turbocharger. ISO 3046-1:2002 specifies the requirements for the declaration of power, fuel consumption, lubricating oil consumption and the test method in addition to the basic requirements defined in ISO 15550.*

## Part two: health and safety legislation, management, and risk assessment

The strict order of priority of all test facility safety systems must be as follows:

*First priority: protect the personnel*
*Second priority: protect the facility*
*Third priority: protect the unit under test*

Formal responsibility for H&S within a large organization will be that of a manager trained to ensure that policies of the company and legal requirements are adhered to by all employees, visitors, and the supervisory organization.

An important feature of the automotive test facility is that, in some circumstances, a potentially hazardous failure of the UUT is to be expected and that an uncontrolled discharge of energy may take place. Therefore as discussed at length in Chapter 3, Test Facility Design and Construction, the concept of "hazard containment" has to be built into, not only into the fabric of the facility but also into its operating procedures.

There are very few, if any, H&S regulations that have been developed exclusively to cover powertrain test facilities; worldwide they come under general laws related to safety at work and environmental protection. Yet the application of these general industrial rules sometimes has unintended consequences and causes operational complications, as in the case of the European ATEX regulations (see Chapter 4: Electrical Design Requirements of Test Facilities), the New Machinery Directive (EN ISO 13849-1), and EN 62061 [1,3]. The demands of EN ISO 13849-1 regulations relevant to the automotive powertrain industry have been absorbed by test organizations, and a set of generally applied best practices has emerged. The main area of difficulty has been a requirement to treat the cell structure as a "machine guard" and therefore to require a dual-processor, "safety" PLC-based system to prevent access to the cell, unless under very specific conditions.

In the numerical safety integrity level (SIL) scoring required within EN 62061, typical powertrain test cells have been graded as SIL level 2 and negotiations with accredited national organizations such as TÜV seem to have arrived at a mutually acceptable level of integration and practices. To allow powertrain component testing to be performed without being made impractical or prohibitively expensive, and in order to maintain our good record of safety, the industrial procedures have tended to be based on established and generally understood best practices. However, where precedent does not exist, as in the use of the new technologies in hybrid and electrical powertrains and vehicles involving large batteries and battery emulation, then renewed vigilance and specific risk analysis is required.

The authors recommend involvement with the industrial forums and the websites of national machinery manufacturing trade associations, many of which tend to give up-to-date advice on detailed compliance with these regulations.

## Considering the common hazards in internal combustion engine, e-motor, and powertrain facilities

The vast majority of "accidents" in automotive test facilities do not result in human injury because of compliance with the rule relating to the test cell having to form a hazard containment and human exclusion box. Reported injuries are very largely confined to those caused by poor housekeeping, such as slipping on fluid-slicked surfaces, tripping on cables, or pipes, falls due to missing floor plates and accidental contact with hot surfaces.

The developing technologies and novel configurations within BEV and HEV have increased the number and types of test now required within an automotive test facility, with a commensurate increase in new hazards, this requires that "traditional" ICE test facility management reassess their risk analyses and working practices. For new entrants into automotive testing it is vital to learnt and adapt existing industrial best practices and build their own safety practices upon them.

Rather perversely, within an EV motor or e-axle test cell, the mounted UUT connected to a battery pack and running at idle speed can appear to be comparatively "safe" and benign when compared with an ICE that is hotter and noisier; it is an appearance that is entirely misleading. The battery pack in particular is currently perceived to represent a significant hazard within a test cell, and in the majority of sites worldwide, they have been either moved out of the cell within their own containment or, more commonly, replaced entirely by using an emulator.

The two most common, serious malfunction incidents experienced in the last 20 years are as follows:

1. *shaft failures*—usually caused by inappropriate system design and/or poor assembly and
2. *fire initiated in the* UUT—in the last 10 years, most commonly caused by fuel leakages from high-pressure (common rail) engine systems, probably the result of poor system assembly or modification.

It follows, therefore, that a high standard of test assembly, and checking procedures together with the design and containment of shafts, plus staff training in the correct actions to take in the case of fire, are of prime importance.

An explosive release into the cell of parts of rotating machinery, other than those resulting from a shaft failure, is rarer than may be supposed; but ICEs do occasionally throw connecting rods, and ancillary units do vibrate loose and throw off drive belts. In these cases the debris and the consequent oil spillage should be contained by the cell structure and drainage system, and humans should, through robust interlocks, good operating procedure, and common sense, be kept outside the cell when running, above idle speed, is taking place.

Incidents of *electric shock* in well-maintained test facilities have been rare, but with the increasing development of hybrid and electric vehicle powertrains, there must be an increasing danger of electric shock and electrical burns.

## The importance of cabinet labeling

The proliferation and the very wide range in rated powers of electrical power sources and distribution systems are giving rise to possible confusion within both the modern test cells and the associated plant rooms. In the opinion of the authors, correct labeling, together with an indication of the "live" status, of many of the "anonymous" electrical panels installed within test facilities needs be improved in order to ensure a safe working environment. Not only should the normal, active or quiescent, status of the facility be considered but that of abnormal conditions when emergency or maintenance personal, unfamiliar with the details of the space, is called rapidly to deal with situations such as flooding or undiscovered source of smoke or during a partial or general power failure.

## Risk analysis

> *Risk can be defined as* the danger of or potential for, *injury, technical failure, financial loss or any combination of the three.*

While H&S managers will concentrate on the first of these, according to the priority set at the start of this chapter, senior managers have to consider all three at the commencement of every new testing enterprise or task.

The legislatively approved manner of dealing with risk management is to impose a process by which, before commencement, a responsible person has to carry out and record a risk assessment. The requirements, relating to judgment and "scoring" of risk level, of the Machinery Directive EN ISO 13849-1, which replaces EN 954-1, are shown in Fig. 2.4.

Risk assessment is not just a "one-off" paper exercise, required by a change in work circumstances; it is a continuous task, particularly during complex projects where some risks may change by the minute before disappearing on task completion.

*Staff involved in carrying out risk assessments need to understand that the object of the exercise is less about describing and scoring the risk but much more about recognizing and putting in place realistic actions and procedures that eliminate or reduce the potential effects of the hazard.*

Both risks of injury (acute), such as falling from a ladder, and risks to health (chronic), such as exposure to carcinogenic materials, should be considered in the risk assessments, as should the risks to the environment, such as fluid leaks, resulting from incidents that have no risk to human well-being.

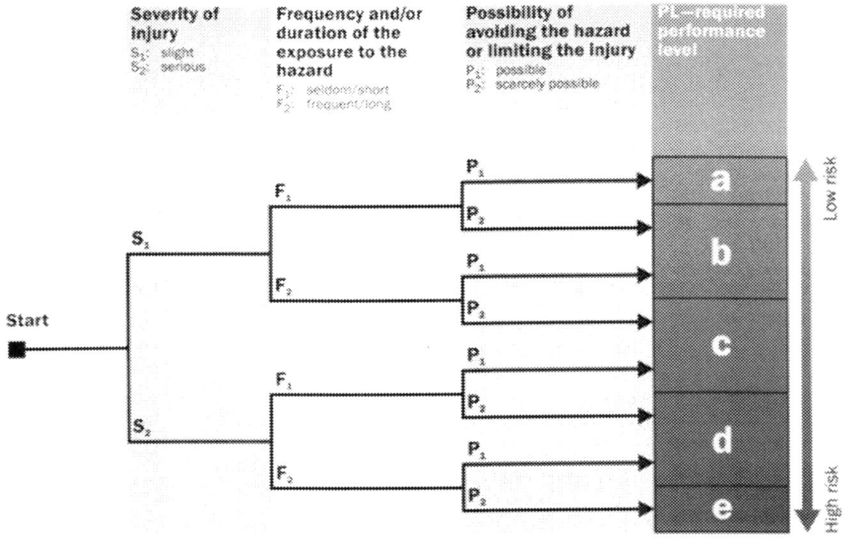


**FIGURE 2.4** The formal classification of risk levels or performance levels as defined in ISO 13849-1:2006.

There are important events within the life cycle of test facilities when H&S processes and risk assessment should be applied:

- planning and prestart stages of a new or modified test facility, both project specific and operational;
- at the change of any legislation explicitly or implicitly covering the facility;
- service, repair, and calibration periods by internal or subcontract staff;
- a significantly different test objects or test routines such as those requiring unmanned running or new fuels; and
- addition of new instrumentation.

*Note concerning subcontractor safety:* The provision of a risk assessment by a subcontractor does not abrogate the responsibility of the Client or Site Management, under whom they work, for health and safety matters related directly or indirectly to the work being done by the subcontractor. The quality of the assessment and the adherence to the processes described therein needs to be checked and monitored. Small contracting companies have been known to use customized templates of risk assessments supplied by their trade Associations and have little knowledge of their detailed contents or the responsibilities they impose.

The formal induction of new staff joining a test facility workforce and the regular review of the levels of training required with its development are important parts of comprehensive QMS, H&S, and environmental policies.

## The special case of management and supervision of university test facilities

The management and operational structures of powertrain test laboratories within universities often differ from those of industrial facilities, as does the levels of relevant training and experience of the facility user group. From casual observation, housekeeping seems to be a particular problem in and around academic automotive test cells where, often because of lack of storage space, it is not uncommon to find workspaces cluttered by stored equipment. Such clutter causes blockage to human access or escape and adds to the facility fire load.

Housekeeping is a matter of primary safety, while physical guarding, which is often given greater management attention, can be of secondary concern.

To qualify for access to the test facility, every student and staff member should have been taken through an appropriate, formal and recorded, facility safety briefing.

The rigorous enforcement and use by the senior manager of the use of test cell logbook, already mentioned, will help overcome the inherent dangers of the sometimes tortuous communication paths in academic organizations and the frequent changing of the student body; it is strongly recommended.

The authors have observed that in both university and government organizations, there is too frequently an organizational fracture between laboratory user groups and their internal facility maintenance group (Estates Department). Such situations and the time, effort, and fund wasting they cause have, from time to time, been a source of frustration and amazement to many contractors involved in facility construction and modification projects. It has been observed that, unless there is close collaboration with a commercial partner, attention to instrument calibration routines in some college test laboratories is lax, so ill prepares students for the rigors of industrial test work.

## Notes concerning determination of cause and effect

Test engineers spend much of their working life determining the difference between cause and effect. Both in isolating the value of design changes observed through test results or trying to find the cause of a system malfunction; test and commissioning staff have to develop both diagnostic skills and a habit of intelligently applied scepticism. All instruments tend to be liars, but even when the data is "true," the cause of an effect observed within complex systems, such as those discussed in this book, can be difficult to determine, even counterintuitive. With so much cause and effect embedded in the software code and design logic, both within the test machinery and the UUT, fault tracing often needs to be a multidisciplinary task and is one good reason for training engineers in mechatronics. Repeatable faults or incidents can be analyzed with relative ease, but spurious faults are a nightmare and as frequently occur in the automotive aftermarket where the common means of curing the fault without discovering its cause, are by module or connection loom replacement.

The Latin "tag" that should be in every test engineer's notebook is "post hoc, ergo propter hoc," meaning "after this, therefore because of this." It has probably been used in the teaching of logic for millennia and is a very tempting logical fallacy much practiced by today. It is an example of *correlation not causation* in which an event following another is seen as a necessary consequence of the former event. Of course, the deduction of causation could be obvious and correct, but we should retain always that attitude of intelligently applied scepticism.

The author of this section, during his years of engine and test facility fault-finding, has found it useful to remember the medical aphorism "When you hear hoof-beats, think horses not zebras," meaning that in the search for the cause of faults, gross errors should be considered before ones of great subtlety.

## End note concerning possible restrictions imposed by health and safety regulations

Quote from Ref. [3]:

> *I am well aware that the strict adherence to H&S laws, as interpreted by some officials, or "barrack-room lawyers," particularly those inexperienced in some of the tasks performed as part of special projects and in particular industries [such as automotive test facilities], can create major program delays and cost increases. However, it is sensible to remember that,* always, *those delays and cost overruns are as nothing when compared with the cost and delays of a major or fatal accident.*

## References

[1] J. Springer, Proceedings of the FISITA 2012 World Automotive Congress: Volume 6, 2012.

[2] C. Mathews, Engineers' Data Book, Wiley-Blackwell, 2012, ISBN-13: 978–1119976226.

[3] T. Martyr, Why Projects Fail, Business Expert Press, 2018, ISBN-13: 978-1-94784-390-5 (paperback).

## Websites

ASAM: Association for Standardization of Automation and Measuring Systems https://www.asam.net/.

ODS: Open Data Services https://www.asam.net/standards/detail/ods/

ISO: http://www.iso.org/iso/home.html.

Chapter 3

# Test facility design and construction

## Chapter Outline



## Introduction

In terms of the new generation of powertrain development cells, those built from around 2015 onward, the configuration of the single, two-quadrant

**Engine Testing. DOI: https://doi.org/10.1016/B978-0-12-821226-4.00003-6**

dynamometer, in-line with an automotive ICE (internal combustion engine), as described in the 1995 edition of this book, maybe considered by some readers as anachronisms. Now for many engineers the unit under test (UUT) has become an "e-axle" (Fig. 3.1) or a small "range extender" device based on a down-sized ICE of conventional or rotary design, and the requirements of the cell and its plant room have changed considerably. While the typical UUT has been shrinking in size, the volume of space required to house the test equipment, their support services, and control systems has increased significantly. In spite of all these changes, all of the fundamental design problems, involved in the physics of rotating machines and the thermodynamics of the test cell spaces, remain to be understood and solved; the first is covered in Chapter 11, Mounting and Rigging Internal Combustion Engines for Test, and the later in Part 1 of this chapter. Housing may now be required for large rectilinear arrays of cabinets and infrastructure to handle battery emulation, multiples of four-quadrant dynamometer power and controllers, in addition the exhaust treatment and ventilation services.

In the case of hybrid ICE-electric systems a full powertrain cell is invariably required. Many facility designers are having to deal with the problems involved in adapting their existing test cells (now individually too small) to deal with the needs of larger, more complex, and more electrical energy-demanding equipment and UUT; in most cases this is difficult, in some

**FIGURE 3.1** A sectioned example of an *e-axle unit* consisting of a PM motor, reduction and differential gearing, and electrical supply inverter unit. This is the UUT that is rapidly replacing the ICE in automotive test facilities. *Reproduced with permission from ZF Passau GmbH.*

practically impossible. But, whatever the size of the test cell it has always to be considered as, and designed to be, a hazard containment box. Any test cell built today should be considered as "machinery" under the broad definition contained in the European Machinery Directive 2006/42/EC. In this sense the cell doors are essentially now the guard for the machine. Any internal guards, such as those for the driveshaft, are essentially supplementary, that is, additional protection to personnel when controlled access to the test cell is required and the UUT is in a "safe," reduced speed mode, or shut down while still hot or electrically alive.

The rapidly increasing electrification of automotive test facilities has brought with it problems in defining the capacity and rating of the main electrical supply. Electrical supply companies are having to come to terms with the rating and types of switchgear and transformers supporting the potential electrical regenerative capacity of modern test cells fitted with AC (alternating current) dynamometers and the electrical energy stored within battery unit test facilities. This is a situation discussed in more depth in Chapter 4, Electrical Design Requirements of Test Facilities.

This role of the test cell as a complex programmable machine has to be kept in mind when considering all aspects of test cell design and its location within its immediate environment.

## Part 1: Calculation of a test facility energy balance

Many problems are experienced in test cells worldwide when the thermodynamics of the cell has not been correctly catered for in the design of cooling

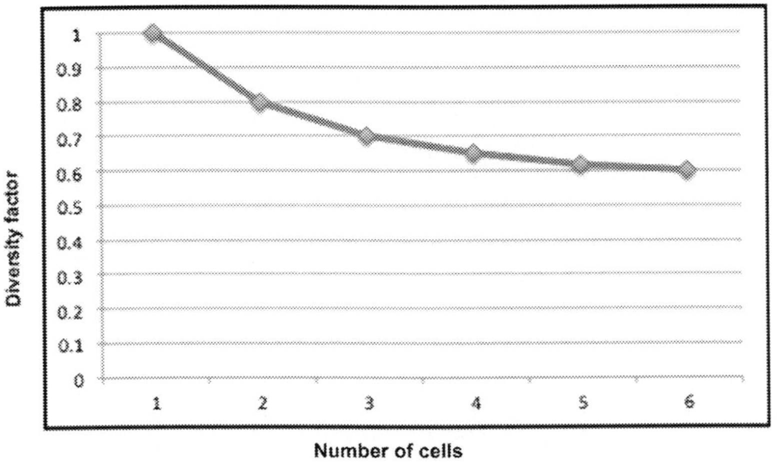


FIGURE 3.2 Diversity factor of the thermal rating of test facility services, plotted against the number of test cells in the facility, based on empirical data from typical automotive test facilities from 1980 to 2010.

systems. The most common problem is high air temperature within the test cell, either generally or in critical areas. The practical effects of such problems will be covered in detail in Chapter 6, Cooling Water and Exhaust Systems, but it is vital for the cell designer to have a general appreciation of the contribution of the various heat sources and the strategies for their control.

Given sufficient detailed information on a fixed engine/cell system, it is possible to carry out a very detailed energy balance calculation. Alternatively, there are some commonly used "rule-of-thumb" calculations available to the cell designer; the most common of these relates to the energy balance of the ICE, which is known as the "30–30–30–10 rule." This refers to the energy balance given in Table 3.1

Like any rule of thumb this is crude, but it does provide a starting point for the calculation of a full energy balance and a datum from which we can evaluate significant differences in balance caused by the UUT itself and its mounting within the cell.

First, there are differences inherent in the engine design. Diesel engines of the size fitted in busses and trucks will tend to transfer less energy into the cell and more into the water than petrol engines of equal physical size.

Second, there are differences in engine rigging in the cell that will vary the temperature and surface area of engine ancillaries such as exhaust pipes. Finally, there is the amount and type of equipment within the test cell, all of which make a contribution to the convection and radiation heat load to be handled by the ventilation system.

Specialist test facility designers have developed their own versions of a test cell's software model, based both on empirical data and theoretical calculation, all of which is used within this book. The version developed by colleagues of the author produces the type of energy balance shown in Fig. 3.3, while Table 3.2 lists just a selection, from an actual project, of the known data and calculated energy flows that such programs have to contain in order to produce such a widely useful diagram.

**TABLE 3.1** **Showing the classic 30–30–30–10 rule of energy distribution in an internal combustion engine test cell.**

| In via | Out via |
|---|---|
| Fuel 300 kW | Dynamometer 30% (90 + kW) |
| | Exhaust system 30% (90 kW) |
| | Engine fluids 30% (90 kW) |
| | Convection and radiation 10% (30 kW) |

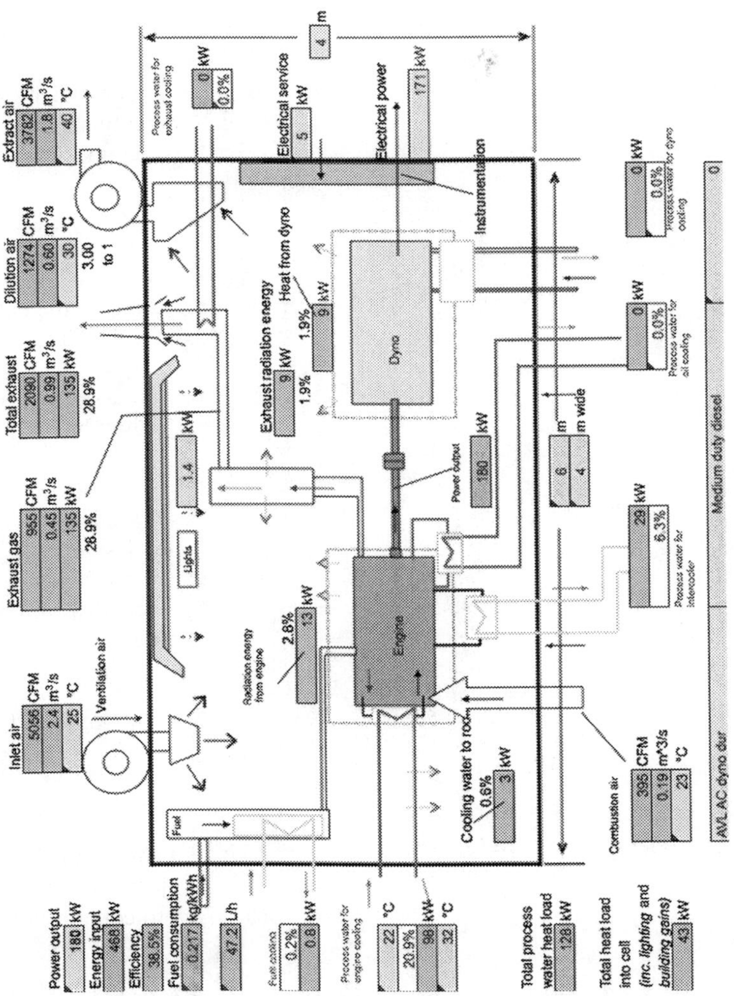


FIGURE 3.3 The spreadsheet output diagram from an ICE test cell thermal analysis and energy flow diagram. It shows all of the energy sources and soaks of a typical single shaft cell except for compressed air which is not considered significant in this case. A selection of the data used to produce the diagram is shown in Table 3.2.

As stated elsewhere in this book the thermodynamic model of engine and powertrain test cells is very different from that of almost every other commercial workspace and requires a great deal of practical experience to get right. Even so, while such tools are extremely useful they cannot be used uncritically as the final basis of design when a range of engines need to be tested or the design has to cover two or more cells in a facility where fluid

**TABLE 3.2 A selection of the known, estimated and calculated data, based on a known engine required to produce an energy and fluid flow diagram as shown in Fig. 3.3.**

| | | | |
|---|---|---|---|
| *Engine and fuel data* | | | |
| Power output | 180 kW | Engine max. power | 180 kW |
| Fuel | Diesel | Primary energy from fuel | 468 kW |
| Calorific value of fuel | 43,000 kJ/kg | *Electricity Output* | |
| Density of fuel | 0.830 kg/L | From AC dyno. | 171 kW |
| *Combustion air* | | *Cooling water loads* | |
| Temp. intake | 23°C | Lube oil HX (N/A) | 0 kW |
| Temp. after compressor | 185°C | Engine jacket HX | 98 kW |
| Temp. after intercooler | 55°C | Intercooler HX | 29 kW |
| Combustion air temp. | 70°C | Chilled Water | |
| *Exhaust gas* | | Fuel cooling | 1 kW |
| Manifold temp. | 650°C | Cell heat loads (radiation from) | |
| Temp. after turbine | 434°C | Engine block | 13 kW |
| Temp. in cell system | 400°C | Exhaust in cell | 9 kW |
| Dilution air temp. | 30°C | Cooling water | 3 kW |
| Temp. at cell exit | 100°C | Dynamometer | 9 kW |
| Dilution air ratio | 3 | Exhaust system | |
| *Plant cooling water* | | Exhaust gas out of cell | 135 kW |
| Glycol content (%) | 50 | | |
| Temperature in | 22°C | | |
| Temperature out | 32°C | | |
| Specific heat capacity | 3.18 kJ/l K | | |
| Density | 1.06 kg/m$^3$ | | |
| *Peak fuel consumption* | | *Exhaust dilution* | |
| Mass rate | 39.1 kg/h | Ratio (kg air/kg exh.) | 3 |
| Volume rate | 47.2 L/h | Mass flow at intake (air) | 2519 kg/h |
| | 0.79 L/min | Density at intake (−50 Pa) | 1.16 kg/m$^3$ |
| Specific fuel consumption | 0.217 kg/kWh | Volume rate at intake | 2164 m$^3$/h |

(*Continued*)

**TABLE 3.2 (Continued)**

| | | | |
|---|---|---|---|
| *Combustion air* | | | 0.60 m³/s |
| Mass rate | 800 kg/h | Total mass flow (air + exh.) | 3358 kg/h |
| Density at 1 bar abs. | 1.19 kg/m³ | Mixture temp. after intake | 123°C |
| Volume rate | 672 m³/h | Mixture density out of cell | 0.95 kg/m³ |
| | 0.19 m³/s | Mixture flow out of cell | 3551 m³/h |
| Exhaust gas | | | 0.99 m³/s |
| Mass rate (air + fuel) | 840 kg/h | | |
| Density, after turbine | 0.739 kg/m³ | | |
| Volume rate after turbine | 1136 m³/h | | |
| Density, out of cell | 0.517 kg/m³ | | |
| Volume rate out | 1622 m³/h | | |
| of closed system | 0.45 m³/s | | |

The production of this type of model has proved very useful in the design of automotive test facilities. When based on the maximum and minimum UUT power outputs the systems "turn-down ratio" can check for feasibility of control. (See Chapter 5, Ventilation and Air Conditioning of Automotive Test Facilities, and Chapter 6: Cooling Water and Exhaust Systems).

services are shared; in those cases the energy diversity factor has to be considered.

To design a multicell test laboratory able to control and dissipate the maximum theoretical power of all its prime movers on the hottest day of the year will lead to an oversized system and possibly poor temperature control at low heat outputs. The amount by which the thermal rating of a facility is reduced from that theoretical maximum is the *diversity factor*. In Germany, it is called the *Gleichzeitigkeits Faktor* and is calculated from zero heat output upward, rather than 100% heat output downward, but the results should be the same, providing the same assumptions are made.

The diversity factor often lies between 60% and 85% of maximum rating but individual systems will vary from endurance beds with a high rating down to anechoic beds with a very low rating.

In calculating, or more correctly estimating, the diversity factor it is essential that the creators of the operational and functional specifications use realistic values of actual engine powers, rather than extrapolations based on possible future developments.

A key consideration is the number of cells included within the system. Clearly, one cell may at some time run at its maximum rating, but it may be considered much less likely that four cells will all run at maximum rating at the same time: the possible effect of this diminution from a theoretical maximum is shown in Fig. 3.2. Note that this graph is based on data collected over the 30 years ending in 2010; since that time modern powertrain cells have been designed for 24 hour running and their UUTs are prerigged to the degree that running time is maximized; therefore in such cases the diversity factor may be in the 85%–90% range.

There is a degree of bravery and confidence, based on relevant experience, required to significantly reduce the theoretical maximum to a contractual specification, but very significant savings in running costs may be possible if it is done correctly.

## Part 2: Test facility workspace design

### Frequency of change of units under test and their handling systems

An important consideration in determining the size and layout of a test cell and its immediate environment is the frequency with which the UUT has to be changed. At one extreme in the automotive sizes are "fuel and lube" cells, where an ICE is virtually a permanent fixture because it is the fluids that are being tested, and at the other extreme are production test cells, where the UUT test duration is a few minutes and the process of rig and derig is fully automated.

The system adopted for transporting, installing, and removing the UUT has to be considered together with the layout and content of the support workshop, any joining corridors, and the position of the control room (see Chapter 11: Mounting and Rigging Internal Combustion Engines for Test, for coverage of UUT pallet systems). All operational specifications for new or modified test facilities should include the intended frequency of change of the UUT in their first draft.

Electrical drive modules, from single motor to double-motor e-axle units, are usually much easier to exchange within their test field than an ICE, which has a greater number and type of service connections. The use, where available, for vehicle/powertrain plugged looms ensures connection integrity and allows fast turnaround.

### Workspaces and their roles

One of the early considerations in planning a new test facility will be the floor space required on two or more levels. Normally, designers have to first consider the vertical layout of the test cell area within the facility; wherever they are located, three functionally defined floor areas will be required:

1. The test cell floor, normally but not always on the ground floor containing a bedplate on which the UUT is mounted. On the same floor will be the control room with direct personnel access between the two.

2. The services floor, housing the plant room, normally but not always on the floor above the cell. It used to be a design principle that the footprint of the cells' individual services matched the cell beneath; but with the proliferation of electrical cabinets serving electrical dynamometers and/or electrical UUT, this is now rarely possible meaning plant-room layout needs careful consideration. Services shared by cells are usually off-set on the Services floor in such a way to minimize cable and duct lengths.
3. The "external" floor, housing equipment such as chiller units, exhaust induced-draught fans, and exhaust outlets that cannot be positioned within a building space. This floor may be on the building roof or may be dispersed outside the building on the ground floor; it is as much an architectural decision as a technical one.

   Considering now the horizontal layout of a test facility, in addition to the three core functional roles defined above, the additional interconnected workspace areas to be considered and whose roles need to have been defined in the operational specification are as follows:
4. The support workshop, UUT rig, and derig areas. Not only will the size and contents depend on the number of cells supported but also in the nature of the units being tested. For a modern hybrid-capable powertrain test facility the following rooms, complete with their own tool and consumables store, within the workshop support complex will be needed, containing:
   a. basic machine and sheet metal workshop,
   b. welding bay equipped for MIG welding of exhaust systems etc.,
   c. electrical and Instrument workshop with multicore connection-cable production facilities, and
   d. test pallet and UUT rigging shop containing a charging and parking bay for test vehicles and any powertrain module transporting truck.
5. A secure storage area required for units to be tested and their rig items.
6. Office space for staff directly involved with the test cell, workshops, also data handling and postprocessing.

## The test cell—the hazard containment box

A modern fully equipped test cell is the most expensive piece of real-estate within any facility; therefore its beneficial use must be maximized by reducing the time taken to carry out tasks such as rigging and derigging the UUT. In particular, there needs to be clear rules concerning the time allowed within the cell for repair and modification of the UUT.

A cramped or overcrowded cell, in which there is no room to move around with ease and safety is a permanent source of danger and inconvenience. As a rule of thumb, in automotive cells, there should be an unobstructed walkway 1 m wide all-round the rigged UUT and its connected dynamometer(s), but in cases where the exhaust tubes form a barrier to a full

360 degrees of access the access walkway is more often horseshoe shaped rather than an oval. The horizontal space required for this 1-m walkway is additional to the space required by the protrusion of wall-mounted cabinets, fuel distribution panels, etc.

In powertrain cells containing large air-sprung bedplates (see Chapter 8, Vibration and Noise) the concrete floor area surrounding the bedplate should provide the walkway.

It is now often necessary, when testing automotive engines, to accommodate most of the exhaust system as used on the vehicle. The variation in engine and exhaust configurations to be used within a single cell may be a significant factor in determining the size of an ICE test cell and the choice of dynamometer. Preplanning activities need to include mock exhaust system layouts and should pay particular attention to the possible need to run some exhausts under or very close to the dynamometer plinth, in which case a physically small permanent magnet motor–based dynamometer such as the AVL DynoULTRA™ may be required.

It must be remembered that much of the plant in a test cell requires regular calibration so there must be adequate access for the calibration engineer and his instrument trolley.

The calibration of dynamometers requires accommodation of a torque arm and calibration weights, usually on both sides of the machine. A classic layout error is to install electrical boxes, trunking, or service pipes that clash with the hanging position of the dynamometer calibration weights.

## Are lifting beams needed on in automotive powertrain test cells?

Before we consider the optimum height of any test cell, we should consider the case for and against the inclusion of a lifting beam within the cell. An in-cell crane is essential when working on large machines such as medium-speed diesels and were often in automotive cells built before 1990, but they come at a construction and insurance cost, particularly in containerized or modular panel-built cells (see later). This added cost comes about because the whole building frame has to be strengthened to allow it to support the crane built to the allowed bending limits required by the insurer's overload capacity. Deep crane beams also lower effective cell height and can make airflow and lighting more difficult to optimize. Besides being one, albeit rather slow, method of engine mounting and demounting, a cell crane is certainly useful during the initial installation of the cell. Subsequently, crane beams can be used for, hopefully, rare, major maintenance and tend to be used as a "sky-hook" for supporting various parts of the rigging looms and instrumentation. Where a powertrain UUT is rigged on a removable pallet or trolley system, which is maneuvered by an electric tug or vehicle handling system, the cost/benefit calculation does not often justify a fixed crane installation.

*Cell roof height.* Most modern automotive cells are between 4 and 4.5 m internal height with the lowest practical height around 3 m.

In addition to any crane system running above head-height, some equipment may be mounted on booms that hinge from the cell wall and carry a box containing an interface panel for internal transducers and signal conditioning (see later).

When design restraints, such as those imposed by some modular cell designs (see modular cells later) restrict the internal roof height of 3 m or below then a customized electrically powered vehicle will have to be employed to go under the UUT, and move it into and out of the cell. In the case of a full powertrain cell, fitted with fixed wheel dynamometers, when being rigged with a complete vehicle, the UUT has to be brought into the cell, without its wheels fitted and be lifted so that the wheel hubs align with the dynamometer centerline; one method of achieving this maneuver is to use a low height, battery powered, four-wheel truck fitted with a lift-and-fall table.

*Note:* In the context of a cell fitted with a tee-slot bed-plate "fixed" wheel-hub dynamometers means that the dynamometers are mounted on a frame that is bolted to the base-plate and on which the axial position of the dynamometer body may be moved in order to engage the articulated flange with the vehicle's wheel hub. This configuration is different from the fully mobile, wheeled, hub-dynamometers used in "flat-floor" test areas (see Chapter 10: Chassis Dynamometers, Rolling Roads, and Hub Dynamometers).

## Best position for power and control cabinets: in an internal combustion engine test cell or plant room?

High powered AC dynamometer drive cabinets are large, heavy, expensive, and contain solid-state devices such as isolated gate bipolar transistors that can be terminally damaged by air polluted with metal dust particles. In the first decade of the 21st century the standard advice was to install AC drive cabinets outside any ICE test cell and in a separate clean, well-ventilated environment within the plant room. Nowadays, cabinets are built for industrial environments and fitted with their own cooling air filtration and circulation systems, in addition, in large units, to water cooling. The problem for the designer therefore is to find the space to house the cabinets array and the swept area of their doors, rather than to worry so much about their environment. In the case of powertrain test cells, the space needed to accommodate the drive cabinets of four-wheel simulating dynamometers may use the entire wall of the powertrain cell's width.

## Seeing and hearing the unit under test: are test cell windows required?

Except in the case of large marine diesel engines, after-market, portable dynamometer stands, and some production test beds, it is a universal practice to have a wall as a physical barrier (the hazard containment) to separate the control space from the running machinery and within that wall to have a window. In addition to using up cell wall space test cell windows cause a number of cell design and operational problems. The degree to which a well-constructed test cell can form a hazard containment box may depend critically on the number and type of windows fitted into the cell walls and doors. The different glass types suitable for cell windows are sold either as "bulletproof" (BS EN1063:2000) or "bandit proof" (BS 5544). There are also quite separate fire-resistant types of glass such as Pilkington Pyrodur. The decision as to whether it is more likely for a large lump of engine (bandit) or a high-speed fragment of turbocharger (bullet) to hit the window has the potential to prolong many an H&S meeting and will not be resolved in this volume. However, good practice would suggest that the glass (and its frame) on the cell side should have the highest fire resistance, and the control room side glass should be the most impact resistant. To achieve good sound attenuation, two sheets of glass with an air gap of some 80 mm is necessary. Many motor-sport or aero-engine cells have three panes of glass with at least one set at an angle to the vertical to minimize internal reflections.

Thanks to modern instrumentation and closed-circuit television, a window between the control room and test cell is rarely absolutely necessary. The use of multiscreen data displays means that many cell windows have become partially blanked out by these arrays.

The choice to have a window or not is often made on quite subjective grounds. Cells in the motor-sport industry all tend to have large windows because they are visited by sponsors and press, whereas original equipment manufacturer (OEM) multicell research facilities increasingly rely on remote monitoring and CCTV. To maintain confidentiality, in test facilities that support several customers and have shared control rooms or a corridor, it is sometimes necessary to fit and operate opaque roller shutter blinds to the windows.

The importance or otherwise of visibility of the UUT from a control room window is linked to the layout and function of the cell. A number of variants of layout are covered later in this chapter.

Even where windows are fitted the experienced operator will be concentrating attention on the indications of instruments and display screens and will tend to use peripheral vision to catch sight of unexpected events in the cell. It is very important to avoid unnecessary visual distractions, such as dangling labels or identity tags that can flutter in the ventilation wind.

Hearing has always been important to the experienced test engineer, who can often detect an incipient failure by ear before it manifests itself through an alarm. Unfortunately, modern test cells, with their generally excellent sound insulation, cut off this source of information and many cells need to have provision for in-cell microphones connected to switchable control room loudspeakers or earphone sockets.

## Test cell flooring and subfloor construction

Whatever the design of the cell structure a substantial concrete floor, or seismic block when fitted (see Chapter 8: Vibration and noise), must be provided with arrangements for bolting down the UUT and the dynamometer(s). A low-cost solution, suitable for ICE testing when prerigged pallets are used, is to precisely level and cast in position, two or more cast-iron T-slotted rails. The machined surfaces of these rails form the datum for all subsequent alignments, and they must be set and leveled with great care and fixed without distortion. The use of fabricated steel-box-beams has proven to be a false economy.

Sometimes complete cast-iron floor slabs containing multiple T-slots are incorporated in the concrete cell floor, but this configuration tends to trap liquids and debris unless effective drainage is provided. All cast-iron floor slabs can become very slippery so good housekeeping practices and equipment are essential.

The air-sprung bedplate discussed in detail in Chapter 8, Vibration and Noise, is the modern standard design. Note that the building construction plan must make arrangements for the maneuvering into place of these large and heavy bedplates before suitable access is closed by building work.

Concrete floors should have a surface finish that does not become unduly slippery when fluids have been spilt; special "nonslip," marine, deck paints are recommended over the more cosmetically pleasing smooth finishes. High-gloss, self-leveling floor surfaces have found wide favor in new factories but when fuels, lubricants, or cooling fluids are spilt on them they become dangerously slippery. Whatever the finish used, it must be able to resist chemical attack from the fluids, including fuels, used in the cell.

Even without the presence of a seismic block, which provides the floor void, it is good practice to provide floor channels on each side of the bed, as they are particularly useful for running fluid services and drain pipes in a safe, uncluttered manner. However, regulations in most countries call for spaces below floor level to be scavenged by the ventilating system to avoid any possibility of the build-up of explosive vapor, thus slightly complicating the cell design (see Chapter 5: Ventilation and Air Conditioning of Automotive Test Facilities).

All floor channels should be covered with well-fitting plates capable of supporting the loads that are planned to run over them. To enable easy removal for maintenance each plate should not weigh more than about 20 kg and be provided with lifting holes. The plates can be cut as necessary to accommodate service connections but have to provide safe footing—accidents caused by insecure channel cover plates are too common. Fuel service pipes, once commonly run into the test cell underground and within the cell in-floor trenches, should nowadays follow modern environmental and safety best practice, and remain in view above ground in cells and be easily accessible.

Apart from the use of floor channels discussed previously, there are two variations in subtest floor design:

- A cellar design where the whole of the test-floor is mounted as an "island" on the top of a table or plinth based on the floor below. The cellar is used to house much of the Service Systems, cables and piping and is a fully ventilated and purged workspace. In these cases, the test cell floor is usually formed by substantial concrete sections rather than the lift-out metal plating used for service channels or the shallow subfloor designs. This type of construction is expensive but gives good results in the case of anechoic test facilities.
- A shallow (c. 1 m) space throughout the whole test cell(s) (but surrounding the plinth) and control room(s) is used to run a substantial proportion of the cable trays carrying power and signal cable looms. The justification for this type of layout is often the wish to hide much of the cable trays and as much as possible of the piped services. There are however a number of significant drawbacks the chief one being that the resulting space is very difficult to work in due to the cramped access. Having to, even infrequently, lift floor plates in order to carry out maintenance work creates trip hazards to other users and any significant spillage of fluids may use cable looms as a convenient conduit.

### *Fluid drainage from test cells and rig areas*

To prevent environmental pollution, any fluids released or spilt in the test cell or rigging workshop has to be contained and held within the facility drainage system rather than leaking into the foul water or land drainage systems. Cell, plant room and garage workshop floors, or any channels cast in them, have to be fitted with drains running into an "intercept tank." As it is hoped that the pollutants, such as oil and fuels spilt are of low volume, such that a common intercept tank for the whole facility can usually be installed. If vehicle washing is carried out in the workshop, then a larger tank or tanks will have to be sized accordingly. A maintenance schedule has to be put in place to inspect and pump out the intercept tanks periodically and after any major incident such as an ICE block failure. Since the flow from cell drains is usually very small it is important that they are fitted with the usual "U Bend" trap that will need occasional flushing with water to prevent evaporation and odors.

## Wall and roof construction of nonmodular test cell systems

Test cell walls are required to meet certain special demands, in addition to those normally associated with an industrial building. In addition to supporting the weight of the services room and roof above them:

- The wall fabric or the frame within which they are built must support the load imposed by in-cell equipment such as any crane installed in the cell, any equipment mounted on, or suspended below, the roof.
- The walls must be of sufficient strength and suitable construction to support wall-mounted instrumentation cabinets, fuel systems, and any equipment carried on booms cantilevered out from the walls.
- They should provide the necessary degree of sound attenuation
- They must comply with requirements regarding fire retention, usually a minimum of 1-hour containment.

In conventional building construction high-density building blocks provide good sound insulation. This feature may be significantly enhanced by filling the voids with dried, dense casting sand after the blocks are laid and before the roof is fitted; however, this lead can to problems when creating wall penetrations after the original construction due to sand and dust leakage from the void above the penetration. Walls of whatever construction usually require some form of acoustic treatment to their internal face, such as 50-mm-thick sound-absorbent panels, to reduce the level of reverberation in the cell. Such panels can be effective in reducing peak noise levels even if some areas are left uncovered for the mounting of equipment.

In non-ICE cells testing electrical powertrain components the easier to clean option of having cast concrete walls that are ceramic tiled may be preferred.

The key feature of test cell roofs, which usually form the floor of the services room above, much disliked by structural engineers, is the number, position, shape, and size of penetrations required by the ventilation ducts and various services. It is vital that the major penetrations are identified early in the facility planning as they may affect the choice of the best construction method. In the author's experience penetrations, additional to those planned and provided during the design stage, are invariably required by one or more subcontractors even when prior discussions have taken place.

Roof construction techniques, in addition to classic form-work supported, cast in situ, concrete slab, include "rib-decking" (Fig. 3.4), which consists of a corrugated metal cell ceiling that provides the base of a reinforced concrete roof that is poured in situ, also hollow-core concrete planking (Fig. 3.5). Each system has inherent advantages and disadvantages. Concrete planking provides an easy and clean erection method but the internal voids in material mean that a substantial topping of concrete screed is required to obtain good sound insulation. The comments above concerning large penetrations are particularly true in this case because major postinstallation modification is extremely difficult.

Suspended ceilings made from fire-retardant materials hung from hangers and a frame fixed into the roof are very rarely fitted in ICE tests cells and only in cases where the "industrial" look of concrete or corrugated metal is

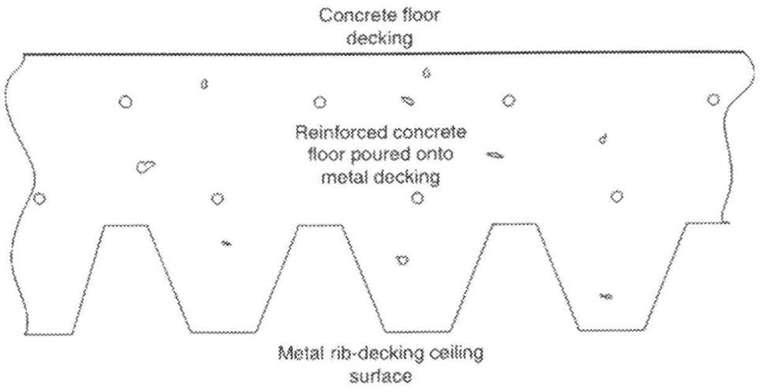


**FIGURE 3.4** Not to scale sketch section through metal rib-decking and poured concrete construction, reinforcing steelwork not shown.

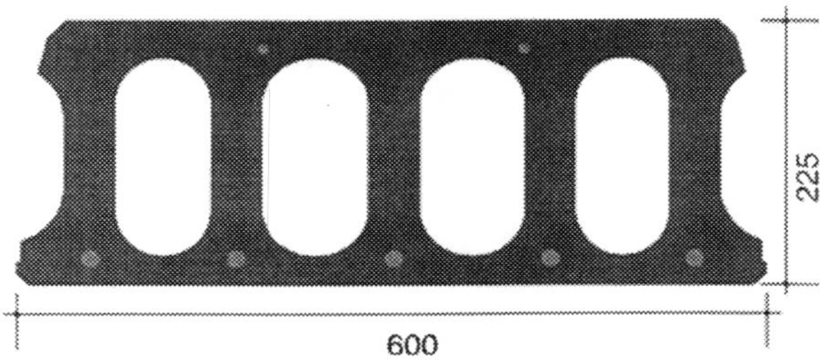


**FIGURE 3.5** A section through a concrete plank which gives a quickly positioned and high strength cell roof. It would normally be topped with a concrete screed. Such construction is difficult to modify postinstallation.

deemed unacceptable, but this has to be shaped around every roof-penetrating service duct and pipe and is usually only financially justified in high-profile research, customer- or sponsor-visited sites.

## *Plant room structure and contents*

Once upon a time, it was usual for the footprint of the plant-room machinery serving an ICE test cell to match that of the cell below. With the increase in the number and size of electrical cabinets housing such things as AC drives, battery emulators, and air-conditioning plant, this is frequently impossible. It is as vital to get right the layout and weight distribution of the services (plant room) floor at the planning stage as that of the cell below. If cell air-conditioning is to be provided by air-handling units (AHUs), these will dominate a large area above the cell's centerline, conveniently requiring other plants to take up space over each side of the cell and use penetrations along the wall lines.

Where an AHU is not used, the ventilation plant (including fans, inlet, and outlet louvers) and ducting (including sound attenuation section) can be difficult to layout in order to gain the best use of floor-space (see Chapter 5: Ventilation and Air Conditioning of Automotive Test Facilities) In addition to the ventilation plant the plant room may have to house part, or the bulk of, the following systems:

- engine or gas-generator exhaust system (see Chapter 8: Vibration and Noise, and for emission equipment housing, see Chapter 17: Engine Exhaust Emissions)
- electrical power distribution cabinets and AC Drive cabinets (see Chapter 4: Electrical Design Requirements of Test Facilities)
- fluid services, including cooling water distribution, chilled water chiller unit and reticulation, fuel pipes, and compressed air system (see Chapter 6: Cooling Water and Exhaust Systems)

In addition to these standard services, space in some facilities needs to include:

- combustion air (CA) treatment unit (see Chapter 6: Cooling Water and Exhaust Systems)
- battery emulator (see Chapter 12: Rigging and Running of Electrical Drive Systems)

**TABLE 3.3** **Examples of the internal dimensions of some modern automotive ICE and powertrain test cells. The tendency is for ever larger space requirements in spite of the automotive UUT 'shrinking'.**

| Call dimensions (actual installation sizes) | Cell category |
|---|---|
| 6.5 m long × 4 m wide × 4 m high | QA test cell for small automotive diesels fitted with eddy-current dynamometer |
| 7.8 m long × 6 m wide × 4.5 m high | ECU development cell rated for 250 kW engines, containing workbench and some emission equipment |
| 6.7 m long × 6.4 m wide × 4.7 m high | Gasoline engine development cell with AC dynamometer, special coolant, and inter-cooling conditioning |
| 9.0 m long × 6 m wide × 4.2 m high (to suspended ceiling) | Engine and transmission development bed with two dynamometers in "tee" configuration. Control room runs along 9 m wall |
| 12 m length × 9 m wide, 2.9 m high. (*roof height enabled by use of low electric truck to transport UUT into and out of cell*) | Full four-wheel powertrain using whole vehicle or powertrain components and HiL configurations |

### *Disabled access*

The authors have been asked about the provision of wheel-chair access to various areas of automotive test facilities. The general rules recently applied in Europe and the United States are all based on local H&S legislation and the legal designation of the various workspaces. Any new test (2020) facility should make all areas that can be zoned as office space wheel-chair accessible, this could include the (ground floor) control room(s), which would, of course, require a suitable escape route in the case of an emergency.

Unless a plant-rooms contains work areas whose function requires, or forms, the normal place of office work for an operative, they do not have to be accessible for wheel-chair users via an elevator—they are deemed maintenance or occasional work areas. In many cases wheelchair access would be limited due to low- and mid-level ventilation ducting.

## Cell and control room lighting

Lighting of all work areas is an H&S critical subject and good publications concerning the legal requirements are widely available in print and on-line [1]. The internal spaces of a test cell and its control room require special consideration.

The typical test cell ceiling may be cluttered with fire-sprinkler systems, exhaust outlets, ventilation ducting, and a lifting beam. The position of lights is often a late consideration, but it is of vital importance, for all roles and viewing methods, to provide even illumination of both the UUT and the in-cell plant without areas of deep shadow without causing glare into the control room.

Lighting technology has advanced considerably in the last decade with the advent of high efficiency LED luminaires which allow many more design possibilities than the past generations of tube and bulb light sources, so the detailed design of a modern lighting system is a matter for the specialist. All lighting units within the cell must be securely mounted so as not to move in the ventilation "wind."

Unless special and unusual conditions or regulations apply, cell lighting does not need to be explosion proof; however, units may be working in an atmosphere of soot- and oil-laden fumes and need to be sufficiently robust and be easily cleaned. Lighting units fitted with complex and flimsy diffuser units are not suitable.

Lighting in the control room has to be arranged such that reading the flat-screen displays is not adversely affected by reflections; if a cell window is fitted the same lack of reflection in the window is important.

*Note of lighting standards*: The "lumen" method of lighting design gives the average level of illumination on a working plane for a particular number of light sources (luminaries) of specified power arranged in a symmetrical pattern.

(*Continued*)

**(Continued)**

Factors such as the proportions of the room and the albedo of walls and ceiling are taken into account. The unit of illumination in the International System of Units is known as the *lux*, in turn defined as a radiant power of 1 $lm/m^2$. The IES Code [1] lays down recommended levels of illumination in lux for different visual tasks. A level of 500 lux in a horizontal plane 500 mm above the cell floor should be satisfactory for most cell work, provided areas of deep shadow are avoided. In special cells where detailed work and in situ inspections take place, the lighting levels may be variable between 500 and 1000 lux.

*Emergency lighting* with a battery life of at least 1 hour and an illumination level in the range 30–80 lux should be provided in both test cell and control room.

## Test cell doors

Note that the following comments on doors apply to both concrete and modular panel constructed cells.

Cell doors have to meet the criteria set for the hazard containment box they for part of and to meet those requirements of noise attenuation and fire containment they are inevitably heavy and require more than normal effort to move them. This is a safety consideration to be kept in mind when designing the cell and its forced ventilation system.

Forced or induced ventilation fans can cause pressure differences across doors, making it dangerous (door flies open when unlatched) or impossible to open a large door. The recommended cell depression for ventilation control is 50 Pa (see Chapter 5: Ventilation and Air Conditioning of Automotive Test Facilities).

Interlocked (see later) test cell doors where operators are permitted to enter when tests are running must be either on slides or be outward opening. The double-walled and double-doored cell shown in Fig. 3.6 is interlocked to prevent entry whenever the cell controls are reset because the inner door opens inward; the unusual double sections are required to contain the very high noise levels of F1 and other race engines.

There are designs of both sliding and hinged doors that are suspended and drop to seal in the closed position. Sliding doors have the disadvantage of creating "dead" space along the wall against which they open.

Doors opening into normally occupied workspaces should be provided with small observation windows and may be subject to regulations regarding the provision of "exit" signs. As mentioned elsewhere and detailed in Chapter 4, Electrical Design Requirements of Test Facilities, all doors into a

**FIGURE 3.6** A double set of well-sealed, interlocked cell doors built as part of a modular panel constructed cell housing a high-noise ICE test bed. *Photo courtesy Envirosound.*

test cell must be interlocked in such a way as to switch the machinery within into an agreed safe state as soon as one of the doors begins to open.

## Modular and containerized test cells

In the context of automotive test cells use of the term "containerized" has become unclear and misused so this edition uses the terms related to the family of "metal box" cells with the following specific meanings:

*Type 1:* A *Containerized* test cell or cell complex is based on one or more, modified but with external dimensions unaltered, ISO Container Units which may be connected (tethered) to service connections and drains but are essentially able to be uplifted and moved to another site. The size and internal space are restricted by the ISO standard for standard or refrigerated shipping containers (see Table 3.4).

*Type 2*: A *modular, all-weather*, test cell or cell complex is constructed in such a way as it may be substantially constructed remote from its final workplace then, in whole or part, can be transported and installed outside on a preprepared site or within a preprepared shell building. Such units may be constructed from ISO containers that have had side or end walls removed so that they can be bolted together to form the building sections of a much larger (non-ISO) construction. The size of its sections is restricted only by the restraints imposed by road transport of the major sections of the disassembled unit.

**TABLE 3.4 Dimensions of standard ISO containers.**

| ISO containers | | 20 ft. | 40 ft. | 40 ft. high-cube | 45 ft. high-cube |
|---|---|---|---|---|---|
| *External* | Length | 6.058 m | 12.193 m | 12.193 m | 13.716 m |
| | | 19′ 10.5″ | 40′ 0″ | 40′ 0″ | 45′ 0″ |
| *Dimensions* | Width | 2.438 m | 2.438 m | 2.438 m | 2.438 m |
| | | 8′ 0″ | 8′ 0″ | 8′ 0″ | 8′ 0″ |
| | Height | 2.591 m | 2.591 m | 2.896 m | 2.896 m |
| | | 8′ 6″ | 8′ 6″ | 9′ 6″ | 9′ 6″ |
| *Interior* | Length | 5.867 m | 12.032 m | 11.989 m | 13.513 m |
| | | 19′ 3″ | 39′ 5.7″ | 39′ 4″ | 44′ 4″ |
| *dimensions* | Width | 2.352 m | 2.352 m | 2.311 m | 2.352 m |
| | | 7′ 8.9″ | 7′ 8.9″ | 7′ 7″ | 8′ 9.6″ |
| | Height | 2.385 m | 2.385 m | 2.667 m | 2.691 m |
| | | 7′ 9.9″ | 7′ 9.9″ | 8′ 9″ | 8′ 9.9″ |
| *Door aperture* | Width | 2.340 m | 2.340 m | 2.286 m | 2.340 m |
| | | 7′ 8.125″ | 7′ 8.125″ | 8′ 9″ | 7′ 8.125″ |
| *Dimensions* | Height | 2.280 m | 2.280 m | 2.565 m | 2.585 m |
| | | 7′ 5.75″ | 7′ 5.75″ | 8′ 5″ | 8′ 5.77″ |

Note: Refrigerated version of ISO containers lose about 10 cm (4″) of internal width and 61 cm (2 ft.) of internal length due to chiller unit and insulation space.

*Type 3*: A *modular building within a building*, which is a steel-framed, panel-wall, and roof-constructed cell unit that is usually assembled on-site, on a flat floor or possible around a preprepared seismically isolated block, inside an existing shell building.

The common advantage over conventional buildings is that of ease and speed of site construction, rather than material cost. A further advantage, when the unit is constructed inside an already working factory, is that "wet trades" are not required in their construction.

*Note.* "Wet trades" include concrete pouring, block laying, and plastering, all of which increase the humidity of the enclosing space and require variable curing times before other work can be carried out in that space. Also included in this category is ceramic tiling which may produce highly abrasive dust which can be destructive if entrained in pipework or machinery.

*Containerized* test cells (Type 1). In all but very specific uses, this type of ICE test cell construction, unless radically modified, is severely compromised by internal size and access restrictions. They have found use as QA test cells or "overflow" test facilities where factory space has been limited of difficult to adapt but the wider industrial use of these ISO "test beds in a box" have had mixed results and have proved to be successful only when built specifically for a very limited range of ICE units. The very restricted volume for services makes them very difficult to adapt, or even maintain, once built. This possibility of future relocation is sometimes quoted as an industrial advantage, but in reality, no really successful example of relocation of a civilian unit is known of by the authors.

In 2020 the ISO container may now be a more viable form of test-cell construction because of the increased requirement to test electrical drive-train components, the commensurate shrinkage in the size of the UUT, and the lack of fluid and exhaust services. In all the latest examples known to the author, such containers are not constructed from "off-the-self" or second-hand ISO units; they retain their size and fixing points but are purpose built with the correct standard of insulation and a damp barrier to inhibit sweating (Table 3.4).

The ability to transport ISO containers using internationally standardized cranes, road and marine carrier systems has meant that ISO standard engine and gas-turbine test facilities have been built for Military use.

One configuration, used for testing overhauled helicopter turbines, has been based on four individual containers as listed:

- one 40-ft. test bed container with bedplate
- one 40-ft. control room and auxiliary generator container
- one 40-ft. plant room and interconnecting rigging parts store
- one 20/40-ft. workshop container.

Some of the cost of this facility was created by the need to make all the services, electrical power, and control systems "plug-compatible" to facilitate rapid mounting and demounting.

The use of the *refrigerated version of the ISO 20-ft. shipping container* (known as "Reefers") have for some years found frequent use in automotive test facilities for providing a cheap and quickly commissioned cold soak cells for cars prior to emission certification testing. They have also been used for cold-start testing where the short duration for the test means that no elaborate exhaust or ventilation system is needed. Standard versions of both 20- and 40-ft. Reefers are rated for road and marine use down to −29°C (−20°F) and some down to −40°C/F. Once chilled down the power consumption of the integrated refrigeration unit will be around 4.5 kW.

### *Battery pack testing using cells based on ISO Reefers or similar modular units*

The designers of cells intended for testing lithium-ion batteries must take into account the hazard levels defined by the European Council of Automotive Research (EUCAR). Any cell carrying out the accelerated thermal and electrical cycling discussed below must have cells designed to deal with incidents up to and including Hazard Level 6 (see Table 3.5).

Recently, refrigerated ISI containers have been used for climatic, particularly low temperature, testing of automotive battery packs (Fig. 3.7). Climatic testing of batteries allows the UUT to be electrically abused under conditions of charging and discharge at temperature extremes; since such abuse can lead to a thermal runaway and fire the physical isolation of a test containment that is relatively expendable makes good sense. The container can be internally compartmentalized to form:

1. The battery pack test cell entered via the double doors and containing a prerigged palletized UUT. The climatic conditioning equipment for that

**TABLE 3.5 The various hazard levels defined by European Council of Automotive Research for the use in the storage and testing of electric vehicle battery packs and cells.**

| Hazard level | Description | Classification criteria |
|---|---|---|
| 0 | No effect | No effect. No loss of functionality. |
| 1 | Passive protection activated | No defect. No Leakage. No venting flame or fire.<br>No exothermic reaction. Battery cell reversibly damaged. Repair of protection device is needed. |
| 2 | Defect/damage | No leakage, no venting flame. No rupture. Battery cell irreversibly damaged. Repair needed. |
| 3 | Leakage Δ mass <50% | No venting flames. No rupture. Weight loss <50% of electrolyte (solvent + salt). |
| 4 | Venting Δ mass ≥50% | No venting flames. No rupture. Weight loss ≥50% of electrolyte (solvent + salt). |
| 5 | Fire or flame | No rupture. No explosion or flying parts. |
| 6 | Rupture | No explosion but flying parts of the active mass. |
| 7 | Explosion | Explosion and disintegration of the battery cell (s). |

FIGURE 3.7 A view of 12 battery test chambers at Millbrook Proving Ground, each able to test automotive battery packs up to 1100 V, 1400 A, 750 kW over a temperature range of −40°C to +90°C. The individual cell construction is based on ISO container construction, being modular all-weather type and equipped with its own power (charge/discharge) control and climatic systems. The large conventional building attached houses the Workshop, Control Room and Offices while the two containers in front are the secure storage for untested batteries, which may be of novel design and chemistries. *Illustration courtesy Millbrook.*

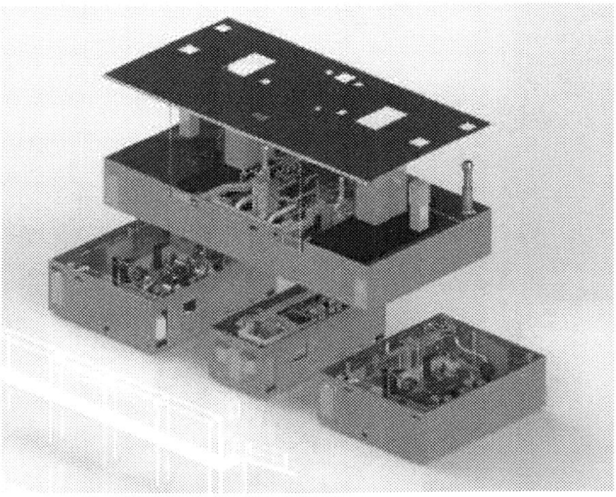

FIGURE 3.8 A complex example of an all-weather modular building consisting of two full powertrain test cells with a central control room and service floor above. All sections were preassembled and tested before trans-European road transport and final assembly on a prepared flat concrete base. Picture courtesy of AVL Tech Centre, UK.

space only will be roof mounted and typically will be capable of providing a test cell temperature between −40°C and +90°C. A high-volume fire suppression system (water mist supplemented with external fire hose connecter) will be supported by a floor drain that would take contaminated water to a drain intercept. Since the units are unmanned the cell will be fitted with high definition video and thermal imaging cameras.

2. A fully ventilated equipment room separated from the cell by a firewall containing both the I/O cabinets, fitted amongst other devices with probably 100 + thermo-couple channels. It will also house the high-powered charging and discharging equipment cabinet currently (2020) typically capable of cycling at up to 800 V and 700 amp.

Such cells can be arranged in multiples but should to have physical separation space between them. A control and monitoring building, usually containing a small UUT preparation workshop is required to be a physically separate entity nearby. (See Chapter 7: Energy Storage, for details.)

## Modular, all-weather (type 2) test cells

Using steel frame (usually rectangular hollow section) and metal clad insulated panels all-weather, custom-built, panel-constructed cell, and control room units have been constructed at several sites around the world.

The justification for their use is as varied as the sites, but generally, it is the idea of minimizing construction and commissioning time on the user's site by shifting as much as possible to the (remote) OEM manufacturing site. By prefabricating, "fitting out" and even testing as much of the test cell as possible before transportation to a prepared concrete base, the final installation site time may be greatly reduced. Since these units will include the cell floor and could include its section of bedplate, the prepared base needs to be recessed to the depth of that floor so that a raised entrance door-sill is avoided. Such a construction can offer the advantage of quickly extending an existing manufacture or test facility without disruption to the main building or the need for substantial excavations. In some locations the "temporary" nature of such a facility may offer advantages in fast-tracking planning and building permits. Experience has shown that the long-term external corrosion resistance has been very variable, and durability requires careful design and choice of enclosure materials particularly when installed in corrosive environments (tropical, saline, etc.). The weather-tight sealing of the long sectional seams involved in this type of design has to be carried out with particular care since once fully assembled the location and stopping of any source of leakage can be problematic.

The detailed design of the services has to be done in such a way as to allow splitting and rejoining at the sectional joints and weather sealing requires a high standard of work at every stage of construction.

The size of the building sections will be determined by the route chosen for transportation and the cost and time required for that loading, transport, and unloading have to be factored into the project plan.

While there are limits of size and weight in most countries for "normal loads" on public roads, the size of "abnormal loads" is limited mainly by the physical restrictions of the chosen road such as bridges and sharp bends amongst buildings. In the United Kingdom, abnormal loads have:

- a weight of more than 44,000 kg
- a width of more than 2.9 m
- a rigid length of more than 18.65 m

Such loads have to be granted special permission and a special-order application must be completed 10 weeks before the scheduled date of the move. Because of the requirement to have, in most cases, police escort, pre-booking of every stage is important. Even in Europe, regulations can differ in different countries so it is vital to employ specialist transport contractors with the relevant knowledge and international certification.

## Modular "building within a building" (Type 3)

There are many successful examples of modular test cells with their associated control rooms and services being installed within a shell building (see Fig. 3.9). In practice, this type of facility has some of the advantages of type 2 constructions with none of the associated problems of weather proofing or transport when they are built up inside the host building from their prefabricated frame and panels.

While many ICE and electrical drive cells, of this construction are simply bolted and sealed on a high-quality flat concrete floor, in the case of special-

**FIGURE 3.9** An example of a flat-floor mounted, modular "building within a building," twin test cell unit with ventilation plant on its roof. *Photo courtesy of Envirosound.*

purpose cells, such as anechoic structures, the test frame will need to be set on preprepared seismic block with subfloor service pipes into the control room (see Chapter 18: Anechoic and Electromagnetic Compatibility Testing and Test Cells).

## Common variations in layouts of multiples of test cells

Test cells of whatever size are often built in side-by-side multiples. A key consideration when designing such layouts is whether or not confidentiality between control rooms and cells needs to be preserved, as may be the case in facilities run by consultancies or government agencies. If this is the case then each cell has to have its own control room, while if this is not a consideration, then control rooms can be shared by one control room positioned between them, or a control corridor can run along the ends of a number of cells as shown in Fig. 3.10. However, it is the chosen method of loading UUT and routes it into and out of the cells that will most influence the detailed layout of the facility floor plan.

It is a general practice in the United States for test cells running liquid- or gas-fueled engines to have an outside wall at one end of the cell and for that wall to be fitted with an emergency exit door and a blast panel (the door may be designed to serve both purposes). This common layout can have a "sandwiched" control room between cells as shown in the exploded view of two modular constructed full powertrain cells shown in Fig. 3.8 and diagrammatically in Fig. 3.12.

The alternative arrangement, shown in Fig. 3.11 of having the control desk at the end of each cell, presents the problem that space has to be found in the control room wall space for a larger than pedestrian door through which the UUT can be transported. The rear wall of the cells shown in Fig. 3.11 may form the outside wall of the building shell and be fitted with an emergency personnel exit door and, where required, a blast relief panel. The exterior of such a wall may also be fitted with a docking station for connecting a bunded, pallet mounted reference fuel barrel unit.

In the case of the layouts shown in both Figs. 3.10 and 3.12, the cells have corridors running at both ends; the "front" of the cells is taken up by a common, undivided, control room or corridor, while the rear corridor is used for transport of rigged ICE mounted on pallets or trolleys. This arrangement of a building also determines that the position of the UUT in the test cell will be at the opposite end from the control room and its window. It requires a larger footprint than that shown in Fig. 3.11, but it increases the control space and keeps it clear of the disruption caused by periodic movement of UUT pallets through the space.

Whereas in the design of "traditional" ICE test cells, the layout tended to be dominated by the logistics of getting the UUT into and out of the test bed, the new designs of cells are dominated by the size of the test equipment rather than that of the UUT. Fig. 3.13 shows part of the internal layout of a

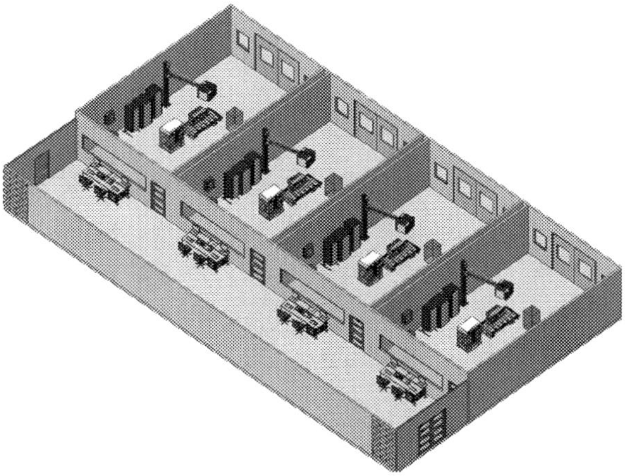

FIGURE 3.10 Cells arranged side-by-side with a common control corridor in front and with UUT access through double doors at the rear.

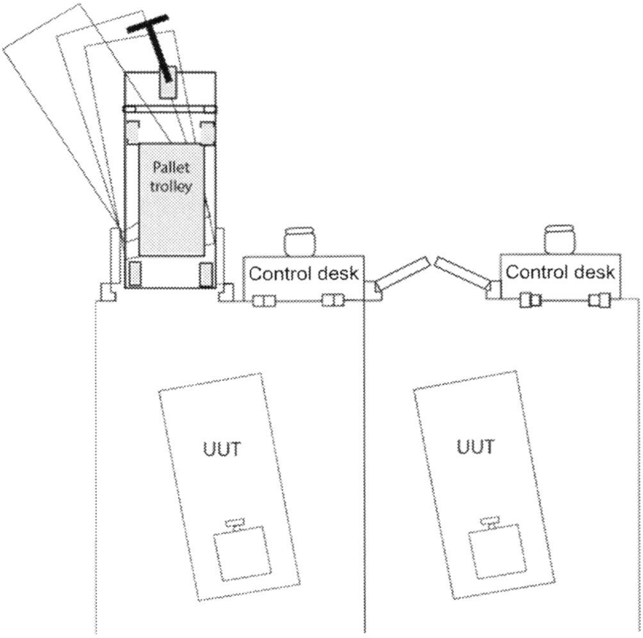


FIGURE 3.11 Sketch of a "side-by-side" arrangement of test cells which can cater for ICE or electric motor testing. The UUT is transported, fully rigged, by trolley through the control corridor reducing the space for a permanent control desk but overall having a smaller footprint than the cells other arrangements.

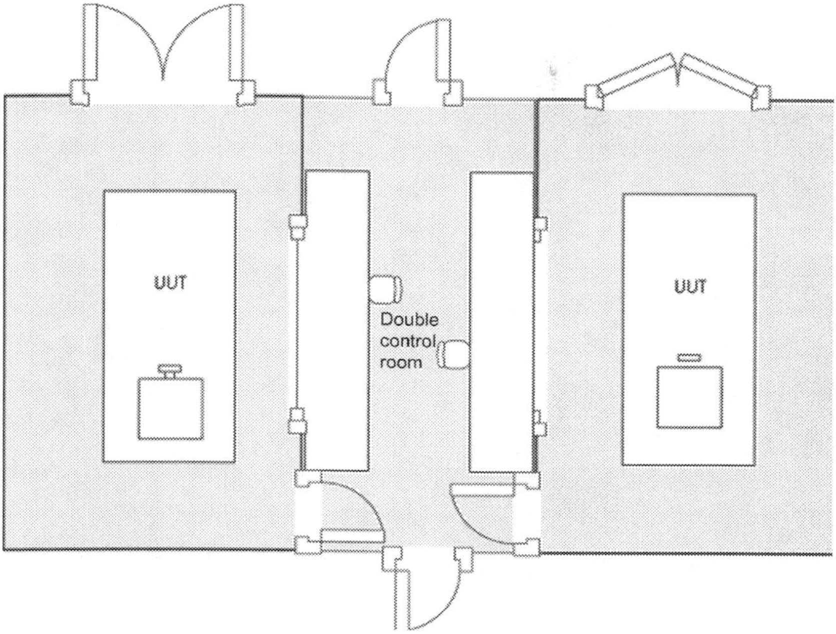


**FIGURE 3.12** A 'back to back' control room arrangement between two cells. This arrangement can be scaled up cater for full powertrain cells as is the case in Fig. 3.8.

**FIGURE 3.13** Photograph showing the layout of a 5-axis transmission test rig where the UUT is small compared to the test equipment and the test duration is a number of weeks. The floor-mounted device in the foreground is part of a gear change robot system.

five-axis cell dedicated to transmission testing, whose layout is similar to, although less configurable than, a full powertrain cell. In this case the UUT is built within the dynamometers and will be subjected to continuous, manned and unmanned, durability cycles. In such cells, palletization of the UUT is not really practical and an overhead crane is an essential requirement.

### *Limitations in the location of test cells*

The prevailing wind direction should be marked on the site plan of any proposed ICE test facility because it is important that they are kept out of any vapor plume from other exhaust systems or industrial processes emitting chemical pollutants; paint shops being a common "worse offender." Roof-mounted ventilation inlets and outlets together with exhausts must be designed and positioned to prevent test cells, very particularly those carrying out emission tests, from ingesting polluted air.

NVH cells need to be as remote as possible from internal and external roadways, railroads, or any source of ground transmitted vibration. Ground vibration from "shaker-tables" is particularly damaging to any rotating equipment having precision rotating element bearings because it may induce damage from brinelling. Ground-borne vibration from such facilities should be well isolated and they should be located remote from all other test cells or precision rotative machinery.

## Alarms, emergency exits, and safety signs

Test facilities may have a hierarchy of several alarms, some of which will require immediate building evacuation, whereas others require those responsible to take immediate remedial action, it is therefore vital that all employees and contractors on site should know what each alarm means, whether or not the cause of the alarm has the potential to affect them and the required response. Within the test cell, control room, and immediate support areas the following design and operational practices are recommended:

- Alarm signals should be at least 10 dB(A) over the background noise of the area in which it is located.
- An acknowledged alarm should reset automatically if the fault that generated it is rectified.
- Alarms should not prevent effective communication within the control room.
- Alarm messages should be presented in a standard format based upon existing conventions, across all cells in the same facility and between all systems.

- A sequential and timed alarm log should be provided for diagnostic purposes. This would normally be a function of the building management system (BMS—see Chapter 13: Test Cell Safety, Control, and Data Acquisition).

The design of the alarm system(s) should prevent masking and flooding of alarms. *Masking* is where one alarm noise masks a similar-sounding alarm preventing the operator from detecting or differentiating the signal.

*Flooding* happens when a system alarms which has a "knock-on" effect on other related systems, the result of which is the triggering of myriad other alarms—flooding the control room with sound.

Note: One author had an experience of such an alarm "flooding" event when the flywheel of truck diesel engine under test shattered. The debris and a flailing connecting shaft cut electrical cables fluid lines and the fire detection system. The resulting cacophony and noise level from multiple alarms was completely disorientating and prevented those present from immediately taking the most appropriate, indeed any, action that would have reduced the considerable consequential damage.

Many test cells and control rooms are visually and audibly confusing spaces, particularly to visitors; good design and clear signage are required to minimize this confusion and thus improve the inherent safety of the facility. Large powertrain test cells are a particular problem because operators, when inside the cell, can be amongst the UUT assemblage and its dynamometers, out of sight of the cell window and with no clear view of the door(s). In such a cell, it is important that there is a clear walkway around the bedplate "clutter" and that each wall bears signage indicating the nearest exit.

The confusing use of multiple signs has become so common, particularly during facility construction and commissioning, as to become a safety hazard in its own right and is now warned against in British Standard 5944.

The number and position of emergency exits within both cell and control room spaces will be determined by national safety legislation and local building codes that will vary from country to country but will be planned to give the shortest route to the building exterior. All doors and emergency exit routes have to be designated with the legislatively required battery-illuminated signs. All emergency doors that are fitted with a "burst bar" opening device open outward from the source of danger and their use must not, by law, be blocked, or inhibited by stored or discarded objects blocking the escape route.

All workers in the facility must be trained and practiced in the required actions in the case of specified emergencies and all visitors should be briefed in the use of emergency exits identified by their host.

## Transducer boxes and wall-mounted booms

The signals from each of the transducers with which the UUT is fitted (Chapter 13: Test Cell Safety, Control, and Data Acquisition) flow through individual cables to a nearby marshaling box. Because of the low level of signals and the need to prevent signal corruption, most signal conditioning and analog-to-digital conversion are carried out by electronics within these transducer boxes, which then send the data to the control computer system via one or more digital cables under the control of a communication bus such as the IEEE 1394 interface (Firewire). These boxes may be either mounted on an adjustable boom cantilevered and hinged from the cell wall or on a pillar mounted from the bedplate but, since they may be housing very expensive electronics, due care needs to be taken in the support and location so that damage is not caused by gross vibration and overheating. Boom-mounted boxes intended to be close to and or immediately above the ICE need to be force-ventilated and temperature monitored. Well-designed booms are made of hollow aluminum extrusions, of the type used in sailing-boat masts, and have a fan at the wall end blowing cooling air through the whole structure.

Boom-mounted transducer boxes can be of considerable size and weight and, because of the convenience of their location, the contents and attachments tend to increase during the process of cell development. Because of this high-cantilevered load, the ability of the wall to take the stresses imposed must be carefully checked; frequently, it is found necessary to build into the wall, or attach to it, a special support pillar.

Fuel booms are often used to take fuel flow and return lines between the ICE on test and wall-mounted conditioning units. These are of light construction and must not be used to attach other instrumentation or cables.

Fig. 3.14 shows part of an advanced powertrain test cell constructed within a modular building (Fig. 3.8). The UUT is a $4 \times 4$ hybrid ICE/electric powertrain. There is *no* shaft connection between the transverse 1.5 L ICE and its 9-speed automatic gearbox driving the front wheels and the 120 kW e-axle driving the rear wheels (foreground in photo).

The powertrain can be run in the cell in all four modes: front wheel drive (ICE only), rear wheel drive (electric vehicle only), all wheel drive with torque splitting or dedicated four wheel drive. This Powertrain-in-the-Loop (PiL)-capable setup uses the latest test bed automation systems that allow for the entire $4 \times 4$ system to run, or for the independent control of each subcomponent when needed. Thus using the same rigging set-up just the e-axle can be run for a motor control only test program or just ICE for an ECU (electronic control unit) test program, or various other power electronics, present or simulated.

The large spot cooling fans are not required during normal full-power running but are used to supplement the cell system and the CA system to produce an ICE system rapid cool-down.

**FIGURE 3.14** "Putting it all together." A view of a full 4 × 4 ICE hybrid powertrain rigged to individual wheel-simulating dynamometers—description in the text below. Most of the power cabinets and battery emulator are on the plant room above. *Reproduced with permission of AVL-TCC (United Kingdom).*

## Part 3: Fire safety and fire suppression in automotive test facilities

### European *At*mospheric *ex*plosion codes applied to engine test cells

*At*mospheric *ex*plosion (ATEX) regulations were introduced as part of the harmonization of European regulations for such industries as mines and paper mills, where explosive atmospheres occur. The engine and vehicle test industry was not explicitly identified within the wording; therefore the European automotive test industry has had to negotiate the conditions under which test cells may be excluded in the few cases where conformity would make operation impossible. As with all matters relating to regulation, it is important for the local and industrial sector-specific interpretation to be checked. The regulation classification of zones is shown in Table 3.6 Areas classified into zones 0, 1, and 2 for gas–vapor–mist must be protected from effective sources of ignition. As has been determined over many years, secondary explosion protection measures such as using EX-rated equipment (even if it was available and, in many cases, it is not) in the ICE test cell makes little sense since the ignition source is invariably the UUT itself. Therefore it is necessary to use primary explosion prevention methods that prevent the space from ever containing an explosive atmosphere covered by ATEX.

**TABLE 3.6** Characteristics of major fire suppression systems.

| US divisions | ATEX zone designation | Explosive gas atmosphere exists | Remark (h/year) |
|---|---|---|---|
| | 0 | Continuously, or for long periods | >1000 |
| 1 = zones 0 and 1 | 1 | Occasionally | 10–1000 |
| 2 = zone 2 | 2 | For a short period only | <10 |

The following primary precautions, which cover both gasoline- and diesel-fueled beds without distinction, are certified in Europe by the relevant body Technischer Überwachungsverein (TÜV):

- The cell space must be sufficiently ventilated both by strategy and volume flow to avoid an explosive atmosphere.
- There has to be continuous monitoring and alarming of hydrocarbon concentration (normally "warning" at 20% of the lower explosive mixture and "shutdown" at 40%).
- Leak-proof fuel piping using fittings approved for use with the liquids contained.
- The maximum volume of fuel "available" in the cell in the case of an emergency or alarm condition is 10 L.

With these conditions fulfilled, the only EX-rated electrical devices that need to be included in the cell design are the gas detection devices and the purge extraction fan.

In the United States the hazardous zone classification system that is most widely utilized is defined by the NFPA Publication 70, NEC, and CEC. They define the type of hazardous substances that is, or may be, present in the air in quantities sufficient to produce explosive or ignitable mixtures. Like most regulations having relevance to our industry, parts of the code are subject to local interpretation by the "authority having jurisdiction" (AHJ), who may be a fire marshal or city planning officer. The codes often refer to the "adequate ventilation" and "sound engineering judgment" being required in classification of industrial spaces. The best and usual practice in the United States often uses changes-per-hour figures in classified zones. ICE test cell subfloor trenches and areas up to 18″ above the floor "where volatile fuels are transferred" require specific ventilation flows. In general, the resulting classifications give similar airflow requirements to European practice, but close working with the local AHJ is advised from the inception and initial planning of any test facility.

These policies continue to be true with the advent of electric and hybrid powertrain testing where batteries are including in the UUT. However, since one mode of failure of an abused Lithium-ion battery is thermal runaway, which does not require an external ignition source to be initiated, some test facilities have opted to remove the battery pack from the main test cell and put it in its own containment outside the facility building. At the time of writing (2020) 90% of test facilities in contact with the authors were using battery emulators in, or adjoining, their test cells rather than the actual vehicle battery pack. Refer to Chapter 5, Ventilation and Air Conditioning of Automotive Test Facilities, for a discussion of detailed design strategies of ventilation systems compliant with ATEX regulations.

## Fire and gas detection and alarm systems for test facilities

There are three separate system design subjects:

- *the prevention of explosion and fire* through the detection of flammable, explosive, or dangerous gases in the cell and the associated remedial actions and alarms;
- *the suppression of fire* and the associated systematic actions and alarms; and
- *detection of gases* injurious to health.

A fuel cell or ICE test cell's gas detection system, supplied by a specialist subcontractor, will be fitted with transducers designed to detect various levels of hydrocarbon vapor proscribed in ATEX regulations.

In facilities testing fuel cells, hydrogen detectors should be fitted in the roof spaces pockets. Alarms for hydrogen concentration levels should be triggered before they approach the lower flammability limit (LFL); a concentration in air *lower* than the LFL means gas mixtures are "too lean" to burn. The lower flammability limit of hydrogen is about 4% by volume.

Depending on the requirements of the risk analysis valid for the facility, some areas within and outside the cell may be fitted with carbon monoxide sensors. This is particularly important if pressurized and undiluted engine exhaust ducting is routed through a building space, a design feature to be avoided if possible.

Gas detection systems must be linked to the cell control system and, where one exists, the BMS, in conformity to local and national regulations.

In order to prevent spurious and expensive shutdowns the gas hazard alarm and fire extinguishing systems, when under operator control, are independent of the test cell's emergency stop circuit system. When under automatic and unmanned control, the hierarchal system of alarms and system activation/shutdown needs to be carefully designed to suit individual facilities and should be specified within the control and safety interlock matrix covered in Chapter 5, Ventilation and Air Conditioning of Automotive Test Facilities.

There is much legislation relevant to industrial fire precautions and a number of British and European Standards that are generally similar to those applied worldwide. In the United Kingdom the Health and Safety Executive, acting through the Factory Inspectorate, is responsible for regulating such matters as fuel storage arrangements and should be consulted, as should the local fire authority.

The choice of audible alarms and their use requires careful thought when designing the facility safety matrix.

Audible fire alarms should have their control system that is independent of and unique in sound to, other alarms, The fire alarms usually use electronic solid-state sounders with multitone output, normally in the range of 800–1000 Hz, or can be small sirens operating in the range of 1200–1700 Hz. Regulations around the world usually state alarms should have an output sound level 10 dB above local ambient; since ambient, even in the control area may occasionally be 80 dB, most fire alarms fitted into engine test cell areas tend to be painfully loud and quickly drive humans out of their immediate area. This may not be a good outcome since if the alarm has been triggered by an *in-cell fire* the operators are best employed to trigger local safety procedures in order to contain the problem rather than abandon the area to its fate. It would seem to be better practice to give obvious visual alarm together with an audible alarm that allows remedial work to be carried out in the control space while triggering the evacuation of other staff from a predetermined area of the building according to practiced procedures.

It is not uncommon for hot, sometimes incandescent, UUT parts to trigger false fire alarms through in-cell detectors, so it is most important that the operators are trained and able to identify and kill such false alarm states and avoid consequential automatic shutdown of unassociated plant, or inappropriate release of a fire suppression system.

## Fire extinguishing systems

Whatever the type or types of fire suppression systems fitted in test cells or available in control rooms, it is absolutely vital that all responsible staff are formally trained and certified in their correct use and that this training is kept up to date with changes in best practice and the nature of the units under test.

Table 3.7 lists the common fire suppression technologies and summarizes their characteristics, which are covered in more detail in the following sections.

### *Microfog water systems*

Microfog or high-pressure mist systems, unlike other water-based fire extinguishing systems, have the great advantage that they remove a lot of heat from the fire source and its surroundings and thus reduce the risk of reignition when it is switched off. The system is physically the smallest available

**TABLE 3.7** Characteristics of major fire suppression systems.

| | Water sprinkler | Inert gas ($CO_2$) | Chemical gases | High-pressure water mist |
|---|---|---|---|---|
| Cooling effect on fire source | Some | Some | None | Considerable |
| Effect on personnel in cell | Wetting | Hazardous/fatal | Minor health hazards | None |
| Effect on environment | High volume of polluted water | Greenhouse and ozone layer depleting | Greenhouse and ozone layer depleting | None |
| Damage by extinguishing agent | Water damage | None | Possible corrosive/hazardous by-products | Negligible |
| Warning alarm time before activation | None required | Essential | Essential | None required |
| Effect on electrical equipment | Extensive | Small | Possible corrosive by-products | Small |
| Oxygen displacement | None | In entire cell space | In entire cell space | At fire source |

and therefore makes its integration within the crowded plant-room space much easier than the gas-based systems.

Microfog systems use very small quantities of water and discharge it as a very fine spray. They are particularly efficient in large cells, such as vehicle anechoic chambers, where they can be targeted at the fire source, which is likely to be of small dimensions relative to the size of the cell.

Other advantages of these high-pressure mist systems are that they tend to entrain the black smoke particles that are a feature of engine cell fires and prevent, or considerably reduce, the need for a major cleanup of ceiling and walls. Such systems are used in facilities containing computers and high-powered electrical drive cabinets with proven minimum damage after activation and enabling a prompt restart.

### *Carbon dioxide*

$CO_2$ has been widely used in the industry for at least 50 years. It is effective at dealing with the types of flammable liquid fires sometimes experienced in test cells which it suppresses by cooling and asphyxiation. It is hazardous to life in confined spaces; breathing difficulties become apparent above a concentration of 4% and a concentration of 10% can lead to unconsciousness or, after prolonged exposure, to death. Therefore a warning alarm period must be given before activation to ensure preevacuation of the cell. $CO_2$ is about 1.5 times denser than air and it will tend to settle at ground level in enclosed spaces.

The discharge of a $CO_2$ flood system is likely to be violent and frightening to those in the region. The pressure pulse will blow out any incompetent blockage of holes made in the cell walls for the transit of cables, etc. The sudden drop in cell temperature causes dense misting of the atmosphere to take place, obscuring any remaining vision through cell windows or cameras.

### *Inergen*

Inergen is the trade name for the extinguishing gas mixture of composition—Argon 50%, Nitrogen 42%, and $CO_2$ 8%.

It works by replacing the air in the space into which it is discharged and taking the oxygen level down to $<15\%$ when combustion is not sustained.

There are other alternative fire suppressants of the same type as Inergen, including pure argon, many of which may be used in automatic mode even when the compartment is occupied, provided the oxygen concentration does not fall below 10% and the space can be quickly evacuated.

With all gaseous systems, precautions should be taken to ensure that accidental or malicious activation is not possible. In particular, with carbon dioxide systems, automatic mode should only be used when the space is unoccupied and in conjunction with alarms that precede discharge.

Total flood systems, of whatever kind, are usable only after the area has been sealed, also:

- They must be interlocked with the doors and special warning signs must be provided.
- Positive indication that the systems are armed should be provided to the operators. There have been cases of systems failing to function because of being accidentally left switched to "maintenance" or "training" modes.

## Handheld fire extinguishers

Portable fire extinguishers are classified to indicate their ability to handle specific classes of fires which are listed in Table 3.8.

Training in the types, use, and positioning of fire extinguishers, on a site as complex as an automotive test facility, needs to be given by an expert with the relevant background the following notes are intended as guidance based on experience.

**TABLE 3.8 Classes of fires for which hand-held fire extinguishers are certified.**

| | |
|---|---|
| Class A fires | Solid combustibles<br>Fires involving solid combustible materials such as wood, textiles, straw, paper, coal etc. |
| Class B fires | Flammable liquids<br>Fires caused by combustion of liquids or materials that liquify such as petrol, oils, fats, paints, tar, ether, alcohol, stearin and paraffin |
| Class C fires | Flammable gases<br>Fires caused by combustion of gases such as methane, propane, hydrogen, acetylene, natural gas and city gas |
| Class D fires | Flammable metals<br>Fires involving combustible metals such as magnesium, aluminum, lithium, sodium, potassium and their alloys. Combustible metal fires are unique industrial hazards which require special fire extinguishers. |
| Class F fires | Combustible cooking media<br>Fires involving particularly hot or deep oil and grease fires, such as deep fat fryers in commercial kitchens or overheated oil pan fires in homes. |
| Electrical fires | Electrical appliances<br>Fires involving electrical appliances such as computers, electrical heaters, stereos, fuse boxes |

Fires within powertrain test cells and their control rooms when the use of handheld extinguishers is possible or safe for the operator are most likely to be electrical, or Class A or minor incidents of Class B.

Class D fires are a special case discussed later.

Most cases of Class A fires known to the author are caused by poor housekeeping such as when oily rags accumulate in-floor trenches shared with exhaust pipes; in these cases, synthetic aqueous film-forming foam (AFFF) fire extinguishers have proved effective even if electrical cabling was involved.

Class B fires can occur suddenly in ICE cells and will invariably require the cell fire system and shut-down process to be triggered rather than any attempt to use handheld equipment. An example was caused by the common fuel rail of a diesel engine splitting and spraying the red-hot turbocharger turbine casing. However, AFFF fire extinguishers are recommended for less dramatic Class B fires.

### *Dry powder (Class D extinguishers)*

Powders are discharged from hand-held devices and are designed for quickly extinguishing of flammable liquids such as petrol, oils, paints, and alcohol; they can also be used on electrical or engine fires. It must be remembered that dry chemical powder does not cool, nor does it have a lasting smothering effect and therefore care must be taken against reignition.

Electrical fires, against which $CO_2$-based extinguishers are recommended, can occur anywhere within a test facility. The danger to humans created by the fire is compounded by the danger of electrical shock therefore in addition to knowing how to deal with the fire both the site staff and the local fire officers need to know how to kill the power supply to the seat of the fire.

*Class D fires involving lithium-ion batteries.* The automotive batteries are developing very quickly and, like any energy storage system, they are potentially hazardous. The dangers of thermal runaway of lithium-based batteries have received lots of publicity but not a great deal of empirical information is available. Dealing with the possibility of fire by the design of the test cell has been covered in this chapter; meanwhile, the following advice is taken from Battery University [2] which is a regularly updated source of information.

> *If a Li-ion battery overheats, hisses or bulges, immediately move the device away from flammable materials and place it on a non-combustible surface. If at all possible, remove the battery and put it outdoors to burn out. Simply disconnecting the battery from charge may not stop its destructive path.*
>
> *A large Li-ion fire, such as in an EV, may need to burn out. Water with copper material can be used, but this may not be available and is costly. Increasingly, experts advise using water even with large Li-ion fires. Water lowers combustion temperature but is not recommended for battery fires containing lithium-metal.*
>
> *When encountering a fire with a lithium-metal battery, only use a Class D fire extinguisher. Lithium-metal contains plenty of lithium that reacts with water*

*and makes the fire worse. The gas released by a venting Li-ion cell is mainly carbon dioxide ($CO_2$). Other gases that form through heating are vaporized electrolyte consisting of hydrogen fluoride (HF) from 20–200 mg/Wh, and phosphoryl fluoride ($POF_3$) from 15–22 mg/Wh. Burning gases also include combustion products and organic solvents. The knowledge on the toxicity of burning electrolyte is limited and toxicity can be higher than with regular combustibles. Ventilate the room and vacate area if smoke and gases are present.*

## Fire stopping of cable and pipe penetrations in cell structure

To maintain the integrity of the "hazard containment box" where ventilation ducts, cables, or cable trunking break through the test cell walls, roof, or floor they must pass through a physical "fire block" to preserve the 1-hour minimum fire containment capability. All wall penetrations carrying cables between control space and cell should be, as a minimum, sealed by using wall boxes having "letterbox" brush seals in steel wall-plates fixed on either side of the central void, the void in which is then stuffed with fire-stopping material. However, poor maintenance or incomplete closure of these firebreaks, commonly caused by the frequent installation of temporary looms, can allow fire or the pressurized extinguishant, such as $CO_2$, to escape explosively into the control room. A hole in the control room wall loosely stuffed with rag not only compromises the building's fire safety but may also invalidate insurance. Alternatively, the more robust Hawke Gland type of through-wall boxes may be used; these devices provide rigid clamping of the individual cables and although the fire block can be disassembled and extra cables added, it is not particularly easy to use after a year or more in service, so it is advisable to build in a number of spare cables at the time of the initial closure.

Intumescent materials are widely used to "fire-stop" test cell penetrations. Often based on a graphite mixture, such materials swell up and char in a fire, thus sealing their space and creating a barrier of poor heat conduction. Intumescent material is available in bags of various volumes that are ideal for stuffing between the ventilation ducting and the roof or wall, since it allows thermal movement without being dislodged. However, it must be held in place, top and bottom, by plates fixed to the concrete to prevent it from being dislodged by any pressure pulses in the cell.

Foams are available to use for sealing pipe and cable ducts, but in all cases "listing and approval for use and compliance" should be sought from local fire authorities and facility insurers before particular products are used.

Plastic "soil pipes" cast into the floor are a convenient way of carrying cables between the test cell and the control room. Several such pipes need to be dedicated to cables of the same type to avoid crosstalk and signal corruption. These ducts should be laid to fall slightly in the direction of the cell to prevent liquid flow into the control room and should have a raised lip of 20 mm or more to prevent drainage of liquid into them but positioned so as not to create a trip hazard. Spare cables should be laid during installation. These pipes can be "capped" by foam or filled with dried casting sand to create a noise, fire, and vapor barrier.

Large power cables may enter a cell through cast concrete trenches cast under the wall and filled with dense dry sand below a floor plate; this method gives both good sound insulation and a fire barrier, and it is relatively easy to add cables later. If coming through the roof the cables should come through individual flameproof glands.

## Part 4: The control room

### Control room design

It is obvious that control room operators' surroundings have a direct relation to their performance; in their vigilance, output, quality, and safety of the people and equipment, they are monitoring. It should be equally obvious that the role and profile of the staff using the control room is of fundamental importance to the design of that workspace and to the operation of the test department.

Gone are the days when most ICE cells were operated by a single "engine driver"; the control rooms of modern powertrain test cells are more likely to be the permanent workspace for at least two engineers/operators spending some 95% of their time there or in the cell. It will also frequently house a floating population of two or three more specialist test engineers spending perhaps 5%–20% in the room during testing; while, during the night, when tests are running "unmanned," there may only be an occasional security staff visit, so the layout has to work for high and low staffing levels. The whole range of operatives needs to see information and have controls relevant to their tasks and function without compromising the view and work of others. There is no "correct" solution, but the layout has to be optimized for the particular roles and needs of each site and there are some general features that should be avoided and established best practice from both the automotive and continuous process industries should be observed.

*Avoid visual and spatial clutter.* The multiple screens that dominate the operators' work space in modern control rooms present the control system designer with the problem of organizing a coherent display giving displayed data the correct priority. The English expression that defines the problem as "not seeing the wood for the trees" is relevant and particularly apposite when cells are upgraded, and screens added haphazardly. A good example of a common screen and control layout can be summarized as follows:

- Screens displaying data directly relevant to the control of the UUT is center middle of the chief operator's position with a screen displaying current and future test sequence stages immediately above or below.
- To the right-hand side of the control desk, screens, and physical control devices related to the control of the test cell services.

- On the left-hand side of the center screens, screens, and physical control devices relating to ancillary test systems such as emission benches, battery monitoring, simulation system outputs.

The same "handed" concept with the position and layout of the screens has to be copied for that of rack or cabinet mounted controls and be common to all cells in the same facility.

In small control spaces, it may be advisable to use articulated support booms for screens that are needed only during particular phases of tests that can be folded away when not required "center-stage."

Recently observed examples of a poor control room layout have all been the consequence of *incremental modifications* as the test facility has adapted to more complex UUT and the integration of new instrumentation. The common features of poor control room development include:

- Operators' work area becoming confined and cramped with controls and displays in positions dictated by available floor or desk space rather than ergonomic or operational considerations.
- Instrumentation that has been added in a piecemeal manner has meant that the cable routes are less than optimum and that signals are vulnerable to electrical noise and therefore data corruption (see Chapter 4: Electrical Design Requirements of Test Facilities).
- Room lighting has been designed for an empty space; it is now creating shadows and reflections on screen and not illuminating the work areas in a manner that is sympathetic to the tasks being carried out.
- Interconnecting wiring exposed and vulnerable with cables joining shared equipment unable to run through dedicated trunking but running taped to floor or walls.

***Established best practices relating to control rooms include*:**

- Under UK law the average noise level within a working space such as the control room must not exceed 85 dB(A) during the length of the working day. Any noise levels regularly approaching this level should require that operators are provided with suitable ear protection and are indicative of a design problem requiring resolution.
- Control rooms rarely seem to be provided with sources of natural daylight because of screen reflection but if they are then adequate means of blocking out or deflecting direct sunlight should be provided.
- The type of lighting should be adequate for the task and it is desirable to provide adjustable lighting for control rooms particularly those that are manned 18–24 hours a day; a lux figure of between 500 and 800 is suggested.
- The temperature and airflow should be adjustable remembering that many control rooms will contain significant heat sources. As a design

guide, a "comfortable" temperature for sedentary work should be between 18.3°C and 20.0°C with airflow between 0.11 and 0.15 m/s. European Standards state 16°C is the minimum temperature for office work and although there is no legal maximum quoted most trade unions use the figure of 30°C.

- In multioperator workstations desks seating should mean that operators are not sitting within each other's "intimate" and responsibility zones. As a guide the minimum spacing distance should be between 300 and 700 mm.
- The degree to which the control room is used as a team discussion area needs to be established at the design stage; if this is the case and team discussions take place during testing, then the seating provided should sufficiently mobile to allow eye contact without disturbing the hierarchy of cell control. The alternative area for team discussion would be remote from the test cell in a room with data displays uploaded to the host computer and subjected to posttest processing.
- It is a sensible practice to provide a two-tray, wheeled trolley for the operators to use, on which to place drinks, store manuals, and other occasionally used items that should not clutter or lead to spillage on the control desk.

If, as is the case in some motorsport or academic facilities, disinterested visitors, under escort, are allowed into the control room, they should occupy an area well to the rear of operations staff positions as shown in Fig. 3.15, which is a control room of a Formula 1 powertrain test facility. It has to

**FIGURE 3.15** A general view of a modern full powertrain test cell control room. The screen above the window is dedicated to steerable CCTV cameras in the cell.

house its two operators and a floating population of specialist development engineers but also, on occasions, VIP visitors. There are several key features to support its roles:

- A large, triple-glazed window and free area behind the desk allow visitors, such as team sponsors, to visit and observe work in progress without disturbing the operators.
- The 12 screen displays are designed to display data in the required format best suited to the subsystem they serve and are placed in the optimum position and lighting to serve that purpose.
- Above the window are placed color CCTV screens displaying pictures from steerable and zoomable in-cell cameras; facility status screens are to the right of the operator's desk.
- The several keyboards and mice are cable attached to the devices they control rather than wireless-enabled to avoid potentially dangerous confusion caused by changing of desk positions.

## In-cell control of the unit under test

The ability to control the UUT and dynamometer load from within the test cell via a "repeater station" installed as wall- or pedestal-mounted boxes, used to be a common feature of automotive test cells, but H&S practices have tended to forbid their use. However, they are very useful in postproduction testing when carrying out some calibration and setting up functions (such as the throttle actuator stroke limits or fuel pump stops), and, provided sensible operational practices and two operators working are enforced, their use may be allowed. Built into any control system that has two manual control stations there has to be a robust safety system that inhibits dual control or control conflicts; therefore the main control desk has to be fitted with a key switch that allows the choice of either "desk control and auto" and "remote and manual." Control follows the single key that can only be used in the operational station while all hard-wired "emergency stop" buttons remain active.

# Part 5: Special-purpose test facilities

## The requirements of a basic after-market internal combustion engine test bed

For at least another decade, there will continue to be a requirement to test the basic function of both petrol and diesel engines, for commercial and military use, outside a vehicle. Engine rebuilders, large bus workshops, and some motor-sport tuners require a cheap, reliable, and safe way to certify the performance of the work and in which there is a requirement to run engines under load, but so infrequently that there is no economic justification for building a permanent test cell.

To provide controlled load in such cases, a "bolt-on" dynamometer can be used (see Chapter 9: Dynamometers: The Measurement and Control of Torque, Speed, and Power, for details). These require no independent foundation and are bolted to the engine bell housing, using an adaptor plate, with a splined shaft connection engaging the clutch plate.

To support the needs of this sort of occasional test work all that is required is an adequately ventilated space, possibly out of doors provided with:

- water supplies and drains with adequate flow capacity for the absorbed power (see Chapter 6: Cooling Water and Exhaust Systems);
- a portable fuel tank and supply pipe of the type sold for large marine outboard motor installations;
- arrangements to take engine exhaust to exterior, if within a building;
- minimum necessary sound insulation or provision of personnel protection equipment; and
- adequate portable fire suppression and safety equipment.

Fig. 9.11 in Chapter 9 shows a typical installation, consisting of the following elements:

- portable test stand for engine fitted with an exvehicular radiator system;
- dynamometer mounted directly to the engine flywheel housing using a multimodel fixing frame;
- control console; and
- flexible water pipes and control cable.

For small installations and short-duration tests the dynamometer cooling water may be simply run from a mains supply, providing it has the capacity and constancy of pressure, and the warm outflow runs to waste. Much more suitable is a simple pumped system, including a cooling tower, operating from a sump into which the dynamometer drains. The control console requires, as a minimum, indicators for dynamometer torque and speed. The adjustment of the dynamometer flow control valve and engine throttle may be by cable linkage or simple electrical actuators. The console should also house the engine stop control and an oil pressure gauge. If required, a simple manually operated fuel consumption gauge of the "burette" type is adequate for this type of installation. The same attention to instrument calibration and data collection has to be observed as in any test facility.

Note that a dynamometer of the bolt-on type may be installed in a truck chassis without removal of the engine, by dropping the propeller shaft and mounting the dynamometer on a hanger frame bolted to the chassis. This technique is useful for testing whole vehicle cooling systems without using a chassis dynamometer.

## Automotive engine production test cells: hot and cold testing

The postproduction loaded hot testing of 100% of petrol engines has become very rare in the mass-market sector and has been replaced by random QA testing and extensive use of in-process test stations. The number of build variants of the "world engines" built by the major OEMs means that only a basic form of any engine model is common to all and physically suitable for a common test system. The hot testing of 100% of diesel engines, particularly commercial truck engines, still takes place and the cells can be highly specialized installations that form part of an automated handling system. The objective is to check, in the minimum possible process time, that the engine is complete, leak-proof, and runs. The whole procedure—engine handling, rigging, clamping, filling, starting, draining, and the actual test sequence—can be highly automated, with interventions, if any, by the operator limited to dealing with fault identification. Typical "floor-to-floor" times for small automotive engines can be between 5 and 8 minutes. Leak detection may be difficult in the confines of a hot-test stand, so it is often carried out at a special (black-light) station following test while the engine is still warm.

The test cell will be designed to read from identity codes on the engine, recognize variants, and to adjust the pass or fail criteria accordingly.

Cold testing along with pressure and rotational testing "in process" of subassemblies, is increasingly being applied to (near) completely built engines. It has considerable cost and operational advantages over hot testing, which requires an engine to be fully built and dressed for running in a test cell with all supporting services. Cold-test areas also have the cost advantage of being built and run without any significant enclosure other than safety guarding and can form an integral section of the engine production line.

Cold-test sequences are of short duration and some potential faults are easier to spot than in hot testing when "the bang gets in the way." The principal technique of cold testing is that the engine is spun at lower than normal running speed, typically 50–200 rpm, by an electric motor with an in-line torque transducer.

Early in both hot and cold test sequences, the "torque to turn" and rate of oil pressure rise figures are used to check for gross assembly errors. The engine wiring loom is usually connected through the ECU connector to a slave unit that is programed to check the presence, connection, and correct operation of the engine's transducers. Thereafter, vibration and noise patterns are recorded and compared to an ever developing and improving standard model. This is a highly automated process that uses advanced computer models and pattern recognition technology. The maximum long-term benefit of cold testing is derived by feeding back-field service data to refine the "pass–fail" algorithms that pick out production faults at their incipient stage.

## End-of-line test station facility layout

In the design phase of either type of production end-of-line (EOL) test facility, a number of fundamental decisions have to be made, including:

- layout, for example, in-line, branch line, conveyor loop with workstations, carousel;
- what remedial work, if any, is to be carried out on the test stand;
- processing of engines requiring minor rectification;
- processing or scrapping and parts recycling of engines requiring major rectification;
- engine handling system, for example, bench height, conveyor and pallets, "J" hook conveyor, automated guided vehicle;
- engines or Motor units rigged and derigged at test stand or remotely;
- storage and recycling of rigging items;
- test-stand maintenance facilities and system fault detection; and
- measurements to be made, handling, and storage of data.

Production testing, hot or cold, imposes heavy wear and tear on engine rigging components which must be considered as consumable items, and which need constant monitoring with spares immediately available. The use of vehicle standard plug or socket components on the rig side has proved to be totally unsatisfactory; these components have to be significantly toughened versions to survive in the EOL environment.

Automatic shaft docking systems may represent a particularly difficult design problem, particularly where multiple engine types are tested, and where faulty engines are cranked or run for periods leading to unusual torsional vibration and torque reversals.

Shaft docking splines need adequate tooth lubrication to be maintained and, like any automatic docking item, can become a maintenance liability if one damaged component is allowed to travel round the system, causing consequential damage to mating parts.

Modular construction and the policy of holding spares of key subassemblies will allow repairs to be carried out quickly by replacement of complete units, thus minimizing production downtime.

## Large and medium-speed diesel engine test areas

As the physical size of the UUT increases, the logistics of handling them becomes more significant; therefore the test area for medium-speed diesel engines is, more often than not, located within the production plant close to the final build area.

Above a certain size, engines are often tested within an open shop in the position in which they have been finally assembled. The dynamometers designed to test engines in ranges about 20 MW and above are small in comparison to the prime mover; therefore the test equipment is brought to the engine rather than the more usual arrangement where the engine is taken to a cell.

Cells for testing medium-speed diesels require access platforms along the sides of the engine to enable rigging of the engine and inspection of the top-mounted equipment, including turbochargers, during test.

Rig items can be heavy and unwieldy; indeed, the rigging of medium-speed diesel engines can be a considerable design exercise in itself. The best technique is to pre-rig engines of differing configurations in such a way that they present a common interface when put in the cell or test shop. This allows the cell to be designed with permanently installed semi-automated or power-assisted devices to connect exhausts, intercooler, and engine coolant piping. The storage of rig adaptors will need careful layout in the rig/derig area.

Shaft connection is usually manual, with some form of assisted shaft lift and location system.

Special consideration should be given in these types of test areas to the draining, retention, and disposal of liquid spills or wash-down fluids.

Large engine testing is always of a duration exceeding the normal working day, therefore running at night or weekends is common and may lead to complaints of exhaust noise or smoke from residential areas nearby. Each cell or test area will have an exhaust system dedicated to a single engine; traditionally and successfully the silencers have been of massive construction built from the ground at the rear of the cells; they should be fitted with smoke dilution cowls and require well-maintained condensate and rain drains to prevent accelerated corrosion.

## Summary

The radical changes being brought about by the advent of e-vehicles, fuel cells, downsized ICE, and hybrid powertrains have brought with them the need to radically redesign some of the test facilities required. Twenty-four-hour running, with 8 hours unmanned, is becoming common and the control of running test cells from geographically remote stations is not uncommon. Large and complex cells are being built to house full 4 × 4 powertrains, with or without the vehicle structure, while "villages" of small identical modular cells are being built for the climatic and electrical testing of battery packs. Cell designers are having to decide whether to modify existing ICE cells, perhaps combining two into one, or to rebuild afresh. New entrants into the automotive test industry are having to meet the test requirements of clients without having the background, established over years, developing the best practices in testing rotational machinery. It is a time of considerable flux when automotive test engineers of both OEMs, academia, tier 1 suppliers, and specialist hardware and software start-up companies should spend as much time as possible exchanging experiences and ideas for the benefit and safety of all.

## References

[1] HSE, Lighting at Work, ISBN 978 0 7176 1232 1 or, a more in-depth treatment, CIBSE, Code for Lighting, 2012, ISBN 9781906846282.

[2] Battery University™, The Illuminating Engineering Society (IES) is a joint sponsor of ASHRAE/IES Standard 90.1 and Standard 189. <https://batteryuniversity.com/>.

Chapter 4

# Electrical design requirements of test facilities

## Chapter outline



Engine Testing. DOI: https://doi.org/10.1016/B978-0-12-821226-4.00004-8

# Introduction

## Note concerning effect of electrical shock on humans

At the time of writing (2020) there is a considerable effort and investment in requipping and modifying existing automotive test facilities, or building new ones to support the development of new electrical energy propulsion systems. The need to work, at all levels of the automotive industry, with high-powered direct current (DC) systems by staff previously only experienced with medium-voltage alternating current (AC) systems gives rise to concerns about personnel safety.

An AC voltage of below 50 VAC (voltage, alternating current) is considered extra-low voltage, while a DC level below 120 VDC (voltage, direct current) is also considered as extra-low voltage; this indicates that the two systems have different effects on the human body and from which one may infer that DC is "safer" than AC.

AC shock through the chest for a fraction of a second may induce ventricular fibrillation at currents as low as 30 mA, while DC shock can stop the heart; in terms of fatalities, both kill, but more milliamps are required of DC current than an AC current at the same voltage. Familiarization with 12 and 24 V lead–acid batteries is not good preparation for dealing with modern battery packs and their associated cabling; training is essential.

The danger of someone suffering from electrical shock or burns is greatly increased by unusual circumstances such as the period post power failure or while a device is "temporally" installed for a non-standard test routine; the habit of risk analysis must be ingrained. Finally the location of system "kill switches" should be obvious to not only staff but any fire department personnel or first responder who may have to attend in an emergency.

Secondary safety measures include the provision of defibrillator pack within easy reach of the test cell and training of the staff in its correct use. Also within easy reach of any cell testing high-voltage drives or batteries should be a large non-conductive S-shape rescue hook for pulling away from the power source anyone immobilized by electric shock.

## Note concerning terms related to "earthing" and "grounding"

In this text and in context with electrical engineering the UK English terms "earth" and "earthing" have generally been used in preference to the US English terms "ground" and "grounding" with which they are, in the context of this book, interchangeable.

## Overview

This chapter, as part of its content, mentions the subject of electromagnetic compatibility (EMC), but only in the context of the automotive test facility, both as a source and a victim of electromagnetic noise. The subject of testing the EMC of the whole vehicle, powertrain units, and their components is treated in Chapter 18, Anechoic and EMC Testing and Test Cells.

The electrical system of a test facility provides the power, nerves, and operating logic, that control the test piece (UUT, Unit Under Test), the test instrumentation, and building services. The power distribution to, and integration of, these many parts falls significantly within the remit of the electrical engineer, whose drawings will be used as the primary documentation in system commissioning and for any subsequent faultfinding tasks.

The theme of system integration is nowhere more pertinent than within the role of the electrical designer, nor is any area of engineering more constrained worldwide by differing national regulations, making integration of international projects fraught with traps for the unwary. On either side of the Atlantic electrical engineers can use similar words that have dissimilar meanings or technical implications. The integration of test cell equipment in Europe with equipment that is built to American standards and vice versa can create detailed compatibility problems. Such problems may not be appreciated by specialists at either end of the project unless they have experience in dealing with each other's practices. Within Europe there are national standards of which electrical engineers, working outside these territories, must be aware. For example, power distribution cabinets built to British standards will not meet several specific layout requirements of the German VDE (Verband Deutscher Elektrotechniker) standards.

## Modern test facilities uses of electrical power

Electrical systems in a classical internal combustion engine (ICE) test cells can be subdivided according to function group and power levels, which are discussed in the following sections.

### Power to cell services and systems

These will be served through a dedicated main AC switchboard with downstream boards per cell or per system. The control of the services dedicated to a cell is usually via a panel in the control room of that cell but will also be incorporated into the building management system (BMS) (see Chapter 13: Test Cell Safety, Control, and Data Acquisition). The test bed engine and dyno controller plus the automation and measurement system all require electrical power at low- or extra-low-voltage levels (according to International Electrotechnical Commission definitions) in the range of 24–240 V, single phase. Often a three-phase supply may be used with the phases separated and balanced with respect to test system load. It may be necessary to have a three-phase system where higher power is needed—for example, heating systems for emissions sample handling.

## Power to an internal combustion engine unit under test

The engine system itself requires electrical power to able to start and operate, as it would when installed into the vehicle. The engine control system requires a low-voltage supply at 12 or 24 V to enable the engine controller and electronics to function. In addition, it is often required to provide a starting system as would be used and installed in-vehicle, this could be either due to the use of a passive load unit with no motoring capability or to be able to test starting performance of the UUT with the same starting motor arrangement as installed in the vehicle. For the latter purpose, 1 or 2 kW of power are required at extra-low voltage, which could mean power supplies, switchgear and cabling with up to DC 200 amp capability.

## Dynamometer power systems

Mostly significant where active, four-quadrant electric machines are used. First-generation active load units were DC machines with thyristor-based drive systems that were notoriously "noisy" electrically. Modern systems are equipped with AC machines using insulated gate bipolar transistor (IGBT) based drive systems that are far more acceptable with respect to electromagnetic interference (EMI) due to higher converter switching frequencies. Power levels are in the range of kilowatts—the network supply must be able to supply and absorb power (often asymmetrically). Dedicated supply transformers are required for the dynamometer and these are normally at 11 kV input supply voltage level.

## Power to hybrid electric vehicles and battery electric vehicles test cells

In addition to the power supplies listed previously for ICE cells, the new requirement for current and future powertrain test environments is the need for handling traction power, DC vehicle systems. These possess different requirements when compared to the above:

- Hybrid propulsion systems: Depending upon the concept being tested, mild hybrid systems may require, in addition to that stated above for a classical engine test cell, a 48 DC volts supply with up to 20 kW capability as well as the ability to emulate a battery system accurately using a power electronic systems.
- Battery electric vehicles (BEV) propulsion systems: The ability to emulate and test, high-voltage battery and motor systems, with DC voltage levels of 400–800 V, 500 kW increasing to over 1000 kW for electric hypercars at the time of writing.

## The electrical engineer's design role

The electrical engineer's prime guidances are the operational and functional specifications (see Chapter 1: Test Facility Specification, System Integration and Project Organization) from which he or she will develop the final detailed functional specification, system schematics, power requirement, distribution, and alarm logic matrices.

Before the electrical engineer can start to calculate the electrical power required in order to produce an electrical distribution scheme, all of the major test and service plant and their electrical loads have to be identified. Before transformers, power distribution boards, control cabinets, and interconnecting cable ways can be located, all of which may have implications for the architect and structural engineer, the general building layout has to be known. The electrical engineer is therefore a key design team member from the start of the project and throughout the iterative design and system integration processes.

Test facilities are, by the nature of their component parts and the low power of their measurement signals, particularly vulnerable to signal distortion caused by various forms of electrical noise. This vulnerability has changed significantly during the previous 25-year lifetime of this publication In the last decades of the 20th century "earth loops" were blamed for every sort of system commissioning delay as existing building infrastructures were adapted to house the latest computerized equipment. Developments in high-power pulse-width modulation (PWM) speed controllers produced a change in the nature of the electromagnetic noise within the test cell environment.

Then as now the only way to minimize or avoid the possibility of signal corruption and cell downtime due to instrumentation errors is by giving special and detailed attention to the standard of electrical installation within a test facility. The electrical designer and installation supervisor must be able to take a holistic approach and be aware of the need to design and install an electrically integrated and mutually compatible system. Laboratory facilities that are built and developed on an "ad hoc" basis often fall foul of unforeseen signal interference or control logic errors within interacting subsystems.

## Power to cell services and systems

Perhaps more than any other aspect of test cell design and construction, the electrical installation is subject to regulations, most of which have statutory force. National differences in the details of regulations can lead to quite significant differences in the morphology of power distribution cabinets and the type of protection devices house within; for this reason it is always a good practice to have such units as motor control panels built in the country of operation, even if schematically designed outside of that country.

Over and above compliance with regulations is the need for all electrical installations within an entire test facility to be of the highest possible standard. The integrity of data produced by the facility relies on the correct separation of uncompromised cables and the high quality of their connections; such work should not be carried out by inexperienced or unsupervised staff. The authors have experience of a significant delay in commissioning test plant that was caused by an electrical contractor, employing an individual subcontractor, experienced in domestic power electrical installation, but suffering from undetected "red-end" color blindness (protanopia). This individual was contracted to install a system containing 40-core, individual-screened, and color-coded signal cables. The lesson is clear: attention to every detail and relevant experience is vital.

It is essential that any engineer responsible for the design or construction of a test cell in the United Kingdom should be familiar with BS 7671, "Requirements for Electrical Installations." There are also many British standards specifying individual features of an electrical system; other countries will have their own national standards. In the United States the relevant electrical standards may be accessed through the ANSI and NEMA websites [1,2].

The rate of change of regulations in the current industrial situation, particularly European regulation, will outpace that of any general textbook. Note that BS EN 60204 is of particular relevance to test cell design since it includes general rules concerning safety interlocking and the shutting down of rotating plant; this is covered in Chapter 9, Dynamometers: The Measurement and Control of Torque, Speed and Power, that also discusses shutting down of UUT/dynamometer combinations.

While the relevant regulations cover most aspects of the electrical installation in test cells, there are several particular features that are a consequence of the special conditions associated with the test cell environment that are not explicitly covered. These will be included within this chapter, as will the special electrical design features required if four-quadrant electrical dynamometers are being included within the facility.

## Physical environment

The physical environment inside a test cell can vary between the extremes used to test the engine or vehicle. The inside of subsystem control cabinets or transducer boxes can become overheated unless they are protected from radiant heat sources and well ventilated. ICE cells, particularly those running diesel engines, may have particulates entrained in the airflow that, over time, block filters fitted to control cabinets. Unlike the case of large powertrain cells, it is not advisable to locate the power cabinets for AC dynamometers inside more cramped engine test cell or in dusty or damp equipment spaces, as they can be particularly vulnerable to ingested dirt and moisture.

The guidelines for acceptable ambient conditions for most electrical plant are as follows:

- The ambient temperature of the control cabinet should be between +5°C and +35°C.

- The ambient temperature for a control room printer is restricted to +35°C and for the PC (personal computer) to +38°C.
- The designed ambient temperature of a nonclimatic test cell should be between +5°C and +45°C.
- The air must be free of abnormal amounts of dust, acids, corroding gases, and salt (use suitable inlet filters).
- The relative humidity at +40°C must not be more than 50% and 90% at +20°C (use anticondensation heating).

If air-conditioning of control room or service spaces is installed in tropical atmospheres of high humidity, the electrical designer must consider protection against condensation within control cabinets and dynamometer drive cabinets. In such conditions, both in operation and during installation, anticondensation heaters should be installed and switched on (see "Electrical cabinet ventilation" section later in this chapter).

The Ingress Protection (IP) rating of electrical devices such as motors and enclosures must be correctly defined in the design stage once their location is known.

In some vehicle and engine cells, particularly large diesel and engine rebuild facilities, it is sometimes the practice for a high-pressure heated water washer to be used, in which case the test bed enclosures should have a rating of at least IPx5 (see Table 4.1).

**TABLE 4.1 Ingress Protection (IP) rating methodology.**

IP54 = IP letter code ______ IP
First digit ______________5 Second digit ______________4

| First digit | | Protection from solid objects | Second digit | | Protection from moisture |
|---|---|---|---|---|---|
| 0 | | Nonprotected | 0 | | Nonprotected |
| 1 | 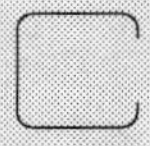 | Protected against solid objects greater than 50 mm | 1 | 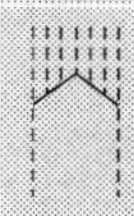 | Protected against dripping water |
| 2 | 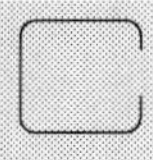 | Protected against solid objects greater than 12 mm | 2 | 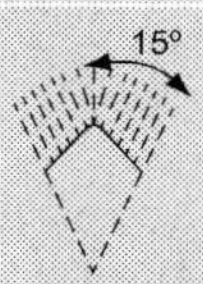  | Protected against dripping water when tilted up to 15 degrees |

(*Continued*)

**TABLE 4.1 (Continued)**

| | | | | | |
|---|---|---|---|---|---|
| 3 | 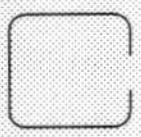 | Protected against solid objects greater than 2.5 mm diameter | 3 | 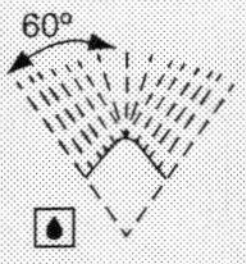 | Protected against spraying water |
| 4 | 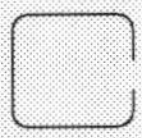 | Protected against solid objects greater than 1.0 mm diameter | 4 |  | Protected against splashing water |
| 5 | 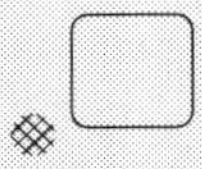 | Dust protected | 5 | 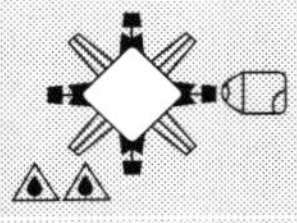 | Protected against water jets |
| 6 | 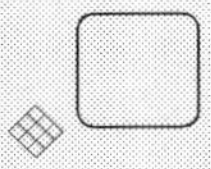 | Dust tight | 6 | 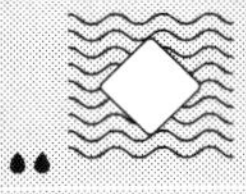 | Protected against heavy seas |
| | | | 7 | 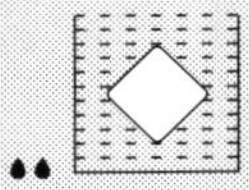 | Protected against the effects of immersion |
| | | | 8 | 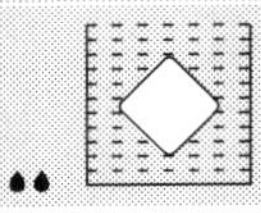 | Protected against submersion |

## Electrical signal and measurement interference

The protection methodology against signal interference in engine test cells has changed significantly over the period since the mid-1990s because of the arrival of EMI in the radio frequencies (RFs) as the dominating source of signal corruption.

The pulse-width-modulating drive technology based on IGBT devices, used with AC dynamometers, has reduced the total harmonic distortion (THD) experienced in power supplies from that approaching the 30% that was produced by first-generation DC thyristor-controlled drive, to less than 5% THD:

$$THD\,(\%) = \sqrt{\sum_{2}^{k}\left(\frac{H_x}{H_f}\right)^2}$$

*where $H_x$ is the amplitude of any harmonic order and $H_f$ is the amplitude of the fundamental harmonic (first order).*

Most sensitive devices should be unaffected by a THD of less than 5% in fact computers and allied control equipment, frequently stipulate that AC sources should have no more than 5% harmonic voltage distortion factor with the largest single harmonic being no more than 3% of the fundamental voltage.

The harmonic distortion produced by thyristor drive–associated DC machines is load dependent and causes a voltage drop at the power supply (see Fig. 4.1).

While IGBT technology, associated with AC dynamometers, has reduced THD that caused problems in facilities fitted with DC dynamometers, it has joined other digital devices in introducing disturbance in the frequency range of 150 kHz to 30 MHz. IGBT-based drive systems produce a common-mode, load-independent disturbance that causes unpredictable flow through the facility earth system. This has meant that previous standard practices concerning signal cable protection and provision of "clean earth" connections, which are separate from "protective earths (PEs)," are tending to change in order to fight the new enemy: EMI in the radio-frequency range. The connection of shielded cables between devices shown in Fig. 4.2 *was* a standard method aimed at preventing earth loop distortion of signals in the shielded cable.

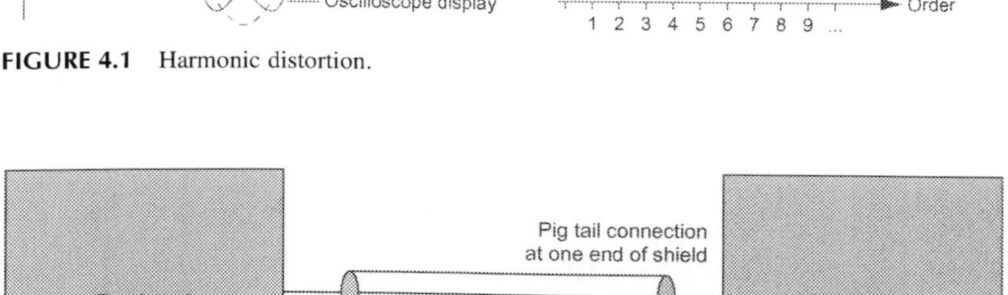

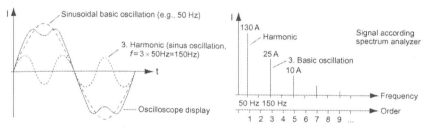


**FIGURE 4.1** Harmonic distortion.

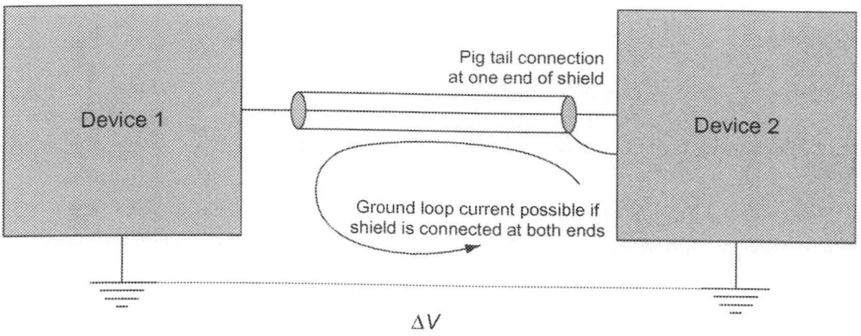


**FIGURE 4.2** Recommended method of connecting shielded cable between devices to prevent ground loop current distortion of signals.

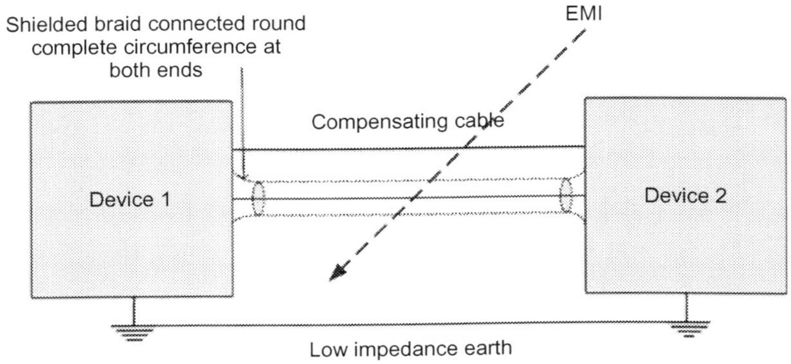


**FIGURE 4.3** Recommended method of connecting shielded cable between devices, in test facilities containing high-powered AC dynamometers with IGBT drives, to prevent EMI and ground loop corruption of signals. *AC*, Alternating current; *EMI*, electromagnetic interference.

*Earth loops* are caused by different earth potential values occurring across a measuring or signal circuit that induces compensating currents.

The single end connection of the cable shield shown in Fig. 4.2 offers no protection from EMI, which requires connection at both ends (at both devices); this requires that in installations containing IGBT drives, the ground loops are defeated by different measures. At the level of shielded cable connection the countermeasure is to run a compensation lead of high surface area in parallel with the shield, as shown in Fig. 4.3. This method also reduces the vulnerability of the connection to external magnetic flux fields.

## Earthing system design

The earthing systems used in many industrial installations are primarily designed as a human protection measure. However, electromagnetic compatibility (EMC) requirements increasingly require high-frequency equipotential bonding, achieved by continuous linking of all ground potentials. This modern practice of earthing devices has significantly changed layouts in which there was once a single PE connection plus, when deemed necessary, a "clean earth" physically distant for the PE.

A key feature of modern EMC practice is the provision of multiple earth connections to a common earthed grid of the lowest possible impedance. The physical details of such connection practice are shown in Fig. 4.4.

To ensure the most satisfactory functioning of these electromagnetic immunity systems, the earthing system needs to be incorporated into a new building design and the specification of the electrical installation. Ideally, the building should be constructed with a ground mat made of welded steel embedded in the concrete floor and a circumferential earth strap.

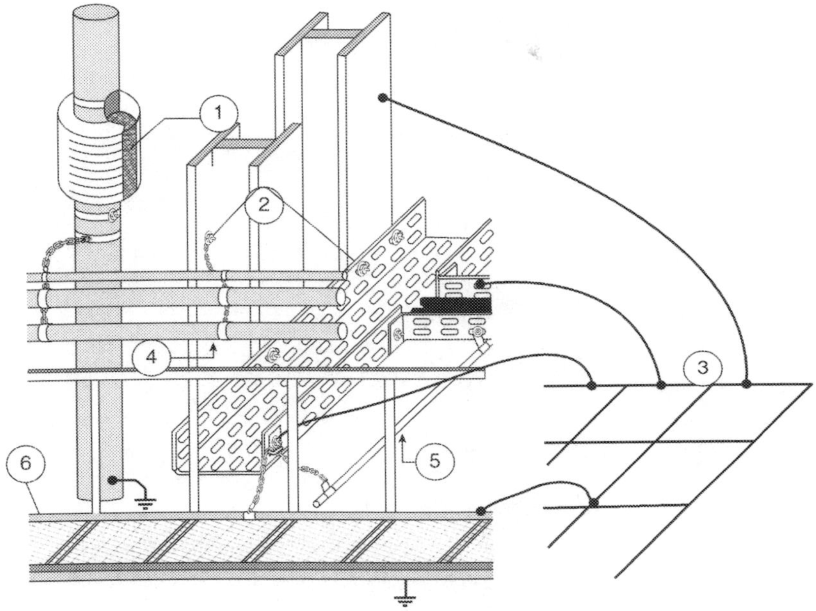


**FIGURE 4.4** Continuous linked network of all ground potentials. (1) Braided conducting strip bridging pipe joints. (2) Bolts welded into the metal building frame that is connected to the reinforcement mat. (3) Foundation earthing mat installed in new buildings and/or runaround grounding strap in existing buildings providing an EMC reference potential. (4) Braided strap linking conductive clamps to service pipes. (5) Protective earth wire. (6) Cable trays, steelwork grid, etc.

## The layout of cabling

Transducer signals are usually "conditioned" as near to the transducer as possible; nevertheless, the resultant conditioned signals are commonly in the range 0–10 VDC or 0–5 mA, very small when compared with the voltage differences and current flows that may be present in power cables in the immediate vicinity of the signal lines.

Cable separation and layout is of vital importance. Practices that create chaotic jumbles of cables beneath trench covers and "spare" cable lengths coiled in the base of distribution boxes are simply not acceptable in modern laboratories. Fig. 4.5 shows an example of a very poor test facility installation.

Coiling of overlong power cables is a common error and it causes inductive interference in control and transducer cables; the more coils, the stronger the effect.

Test facilities will contain some or all of the following types of wiring, which have distinctly different roles and which must be prevented from interfering with each other:

- power cables, mains supply ranging from high-power wiring for dynamometers through three-phase and single-phase distribution for services

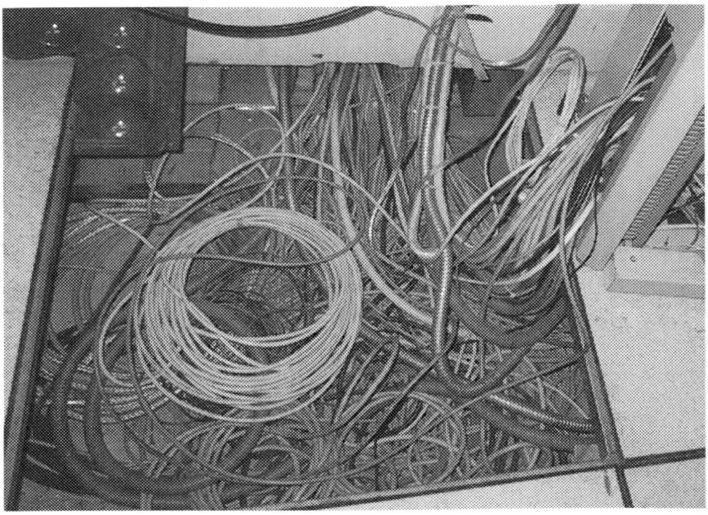

**FIGURE 4.5** The real-life horrors of a bad electrical installation revealed under a control room floor. Power cables mixed with signal cables in close parallel proximity, overlong cables coiled and producing magnetic interference, A fault-finding nightmare!.

and instruments to low-power supplies for special transducers, also high current DC supply for starter systems;

- (orange) power cables forming part of the hybrid or electric vehicle (EV) powertrain made compliant with ECE-R100;
- Control cables for inductive loads, relays, etc.; and
- Signal cables:
  - digital control with resistive load;
  - Ethernet, RS232, RS422, IEEE1394;
  - bus systems such as controller area network (CAN), CAN-FD, Flexray, EtherCAT, and other industrial fieldbus systems;
  - measuring cables associated with transducers and instrumentation transmitting analog signals; and
  - extra-low voltage, DC supplies

In the following paragraphs the common causes of signal interference are identified and practices recommended to avoid the problems described.

*Inductive interference* is caused by the magnetic flux generated by electrical currents inducing voltages in nearby conductors. Countermeasures include the following:

- Do not run power cables close to control or signal cables (also see capacitive interference next). Use either segmented trunking or different cable tray sections.
- Use twisted pair cables for connection of devices requiring supply and return connection (30 twists per meter will reduce interference voltage by a factor of 25).
- Use shielded signal cables and connect the shield to earth at both ends.

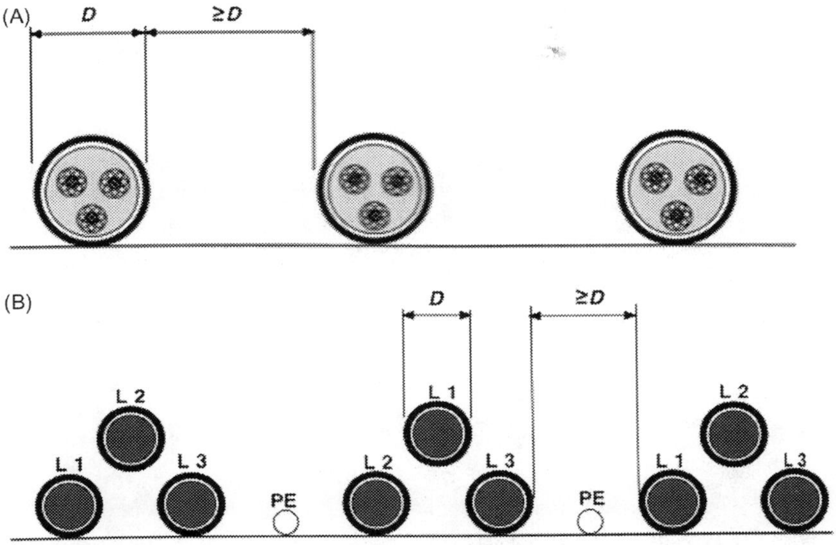


**FIGURE 4.6** (A) Recommended spacing of multicore power cables in a trench or tray. (B) Recommended spacing orientation and layout of individual phase power cables with PE cables between each bunch (only two shown).

Fig. 4.6 shows a suggested layout of power cables running in an open cable tray. The same relationship and metal division between cable types may be achieved using segmented trunking, trays, or ladders. It is important that the segments of such metallic support systems are connected together as part of the earth bonding system; it is not acceptable to rely on metal-to-metal contact of the segments.

Note that cables of different types crossing through metal segments at 90 degrees to the main run do not normally cause problems.

*Capacitive interference* can occur if signal cables with different voltage levels are run closely together. It can also be caused by power cables running close to signal lines. Countermeasures include the following:

- Separate signal cables with differing voltage levels.
- Do not run signal cables close to power cables.
- If possible use shielded cable for venerable signal lines.

*EMI* can induce both currents and voltages in signal cables. It may be caused by a number of "noise" transmitters ranging from spark plugs to mobile communication devices.

RF noise, produced by motor drives based on IGBT devices, is inherent in the technology with inverter frequencies of between 3 and 4 kHz and is due to the steep leading and failing edge of the 500 V pulses that have a voltage rise of around 2 kV/μs. There may be some scope in varying the pattern of frequency disturbances produced at a particular drive/site combination by adjustment of the pulsing frequency; the supplier would need to be consulted.

Following are the countermeasures for AC drive–induced noise:

- Choice and layout of both motor/drive cabinet and control cabinet to supply cables is very important:
- Motor cable should have three multicore power conductors in a symmetrical arrangement within a common braided and foil screen. The bonding of the screen at the termination points must contact 360 degrees of the braid and be of low impedance.
- The power cables should contain symmetrically arranged conductors of low inductivity within a concentric PE conductor.
- Single-core per-phase cables should be laid as shown in Fig. 4.7.
- Signal cables should be screened.
- Signal cables should be encased in metal trunking or laid within metal cable tray.

The layout of power cables of both multicore and single core is shown next as following recommended practice.

*Conductive coupling interference* may occur when there is a supply voltage difference between a number of control or measuring devices. It is usually caused by long supply line length or inappropriate distribution layout. Countermeasures include the following:

- Keep supply cables short and of sufficient conductor size to minimize voltage drop.
- Avoid common return lines for different control or measurement devices by running separate supply lines for each device.

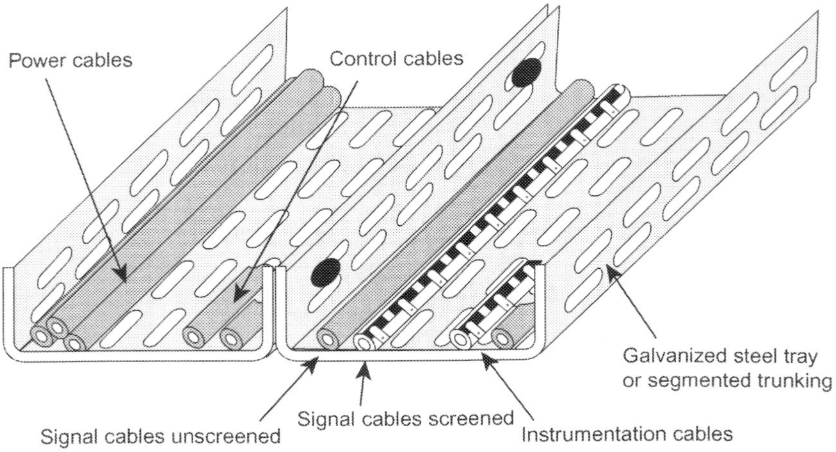


**FIGURE 4.7** Segmented trays or trunking and separation of types of cable.

The same problem may occur when two devices are fed from different power supplies.

To minimize interference in analog signal cables, an equipotential bonding cable or strap should connect the two devices running as close as possible to the signal cable. It is recommended that the bonding cable resistance be less than 1/10th of the cable screen resistance.

## Electrical cabinet ventilation

Instrument errors caused by heat are particularly difficult to trace, as the instrument will probably be calibrated when cold, so the possibility of such damage should be eliminated in the design and installation phases. Many instrumentation packages produce quite appreciable quantities of heat and if mounted low down within the confined space of a standard 19-in. rack cabinet, they may raise the temperature of the apparatus mounted above them over the generally specified maximum of 40°C.

Control and instrument cabinets should be well ventilated, and it may be necessary to supplement individual ventilation fans by extraction fans high in the cabinet. Cabinet ventilation systems should have filtered intakes that are regularly changed or cleaned.

Special attention should be given when signal conditioning and complex instruments are within a boom-mounted box that may be positioned above an ICE under test; then forced ventilation must be provided and particular attention must be paid to the choice of equipment installed. Some signal conditioning modules are specially designed to run in such situations and at temperatures of up to 60°C. It is recommended that thermal "telltale" strips are installed in boom boxes and in cabinets as a good maintenance device and aid to faultfinding.

## Security of supply

The electrical system designers must carefully consider the possible consequences of a failure of the electrical supply either to part of the system due to protection device operation or an incoming grid failure. Particular dangers can arise if the incoming power is restored unexpectedly after a short duration mains supply failure. While major grid failures may be very uncommon in industrial areas of Europe and America, they may not be in countries with less developed intrastructure where modern test plant is being installed, local knowledge is essential and special design features may be required. It is usual practice to design systems as normally deenergized and for the cell control logic to require that equipment, such as solenoid operated fuel valves be individually physically reset after mains failure as part of an orderly operator initiated restart procedure.

The installation of an uninterrupted power supply for critical control and data acquisition devices is strongly recommended. Such a relatively inexpensive piece of low maintenance equipment need only provide power for the short time required to save or transfer data and bring instrumentation into a safe condition.

## Power to an internal combustion engine under test

A power supply is often required to supply DC motors that are required to start or "crank" the ICE. Voltage is DC, less than 50 V (12 or 24 V), with currents in the range of hundreds of amps. In a vehicle, this is normally provided by a starter battery, which can also be installed at the test environment to provide the required power. However, this presents some issues with respect to test cell safety (i.e., starter batteries located inside test cells, often containing acid electrolyte as liquid or gel, possibly giving off volatile fumes during operation)—and also the practicality of charging and maintaining this battery. This latter issue often depends if the engine has a generator mounted and if it can be used—as some tests specify the use of ancillary equipment that apply parasitic loads on the engine (and some do not).

Another possible option is the use of low-voltage power supplies for starting and charging. Electrical power supply units capable of supplying extra-low-voltage DC power can be considered a safer option in the test cell when compared to a battery. It is advisable to plan for the use of this equipment at an early stage so that the device can be engineered into the overall electrical system, as opposed to mobile devices being used for engine cranking and power.

The electronic control unit of the UUT monitors the supply voltage provided, as certain control parameters are sensitive to voltage. Therefore a high-quality DC supply must be provided by the facility-based system. Often these facility devices are interfaced to the test cell controller such that output voltage can be defined and controlled, in order to monitor UUT system behavior to a fluctuating voltage power supply. Such power supplies need to be selected carefully as DC ripple is a significant factor that tends to be a function of price and quality of the selected device. Relatively small ripple currents can cause significant issues with respect to supply quality monitoring, also where DC power supplies are used for charging batteries, voltage ripple must be kept within limits to prevent excessive heat generation in the battery that will cause premature failure. Note this is a general issue when using DC power supplies connected to battery systems due to the fact that any battery has internal impedance (or resistance) and the ripple current flowing into a battery at normal float charge conditions can cause considerable heating due to simple power dissipation effects (based on $I^2R$ losses).

A typical DC supply for use in ICE test environments would require a voltage range from 0 to 50 V, with a current supply capability of 20–100 amp. Typically, peak-to-peak ripple will vary depending upon the quality of the chosen device from 100 mV (worst) to 10 mV (best). Where batteries are connected then a maximum allowable ripple current can be approximated based upon the battery capacity.

## Dynamometer power systems

Careful consideration should be made when integrating an AC drive system into an engine test facility and the whole scheme should be discussed with the supply company because of the implications of the facility becoming a transient power exporter (see the "Electrical supply connection" section).

Where possible it is most advisable to provide a dedicated electrical supply to an AC drive, or number of AC drives.

There needs to be a clear understanding of the status of the existing electrical supply network, and the work involved in providing a new supply for an AC dynamometer system. It is important to calculate the correct rating for a new supply transformer. This is a highly specialized subject and the details may change depending on the design of the supply system and AC devices, but the general rule is that the "mains short-circuit apparent power ($S_{SC}$)" needs to be at least 20 times greater than the "nominal apparent power of one dynamometer ($S_N$)." Hence, $S_{SC}/S_N > 20$.

In a sample calculation based on a single 220 kW dynamometer, the following sizing of the supply transformer may be found:

$P_N = 220$ kW nominal power, cos $\Pi = 1$, overload 25%

$S_N = P_N \times \cos \Pi = 220$ kVA

$S_{SC} = S_N \times 20 = 4.400$ kVA

Standard transformer (ST) ~ $U_{SC} = 6\%$

ST $= S_{SC} \times 0.06 = 264$ kVA, $R$ transformer $= 315$ kVA

ST $= S_N \times 1.25 = 275$ kVA, $R$ transformer $= 315$ kVA.

In the case of multiple dynamometer installation, some manufacturers in some conditions will allow a reduction in the $S_{SC}/S_N$ ratio; this can reduce the cost of the primary supply.

Shown below is the basic calculation of power supply transformer in the case of 3 × 220 kW dynamometers:

$P_N = 220$ kW nominal power, cos $\Pi = 1$, overload 25%, diversity 1, 0.8

$S_N = 3 \times P_N \times \cos \Pi = 660$ kVA

$S_{SC1} = S_N \times 20 = 13{,}200$ kVA in the case of no reduction for multiple machines

$S_{SC2} = S_N \times 10 = 660$ kVA in the case of ratio reduction to minimum

ST ~ $U_{SC} = 6\%$

ST $= S_{SC1} \times 0.06 = 792$ kVA, $R$ transformer $= 800$ kVA

ST $= S_{SC2} \times 0.06 = 396$ kVA, $R$ transformer $= 400$ kVA

ST $= S_N \times 1.25 \times 0.8 = 660$ kVA, $R$ transformer $= 800$ kVA.

### Supply interconnection of disturbing and sensitive devices

An infinite variation of transformer and connection systems is possible, but Fig. 4.8 shows the range from poor (A) to a recommended layout (C). The worst connection scheme is shown in Fig. 4.8A, where both the sensitive and

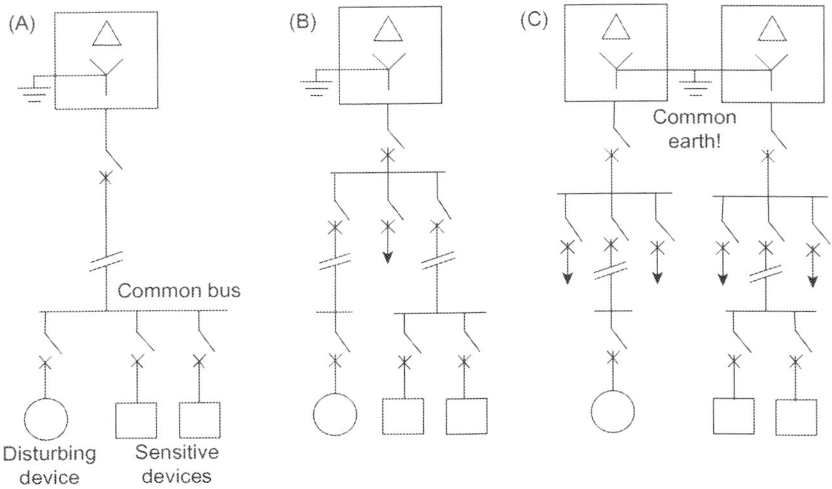


**FIGURE 4.8** Different power connection layouts, ranging from poor (A) to recommended (C).

a disturbing device, such as a PWM drive, are closely connected on a common local bus. Fig. 4.8B shows an improved version of Fig. 4.8A, where the common bus is local to the transformer rather than the devices. The best layout is based on Fig. 4.8C, where the devices are fed from separate transformers that share a common earth connection.

## Power cable material and bend radii

The wide use of AC motors and dynamometers in modern test cells has increased the need for the careful choice of power cables, all of which must be highly resistant to gasoline, oils, and hydraulic fluids but normally need not be gasoline or oil "proof" or "fast." The conductor copper must be stranded rather than massive and the bending radius should not exceed the following figures:

| | |
|---|---|
| Fixed cables: | 15 times the cable outside diameter |
| Moving or movable cables: | 10 times the cable outside diameter |

## Electrical power supply specification

Most suppliers of major plant will include the required details of power supply conditions their equipment will accept; in the case of AC dynamometers this may require a dedicated isolating transformer to provide both isolation and the required supply voltage.

**TABLE 4.2** A typical mains power specification in the United Kingdom.

| Voltage | 230 VAC −6% + 10% |
|---|---|
| Frequency | 0.99−1.01 of nominal frequency 50 Hz continuously, 0.98−1.02 of nominal frequency 50 Hz short time |
| Harmonic distortion | Harmonic distortion is not to exceed 10% of total r.m.s. voltage between the live conductors for the sum of the second through to fifth harmonics, or 12% max of total r.m.s. voltage between the live conductors for the sum of the sixth through to the 30th harmonics |
| Voltage interruption | Supply must not be interrupted or at zero voltage for more than 3 ms at any time in the supply cycle and there should be more than 1 s between successive interruptions |
| Voltage dips | Voltage dips must not exceed 20% of the peak voltage or the supply for more than one cycle and there should be more than 1 second between successive dips |

The IEEE 519-2014 standards are applicable to supplies for AC automotive dynamometers.

The voltage used in America and Canada is based on 120 VAC and a frequency of 60 Hz, a useful text on compliancy with this system is found at [3].

The voltage used throughout Europe (including the United Kingdom) has been harmonized since January 2003 at a nominal 230 V 50 Hz. It was formerly 240 V in the United Kingdom and 220 V in the rest of Europe.

The new "harmonized voltage limits" in most of Europe, which were the former 220 V nominal countries. are now:

230 V −10% + 6% or between 207.0 and 243.8 V

While in the United Kingdom, which was formerly 240 V nominal, they are:

230 V −6% + 10% or between 216.2 and 253.0 V

To cope with both sets of limits all modern equipment must therefore be able to accept 230 V ± 10%, that is, 207−253 V.

A typical mains power specification in the United Kingdom is shown in Table 4.2.

## Electrical supply connection

Ultimately, the test facility must be connected to a mains power supply network, the robustness of which can vary from country to country and from urban to rural areas. For an industrial facility (non-Automotive test), where the supply does not involve exporting of electrical power back into the network—then the discussion with the supplier or network operator is normally quite straightforward and involves characterization of the required load and duty cycle, such that hardware can be specified accordingly. Modern automotive test facilities, with active dynamometers, present a challenge in this respect. The energy created by the UUT is dissipated as electrical energy into the electrical network; battery test facilities can also export pulses of energy. In theory this sounds acceptable and efficient but in practice the conversation with the electrical supplier is more complicated because the energy that is generated by the test object is not predictable, and highly transient. Therefore it is often the case that the supplier sees the test facility as a disturbance and not a potential source of energy.

It is often the case, at least in the United Kingdom, that vehicle test facilities are located away from the main power supply infrastructure, in rural locations. This generally means that there will be a natural constraint on what can be handled by the surrounding network because the network was not originally designed to pass any significant industrial load. Also, when much of the network was built, the power would have flowed in one direction only, that is, from the power station through the high-voltage network to the distribution network through to the end user. The connection of a test facility that is exporting significant power means that there is distributed generation, at the various voltage levels on the local network and there can be reverse power flow in some parts of the network. The cost of removing this constraint and the technical issues around it can be a significant source of delay and cost when planning significant upgrades to a facility.

When connection to the network is being considered, it is necessary to involve the key persons from the supply and distribution network company. Distribution managers and planners are able to understand the exact constraints for a particular piece of network—as that is part of their "patch" and responsibility. If the network limitations are exposed due to the supply infrastructure, and upgrades are not possible, then other technical options, located within the test facility itself must be considered and approved by the electrical infrastructure stakeholders. Depending upon the nature of the problem, this could include active mains compensation systems to improve the facility supply quality when generating/exporting under highly transient UUT operating conditions. Resistive load banks can be deployed to limit the export capacity and dissipate excess energy as waste heat. The latter are often considered environmentally unfriendly but nevertheless, many facilities still require them, even just as a backup, due to the complications of supply network connections and the extended discussions that often occur around this topic.

In general, the electrical engineer for the facility must define the expected maximum export capacity under typical working conditions. A network simulation can provide not only typical values but also duty cycles and boundary conditions for the infrastructure before it is even built. With this data the distribution operator can state the maximum export/import capacity of the local network is and then provide a cost estimate for reinforcement of the local network. This cost can then be compared to the costs of an export limiting scheme.

## Test cell's electrical roles: importing and exporting

An important factor to consider at the time of writing is the effect of electrification of powertrains that now form a UUT with a complex mix of generators and consumers as follows:

- ICE—exporter, through AC dynamometers
- battery emulator—consumer, and exporter when testing hybrid drives
- battery tester—exporter and consumer
- inverter testing—exporter and consumer
- DC motor system—exporter and consumer

As a modern powertrain contains a combination of the above mentioned elements, it is highly probable that any new test facility or significant upgrade to any facility will include a combination of all of these systems. When an electric machine is both the UUT and the dynamometer, then power can be circulated around the system with only makeup for losses being imported and no need to export power to the supply network. This type of circuit is shown in Fig. 9.13 and simplifies the infrastructure requirements it also reduces the running costs involved. Power circulation can be considered at network level but also at test system level via the interconnection of inverter and drive systems at DC link level.

The planning of any new facility or facility upgrades should consider the power density of developing powertrains. At the time of writing manufacturers are now also producing vehicles with electric traction systems that have power levels in the range of 0.5–1 MW for both cars and commercial vehicle applications. These power levels are likely to have a significant impact on the supply network.

## Power to Mild-hybrid, hybrid electric vehicles and battery electric vehicles test cells

### Mild-hybrid electric vehicle (mHEV)

It should be possible for an existing ICE test cell to be adapted to test a 48 V mild hybrid powertrain without major structural modification but it requires a 48 V power and recuperation and storage system with all the control and safety systems that come with it.

The main issue for this system is to manage the electrical current that is in the range of 200–400 amps. The main consequence of the physical layout in the cell is the potential for considerable voltage drop along extended cable lengths, together with the robustness and quality of the required switchgear to control and isolate such DC supplies. Voltage drops have to be dealt with by ensuring that control systems control the supply voltage at the point of delivery in the cell, rather than at the power distribution cabinet remote from the UUT.

## Hybrid electric vehicles and plug-in hybrid electric vehicles

The higher electrical powers of full hybrid electric vehicles (HEV) and plug-in hybrid electric vehicles (PHEV) systems together with their mechanical complexity require a completely new design of both the cell structure and the electrical supplies. Existing ICE cells, providing they are sufficiently large and equipped with an AC dynamometer system may be adaptable to some HEV configurations without any changes to the major services, with the exception of the electrical power system, which will have to undergo a complete upgrade. Whereas the design and layout problems faced by the electrical engineer designing BEV and HEV systems are similar, the mechanical system designers have a completely different set of options. Decommissioned ICE cells can often provide adequate spaces for testing e-axles or their component parts (e.g., Fig. 9.14).

## Battery electric vehicles

For BEV test environments, voltages and power densities are much higher and high-voltage DC supplies have quite specific requirements with respect to safety, switchgear and cabling. For component tests of motors and inverters—power can be recirculated and this reduces the overall consumption of electrical energy but the complexity involved in recirculation of energy within the facility can be a consideration. A typical BEV may be equipped with several batteries and energy sources to provide power to multiple consumers—pumps, accessories, thermal management, etc. The considerable number of these consumers and their synchronized operation can cause significant system voltage ripple in the high-voltage system. This has an effect of increasing the current carried by the complex cable structure (i.e., the shielding) and can cause issues via overloading and failure of the protective shield and cable glands in the electrical system architecture of the powertrain wiring system. It is important that the cell power system for BEV application does not multiply this ripple effect and enables the measurement and characterization of such phenomena in the development phase of the BEV propulsion system, as this is a known failure mode in the operation of electrified powertrains in service.

## Considerations in converting internal combustion engine to electric motor assembly test cells

As existing ICE test facilities are being modified and new ones are being built in order to deal with e-vehicle powertrain units (Fig. 4.9), two distinctly

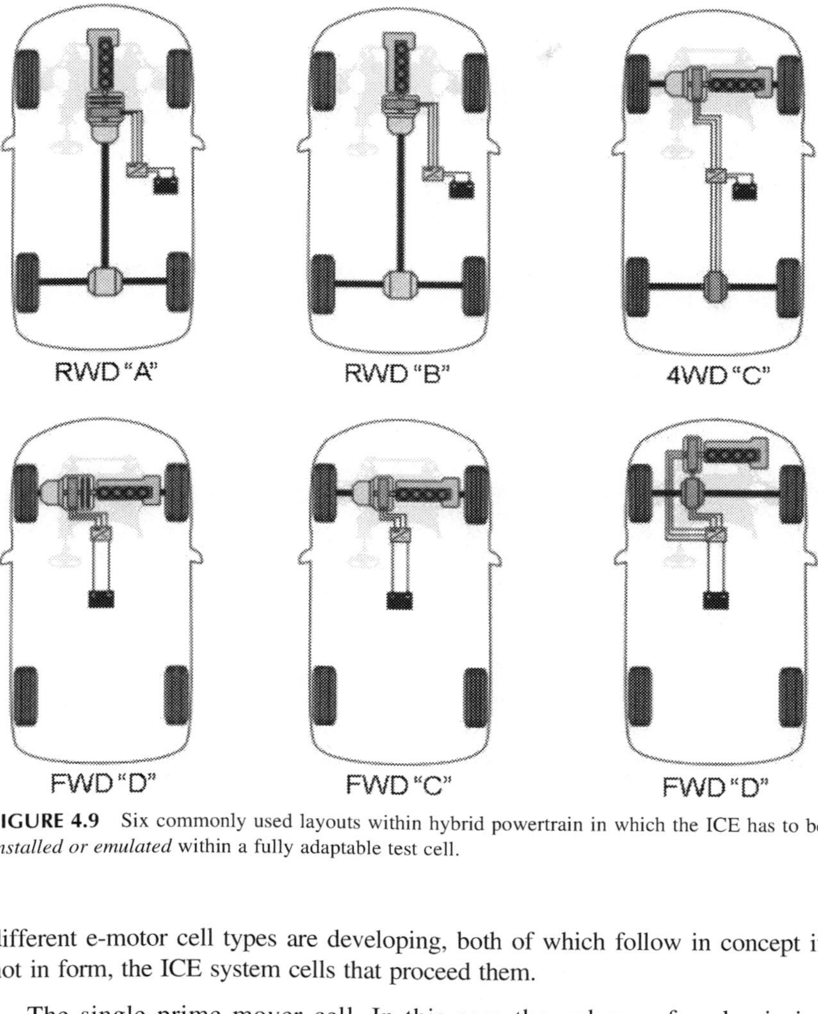


**FIGURE 4.9** Six commonly used layouts within hybrid powertrain in which the ICE has to be *installed or emulated* within a fully adaptable test cell.

different e-motor cell types are developing, both of which follow in concept if not in form, the ICE system cells that proceed them.

- The single prime mover cell. In this case the volume of a classic in-line ICE/dynamometer cell is usually adequate and conversion to a mHEV or traction motor cell is possible. The ventilation load should be lower so no modification is required but additional space will be required in the plant room for the chosen source of DC power. A key difference between ICE and motor test stands is that the motor type is frequently packaged with a climatically controlled housing. This is made easy by the lack of fluid and exhaust connections that would be required in an ICE test stand.
- The part-powertrain cell where the UUT is for a BEV and can be in more than one e-axle configuration does not have a direct equivalent in ICE

cells of the past and, in the example shown next, requires the housing of three AC dynamometers together with the UUT mounting plinth. In the example shown in Fig. 12.1 the pallet mounted UUT is directly connected to an ICE simulating dynamometer and the two output shafts are each connected to a wheel simulation dynamometer. This type of pallet and dynamometer unit, using in vehicle mounting fixings, forms the rigid fixed node on a large bedplate that allows different configurations of powertrain with up to five dynamometers. In the system shown the wheel dynamometers are mounted on height-adjustable frames. Such a cell, free from the requirement of running an ICE, can house most of the power-electric cabinets within its enclose.

When full hybrid powertrains are required they tend to require to be new-built and while they share the large sectional bedplate that allows multiple configurations due to the complexity of the services supporting both ICE and HEV systems, they tend to be rather less quickly adaptable and less likely to be able to house the power electrical systems. Fig. 3.13 gives a good idea of a fully adaptable high-powered ICE/HEV powertrain cell.

The main impact of e-drive propulsion testing is the need for a DC power source, at voltage levels considerably higher than encountered up to now with conventional propulsion system test environments. As a battery can only store energy in DC form, and as the power requirements are in the range of 300–600 kW for traction with DC voltages up to 800 V and currents up to 900 amp, this creates new challenges for the test cell design, where integration of high-powered DC systems are required. These challenges can be summarized as:

- DC switchgear and cabling technology has quite different requirements to high-powered AC—breakers and circuit protection devices have different classifications and performance requirements.
- Personal safety requirements around high-powered DC supply systems, energy storage systems, and the use of permanent magnet (PM) motors are significantly different so specific measures and safety constraints for those working in and around these systems are required.
- Battery technology is sensitive to temperature and the risk of "thermal runaway" means that many operations are using battery emulation in their place. There are three different strategies emerging in general practice:
  - Eschewing the in-cell use of the vehicle battery pack an emulator is used which, due to its bulk, may be positioned outside the cell and even shared between cells.
  - While requiring a battery pack to be part of the system under test, it is housed separately from the main cell, close but outside in its own isolatable housing.

- When the battery forms an integral part of the system under test, it is installed within the test cell but mounted with multiple temperature measurement and thermal camera monitoring.
- Cabling and voltage drop. Voltage levels must be maintained as required by the test regime and the system, interfaces and cabling must be capable of the required level of performance with respect to load, temperature, ingress and humidity, as well as handling. In order to connect and reconnect quickly to any part of the EV power system the distractive cables and plugs, compliant with ECE-R100, should be used (see Fig. 12.2).

High-powered DC and AC systems in a vehicle and their emulation devices have the potential to create a considerable risk to other equipment due to RF interference and electromagnetic compatability (EMC).

Battery emulators should comply with, as a minimum, the following standards:

- EN 61000-6-4 covering EMC emission;
- EN 61000-6-2 covering EMC immunity; and
- EN 50178, ISO 12405-4:2018 and EN 60439-1 covering general standards for batteries and switchgear.

See Chapter 18, Anechoic and EMC Testing and Test Cells, for detailed coverage of EMC testing which is an important parallel activity to the development of the electric powertrain.

High power test systems and infrastructure are not new topics, but the staff and management of existing vehicle powertrain development facilities, which are being upgraded or created, now need to consider an electrical power flow requirement, which has a different pattern to a traditional ICE or powertrain test cells (that are developing units for transportation or vehicle applications). Large test cells for power generation, marine and wind turbine applications already exist and overcome all of the problems mentioned. The main issue is that the power density of the UUT in these latter applications has not evolved in the same way as that for a vehicle powertrain, that is now transitioning from ICE to electric power concepts, and that high capacity energy storage devices (batteries up to 100 kWh at the time of writing) for vehicle propulsion systems are a new consideration in the test environment that necessitate fresh work processes and training.

## Cell heat load from e-motors

In the absence of a detailed analysis the heat load put into the cell ventilation air by air-cooled traction motors and e-axles has been assumed to be akin to that from AC dynamometers at around 0–15 kW/kW of power output.

There are some traction motors that are fluid cooled and there now exist a range high torque-density motors, currently used in electric racing cars, for which an extremely efficient cooling system is required for the heat extraction from the motor. In these designs the stator winding is cooled with forced 3M Fluorinert oil or similar, which is retained in the stator slots region with a carbon fiber tube, while the end windings are potted. The volume flow rate is variable between 5 and 15 L/min allowing a winding temperature below 120°C. The test cell has to be equipped with a closed system for the cooling oil supply and treatment (see also Chapter 12: Rigging and Running of Electrical Drive Systems).

## General consideration for battery emulation and direct current power sources and sinks

The electrical engineering design team involved with hybrid and/or electric powertrain testing has to find space for and integrate into their systems one or more grid-connected, bidirectional, fully controllable DC power sources or separate modules that when combined are of fulfilling the roles of testing batteries, supercapacitors and fuel cells, inverter systems, as a programmable DC power source and sink and also test e-motors and e-powertrains as a ISO 12405-4:2018 compliant battery emulator.

Such full battery emulating devices, holding all the required circuitry and contactors, that are currently available as package products will range from around 75 up to upward of 350 kW are physically large three- or four-bay cabinets with the higher power units probably requiring a water cooling supply.

It is therefore sensible that they are outside the test cell and matched with an interface panel in the cell that provides short EV-standard plugged cables to the UUT.

There are a wide range of DC power sources, both unidirectional and bidirectional, matched with battery emulators being made around the world and choosing the most suitable package or individual modules requires careful matching with the users test regimes and products.

Future-proofing the electrical simulation and emulation environment should be considered in the first facility design. Capacity for energy flow to and from the network is likely to increase, along with voltage levels. This must be accounted for in the initial specification of equipment along with the plant room space required and the subsequent cost trade-off. A modular hardware network allows stacking of systems in parallel in order to provide flexibility, but the overall infrastructure capability must be capable in the first place. The modular approach is an important topic for the electrical engineer and planner—as it facilitates a multi application environment.

## Recording the evolution of the test cell electrical system

Once a test cell is commissioned and handed over for productive use, and as-built drawings have been delivered, the implementation and evolution phase will start. Over time this often means additions or removal of devices, requiring modifications to the electrical system and subsystems, which may be undocumented. This occurs most often incrementally, where small changes and adaptions to the special use or application of the cell take place, when test equipment is changed or modified and the electrical interfaces have also be adapted. Often the owner or user of the cell sees these small changes as insignificant and in isolation they may be so. In the real world though, these small changes have a cumulative effect over time where the test cell as delivered schematics no longer reflect the current "as now built" situation. At the time it is thought that the cost and effort of updating drawings may be an unnecessary expense adding no real value. However, this point of view is a mistake—it is willful neglect of the facility and can cause additional, greater costs further down the line where upgrades and extensions are necessary in order to maintain the test cell productivity and efficiency.

Therefore a disciplined approach is required to the integration of new equipment or update of existing equipment in a working test environment. Drawings are normally provided electronically, with hard copies as an option. The electronic versions are normally not editable by the customer, but with the hard copy version, manual markup is always possible. Therefore this is a good justification for an old-school approach to manually markup hard copy drawings whenever a significant change occurs. The definition of the threshold for a markup activity is the responsibility of facility's technical management or owner. Simply changing a sensor or cable can be excused. But a significant change, such as the installation of a new instrument with a new interface cable, or the replacement of a section of multiple cables due to damage, or upgrade requirements must be recorded along with the drawing markup notes should be made in the test cell log with cross-referencing to the relevant schematic.

Another challenge in this respect is the documentation relating to mobile equipment, it is often the case that certain measurement devices are based on a trolley and shared between one, two, or more cells. Interface panels to the test cell control system are provided and recorded in the schematics of the overall system. Normally though, the mobile equipment often does not have its own electrical subsystem or interface wiring documented (often seen as overkill). Consequently, such mobile system can evolve in an undocumented way and it is the facility management's responsibility to ensure that such devices are included in the schematic or document maintenance philosophy.

## European safety standards and Conformité Europeen marking

The United Kingdom, via the European Community has, since 1985, been developing regulations to achieve technical harmonization and standards to permit free movement of goods within the Community. There are currently four directives of particular interest to the builders and operators of engine test facilities (see Table 4.3).

The use of the "CE mark" (abbreviation for "Conformité Europeen") implies that the manufacturer has complied with all directives issued by the European Economic Community (EEC) that are applicable to the product to which the mark is attached.

There may be some confusion as to the difference between CE marking and the e-mark, the latter being part of the vehicle homologation requirements for type-approved vehicles and vehicle components sold in the EU. "E" and "e" type approval marks are required on devices related to all safety-relevant functionality of the vehicle. Such devices will have to be e-mark certified by a certifying authority; unlike most CE marking, manufacturers cannot self-declare and affix an e-mark (see the "EU Type Approval "E" and "e" Marking" section in Chapter 18: Anechoic and EMC Testing and Test Cells).

An engine test cell must be considered as the sum of many parts. Some of these parts will be items under test that may not meet the requirements of the relevant directives. Some parts will be standard electrical products that are able to carry their individual CE marks, while other equipment may range from unique electronic modules to assemblies of products from various manufacturers. The situation is further complicated by the way in which electronic devices may be permanently or temporally interconnected.

If standard and tested looms join units belonging to a "family" of products, then the sum of the parts may comply with the relevant directive.

**TABLE 4.3** EC directive for Conformité Europeen marking relevant to powertrain test facilities.

| Directive | Reference | Optional | Mandatory |
|---|---|---|---|
| Electromagnetic compatibility | 89/336/EEC | January 1, 1996 | |
| Machinery | 98/37EC | January 1, 1993 | January 1, 1995 |
| Low voltage | 73/23/EWG plus 93/68/EEC | January 1, 1995 | January 1, 1997 |
| Pressure equipment | | November 1999 | May 2002 |

If the interconnecting loom is unique to the particular plant, the sum of the CE marked parts may not meet the strict requirements of the directive.

It is therefore not sensible for a specification for an engine test facility to include an unqualified global requirement that the facility "be CE marked." Some products are specifically excluded from the regulations, while others are covered by their own rules; for example, the directive 72/245/EEC covers radio interference from spark-ignition vehicle engines. Experimental and prototype engines may well fail to comply with this directive, just as hardware-in-the-loop tests running unproven simulation models may not comply in every respect with some strict safety standards. These are possible examples of the impossibility of making any unqualified commitment to comply in all respects and at all times with bureaucratic requirements drawn up by legislators unfamiliar with our test industry.

There are three levels of CE marking compliance that can be considered:

- All individual control and measuring instruments should individually comply with the relevant directives and bear a CE mark.
- "Standardized" test bed configurations that consist exclusively of compliant instrumentation, are configured in a documented configuration, installed to assembly instructions/codes, and have been subjected to a detailed and documented risk analysis may be CE marked.
- Project-specific test cells. As stated previously, the CE marking of the complete hybrid cell at best will require a great deal of work in documentation and at worst may be impossible, particularly if required retrospectively for cells containing instrumentation of different generations and manufacturers.

The reader is advised to consult specific "health and safety" literature, or that produced by trade associations, if in doubt regarding the way in which these directives should be treated.

## Safety interaction matrix

A key document of any integrated system is some form of safety or alarm interaction matrix, and its first draft is usually the responsibility of the electrical engineer having system design authority. It is the documentary proof that a comprehensive risk analysis has been carried out (see Table 4.4 for an example of a recommended matrix layout). To be most effective the base document should be verified between the electrical design engineer, the system integrator, and user group. It is the latter that have to decide, within a framework of safety rules, upon the secondary reactions triggered in the facility by a primary event.

The control logic of the BMS has to be integrated with that of the test control system; if the contractual responsibility for the two systems is split then the task of producing an integrated safety matrix needs to be allocated and sponsored.

**TABLE 4.4** Example test cell safety matrix.

| Test cell safety matrix for exampled faults below | Cell control system | 400 VAC electrical supply | Test cell power sockets | Ventilation fans | Ventilation system dampers | Combustion air system | AC dynamometer | Engine control | Test cell incoming fuel solenoids | Fuel conditioning unit | Compressed air system |
|---|---|---|---|---|---|---|---|---|---|---|---|
| Controlled by: | CS | ESR | ESR | BMS | BMS | BMS | CS | ESR | ESR-DO | ESR | ESR-DO |
| Main building fire panel | NO reaction | Enabled | Enabled | BMS shuts down service after 4 min | Manual shutdown of test cell before evacuating the facility | | | | | | |
| Test cell fire system (automatic) | Stop | Enabled | Disabled | Disabled | Closed | Disabled | Stop | Stop | Closed | Power off | Closed |
| Level 1 HC/CO gas alarm | Message display | Enabled | Enabled | Vent to purge | Open | Enabled | Power on | Enabled | Open | Power on | Open |
| Level 2 HC/CO gas alarm | Fast stop | Enabled | Enabled | Vent to purge | Open | Power off | Regen. stop | Stop | Closed | Power off | Open |
| Emergency stop (Cat 1) | Fast stop | Enabled | Disabled | Disabled | Closed | Disabled | Regen. stop | Stop | Isolated | Power off | Closed |
| Fast stop—via control desk button | Fast stop | Enabled | Enabled | Freeze vent | Open | Enabled | Regen. stop | Stop | Open | Power on | Open |
| Test cell doors opened | Fast stop | Enabled | Enabled | Freeze vent | Open | Enabled | Regen. stop | Stop | Open | Power on | Open |
| Test sequence engine alarm | Message display | Enabled | Enabled | Enabled | Open | Enabled | Power on | Enabled | Open | Power on | Open |
| Engine stop (automatic) | Stop | Enabled | Enabled | Enabled | Open | Enabled | Power on | Stop | Open | Power on | Open |

*BMS*, Building management system; *CO*, carbon monoxide; *CS*, calibrated span; *ESR*, Emergency stop relay; *HC*, hydrocarbon.

## References


[1] The National Electrical Manufacturers Association, <http://www.nema.org/stds/>.

[2] The American National Standards Institute (ANSI), <www.ansi.org>.

[3] Lisa M. Benson, Karen Reezek, A Guide to United States Electrical and Electronic compliance requirements, National Institute of Standards and Technology ref: NISTIR 8118r1. Available FOC from https://doi.org/10.6028/NIST.IR.8118r1.


## Further reading


BS 7671:2008. Requirements for Electrical Installations: IEE Wiring Regulations, eighteenth ed., ISBN: 978-0-86341-844-0.

BS EN 60204-1:2006 þ A1. Safety of Machinery Electrical Equipment of Machines, 2009.

J. Goedbloed, Electromagnetic Compatibility, Prentice Hall, New Jersey, 1992. ISBN: 0-13249293-8.

Institution of Engineering and Technology, Electromagnetic compatibility and functional safety, latest versions available through <www.theiet.org/factfiles>.

J. Middleton, The Engineer's EMC Workbook, Marconi Instruments Ltd, Sterenage, 1992. ISBN: 0-95049413-5.

T. Williams, EMC for Product Designers, Newnes, Oxford, 1992. ISBN: 0-75061264-9.

Chapter 5

# Ventilation and air-conditioning of automotive test facilities

## Chapter outline



**Engine Testing. DOI: https://doi.org/10.1016/B978-0-12-821226-4.00005-X**

## Introduction

This chapter and the next are concerned with the thermodynamics of the automotive powertrain test facility; how the units under test (UUT) within the cells and their workspaces are cooled or heated, and how the creation of an explosive atmosphere is prevented.

*Note:* The general principles covering the avoidance of a build-up of an explosive atmosphere is covered in detail within Part 3 of Chapter 3 'Test facility design and construction' while detailed features required by ATEX regulations are covered here.

In almost all automotive internal-combustion engine (ICE) test cells, the vast majority of the energy comes into the system as highly concentrated chemical energy entering the cell by way of pipes via the smallest penetrations in the cell wall. It leaves the cell as lower grade heat energy through the largest penetrations: the ventilation ducts, engine exhaust, and the cooling water pipes.

In the case of cells testing an ICE and fitted with electrically regenerative dynamometers, almost one-third of the energy supplied by fuel will leave the cell as electrical.

In the case of EV (Electric Vehicle) and hybrid powertrain testing, the energy flows are more complicated and variable, particularly when some of the electrical energy is recirculating within the test system. In general, when converting a cell from testing only ICE units to testing hybrid arrangements or pure electric drive units, there is little need to modify the ventilation system since the logic of its control will be the same and the thermal load will range from similar to significantly lower.

Since the ICE puts more heat energy into its test cell than any other automotive component, we will use it as the prime example in the following content.

## Part 1: The basics and the heat capacity of cooling air

By definition the test cell environment is mainly controlled by regulating the quantity, temperature, and in some cases the humidity, of the air passing through it. Air is not the ideal heat transfer medium: it has low density and low specific heat; it is transparent to radiant heat, and its ability to cool hot surfaces is much inferior to that of liquids.

The main properties of air, of significance in air-conditioning, may be summarized as reviewed next:

The gas equation:

$$p_a \times 10^5 = \rho R(t_a + 273) \tag{5.1}$$

where $p_a$ is the atmospheric pressure (bar), $\rho$ is the air density (kg/m$^3$), $R$ is the gas constant for air which equals 287 J/kg K, and $t_a$ is the air temperature (°C).

Under conditions typical of test cell operation with an air temperature of 25°C (77°F) and standard atmospheric pressure, the density of the air from the previous equation above is:

$$\rho = \frac{1.01325 \times 10^5}{287 \times 298} = 1.185\ \text{kg}/m^3$$

which is about 1/850th that of water.

The specific heat, at constant pressure, of air at normal atmospheric conditions is approximately:

$$C_p = 1.01\ \text{kJ}/\text{kg K}$$

which is less than one-quarter of that of water.

The airflow necessary to carry away 1 kW of power with a temperature rise of 10°C is:

$$m = \frac{1}{1.01 \times 10} = 0.099\ \text{kg}/\text{s} = 0.117\ \text{m}^3/\text{s} = 4.132\ \text{ft}^3/\text{s}$$

All these indicate that as much heat as possible should be taken out of the cells system via the cooling water systems.

## Sources of heat in the test cell

### The internal-combustion engine

A worked example next shows the calculations that can be used to estimate the heat load put into the cell from an ICE, meanwhile there are many "rules of thumb" used in the industry. One authority quotes a maximum of 15% of the heat energy in the fuel, divided equally between convection and radiation. This would correspond to about 30% of the power output of a diesel engine and 40% in the case of a gasoline engine.

In the experience of the authors, a figure of 40% (0.4 kW/kW engine output) represents a safe upper limit to be used as a basis for design for water-cooled engines. This is divided roughly in the proportion 0.1 kW/kW engine to 0.3 kW/kW exhaust system; it is thus quite sensitive to exhaust layout and insulation.

In the case of an air-cooled engine, the heat release from the engine will increase to about 0.7 kW/kW output in the case of a diesel engine and to about 0.9 kW/kW output for a gasoline engine. That proportion of the heat of combustion, that passes to the cooling water in a water-cooled engine, in an air-cooled unit must of course pass directly to the surroundings. High-powered air-cooled engines, such as aeronautical radial engines, tend to be tested on thrust and torque measuring cradles in open-ended buildings. Cooling is provided by a flow of air, propelled by the airscrew that also provides engine load.

### The dynamometer

A water-cooled dynamometer, whether hydraulic or eddy current, runs at a moderate temperature and heat losses to the cell are unlikely to exceed 5% of power input to the brake. When AC and DC machines are air cooled, and heat loss into the cell is around 6%–10% of power input, the contribution to cell air heating of the water-cooled versions of AC dynamometers are usually taken as negligible for these calculations.

### Other sources of heat

AC drive cabinets generate heat that may prove significant if they are located in control spaces. The heat output from these drive cabinets can vary with the manufacturer and the work cycle; many of the higher power units can be liquid cooled but for general ventilation calculations, a figure of 5% of the maximum rated power of the connected dynamometer can be used. In all cases, with AC drives and particularly with devices such as battery emulator cabinets, the manufacturer's specification must be obtained.

The heat generated by e-axles and their components depends upon the test bench design and its installation space, but it will always be less than an ICE of the same power rating. It is now general practice for test benches capable of full power testing of traction motors to mount the UUT on an insulated pallet and be fitted with a sliding insulated cover so it is enclosed in a small chamber wherein the temperature may be controlled between −40°C and 120°C. Such test bench specific, stand-alone systems reduce the need from supplementary ventilation or spot-fans in the main cell chamber but require their own services.

Effectively all the electrical power to lights, fans, and instrumentation in the test cell will eventually appear as heat transmitted to the ventilation air. The same applies to the power taken to drive the forced-draught inlet fans: this is dissipated as heat in the air handled by the fans. The secondary heat exchangers used in coolant and oil temperature control will similarly make a contribution to the total heat load; here long lengths of unlagged pipework holding the primary cooling fluid will not only add heat to the ventilation load but will also adversely affect control of the fluid temperature.

Recommended values as a basis for the design of the ventilation system are given in Table 5.1. In all cases, they refer to the maximum rated power output of the engines to be installed.

**TABLE 5.1** Heat transfer to test cell ventilation air; based on empirical data.

| Heat source | kW/kW power output |
|---|---|
| Engine, water cooled | 0.1 |
| Engine, air cooled | 0.7–0.9 |
| Exhaust system (manifold and silencer) | 0.3 |
| Hydraulic dynamometer | 0.05 |
| Eddy-current dynamometer | 0.05 |
| AC or DC dynamometer air blast cooled | 0.15 |

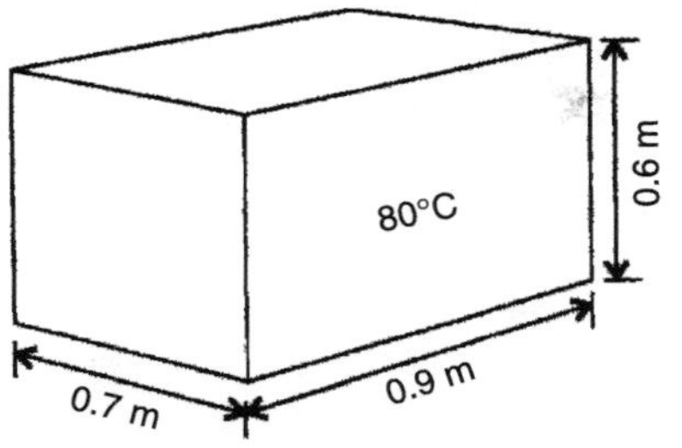


**FIGURE 5.1** Simplified model of example 100 kW ICE, for analysis of heat transfer to surroundings.

## Worked example of heat transfer into the test cell from the internal-combustion engine body and exhaust

It is useful to gain a feel for the relative significance of the elements that make up the total of heat transferred from a running ICE system into its surroundings by considering rates of heat transfer from bodies of simplified form under test cell conditions.

Consider a body of the shape sketched in Fig. 5.1. This might be regarded as roughly equivalent, in terms of projected surface areas in the horizontal and vertical directions, to a gasoline engine of perhaps 100 kW maximum power output, although the total surface area of the engine could be much greater. Let us assume the surface temperature of the body to be 80°C and the temperature of the cell air and cell walls as 30°C.

Heat loss occurs as a result of two mechanisms: natural convection and radiation. The rate of heat loss by natural convection from a vertical surface in *still air* is given approximately by:

$$Q_v = 1.9(t_s - t_a)^{1.25}\ W/m^2 \tag{5.2}$$

The total area of the vertical surfaces of Fig. 5.1 equals 1.9 m$^2$. The convective heat loss is therefore:

$$1.92 \times 1.92(80-30)^{1.25} = 480\ \text{W}$$

The rate of heat loss from an upward-facing horizontal surface is approximately:

$$Q_h = 2.5(t_s - t_a)^{1.25}\ W/m^2 \tag{5.3}$$

which in this example gives a convective loss of:

$$0.63 \times 2.5\ (80-30)^{1.25} = 210\ \text{W}$$

The heat loss from a downward-facing surface is about half that for an upward-facing surface, giving a loss of approximately 110 W.

We thus calculate a rough estimate of the total convective loss as 800 W; if the surface temperature was raised to 100°C, this would increase to about 1200 W. However, this is the heat loss in still air, and the air within an ICE test cell will always in a ventilation flow which means that the heat losses into that air will be far greater.

*It is vital, when considering problems of cooling heat sources in a test cell, to remember that: an increase in cooling air velocity greatly increases the rate of heat transfer into that air and that this effect may aggravate, rather than solve, the problem of high temperatures within a cell.* (See "Use of spot fans" later in this chapter.)

As a rough guide, doubling the velocity of airflow past a hot surface increases the heat loss by about 50%. The air velocity due to natural convection in our example is about 0.3 m/s.

An air velocity of 3 m/s would be moderate for a test cell with ventilating fans producing a vigorous circulation, and such a velocity past the body of Fig. 5.1 would increase convective heat loss fourfold, to about 3.2 kW at 80°C and 4.8 kW at 100°C.

The rate of heat loss by radiation from a surface depends on the emissivity of the surface (the ratio of the energy emitted to that emitted by a so-called black body of the same dimensions and temperature) and on the temperature difference between the body and its surroundings. Air is essentially transparent to radiation, which mainly heats up the surfaces of the surrounding metalwork and cell; this heat must subsequently be transferred to the cooling air by convection or conducted to the surroundings of the cell.

Heat transfer by radiation is described by the Stefan–Boltzmann equation, a form of which is as follows:

$$Q_r = 5.77\varepsilon\left[\left(\frac{t_s+273}{100}\right)^4 - \left(\frac{t_w+273}{100}\right)^4\right] \tag{5.4}$$

A typical value of emissivity ($\varepsilon$) for machinery surfaces would be 0.9, $t_s$ equals the temperature of the hot body (°C), and $t_w$ is the temperature of an enclosing surface.

## Heat transfer into the test cell from the exhaust system

The other significant and highly variable source of heat loss associated with the ICE is the exhaust system. In the case of turbocharged engines, this can be particularly significant. Assume in the present example that exhaust manifold and exposed exhaust pipe are equivalent to a cylinder of 80 mm diameter × 1.2 m long at a temperature of 600°C, surface area 0.3 m$^2$.

Heat loss at this high temperature will be predominantly by radiation and equal to:

$$0.3 \times 5.77 \times 0.9 \left[ \left( \frac{873}{100} \right)^4 - \left( \frac{303}{100} \right)^4 \right] = 8900 \text{ W}$$

from Eq. (5.2).

The convective loss is:

$$0.3 \times 1.9 \times (600 - 30)^{1.25} = 1600 \text{ W}$$

From this, it may be seen that the heat losses into the cell from an unlagged exhaust system heavily outweigh those from the engine block. To reduce the heat load in the cell from an exhaust system, it is necessary to insulate (lag) those parts of the system that form part of the cell's permanent installation. However, those parts of the exhaust that are ex-vehicle and form part of the engine's rigging system will usually have to be left unlagged but guarded from accidental contact by operators.

## Heat transfer into cell air from the walls

Most of the heat radiated from the engine and exhaust system will be absorbed by the cell walls and ceiling, also by instrument cabinets and control boxes, and subsequently transferred to the ventilation air by convection.

If we take the example of the "engine" and "manifold" and consider them to be installed in a test cell 6 m long × 5 m wide × 4 m high, then the total wall area is 88 $m^2$. Assuming a wall temperature 10°C higher than the mean air temperature in the cell and an air velocity of 3 m/s, the rate of heat transfer from wall to air is in the region of 100 $W/m^2$, or 8.8 kW for the whole wall surface, roughly 90% of the heat radiated from engine and exhaust system; the equilibrium wall temperature is perhaps 15°C higher than that of the air (Table 5.2).

The previous worked example is not an attempt to make a detailed, using exact values of surface areas and temperatures, but a simplified treatment to help one to clarify the principles involved.

**TABLE 5.2** Heat losses from the bodies in Fig. 5.1.

| Heat source | Convection (kW) | Radiation (kW) |
|---|---|---|
| Engine, jacket, and crankcase at 80°C | 3.2 | 1.2 |
| Engine, jacket, and crankcase at 100°C | 4.8 | 1.8 |
| Exhaust manifold and tailpipe at 600°C | 1.6 | 8.9 |

**TABLE 5.3** Estimated heat transfer into test cell ventilation air.

| Heat source | kW/kW power output |
|---|---|
| Engine, water cooled | 0.1 |
| Engine, air cooled | 0.7–0.9 |
| Exhaust system (manifold and silencer) | 0.3 |
| Hydraulic dynamometer | 0.05 |
| Eddy-current dynamometer | 0.05 |
| AC or DC dynamometer air blast cooled | 0.15 |

## Calculation of ventilation load

The first step is to estimate the various contributions to the heat load from engine, exhaust system, dynamometer, lights, and services. This information should be summarized in a single flow diagram. Table 5.3 shows typical values.

Heat transfer to the ventilation air is to a degree self-regulating: the cell temperature will rise to a level at which there is an equilibrium between heat released and heat carried away. The amount of heat carried away by a given airflow is clearly a function of the temperature rise ($\Delta T$) from inlet to outlet.

If the total heat load is $H_L$ (kW), then the required airflow to achieve the required $\Delta T$ (°C) is:

$$Q_A = \frac{H_L}{101 \times 1.185\Delta T} = 0.84\frac{H_L}{\Delta T}\ \mathrm{m^3/s} \tag{5.5}$$

**Formula Eq. (5.5) will be found very useful for general ventilation air-flow calculations and is worth memorizing; of course, the main difficulty in its use is the realistic estimation of the value of the heat load $H_L$.**

A temperature rise $\Delta T = 10°C$ is a reasonable basis for design. Clearly, the higher the value of $\Delta T$, the smaller the corresponding airflow. However, a reduction in airflow has two influences on general cell temperature: the higher the outlet temperature, the higher the mean level in the cell, while a smaller airflow implies lower air velocities in the cell, calling for a greater temperature difference between cell surfaces and air for a given rate of heat transfer.

## Part 2: Ventilation strategies

Given sufficient detailed information on a fixed ICE or hybrid cell system, it is possible to carry out a very detailed energy balance calculation but there are some commonly used "rule of thumb" calculations available to the cell designer; the most common of these relates to the energy balance of an ICE, which is known as the "30–30–30–10 rule." This refers to the energy balance where the percentages quoted are that of the fuel energy coming into the cell. Thus the first listed 30% is the engine's rated power going out of the cell via the dynamometer. The second 30% is the energy transferred into the engine's, and therefore the cell's, cooling system, the third is the energy going into the cell air, and the final 10% represents the energy going into oil cooling and ancillary systems. Like any rule of thumb this is crude, but it does provide a starting point for the calculation of a full energy balance from which we can evaluate significant differences in balance caused by the engine itself and its rigging within the cell.

First, there are differences inherent in the engine design. Diesels will tend to transfer less energy into the cell air than petrol engines of equal physical size. From past experience, testers of rebuilt bus engines, which have both vertical and horizontal configurations, have noticed that different engines, with the same nominal power output, will show quite different distribution of heat into the test cell air and cooling water. Second, there are differences in engine rigging in the cell that will vary the temperature and surface area of the exhaust pipes. Finally, there is the amount and type of equipment within the test cell, all of which makes a contribution to the convection and radiation heat load to be handled by the ventilation system.

Specialist designers have developed their own versions of a test cell software model, based both on empirical data and theoretical calculation, all of which is used within this book. The version developed by colleagues of the author produces the type of energy balance shown in Chapter 3, Test Facility Design and Construction, Fig. 3.3 and Table 3.2 lists just a selection, from an actual project, of the known data and calculated energy flows that such programs have to contain in order to produce Fig. 3.3.

Such tools are extremely useful but cannot be used uncritically as the final basis of design when a range of engines needs to be tested or the design has to cover two or more cells in a facility where fluid services are shared; in those cases the energy diversity factor has to be considered.

## The application of a diversity factor when specifying shared services

To design a multicell test laboratory that is able to control and dissipate the maximum theoretical power of all its prime movers on the hottest day of the year will lead to an oversized system and possibly poor temperature control at low heat outputs. Oversized control valves loose "control authority" at low heat inputs. The amount by which the thermal rating of a facility is reduced from that theoretical maximum is the *diversity factor*. The diversity factor often lies between 60% and 85% of maximum rating, but individual systems will vary from endurance beds with high rating down to anechoic beds with very low rating.

A key consideration is the number of cells included within the system. Clearly one cell may at some time run at its maximum rating, but it may be considered much less likely that four cells will all run at maximum rating at the same time: the possible effect of this diminution from a theoretical maximum was is discussed in Chapter 3, Test Facility Design and Construction, and is shown in Fig. 3.2. There is a degree of bravery and confidence, based on relevant experience, required to significantly reduce the theoretical maximum to a contractual specification, but very significant savings in capital and running costs may be possible if it is done correctly. Of course, it is essential that the creators of the operational and functional specifications use realistic values of actual ICE or motor powers to be tested, rather than extrapolations based on possible future developments. "Future proofing" may be better designed into the facility by planning for possible incremental addition of plant rather than oversizing at the beginning.

## Ventilation strategies

The purpose of air-conditioning and ventilation is the maintenance of an acceptable and specified atmospheric environment in an enclosed space. This is a comparatively simple matter where only human activity is taking place but becomes progressively more difficult as the rate of change and size of energy flows into and out of the space increase. An ICE test cell represents perhaps the most demanding environment encountered in industry. Large amounts of power will be generated in a comparatively small space, surfaces at high temperature are unavoidable, and large flows of cooling water, air, and electrical power have to be accommodated, together with rapid variations in thermal load. In addition to the rapid variation in heat-load, the ventilation system has to play an important role in the maintenance of a non-explosive atmosphere within the test cell, not only those testing ICE but also hydrogen fuel cells and high-power electrical battery systems. Such requirements are dealt with by specifically designed vapor purge systems and by ensuring sufficient airflow through the cell even at times of low thermal load.

Much of this chapter covers the most commonly used method of removing machinery-generated heat from the test cell, by forced ventilation using ambient (outside) air that provides a balanced pressure in the closed-cell system. An alternative pressure-balanced system recirculates some or all of the cell air through a conditioning system such as an air handling unit (AHU). Open-cell ventilation systems use fans that draw air through the cell that enters through ground-level grilles, usually in an external wall; these can be satisfactory in more limited test conditions than a balanced system and suffer from high noise breakout.

The final choice will be influenced by the range of ambient conditions at a geographic location, building space restraints, and the type of testing being carried out.

Since the performance of any type of ICE depends, to a large extent, by the condition, temperature, pressure, and humidity of both ingested and surrounding air, this rather mundane aspect of facility design is vital to both the quality of test results and the safety of the test facility. It may be of interest to readers to know that the problems relating to the ventilation of ICE test cells and/or the spaces used to house high-powered electrical cabinets have out-numbered all others dealt with by one of the authors during 30 years of consultancy work—so it is a subject worthy of study.

**TABLE 5.4 "Rule of thumb"-suggested air changes per hour in listed spaces.**

| | | | |
|---|---|---|---|
| **Banks** | 2–4 | **Laboratories** | 5–10 |
| Cafés | 10–12 | Offices | 5–7 |
| Shower rooms | 15–20 | Toilets | 6–10 |
| Garages | 6–8 | Factories | 8–10 |

*Note:* Some guides on ventilation of inhabited spaces use, as a measure of ventilation requirements, "room volume changes per hour," in this edition it is a measure used only in relation to safety guidelines or legislation where it may be used; Table 5.4 is shown as an example.

## Air handling units

Recirculation of the majority, or variable proportions of test cell air through a temperature-conditioning AHU can have operational advantages over a forced "in-and-out" ambient ventilation systems, particularly in cold climates. They may reduce the problem of engine noise breakout through the external air ducting and conveniently provide workshop mode air-conditioning. Since closed systems will recirculate entrained pollutants, a separate combustion air system is recommended, but a fresh air makeup source will always be required to prevent buildup and to balance the loss through the low-level purge.

Designs are usually based on packaged units, supplied with building services, such as chilled and medium-pressure hot water supply. The units used must be designed with filtration able to deal with air contaminated with the type of particulates produced during some types of engine testing. They are normally mounted above the cell in the services room and aligned on the long axis of the cell to minimize ducting. A rarer alternative arrangement, internal within high-roofed cells such as chassis dynamometer facilities, uses ceiling-mounted units that are based on direct evaporative cooling and electrical heating units.

AHUs, particularly the direct evaporative type, can be energy efficient but a detailed and site-specific calculation of initial and running costs is needed to make a valid comparison.

## Purge fans: safety ventilation system requirements to reduce explosion risk

In closed cells using volatile fuels or in any work or storage area in which any dangerous buildup of fumes can occur, the ventilation system will have to incorporate a purge fan, the purpose of which is to remove heavier-than-air and/or potentially explosive fumes from the lowest points of the cell complex. An open cell system fitted with (closable) ground-level air inlet grilles will not be immune from such vapor buildup if it has floor trenches, so the same purge requirement has to be met.

In ICE cells the purge fan plays an essential role in reducing the fire and explosion and is the only part of the ventilation system which must be manufactured to meet the requirements of the ATEX Directive, IEC standards and EN14986 (see Chapter 2: Quality and Health and Safety Management), these require that both motor and fan are inherently "nonsparking."

The purge system has to be integrated into the test cell control system in such a way as to ensure, on cell start-up, it has run for a minimum of 10 minutes, with no hydrocarbon sensors showing an alarm state, before engine ignition, and therefore fuel inlet valves are allowed to be energized. The rating of the purge fan may be covered by local legislation but as a minimum should provide for 30 air changes of the enclosed cell space per hour. The purge suction duct should extend to the lowest point of the cell, which may include any services trench system. In some older (pre-ATEX) designs, usually multicell installations, the purge fan system is also used for providing exhaust dilution.

## The special case of ventilation of test areas using hydrogen

The small enclosed space of the "traditional" ICE test cell confined within a larger building is generally unsuitable for testing hydrogen-based fuel cell systems. Any building space housing hydrogen devices has to be ventilated such that any hydrogen leak is diluted to 25% of the lower flammable level that is 1% by volume of air. Forced ventilation systems should not be used for the evacuation of hydrogen rather, since it is lighter than air it should be allowed to escape (vented) directly from the building to the outside by opening ducts in the roof. Air inlet openings should be located at floor level in the exterior walls, but hydrogen gas detector sensors must be installed at high level. It is recommended (29 CFR 1910.106.) that inlet and outlet openings should have a minimum total area of 0.003 $m^2$ per 1 $m^3$ of room volume, or 1 $ft.^2$ per 1000 $ft.^3$ of room volume, and that inlet ventilation systems should provide a normal air exchange of 0.3 $m^3$ of air per minute per 1 $m^2$ of solid floor space, or 1 $ft.^3$ of air per minute per 1 $ft.^2$ of solid floor space.

The building must be such that the upper parts of the structure and the underside of the roof must not be formed with pockets that could contain the gas and suspended ceilings must not be fitted. As with all other types of test cells, the building's design should prevent recirculation of vented gasses into another system's inlet.

## Part 3: Design of ventilation ducts and distribution systems

### Pressure losses

The layout of ventilation ducting in the services space above the cell is made more difficult if headroom is constricted and the ducts have to run at the same level as installed plant such as drive cabinets or combustion air treatment plant. The space layout needs careful, advanced planning to obtain an optimum balance between the use and access of floor-mounted equipment and tortuous, energy-wasting, noisy ducting.

The velocity head or pressure associated with airflowing at velocity $V$ is given by:

$$P_v = \frac{\rho V^2}{2} \text{ Pa}$$

This represents the pressure necessary to generate the velocity. A typical value for $\rho$, the density of air, is 1.2 kg/$m^3$, giving:

$$P_v = 0.6V^2 \text{ Pa}$$

The pressure loss per meter length of a straight duct is a fairly complex function of air velocity, duct cross section, and surface roughness. Methods of derivation with charts are given in any of the recommended texts at the end of this chapter. In general, the loss lies within the range 1–10 Pa/m, the larger values corresponding to smaller duct sizes. For test cells with individual ventilating systems, duct lengths are usually short and these losses are small compared with those due to bends and fittings such as fire dampers.

The choice of duct velocity is a compromise depending on considerations of size of ducting, power loss, and noise. If design air velocity is doubled by the size of the ducting being reduced, the pressure losses are increased roughly fourfold, while the noise level is greatly increased (by about 18 dB for a doubling of velocity). Maximum duct velocities recommended are given in Table 5.5.

It is general practice to aim for a flow rate of 12 m/s through noise attenuators built within ducting [1].

**TABLE 5.5 Maximum recommended duct velocities.**

| Volume of flow ($m^3/s$) | Maximum velocity (m/s) | Velocity pressure (Pa) |
|---|---|---|
| <0.1 | 8–9 | 38–55 |
| 0.1–0.5 | 9–11 | 55–73 |
| 0.5–1.5 | 11–15 | 73–135 |
| >1.5 | 15–20 | 135–240 |

The total pressure of an airflow $p_t$ is the sum of the velocity pressure and the static pressure $p_s$ (relative to atmosphere):

$$P_t = P_s + P_v = P_s + \frac{PV^2}{2}$$

The design process for a ventilating system includes the summation of the various pressure losses associated with the different components and the choice of a suitable fan to develop the total pressure required to drive the air through the system.

## Ducting and fittings

Various codes of practice have been produced covering the design of ventilation systems.

Galvanized sheet steel is the most commonly used material, and ducting is readily available in a range of standard sizes in rectangular or circular sections. Rectangular section ducting has certain advantages: it can be fitted against flat surfaces and expensive round-to-square transition lengths for connection to components of rectangular section such as centrifugal fan discharge flanges, filters, and coolers are avoided.

Spiral wound galvanized tubing has been used satisfactorily and cost-effectively in workshop and garage space ventilation schemes but has not proved suitable for test cell work and must not be used in systems carrying diluted engine exhaust, where it suffers from condensate accelerated corrosion.

Once the required air-flow rate has been settled and the general run of the ducting decided, the next step is to calculate the pressure losses in the various elements in order to specify the pressure to be developed by the fan. In most cases a cell will require both a forced-draught fan for air supply and an induced-draught fan to extract the air. The two fans must be matched to maintain the cell pressure as near as possible to atmospheric environment. For control purposes, cell pressure is usually set at 50 Pa below atmospheric conditions, which gives the safest set of conditions concerning door pressurization and fume leakage.

The varied and many components such as miter bends, swept bends, and rectangular ducts all have characteristic pressure losses usually given as its $K_e$ factor in manufacture's data sheets. The pressure loss per unit or length is therefore given by:

$$\Delta P_t = K_e \frac{\rho V^2}{2}$$

Information on pressure losses in plant items such as ducts of different dimensions, bends, hoods are available in texts such as the "CIBSE Guide B2: Ventilation and Ductwork 2016" or at the website section (Heating, Ventilation, and Air Conditioning (HVAC) of the US Department of Energy. Pressure loss data for filters, heaters, and coolers are generally provided by the manufacturer. See also [1].

The various losses are added together to give the cumulative loss in total pressure (static pressure + velocity pressure), which determines the required fan performance.

## Inlet and outlet ducting

Test cell ducts have to be designed so that noise created in the test cell does not break out into the surrounding environment. The problem has to be solved by the use of special straight noise attenuating sections of duct that are usually of a larger section and that should be designed in such a way as to give an air velocity around an optimum value of 12 m/s (see next section).

The arrangement of air inlets and outlets calls for careful consideration if short-circuiting and local areas of stagnant air within the cell are to be avoided. A simple, if rather messy, test to check if there are such "dead" areas in the airflow is to throw a hand-full of polystyrene beads into the center of the cell and observe their paths and distribution.

There are many possible layouts based on combinations of high-level, low-level, and above-engine direction ducts; four commonly used layouts are shown below.

Fig. 5.2 shows a commonly used design in cells without large subfloor voids; designers have to ensure that the airflow is directed over the engine rather than taking a higher level shortcut between inlet and outlet. In this and the next design, the purge of floor-level gases would be carried out through the exhaust gas dilution duct (see comments concerning exhaust systems in Chapter 6: Cooling Water and Exhaust Systems).

Fig. 5.3 shows an alternative layout where the intake air comes into the cell at low level, either by forced ventilation, as shown, or through an attenuated duct drawing ambient air from outside. In both the cases the air is drawn over the engine and exits at high level.

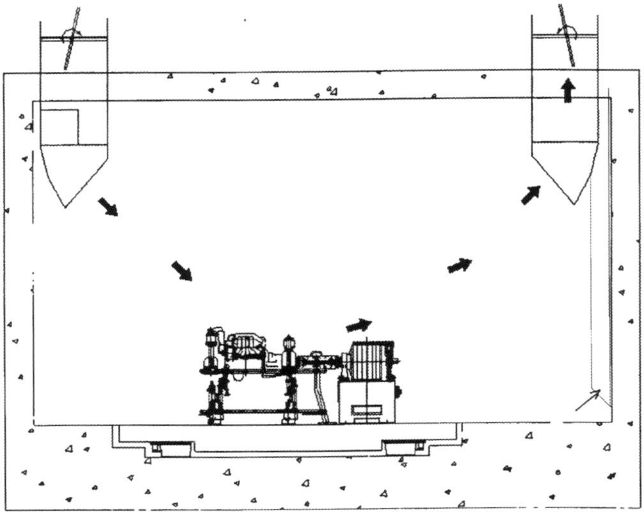

**FIGURE 5.2** Variable-speed fans in a pressure-balanced system using high-level inlet and outlet ducts plus supplementary purge duct at floor level.

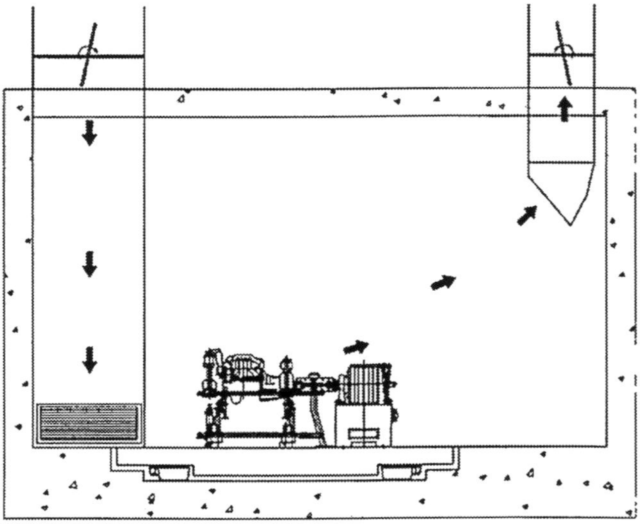

**FIGURE 5.3** Low-level inlet and high-level outlet ducted system; the low-level duct can either be part of a balanced fan system as in Fig. 5.2 or drawing outside air through an inlet silencer. Purge duct has been omitted from the sketch but needs to be fitted as in Fig. 5.2.

An inlet hood above the engine, as shown in Fig. 5.4, may inhibit the use of service booms and UUT access.

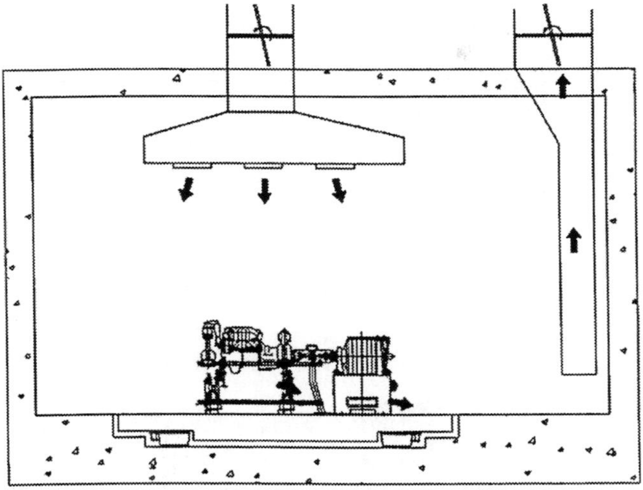

FIGURE 5.4 A balanced fan system as in Fig. 5.2, but inlet ducted through nozzles from an overengine plenum and extracted at low level.

Where a substantial subfloor space is present, the ventilation air can be drawn out below floor level and over the engine from an overhead inlet ventilation plenum.

It will be clear by now that the choice of system layout has a major influence on the layout of equipment both in the cell and the service space above.

Where an AHU is not fitted but heat energy needs to be conserved in the ventilation air, then a recirculation duct, complete with flow control dampers, can be fitted between the outlet and inlet ducts above the fans. This can be set to recirculate between 0% and 80% of the total airflow.

The use of "eyeball" *nozzles* in the ventilation entry duct allows the airflow into the cells to be split up into individually directed jets. This feature certainly helps in preventing areas of the cell from becoming zones of comparatively still air, as may happen with single outlet entry ducts. However, once the optimized flow has been discovered and unless major movements of plant are made within the cell, the nozzles need to be left fixed to prevent casual redirection, otherwise significantly varied airflows may result in subtle variations in engine performance that will affect repeatability of tests. Due to the distance between such nozzles and the UUT, the problems related to spot fans discussed next should not arise.

## The use of "spot fans" for supplementary cooling of unit under test or other in-cell heat sources

Spot fans must be used with care to avoid unintended consequences. A common error made by operators facing high cell temperatures is to bring in

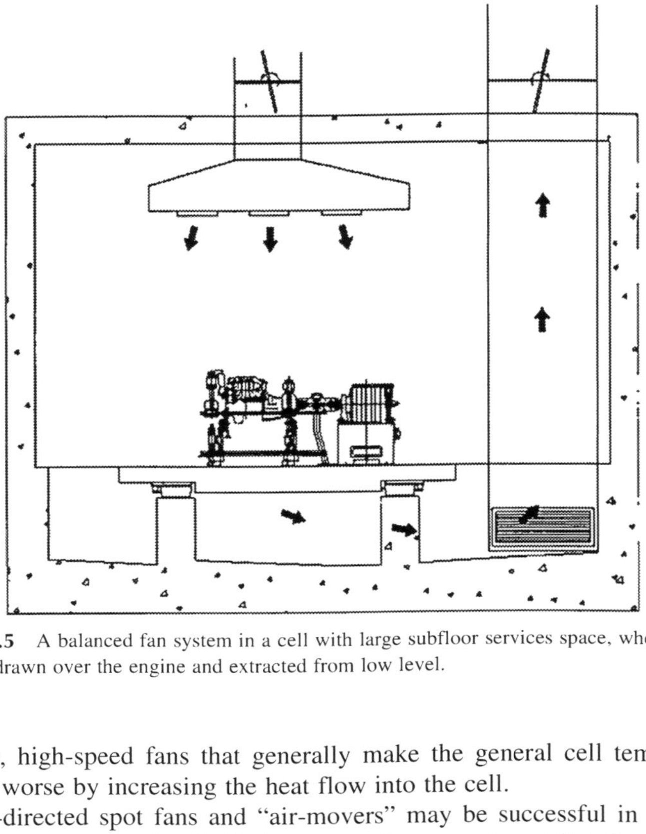

**FIGURE 5.5** A balanced fan system in a cell with large subfloor services space, where ventilation air is drawn over the engine and extracted from low level.

auxiliary, high-speed fans that generally make the general cell temperature problem worse by increasing the heat flow into the cell.

Well-directed spot fans and "air-movers" may be successful in reducing *localized* hot spots on the UUT, but it must be remembered that the cell ventilation system has to cope with the extra heat load created. Spot fans can be usefully employed in large powertrain test cells to provide cooling directly to modules such as motors and transmission units that would normally be cooled by airflow from the vehicles motion.

It should also be understood that a stream, or eddy, of hot air can have unpredictable and undesirable effects on the output of an ICE as it can raise the temperature of the engine inlet air. In addition, cooling or heating airflows can upset the calibration of force transducers that are sensitive to a $\Delta T$ across their surfaces. So that spot fans are stopped in the case of a fire, the switched power outlets supplying them must be included in the emergency stop chain.

## Air movers

A precise way to spot-cool "hot spots" within the cell or UUT-mounted devices than a mobile fan is to use an air-mover of the type made by companies such

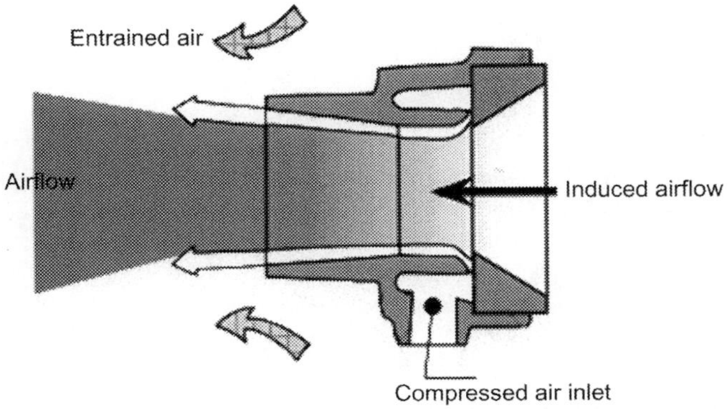


**FIGURE 5.6** Diagrammatic representation of a section through an "air-mover" or "air-flow amplifier" using a compressed air feed for power. These units may be used for spot cooling with more directional precision than a spot fan.

as Brauer. These devices use a supply of compressed air to induce and entrain surrounding air in an accelerated jet up to some 100 times the free volume of the compressed supply. Fig. 5.6 shows in diagrammatic form these small and relatively quiet spot-cooling devices.

## Fire dampers within inlet and outlet ducting

The design of all ducting bringing air into, or taking air out of, the cell *must obey the rule of hazard containment*; therefore fire dampers must be fitted in all ducts at the cell boundary. These devices act to close off the fire from any fresh air supply and form a fire barrier.

The two types commonly used are:

- *Normally closed, motorized open.* These are formed by stainless steel louvers within a special framework having their own motorized control gearing and closure mechanism. They are fitted with switches allowing the building management system or cell control system to check their positional status.
- *Normally open, spring closed through release of a thermal-link-retained steel curtain.* These only operate when the air-flow temperature reaches their release temperature and have to be reset manually. Care must be taken by when working in the immediate area of these spring closed shutters to avoid injury through an unintended release.

An advantage of the motorized damper is that when the cell is not running or being used as a workshop space during engine rigging, the flow of outside (cold) air can be cutoff.

## Ventilation duct silencers

Noise from the cell, quite apart from that of an ICE exhaust and the ventilation fans, will tend to "break out" via any penetration or gap (ill-fitted doors, etc.) in the cell structure so the large ventilation ducts are a significant pathway. The classic method of minimizing noise disturbance in the area surrounding a facility is to direct noise vertically through roof-mounted stacks, but this is often a partial solution at best.

Both inlet and outlet air ducting may need to be fitted with attenuators or duct silencers. The effectiveness of such attenuating duct sections is roughly proportional to their length, a commonly made standard found in test cell systems being 2.4 m long. Their internals usually consist of baffles of glass cloth or wool encased in perforated galvanized steel sheet and work on the same principle as the absorption mufflers described in Chapter 6, Cooling Water and Exhaust Systems. While attenuation above 500 Hz is usually good, it reduces at the lower frequencies. The balance between building space, noise spectrum, and resistance to airflow is complicated to achieve and requires expert site-specific design work.

## External ducting of ventilation systems

The external termination of ducts can take various forms, often determined by architectural considerations but always having to perform the function of protecting the ducts from the ingress of rain, snow, sand, and external debris such as leaves. To this end the inlet often takes the form of a motorized set of louvers, interlocked with the fan start system, immediately in front of coarse then finer, filter screens.

In some cases, in temperate climates, the whole service space above the cell may be used as the inlet plenum for an air-handling unit. This design strategy uses more space than the vertical stack designs but allows more freedom in the position of the inlet louvers and overcomes the frequently faced problem of architectural objections to an "industrial" style of building profile.

In exceptionally dusty conditions such as exist in arid desert areas, the external louvers should be motorized to close and be fitted with vertical louver sections to prevent dust buildup on horizontal surfaces.

## Control of test cell ventilation systems

The simplest practical ventilation control system is that based on a two-speed extraction fan in the roof of the cell drawing outside (ambient) air through a low-level duct at one end of the cell. The fan runs at low speed during times when the cell is used as a workshop and at high speed during engine running. In this case the cell pressure is ambient unless flow through the inlet filter is restricted.

Closed cells require both inlet and extract fans to be fitted with variable speed drives and a control system to balance them. There are several alternatives, but under a commonly used control strategy operating with a nonrecirculation system, one fan operates under temperature control and the other under pressure. Thus with both fans running and with a cell pressure of about 50 Pa below ambient, when the cell air temperature rises, the extract fan control reacts to increase speed and increase airflow; this speed increase creates a drop in cell pressure, which is detected by the control system, so the inlet fan speed is increased until equilibrium is restored. The roles of the fans are somewhat interchangeable, the difference being that in the case described previously the control transients tend to give a negative cell pressure and if the control roles are reversed it gives a, less desirable, positive cell pressure.

The fans controlled in the way described previously must have the capacity to deal with the additional flows in and out of the cell by way of purge and combustion airflows where these systems are fitted. The control system also has to be sufficiently damped to prevent surge and major disturbance in the case of a cell door being opened.

## Ventilation fans: designs and features

Methods of testing fans and definitions of fan performance are given in British Standard 848-7:2003. The subject is not entirely straightforward and requires specialist advice during the design phase of a test facility ventilation system; the classes of fans likely to be used will include the following:

1. *Axial flow fans.* For a given flow rate an axial flow fan is considerably more compact than the corresponding centrifugal fan and fits very conveniently into a duct of circular cross-section. The fan static pressure per stage is limited, typically to a maximum of about 600 Pa at the design point, while the fan dynamic pressure is about 70% of the total pressure. Fan total efficiencies are in the range of 65%–75%. Axial flow fans mounted within a bifurcated duct, so that the motor is external to the gas flow, are a type commonly used in individual exhaust dilution systems. An axial flow fan is a good choice as a spot fan, or for mounting in a cell wall without ducting, but should not be used in highly contaminated airflows or diluted exhaust systems. Multistage units are available but tend to be fairly expensive and rarely used in ventilation systems.
2. *Centrifugal fans, flat blades, backward inclined.* This is probably the first choice in most cases where a reasonably high pressure is required, as the construction is cheap and efficiencies of up to 80% (static) and 83% (total) are attainable. A particular advantage is the near immunity of the flat blade to dust collection. Maximum pressures are in the range of 1–2 kPa.
3. *Centrifugal fans, backward curved.* These fans are more expensive to build than the flat-bladed type. Maximum attainable efficiencies are 2%–3% higher, but the fan must run faster for a given pressure and dust tends to accumulate on the concave faces of the blades.

4. *Centrifugal fans, airfoil blades.* These fans are expensive and sensitive to dust but are capable of total efficiencies exceeding 90%. There is a possibility of discontinuities in the pressure curve due to stall at reduced flow. They should be considered in the larger sizes where the savings in power cost are significant.
5. *Centrifugal fans, forward curved blades.* These fans are capable of a delivery rate up to 2.5 times as great as that from a backward inclined fan of the same size, but at the cost of lower efficiency, unlikely to exceed 70% total. The power curve rises steeply if flow exceeds the design value.

### Fan noise

Taken in isolation, within a ventilation system and services room the fans are usually the main source of system noise. The noise generated varies as the square of the fan pressure head, so that doubling the system resistance for a given flow rate will increase the fan sound power fourfold, or by about 6 dB. As a general rule, to minimize noise, ventilation fans should operate as close to the design point as possible. Dust accumulation on the outlet fan's impeller may spoil its balance and greatly increase the noise through vibration transmitted into the ducting.

## Part 4: Air-conditioning

Most people associate this topic with comfort levels under various conditions in places of human habitation and work but there are special applications in powertrain testing. Levels of air temperature and humidity, along with air change rates, are laid down by various national statutes; in the United Kingdom, these are covered by BS5720:1979. In the United States the Office of Safety and Health Administration (OSHA) is the key agency that monitors workplace regulations nationwide. OSHA's recommendations for workplace air treatment set federal standards for temperature and humidity levels: the minimum temperature for indoor workplaces is 68°F and the maximum 76°F. The acceptable range for indoor humidity is between 20% and 60%.

Such regulations must be observed with regard to the test facility's control rooms. However, the conditions in an engine test cell are far removed from the normal, and justify special treatment, which follows.

Two properties of the ventilating air entering the cell (and, more particularly, of the induction air entering the engine) are of importance: the temperature and the moisture content. Air-conditioning involves four main processes:

- heating the air
- cooling the air

- reducing the moisture content (dehumidifying)
- increasing the moisture content (humidifying)

The study of the properties of moist air is known as *psychrometry* [2]. It is treated in many standard texts on the subject of HVAC and only a very brief summary will be given here.

Air-conditioning processes are represented on the psychrometric chart in Fig. 5.7. This relates the following properties of moist air:

- *The moisture content or specific humidity*, $\omega$ kg moisture/kg dry air. Note that even under fairly extreme conditions (saturated air at 30°C), the moisture content does not exceed 3% by weight.
- *The percentage saturation or relative humidity*, $\varphi$. This is the ratio of the mass of water vapor present to the mass that would be present if the air were saturated at the same conditions of temperature and pressure. The mass of vapor under saturated conditions is very sensitive to temperature. A consequence of this relationship is the possibility of drying air by cooling. As the temperature is lowered, the percentage saturation increases until at the dew point temperature the air is fully saturated and any further cooling results in the deposition of moisture.

| Temperature (°C) | 10 | 15 | 20 | 25 | 30 |
|---|---|---|---|---|---|
| Moisture content $\omega$ (kg/kg) | 0.0076 | 0.0106 | 0.0147 | 0.0201 | 0.0273 |

- *The wet- and dry-bulb temperatures.* The simplest method of measuring relative humidity is by means of a wet- and dry-bulb thermometer. If unsaturated airflows past a thermometer having a wetted sleeve of cotton around the bulb, the temperature registered will be less than the actual temperature of the air, as registered by the dry-bulb thermometer, owing to evaporation from the wetted sleeve. The difference between the wet- and dry-bulb temperatures is a measure of the relative humidity. Under saturated conditions the temperatures are identical, and the depression of the wet-bulb reading increases with increasing dryness. Wet- and dry-bulb temperatures are shown in a psychometric chart.
- *The specific enthalpy of the air.* This is the amount of heat in the air. In the United States the enthalpy of air is measured in BTUs per pound of air, in the rest of the world it is measured in kilojoule per kilogram (kJ/kg) of air.

The specific enthalpy of *moist air* must include both the sensible heat and the *latent heat of evaporation* of the moisture content. The specific enthalpy of moist air is:

$$h = 1.01t_a + \omega(1.86t_a + 2500) \tag{5.6}$$

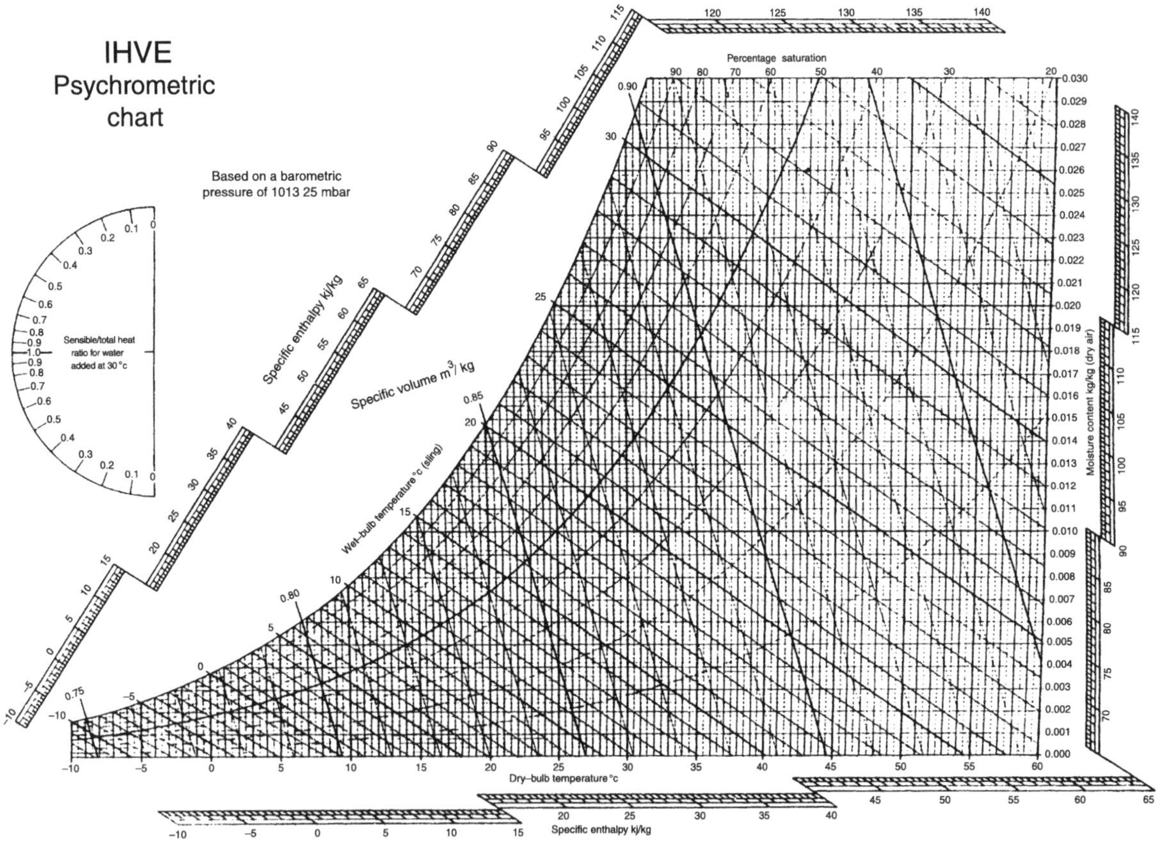


FIGURE 5.7 A version of a psychrometric chart (in SI units).

the last two terms represent the sum of the sensible and latent heat of the moisture. Taking the example of saturated air at 300°C:

$$h = 1.01 \times 30 + 0.0273(1.86 \times 30 + 2500) = 30.3 + 1.5 + 68.3 = 100.1\ \mathrm{kJ/kg}$$

The first two terms represent the sensible heat of air plus moisture and it is apparent that ignoring the sensible heat of the latter, as is usual in air cooling calculations, introduces no serious error. The third term, however, representing the latent heat of the moisture content, is much larger than the sensible heat terms. This accounts for the heavy cooling load associated with the process of drying air by cooling: condensation of the moisture in the air is accompanied by a massive release of latent heat.

Increasing moisture content is achieved either by spraying water into the air stream (with a corresponding cooling effect) or by steam injection. With the appearance of Legionnaires' disease, the latter method, involving steam that is essentially sterile, is recommended.

## Legionnaires' disease

This disease is a severe form of pneumonia and infection is usually the result of inhaling water droplets carrying the causative bacteria (*Legionella pneumophila*). Factors favoring the organism in water systems are the presence of deposits such as rust, algae, and sludge, a temperature between 20°C and 45°C, and the presence of light. Clearly all these conditions can be present in systems involving cooling towers and some types of humidifiers.

Preventive measures include the following:

- Treatment of water with scale and corrosion inhibitors to prevent the buildup of possible nutrients for the organisms.
- Use of suitable water disinfectant such as chlorine (1–2 ppm) or ozone.
- Steam humidifiers are preferable to water spray units.

If infection is known to be present, flushing, cleaning, and hyperchlorination are necessary. If a system has been out of use for some time, heating to about 70°C–75°C for 1 hour will destroy any organisms present.

*This matter should be taken seriously. This is one of the few cases in which the operators of a test facility may be held criminally liable for injury caused to the general public, with consequent ruinous claims for compensation.*

## Effects of humidity: a warning

Electronic equipment is extremely sensitive to moisture. Large temperature changes, when associated with high levels of humidity, can lead to the deposition of moisture on components such as circuit boards, with disastrous results.

This situation can easily arise in hot weather: the plant cools down overnight and dew is deposited on cold surfaces. Some protection may be afforded by continuous air-conditioning. The use of chemical driers is possible; the granular substances used are strongly hygroscopic and are capable of achieving very low relative humidity. However, their capacity for absorbing moisture is of course limited and the container should be removed regularly for regeneration by a hot air stream.

The operational answer, adopted by many facilities, is to leave vulnerable equipment switched on in a quiescent state while not in use, or to fit anticondensation heaters at critical points in cabinets. In the period of facility construction, between delivery of equipment and commissioning, it is vital to fit an auxiliary power supply to run a rudimentary set of anticondensation heaters. Before their use was banned in the United Kingdom, a chain of 60 W incandescent light bulbs was the favored method used but small domestic cupboard heaters can be used instead.

## Part 5: Treatment of internal-combustion engine combustion air and climatic test cells

The condition of the combustion air (its pressure, temperature, humidity, and purity) on engine performance and variations in these factors can have a very substantial effect on performance. In an ideal world the ICE would all be supplied with air in "standard" conditions. In practice, there is a trade-off between the advantages of such standardization and the cost of achieving it.

For routine (non-emissions) production testing, variations in the condition of the air supply are not particularly important, but the performance recorded on the test documents, requiring some degree of correlation with other test beds or facilities, should be corrected to standard conditions.

The simplest and most widely used method of supplying the combustion air is to allow the engine to take its supply directly from the test cell atmosphere. The great advantage of using cell air for vehicle engines in particular is that rigging of complex "in-vehicle" air filtration and ducting units is straightforward, although it may be practical to "flood" the air filter with an excess of treated air (see next).

The major disadvantage of drawing the air from within the cell is the uncontrolled variability in temperature and quality arising from air currents and other disturbances in the cell. These can include contamination with exhaust and other fumes and may be aggravated by the use of spot fans.

For research and development testing, and particularly critical exhaust emissions work, it is necessary that, so far as is practicable, the combustion air should be supplied with the minimum of pollution and at constant conditions of temperature, pressure, and humidity. To create a realistic operational specification for temperature and humidity control of combustion air, requires that the creator needs to have a clear understanding of a psychometric chart (Fig. 5.7). The implications of requiring a large operational envelope plus a clear understanding of the temperature and humidity points required by the planned testing regimes are vital. It is recommended that cell users mark on a psychometric chart the operating points prescribed by known (emission homologation, etc.) tests or investigate the specification of available units before producing specifications that may impose unnecessarily high energy demands on the requested unit.

While the degree of pollution ingested from the outside air is a function of the choice of site, the other variables may be controlled in the following order of difficulty:

- temperature only
- temperature and humidity
- temperature, humidity, and pressure
- temperature and dynamic pressure control during transient engine conditions

## Temperature-only control, flooded inlet combustion air units

The simplest form controls air temperature only and supplies an excess of air via a flexible duct terminating in a trumpet-shaped nozzle close to the engine's inlet; this is known as a "flooded" inlet. The unit should be designed to supply air at a constant volume, calculated by doubling the theoretical maximum air demand of the engine. Because the flow is constant, good, steady-state, temperature control can be achieved; the excess air is absorbed into the cell ventilation inlet flow and compensated for by the ventilation temperature and pressure control.

The normal design of such units contains the following modules:

- inlet filter and fan;
- chiller coils;
- heater matrix; and
- insulated duct, taken through a fire damper, terminating in flexible section and bell-mouth.

By holding the chiller coils' temperature at a fixed value of say 7°C and heating the air to the required delivery temperature, it is possible to reduce the effect of any humidity change during the day of testing without installing full humidity control.

Switching combustion air units on and off frequently will cause the system control accuracy to decline and therefore it is sensible to use the delivered air in a space heating role when the engine is not running, allowing the main ventilation system to be switched off or down to a low extract level. An energy-efficient comfort ventilation regime can be designed by using the combustion air balanced with the purge fan extraction.

## Humidity-controlled combustion air units

Systems for the supply of combustion air in which both temperature and humidity are controlled can be expensive, the cost increasing with the range of conditions to be covered, and the degree of precision required. They are also energy intensive, particularly when it is necessary to reduce the humidity of the atmospheric air by cooling and condensation. Before introducing such a system, a careful analysis should be made to ensure that the operational envelope proposed is achievable from a standard unit and the services available to it.

It is sensible to consider some operational restraints on the scheduling of tests having different operational levels of humidity in order to avoid condensation buildup in the system. Corrugated ducting should not be used in systems providing humid air, since they tend to collect condensate that has to dry out before good control can be restored.

To indicate the energy required and the complexity of the task, we can take the example of a 250 kW diesel engine with a calculated air consumption of 1312 kg/h corresponding to a volumetric flow rate of about 0.3 $m^3/s$.

The necessary components of a suitable air-conditioning unit for attachment to the engine air inlet are shown diagrammatically in Fig. 5.8 and may comprise in succession:

1. air inlet with screen and fan
2. heater, either electrical or hot water
3. humidifier, comprising steam or atomized water injector with associated generator
4. cooling element, with chilled water supply
5. secondary heating element

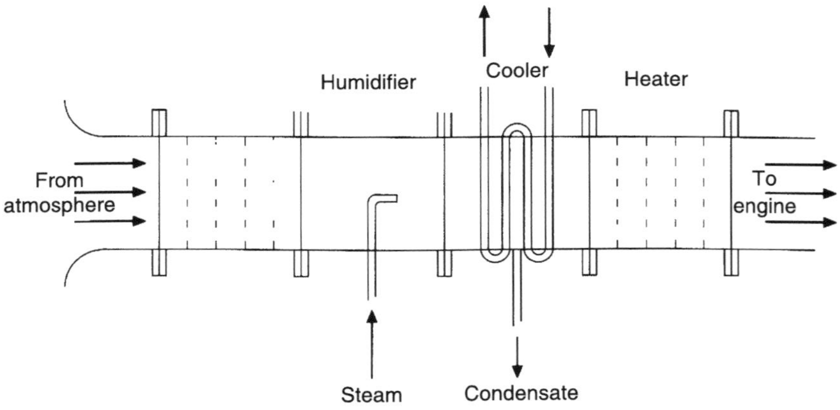


**FIGURE 5.8** Schematic of typical combustion air system contents with inlet filter not shown.

Note that there are two heater elements, one before and another after the humidifier. The first element is necessary when it is required to humidify cold dry air, since if steam or water spray is injected into cold air supersaturation will result and moisture will immediately be deposited. On the other hand, it is commonly necessary, in order to dry moist air to the desired degree, to cool it to a temperature lower than the desired final temperature and reheat must be supplied downstream of the cooler. The internal dimensions of the duct in the present example would be approximately 0.3 m × 0.3 m.

A typical standard unit available for this type of duty with delivery of 1600 $m^3/h$ at temperature range of 15°C–30°C and with option of humidity control 8–20 g/kg would have dimensions of about 2755 mm long × 1050 mm width × 2050 mm in height. It would have a maximum power rating of around 68 kW and would need to be supplied with both chilled water and demineralized water for humidifier.

It will be observed that the energy requirements, particularly for the chiller and from separate chilled water supply, are quite substantial. In the case of large ICE and gas turbines, any kind of combustion air-conditioning is not really practicable, and reliance must be placed on correction factors.

Any combustion air supply system must be integrated with the fire alarm and fire-extinguishing system and must include the same provisions for isolating individual cells as the main ventilation system.

## Pressure–temperature–humidity-controlled combustion air units

A close coupled air supply duct attached to the engine is necessary in special cases such as anechoic cells, where air intake noise must be eliminated, at sites running legislative tests that are situated at altitudes above those where standard conditions can be achieved, and where a precisely controlled conditioned supply is needed.

Since ICE air consumption varies more or less directly with speed, a parameter can vary more rapidly than any air supply system can respond, and since it is essential not to impose external, supply-system induced, pressure changes on the engine air, some form of excess air spill or engine bypass strategy has to be adopted to obtain pressure control with any degree of temperature and humidity stability.

For testing at steady-state conditions, the pressure control system can operate with a slow and well-damped characteristic, typically having a stabilization time, following an engine speed set point change, of 30 seconds or more. This is well suited to systems based on regulating the pressure of an intake plenum fitted with a pressure-controlling spill valve/flap.

The inlet duct must be of sufficient size to avoid an appreciable pressure drop during engine acceleration, and it needs to be attached to the engine inlet in such a way as to provide a good seal without imposing forces on the engine.

It is sensible to encapsulate or connect to the normal engine air inlet filter, if it can sustain the imposed internal pressure imposed by the supply system. If the filter has to be encapsulated, it often requires a bulky rig item that has to be suspended from a frame above the engine.

Various strategies have been adopted for fast pressure control, the best of which are able to stabilize and maintain pressure at set point within <2 seconds of an engine step change of 500 rpm.

Control of inlet pressure only does not simulate true altitude; this requires that both inlet and exhaust are pressurized, or depressurized, at the same pressure, a condition that can be simulated using a dynamic pressure control system patented by AVL List GmbH, as shown in Fig. 5.9. The largest of such systems typically have a maximum flow of 2500 $m^3/h$ and performance figures of:

- inlet air temperature range: −10°C to +40°C with an accuracy of ±0.4°C
- inlet pressure range: barometric +50 to −100 mbar with an accuracy of ±1 mbar
- relative humidity range: 10%–90% with an accuracy of ±5%

and find use in modern powertrain cells in addition to motor sports such as the World Rally Series where events are held worldwide at site of different altitudes. The terms "dynamic" and "transient" when applied to air-conditioning systems need to be qualified and some common sense applied with consideration being given to the condition of the air being drawn into the conditioning unit. Normal conditioning units should have a settling time after a 20% step change of temperature of around 1 minute perhaps half that for pressure, for humidity there are too many variables to give precise figures but clearly it takes longer for humidity to settle to a new figure and times will differ between increase and decrease. For their dynamic system AVL quotes <3 seconds for adjustment time of pressure and <60 seconds for temperature; while the first is important for real-life simulation, the latter is probably not.

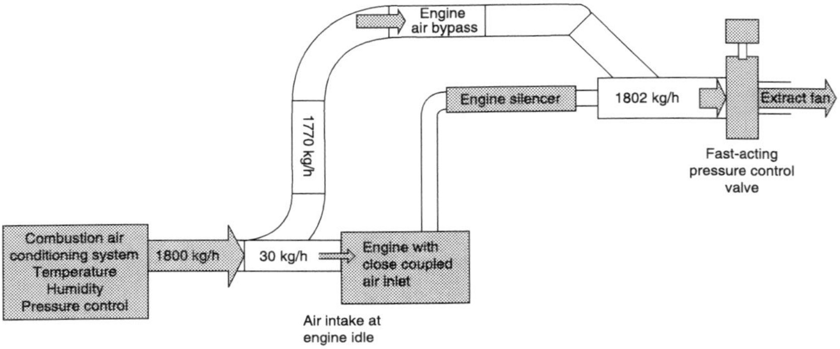


**FIGURE 5.9** Schematic diagram of an AVL design of a dynamic combustion air pressure control system—example flows shown are of a 1600 WRC engine at idle condition.

### *Important system integration points concerning combustion air systems*

Combustion air units of all types will require condensate drain lines with free gravity drainage to a building system; backup or overspill of this fluid can cause significant problems. Those fitted with steam generators will require both condensate and possibly steam "blow-down" lines piped to safe drainage. Condensate drains, if taken through the external wall of the building as is the case of domestic overflow pipes of the same size, can be blocked by ice in cold weather and cause units to flood.

*If steam generators are fitted the condition of feed water will be critical and suitable water treatment plant needs to be installed, particularly where supply water is hard.*

## Climatic testing internal-combustion engine powertrains

Cold chambers built exclusively for powertrain testing are quite rare, since much of the research work in cold starting and operation in extreme environments tends to be carried out in whole vehicle systems (see next). However, specialist research by companies involved in lubricant, and materials research has climatic cells with controlled environments in the range between +40°C and −40°C.

Because climatic cells, particularly when working at low temperatures, require a lot of energy to run, it is sensible to reduce the volume of the chamber containing the UUT to the minimum viable volume to carry out its low-temperature tasks. For relatively short duration cold-start work, a common method is to circulate cold fluids around the engine while it is encased in a demountable, insulated "tent" fed by a supply of refrigerated air. There are facilities where the "boxed" or "tented" UUT can be subjected to a surrounding temperature with coolant circulation in the range of −70°C and up to +180°C; the key to maintaining reasonable energy use at these temperatures is that the air and coolant volumes are kept low.

A common technique is to use a modified refrigerated ISO container (see containerized cells in Chapter 3: Test Facility Design and Construction); this allows the engine to be "cold-soaked" down to the temperature required for legislative cold-start emission tests. In the case of electric vehicle battery pack testing small custom-built test cells such as that shown in Chapter 3, Test Facility Design and Construction, Fig. 3.7 meet all the requirements of an energy efficient space.

For general requirements for special materials and design features required to be built into test equipment operating in low-temperature environments, see the relevant section in Chapter 10, Chassis Dynamometers, Rolling Roads, and Hub Dynamometers, on chassis dynamometers.

## Climatic cells for electrical powertrain units and batteries

Climatic test cells for electric drive axles or motors can be physically smaller and less complex than those testing ICE, and the thermal loads will be lower, while the working range will have a higher maximum temperature. Because the UUT is small, it has become the modern practice to form a thermal chamber, with a moveable cover, that only houses the UUT as shown in Fig. 5.10. The pallet and chamber base-plate system must be suitably insulated and test, chamber opening, and climatic control systems fully integrated.

Climatic cells for battery testing (Fig. 3.7) whose temperature range needs to be in the range of $-40°C$ up to $+90°C$ normally form just one part of a complete battery pack performance testing cell which have several unique features. Since thermal and a full range of electrical cycling that may constitute abuse testing will be carried out in such spaces, they are often built as separate discrete units to minimize consequential damage to other cells in the case of thermal runaway. Such requirement makes the use of highly specialized, containerized, or modular constructions appropriate. Typically they will contain a mid-space firewall and door between the climatic test section and the space housing the battery charging and discharging equipment and the climatic equipment on the roof. The primary fire suppression system should be water mist but often the bunded bottom two-thirds of the test space is built so it can be flooded then drained to an intercept tank. Test control and data analysis needs to be centralized for several such test chambers in a separate building that may also house a workshop for

**FIGURE 5.10** The climatic chamber cover that slides away from the UUT to allow access. This chamber has a maximum controlled temperature range of $-35°C$ to $+120°C$ and a controlled humidity range of between 10% and 95% ± 1% RH (Relative Humidity) at $+10°C$ and $+95°C$. *UUT*, Unit under test. *Photo courtesy ZF Test Systems.*

battery-pack test preparation. Batteries delivered for testing that may be of unknown condition and untested chemistry need to be stored in a bunded floodable, and constantly monitored external storage container.

## Climatic test cells for vehicles

There is a requirement for this kind of facility for development work associated with various aspects of vehicle and powertrain performance under extreme climatic conditions. The accelerating development of electrical powertrains and energy storage in batteries, which are particularly temperature sensitive, has increased the need for whole vehicle climatic testing. Other subjects include vehicle climatic control and solar loading and for ICE vehicles, cold starting, fuel waxing, etc.

The design of air-conditioning plants for combustion air has been discussed briefly earlier in this chapter, but the design of an air-conditioned chamber for the testing of complete vehicles is a very much larger and more specialized problem. Fig. 5.11 is a schematic drawing of such a chamber, built in the form of a recirculating wind tunnel so that the vehicle on test, mounted on a chassis dynamometer, can be subjected to oncoming airflows at realistic velocities, and in this particular case over a temperature range from +40°C to −30°C.

The outstanding feature of such an installation is its very large thermal mass and, since it is usually necessary, from the nature of the tests to be performed, to vary the temperature of the vehicle and the air circulating in the tunnel over a wide range during a prolonged test period, this thermal mass is of prime significance in sizing the associated heating and cooling systems.

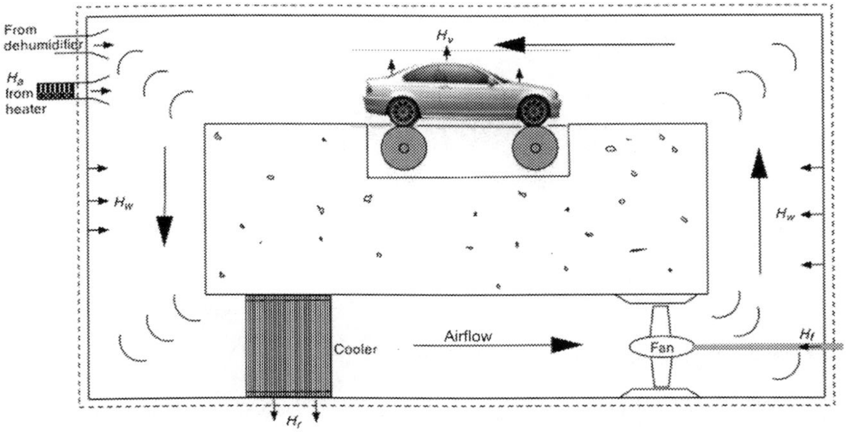


**FIGURE 5.11** Schematic of climatic wind tunnel facility fitted with a chassis dynamometer.

The thermodynamic system of the chamber is indicated in bulleted list next and, using actual site data, the energy inflows and outflows during a particular cold test.

- $C_a$—Air content of chamber, return duct, etc. volume approximately 550 m$^3$, approximate density 1.2 kg/m$^3$, $C_p = 1.01$ kJ/kg K, thermal mass = 550 × 1.2 × 1.01 = 670 kJ/°C.
- $C_v$—Vehicle, a commercial vehicle, weight 3 t, specific heat of steel approximately 0.45 kJ/kg K thermal mass, say 3000 × 0.45 = 1350 kJ/°C.
- $C_e$—Fan, cooler matrix, internal framing, etc. estimated at 2480 kJ/°C.
- $C_s$—Structure of chamber. This was determined by running a test with no vehicle in the chamber and a low fan speed. The rate of heat extraction by the refrigerant circuit was measured and a cooling curve plotted. Concrete is a poor conductor of heat and the coefficient of heat transfer from surfaces to air is low. Hence, during the test a temperature difference between surfaces and wall built up but eventually stabilized; at this point the rate of cooling gave a true indication of effective wall heat capacity. The test showed that the equivalent thermal mass of the chamber was about 26,000 kJ/°C, much larger than the other elements. Concrete has a specific heat of about 8000 kJ/m$^3$ °C, indicating that the "effective" volume involved was about 26,000/8000 = 3.2 m$^3$.

The total surface area of the chamber is about 300 m$^2$, suggesting that a surface layer of concrete about 1 cm thick effectively followed the air temperature. The thermal capacity (thermal mass) of the various elements of the system is defined in terms of the energy input required to raise the temperature of the element by 1°C. Adding these elements together, we arrive at a total thermal mass of 30,500 kJ/°C and since 1 kJ = 1 kW s, we can derive the rate at which the temperature of the air in the chamber may be expected to fall for the present case. This shows that the "surplus" cooling capacity available for cooling the chamber and its contents, $H_w$, amounts to 221 kW. This could be expected to achieve a rate of cooling of 221/30,500 = 0.0072°C/s, or 1°C in 2.3 minutes. However, this is the final rate when the temperature difference between walls and air had built up to the steady-state value of about 20°C. The observed initial cooling rate is much faster, about 1.5°C/min.

Heating presents less of a problem, since the heat released by the test vehicle and the fan assists the process rather than opposing it. A further effect is associated with the moisture content of the tunnel air. On a warm summer's day, this could amount to about 10 kg of water and during the cooling process this moisture would be deposited mostly on the cooler fins, where it would eventually freeze, blocking the passages and reducing heat transfer. To deal with this problem, it is necessary to include a dehumidifier to supply dried air to the tunnel circuit. Note that massive concrete constructions take several months to "dry out" under average temperate conditions; therefore during the early commissioning of installations like the one described previously, increased humidity and freezing of the cooler matrix can be a significant problem.

Some occasionally met, climatically induced, problems can be expensive to repeat and deal with out of the areas in which they may occur. Vapor locking in fuel lines can disable a gasoline-fueled vehicle if they are parked, while hot, for a short time on very hot roadway, particularly at high altitude. Using the so-called Death Valley (which is at low altitude) parking test might be most cost effectively carried out on suitable geographic site rather than being simulated in a chamber.

## Automotive wind tunnels

At the extreme end of any discussion concerning test facility ventilation, the design and operation of automotive wind tunnels deserve mention. This is a specialist field and students needing to study the subject in detail are recommended to read Aerodynamics of Road Vehicles detailed at the end of the chapter. Aerodynamic testing is receiving increasing investment, not only because of the continued need to use less fuel through reducing a vehicle's drag coefficient ($C_d$), but because of the more complex cooling flow requirements of the new hybrid and electric vehicles which generally do not have the conventional front-facing radiator-grill air inlets. The energy required to move air in a laminar flow through an enclosed space in which the test object is placed rises almost exponentially with the speed of flow and the volume moved. However, much of the work is scalable; therefore some of the tunnels needing to work at the highest speeds are, for example, one-third or one-half scale. In spite of this, full-size tunnels requiring 3.5 MW fans and operating at over 250 kph have been built. Designs are both closed designs such as that of climatic versions and semiopen designs where the air is normally sucked through the test space rather than blown. Full-sized, light-vehicle tunnels have test area spaces that vary considerably in size but typically would need to be around 16 m long × 8 m wide × 5 m high. A particular design feature of automotive wind tunnels, not shared with aeronautical facilities, is that the test vehicle has to be in contact with a flat road surface with which it has an ever-changing force relationship, much of which depends on the aerodynamic forces working upon the vehicle. These forces, all transmitted through the road/tire contact points, have to be measured by a "wind tunnel balance" on which the vehicle sits. This balance system is made up of strain gauge flexures that are designed to respond to load. They measure multiple loads and moments by the use of multiple individual flexures, each one measuring load on just one axis. The balance is frequently mounted on a turntable so that phenomena such as side-wind stability can be measured.

One problem in real-life simulation within an automotive wind tunnel is that the *real*, and aerodynamically important, situation of the air being stationary relative to the road while moving relative to the vehicle is expensive to simulate; this requires a "moving-ground" installation that a roller-based chassis dynamometer does not provide. However, moving-ground or "flat-track" chassis dynamometers have been made and are briefly described in Chapter 10, Chassis Dynamometers, Rolling Roads, and Hub Dynamometers, of this book.

While the forces resulting from airflow can be measured, the means by which they are created is difficult to visualize since air is invisible. Various means of air-flow visualization are used, from the simple "smoke wand" that vaporizes oil or propylene glycol to produce a dense white vapor stream, to jets of helium-filled microbubbles, all of which need to be supported by motion-capture cinema-photography units.

## Summary

The design of the ventilation system for a test cell is a major undertaking, and a careful and thorough analysis of expected heat loads from UUT plus any exhaust system, dynamometer, instrumentation, cooling fans, and lights is essential if subsequent difficulties are to be avoided. Once the maximum heat load has been determined, it is recommended that a diversity factor is applied so as to arrive at a realistic rating for the cell services.

Any cell using gasoline, gaseous or other volatile fuels should include within the ventilation system a purge system to remove potentially explosive vapor.

Ventilation systems must be designed so that the cell forms a hazard containment box requiring that they are shut down and isolated so as not to support or "feed" a fire within the cell.

The choice between full-flow ventilation and air-conditioning has to be made after a detailed review of site conditions, particularly building space, and cost.

If a separate treated air supply for an ICE is to be used, careful consideration has to be made concerning its type and rating in order to keep costs and energy requirements in line with actual test requirements.

The ventilation of the control room should ensure that conditions there meet normal office standards.

Specialist facilities for climatic and aerodynamic work require designers and contractors that have the relevant experience in the automotive sector.

## Notations

| | |
|---|---|
| **Atmospheric pressure** | $p_a$ (bar) |
| **Atmospheric temperature** | $t_a$ (°C) |
| **Density of air** | $\rho$ ($m^3$) |
| **Gas constant for air** | $R = 287$ J/kg K |
| **Specific heat of air at constant pressure** | $C_p$ (J/kg K) |
| **Mass rate of flow of air** | $m$ (kg/s) |
| **Rate of heat loss, vertical surface** | $Q_v$ ($W^2$) |
| **Rate of heat loss, horizontal surface** | $Q_h$ ($W^2$) |
| **Temperature of surface** | $t_s$ (°C) |
| **Rate of heat loss by radiation** | $Q_r$ ($W/m^2$) |
| **Ventilation air temperature rise** | $\Delta T$ (°C) |
| **Total heat load** | $H_L$ (kW) |
| **Ventilation air-flow rate** | $Q_A$ ($m^3/s$) |
| **Fan air-flow rate** | $Q_F$ ($m^3/s$) |
| **Velocity of air** | $V$ (m/s) |
| **Velocity pressure** | $p_v$ (Pa) |
| **Static pressure** | $p_s$ (Pa) |
| **Total pressure** | $p_t$ (Pa) |
| **Pressure loss** | $\Delta p_t$ (Pa) |
| **Static air power** | $P_{sF}$ (kW) |
| **Total air power** | $P_{tF}$ (kW) |
| **Shaft power** | $P_A$ (kW) |
| **Moisture content** | $\omega$ (kg/kg) |
| **Relative humidity** | $\varphi$ |
| **Specific enthalpy of moist air** | $h$ (kJ/kg) |
| **Cooling load** | $L_C$ (kW) |
| **Heating load** | $L_H$ (kW) |
| **Emissivity** | $\varepsilon$ |
| **Pressure loss coefficient** | $K_e$ |

## References

[1] A. Vedavarz, S. Kumar, M.I. Hussain, HVAC: Heating, Ventilation & Air Conditioning Handbook for Design & Implementation, Industrial Press, Inc., 2007. ISBN-13: 978-0831131630.

[2] E.G. Pita, Air Conditioning Principles and Systems: An Energy Approach, Pearson Education, 2001. ISBN13 9780130928726.

Chapter 6

# Cooling water and exhaust systems

## Chapter Outline



## Part 1: Cooling water supply systems fundamentals

*Note*: The introductory section of the previous chapter is also relevant to the design of water systems.

The cooling water (CW) system for any heat engine or electric motor powertrain test facility has to provide water of suitable quality, temperature, and pressure to allow sufficient volume to pass through the heat exchanging equipment in


**Engine Testing. DOI: https://doi.org/10.1016/B978-0-12-821226-4.00006-1**

order for it to have the adequate cooling capacity. The water pressure and flow have to be sufficiently constant to enable the devices supplied to maintain control of temperature of the unit under test (UUT) and all installed water-cooled equipment. As recommended in Chapter 3, Test Facility Design and Construction, at an early stage in the design of a new test cell, an energy balance diagram similar to Fig. 3.3 should be drawn up to show all the heat energy flows: air, fuel, water, exhaust gas, electricity, into and out of the cell. This holistic approach will help in sizing the total water-cooling system as part of the facility design. While water required for fire-suppression (sprinkler) systems are not included in the energy balance they must be included in the piping schedules, cell penetrations requirements, and pressure testing time-line.

It is essential for purchasers of water-cooled plant to carefully check the inlet water temperature specified for the required performance, since the higher the facility cooling water inlet temperature supplied the less work the device will be capable of performing before the maximum allowable exit temperature is reached.

In this book the water supplied by the test facility is considered as the *primary* circuit, while the fluid flowing through closed circuits within the UUT (oil, engine coolant, etc.) are the *secondary* circuits. Chilled water circuits are treated as secondary. There are several different terms used in the English language for the same primary circuit fluid, including mains water, plant water, factory water, and raw water.

Most automatically controlled heat exchangers in the powertrain test industry have valves controlling the flow of the secondary fluid through a heat exchanger that is fed by a constant flow from the primary circuit, as is shown later in this chapter.

## The basics: Properties of water

Water is an ideal solvent that is only met in its chemically pure form in laboratories. Unless stated otherwise, this chapter refers to "water" as the treated, potable liquid generally available in city supplies within a developed industrial country. Water is the ideal liquid cooling medium. Its specific heat is higher than that of any other liquid, roughly twice that of liquid hydrocarbons. It is of low viscosity, widely available and, provided it does not contain sufficient quantities of dissolved salts that make it aggressive to some metals, it is relatively noncorrosive. The specific heat of water is usually taken as $C = 4.1868$ kJ/kgK. This is the value of the "international steam table calorie" and corresponds to the specific heat at 14°C. The specific heat of water is very slightly higher at each end of the liquid phase range from 0°C to 95°C, but these variations may be neglected in general industrial use.

Use of the antifreeze (ethylene glycol, $H_2H_6O_2$) as an additive to water permits its operation as a coolant over a wider range of temperatures. A

50% by volume solution of ethylene glycol in water permits operation down to −33°C. Ethylene glycol also raises the boiling point of the coolant such that a 50% solution will operate at a temperature of 135°C with a pressurization of 1.5 bar. The specific heat of ethylene glycol is about 2.28 kJ/kgK and, since its density is 1.128 kg/L, the specific heat of a 50% by volume solution is:

$$(0.5 \times 4.1868) + (0.5 \times 2.28 \times 1.128) = 3.38\ \text{kJ/kgK}$$

or 80% of that of water alone. Thus the circulation rate must be increased by 25% for the same heat transfer rate and temperature rise. The relation between flow rate, $q_w$ (L/h), temperature rise, $\Delta T$, and heat transferred to the water is:

$$4.1868 q_w \Delta T = 3600H$$
$$q_w \Delta T = 860H$$

where $H$ is the heat transfer rate in kW, so to absorb 1 kW with a temperature rise of 10°C, the required flow rate is 86 L/h.

## Water quality required in a primary cooling water circuit

At an early stage in planning a new test facility, it is essential to ensure that a sufficient supply of water of appropriate quality can be made available. Control of water quality, which includes the suppression of bacteria, algae, and slime, is a complex matter and it is advisable to consult a water treatment expert who is aware of local conditions. If the available water is not of suitable quality, then the project must include the provision of a water treatment plant. Most dynamometer manufacturers publish tables, prepared by a water chemist, which specify the water quality required for their machines.

## Solids in water

Circulating water should be as free as possible from solid impurities. If water is to be taken from a river or other natural source, it should be strained and filtered before entering the system. Raw surface water usually has significant turbidity caused by minute clay or silt particles that are ionized and may only be removed by specialized treatments (coagulation and flocculation). Other sources of impurities include drainage of dirty surface water into the sump, windblown sand entering cooling towers, and casting sand from engine water jackets. Hydraulic dynamometers are sensitive to abrasive particles and accepted figures for the permissible level of suspended solids are in the range 2−5 mg/L. Seawater or estuarine water is used for testing large marine prime-movers with dynamometers and heat exchangers fitted with internals made from special stainless steels and marine bronzes, but it is not to be recommended for standard automotive equipment.

## Water hardness

The hardness of water is a complex property. There is a general subjective understanding of the term related to the ease of which soap lather can be created in a water sample, but the quality is not easy to measure objectively.

*Hard water, if its temperature exceeds about 70°C, may deposit calcium carbide "scale," which can be very destructive to all types of dynamometer and heat exchangers.*

A scale deposit greatly interferes with heat transfer and commonly breaks off into the water flow, when it can jam control valves and block passages. Soft water may have characteristics that cause corrosion, so very soft water is not ideal either.

Essentially, hardness is due to the presence of divalent cations, usually calcium or magnesium, in the water. When a sample of water contains more than 120 mg of these ions per liter, usually expressed in terms of parts per million (ppm) of calcium carbonate, $CaCO_3$, it is generally classified as a hard water.

There are several national scales for expressing hardness, but at present no internationally agreed scale:

*American and British*: 1° US = 1° UK = 1 mg $CaCO_3$ per kg water = 1 ppm $CaCO_3$

*French*: 1°F = 10 mg $CaCO_3$ per liter water

*German*: 1°G = 10 mg $CaCO_3$ per liter water

*1°dH* = 10 mg CaO per liter water = 1.25° English hardness

*Note*: The old British system, still listed in some documentation was 1°Clarke = 1 grain per Imperial gallon = 14.25 ppm $CaCO_3$.

For example: the requirements for Froude water-brake dynamometers were usually specified as within the range 2–5°Clarke (30–70 ppm $CaCO_3$).

## The pH value of cooling water and chemical additives

Water may be either acid or alkaline/basic. Water molecules, HOH or $H_2O$, have the ability to dissociate, or ionize, very slightly. In a perfectly neutral water, equal concentrations of $H^+$ and $OH^-$ are present. The pH value is a measure of the hydrogen ion concentration: its value is important in almost all phases of water treatment, including biological treatments. Acid water has a pH value of less than 7.07 and most dynamometer and test equipment manufacturers call for a pH value in the range 7–9; the ideal is within the range 8–8.4.

The preparation of a full specification of the chemical and biological properties of a given water supply is a complex matter. Many compounds—phosphates, sulfates, sodium chloride, and carbonic anhydride—all contribute to the nature of the water, the anhydrides in particular being a source of dissolved oxygen that may make it aggressively corrosive. This can lead to such problems as the severe roughening of the loss plate passages in eddy-current dynamometers and plate heat exchangers that have not been treated to prevent the effect. Some forms of Cooling Water (CW) treatment or "dosing" can themselves cause problems in the special circumstances of powertrain testing. The narrow passages in eddy-current dynamometer loss plates are particularly liable to blockage arising from the use of inappropriate chemicals used in some water treatment regimes. Water treatment specifications should include the fact that, if used with water brakes, the treated water will be subjected to highly centrifugal regimes and local heating that may cause some degrading of the solution.

Chemical treatment of recirculating CW systems is normally required for minimizing corrosion, scaling, and fouling. Treatment programs have to be customized for each system depending on the quality of make-up water and the system configuration; this will usually require local specialist help. A combination of dispersants, biocides, and corrosion inhibitors will be required in most cases to control the quality of the circulated water. Depending on the system's design and size, water treatment chemicals are fed by small chemical metering pumps on timed or constant feed. Treatment programs must be evaluated by regular testing.

A recirculating closed system should include a small constant bleed-off to drain balance by a make-up supply to prevent deterioration of the water by the concentration of undesirable compounds. A bleed rate of about 1% of system capacity per day should be adequate. If no bleed-off is included, the entire system should be periodically drained, cleaned out, and refilled with freshwater [1].

## Special case of large marine engine dynamometers

In the testing of large marine engines, which invariably require a water-brake dynamometer (see Fig. 9.11) using large quantities of cold water, estuarine or seawater may be used as a primary cooling medium.

In all test facility designs consideration should be given to the consequences of a power failure. In the case of large marine engine testing the possible effects of a sudden failure in the water supply can be very serious. Consider the failure of the water supply to a hydraulic dynamometer absorbing over 10 MW from a marine propulsion diesel engine operating at a full speed of 120 rpm. Even if the shutdown system operates immediately the fault is detected the engine system will take some time to come to rest, during which the dynamometer will have emptied and be operating on a mixture of air and water vapor, with the strong possibility of serious overheating. Therefore, in the case of any large engine test facility, some provision for a gravity feed of water in the event of a sudden power failure is advisable.

In general, the supply pressure to hydraulic dynamometers should be stable otherwise the control of the machine will be adversely affected. This implies that the supply pump must be of adequate capacity and having as flat as possible pressure–volume characteristic in the normal operating range.

## Required flow rates for test cell modules and unit under test

A wide variety of automotive test devices are required to be kept at optimum temperature and use water as a cooling medium; in addition to the commonly used in-cell oil and fuel control units, some high-power electrical drive and control units require a chilled water supply for their integral heat exchangers. Many of these devices will be supplied with the manufacturer's specification of water pressure, flow, and temperature required to keep the devices below their maximum operating temperature. In practice the most common problem met by operators is a higher than specified water inlet temperature that leads to a high exit temperature and a reduced performance band. Possible causes are effects of other facility cooling loads in the primary circuit, high ambient temperatures affecting cooling towers or inadequately sized circuit sumps.

For cooling an internal combustion engine (ICE) rigged in a test cell, and in the absence of a specific requirement, it is good practice, for design purposes, to limit the temperature rise of the cooling medium through the engine water jacket to about 10°C. For cooling a dynamometer the flow rate must be determined by the maximum permissible CW outlet temperature, since it is important to avoid the deposition of scale (temporary hardness) on the internal surfaces of the machine or localized boiling.

Eddy-current dynamometers, in which the heat to be removed is transferred through the loss plates, are more sensitive in this respect than hydraulic machines, in which heat is generated directly within the CW. Provided the carbonate hardness of water does not exceed 50 mg CaO/L the maximum discharge (leaving) temperatures are:

- eddy-current machines 60°C;
- hydraulic dynamometers 65°C–70°C (see note below); and
- for greater hardness values, limit temperatures to 50°C.

Note: *When the discharge temperature reaches the maximum figure, some machines can experience internal flash boiling which will be detected by hearing a distinctive "crackling" noise. Running in this condition will cause destructive cavitation damage to the working chamber of the dynamometer.*

Approximate cooling loads per kilowatt of engine power output are shown in Table 6.1.

Corresponding flow rates and temperature rises are as shown in Table 6.2.

## Cooling systems for automotive traction motors

The increasing power density of automotive propulsion motors, and their location within a compact subassembly in the vehicle, is meaning that most are being designed to have dedicated cooling systems which must be accommodated and

**TABLE 6.1** Estimated cooling loads for water cooling circuit per kilowatt of engine power output.

| Heat source | Output (kW/kW) |
|---|---|
| Automotive gasoline engine, water jacket | 0.9 |
| Automotive diesel engine, water jacket | 0.7 |
| Medium-speed marine diesel engine | 0.4 |
| Automotive engine oil cooler | 0.1 |
| Hydraulic or eddy-current dynamometer | 0.95 |

**TABLE 6.2** Estimated temperature rise and flow rates corresponding to the cooling loads shown in Table 6.1.

| Heat source | In (°C) | Out (°C) | *l* (kWh) |
|---|---|---|---|
| Automotive gasoline engine | 70 | 80 | 75 |
| Automotive diesel engine | 70 | 80 | 60 |
| Medium-speed marine diesel engine | 70 | 80 | 35 |
| Automotive oil cooler | 70 | 80 | 5 |
| Hydraulic dynamometer | 20 | 68 | 20 |
| Eddy-current dynamometer | 20 | 60 | 20 |

monitored in the test cell. Although Grade F insulation has a maximum allowable operating temperature of 155°C, this high temperature can cause accelerated insulation failure. Moreover, the remanence and coercivity of the rare earth magnets often used are inversely proportional to the temperature, which means that, over time, it may lead to partial demagnetization at higher temperatures.

Several cooling strategies are being used and the details continually develop:

- Surface cooled electric motors are provided with a separate cooling jacket; the coolant may be either water or oil. To improve cooling the mantles are finned and generally manufactured from bronze alloy or aluminum.
- Internally cooled electric motors wherein a cooling oil is pumped through cooling slots cut directly into the stator laminations.
- Electric motors cooled by a self-contained, phase-change cooling system, in which the coolant is circulated in such a way a portion is transformed into vapor upon heating then condensed.

- Hybrid cooling systems often consisting of water-jacketed motors with forced air rotor cooling.

Oil companies are developing oils and "e-fluids" to operate as cooling fluids for EV and HEV Battery Packs as well as geared motor, lubrication, and cooling systems.

The strategy for the use of different types of motor cooling system within the test process has to be carefully considered particularly when the motor is being subjected to climatic extremes. It is vital within the test process that the fluids used are chosen, stored, used, and recorded with the same rigorous QA procedures as fuels used in emission testing.

## Types of test cell cooling water circuits

Test cell CW circuits may be informally classified as follows, with increasing levels of complexity:

1. Direct mains water supplied systems containing a portable dynamometer and cooling column that allow heated water to run to waste.
2. Sump or tank-stored water systems that are "open," meaning at some point in the circuit water runs back, under gravity and at atmospheric pressure, into the sump by way of an open pipe. These systems normally incorporate self-regulating water/fluid cooling modules for closed engine cooling systems filled with special coolant/water mix and, if required, for oil temperature control. They commonly have secondary pumps to circulate water from the sump through evaporative cooling towers when required.
3. Closed pumped circuits with an expansion, pressurization, and make-up units in the circuit. Such systems have become the most common as most modern temperature control devices and electrical dynamometers which, unlike water brakes, do not require gravitational discharge. Closed water-cooling systems are less prone to environmental problems such as the risk of Legionnaires' disease.
4. Chilled water systems (those supplying water below ambient) are almost always closed, although some may contain an unpressurized, cold, buffer tank.

## Direct mains water to waste cooling

These are systems used in small scale after-market repair and tuning facilities that often use "bolt-on" portable dynamometers of the Go-Power or Piper type and/or Engine Cooling Columns of the type shown in Fig. 6.1. In such systems the raw water is directly involved in the heat extraction from the process rather than from an intermediate heat exchanger. While most of these systems run mains supplied water to waste, it is possible to circulate the water through a gravity feed sump. One drawback of these systems is that the coolant is mains supply water and therefore cannot realistically be dosed with any additives.

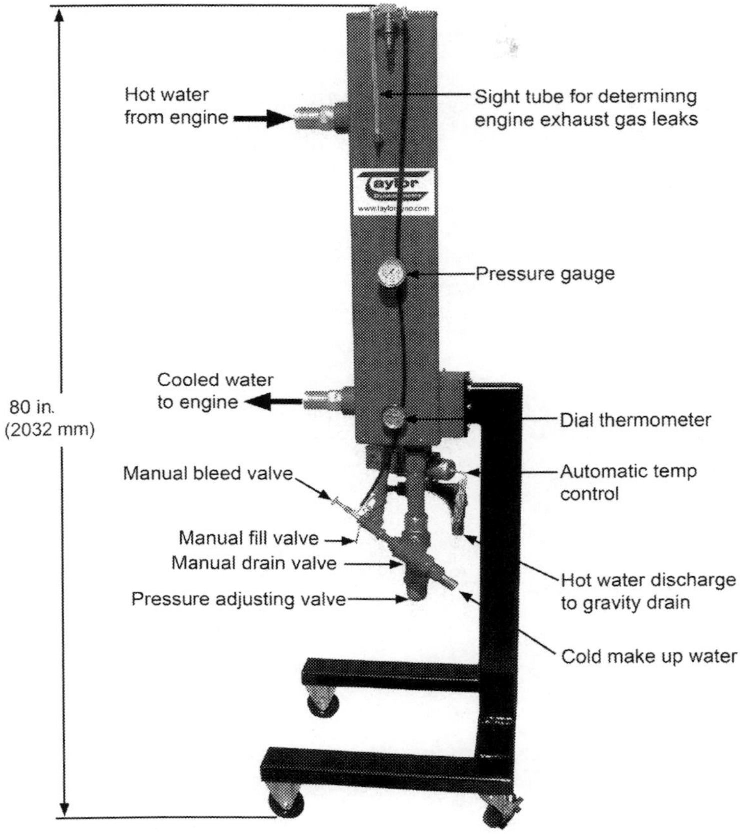


**FIGURE 6.1** Engine cooling column used widely in the United States, often in conjunction with direct engine-mounted dynamometers (see Chapter 10: Chassis Dynamometers, Rolling Roads, and Hub Dynamometers). *Reproduced with the permission of Taylor Dynamometer, Milwaukee.*

## Open plant water-cooling circuits

These systems are "open" because they all discharge into a sump which is at ambient atmospheric pressure. An essential feature is that they store water in a sump lying below floor level from which it is pumped through the various heat exchangers and a cooling tower circulation system. The sump is normally divided into hot and cold areas by a partition weir wall (see Fig. 6.2).

Water is circulated from the cool side and drains back into the hot side. When the system temperature reaches the control maximum, it is pumped from the hot sump and through the cooling tower before draining back into the cool side. A rough rule for deciding sump capacity is that the water should not be turned over more than once per minute; within the restraints of cost the biggest available volume gives the best results.

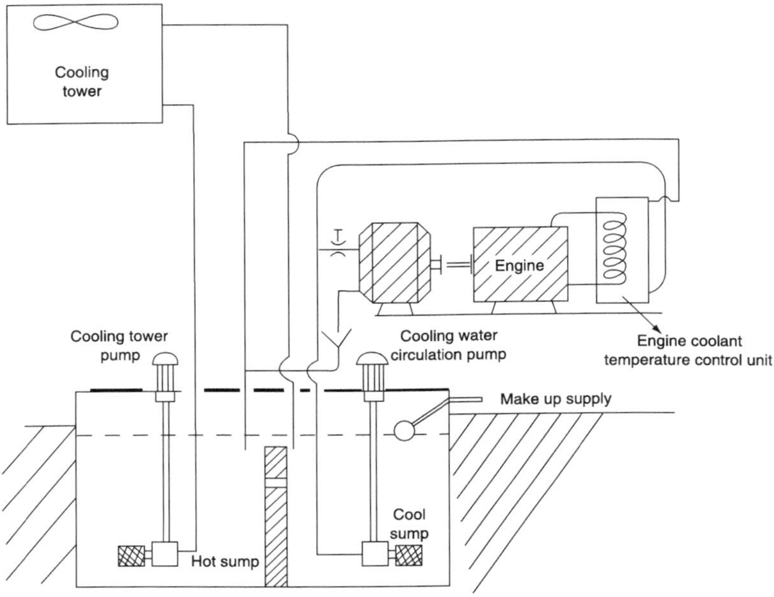


**FIGURE 6.2** Diagram of an open cooling water system incorporating a cooling tower and a below-ground, partitioned sump.

Note: *In the type of system shown in* Fig. 6.2, *sufficient excess sump capacity, above the normal working level, must be provided to accommodate the total drain-back volume from pipework, engines, and heat exchangers upon system shutdown.*

The arrangement shown diagrammatically in above is a classic arrangement of which thousands of similar systems are installed worldwide, but care has to be taken to keep debris such as leaves or flood water "wash-off" from entering the system. All ground-level sumps should have the top surface raised at least 100 mm above ground level to provide a lip that prevents flooding from groundwater.

Since the sump is open to the atmosphere, there is a continuous loss of water due to evaporation plus the required bleed-rate of about 1% of system capacity per day, mentioned above; therefore make-up from the mains water supply needs to be controlled by a float valve. It is important to minimize air entrainment in the pump suction; therefore the minimum level of the sump when the pressure and return lines are full should be sufficient to discourage the formation of an air-entraining vortex. The return flow should be by way of a submerged pipe fitted with an air vent.

A sensible design feature at sites where freezing conditions are experienced is to use pumps submerged in the sump so it can be ensured that, when not being used, the majority of pipework will be empty.

## Closed plant water-cooling circuits

The design and installation of a closed water supply for a large test installation is a specialist task and not to be underestimated, but in principle, it has many of the features of a closed domestic central heating system. However, the industrial version may require the inclusion of a large number of test and flow balancing valves, together with air bleed points, stand-by pumps, and filters with change-over arrangements. By definition these closed systems have no sump or gravity draining from any module within the circuit; such systems do not suffer from the evaporative losses of an open system and are less prone to contamination. Typically, it uses one or more pumps to force water through the circuit, where it picks up heat that is then dispersed via closed-circuit cooling towers, before the water is returned directly to the pump inlet.

It is vital that air is taken out and kept out of the system. The whole pipe system must be provided with the means of bleeding air out at high points or any trap points in the circuit. To achieve proper circulation and cope with thermally induced changes in system volume, also to make up for any leakage, a closed system has to be fitted with an expansion tank, plus some means of "make-up" and pressurization. These requirements can be met by using a form of compressed air/water accumulator connected to a pressurized make-up supply of treated water.

"Balancing" of water systems is the procedure by which the required flow, through discrete parts of the circuit, is fixed by the use of pressure-independent "flow-setter" valves having test points fitted for commissioning purposes. Valves are required for each subcircuit because they have their own particular thermal load and resistance and, therefore, require a specific primary system flow rate. The final balancing of closed cooling systems can be time-consuming, particularly if a facility is being brought into commission in several phases, meaning that the complete system will have to be rebalanced at each significant system addition.

None of the devices fitted within a closed plant water system should have "economizer" valves that themselves regulate the flow of the primary (plant) water, since that variation may continually unbalance the system. To avoid such unbalancing, devices in the circuit, the temperature control valves, should work by regulation of secondary fluid and have constant primary (plant water) flow.

*Freezing protection* must be considered by the facility designer, even if the whole system is within a building. Closed, pressurized water systems can be filled with an ethylene glycol and water mix to prevent freezing but, as mentioned at the beginning of this chapter, the cooling efficiency of the mixture is inversely proportional to the concentration of glycol. It should also be remembered that some materials in seals and pipes, such as natural rubber derivatives, may deteriorate and fail if exposed for long periods to glycols at elevated temperatures.

The pipework of open water systems can be trace heated. Such systems consist of special heating tape being wound around the pipes in a long-pitched spiral under insulating material. The control is usually via one or more thermostatic switches that energize the heating wire when the air temperature at key points of the system reaches 5°C.

## Engine coolant temperature control: Radiators and cooling columns

The use of "vehicle type" radiators within an engine test cell, relying on the engine's water pump for circulation, allows the use of a quite cheap, small, closed-circuit system containing small volumes of special coolants. The problems usually encountered are concerned with getting sufficient airflow through the radiator and around the engine from one or more electrically powered fans to control the coolant temperature. Consequentially, there is then the problem of dealing with the high heat-flow into a closed test cell space; therefore such systems are more suitable for "open-air" (car-port type) test sites.

If special engine coolants are not required, a cooling column is a simple and economical solution, they are commonly used in the United States (see Fig. 6.1) and are portable and easily located close to the engine under test.

The column allows the engine outlet temperature to run up to its set control level; at this operating point a thermostatic valve opens, allowing cold water to enter the bottom of the column and the same volume of hot water to run to waste or the sump from the top. The top of the column is fitted with a standard automotive radiator cap for correct engine pressurization and use when filling the engine circuit. Whether or not the engine under test is fitted with its own thermostat, precise control of coolant temperature is not easily achieved with such a "passive" service module unless it is closely designed to match the thermal characteristics of the engine with which it is associated. In this context "passive" means that the coolant service module is not fitted with either a coolant circulating pump or its own coolant heater.

## Closed engine coolant and oil temperature control modules

The inclusion in the test engine of its own coolant thermostat can cause problems and will depend on the design of experiment and the rigging of the engine system within the cell. Where the thermostat body has to be present due to the physical layout of the pipework but is not required to function, because an active (pumped and heated) coolant control unit is being used, it can be simply rendered ineffective by drilling holes in the diaphragm.

Coolant temperature control units can be adapted to control either the temperature of the coolant entering or leaving the engine; it is the position of the PRT sensor (Platinum resistance thermometry) giving feedback to the controller that determines what is controlled. Control of the "coolant in" temperature is always superior to that of "coolant out"; typical control ranges and limits figures for modern plant are listed in Table 6.3.

Even with closed systems, it may be difficult to achieve stable temperatures at light load. The instability of temperature control is increased if the engine is much smaller than that for which the cooling circuit is designed. The capacity of the heat exchanger is the governing factor [2].

**TABLE 6.3** Typical control ranges and limits for three types of coolant control units.

| Device type | Control range and tolerance |
|---|---|
| Pressurized engine coolant, passive cooling only. Steady state at engine inlet. | 70–125 ± 1°C |
| Engine oil, passive, cooling only. Steady state at engine outlet. | 70–145 ± 2°C |
| Pumped, heated, and cooled with engine bypass flow. Dynamic performance is superior to other types. | 70–145 ± 0.75°C |

There are many passive, closed system engine coolant temperature control units on the market, most working on the principle of a closed-loop, control valve controlling flow of coolant through a heat exchanger, and they can be broken down into the following types:

- type 1: mobile pedestal type (Fig. 6.4)
- type 2: special engine pallet-mounted systems
- type 3: user-specific, wall-mounted systems (Fig. 6.5)
- type 4: complex, heated and pumped, fixed pedestal type with conditioned fluid bypass close to the engine

Fig. 6.3 shows a simplified schematic of a typical service module incorporating heat exchangers for jacket coolant and lubricating oil of the circuit, while Fig. 6.4 is an illustration of the same device. The commercially available combined header tank and heat exchanger is a particularly useful feature. This has a filler cap and relief valve and acts in every way as the equivalent of a conventional engine radiator and ensures that the correct pressure is maintained.

For ease of maintenance, it should be possible to withdraw exchanger tube stacks without major dismantling of the system, and simple means for draining both oil and coolant circuits should be provided. The most usual arrangement is to control the temperature by means of a three-way thermostatically controlled valve in the engine fluid system (Fig. 6.5).

As discussed earlier, the alternative, where temperature is controlled by regulating the primary CW flow, will work, but gives an inherently lower rate of response to load changes. Types 2 and 3 listed earlier are often designed and built by the user. The pallet-mounted systems, which may use specific exvehicle parts for such items as the header tank and expansion vessel, allow a high degree of prerigging to take place outside the test cell and have a low thermal inertia.

Type 4 is the most complex and incorporates a coolant circulation pump, heaters, fluid-to-fluid cooler, and complex control strategies to deal with low

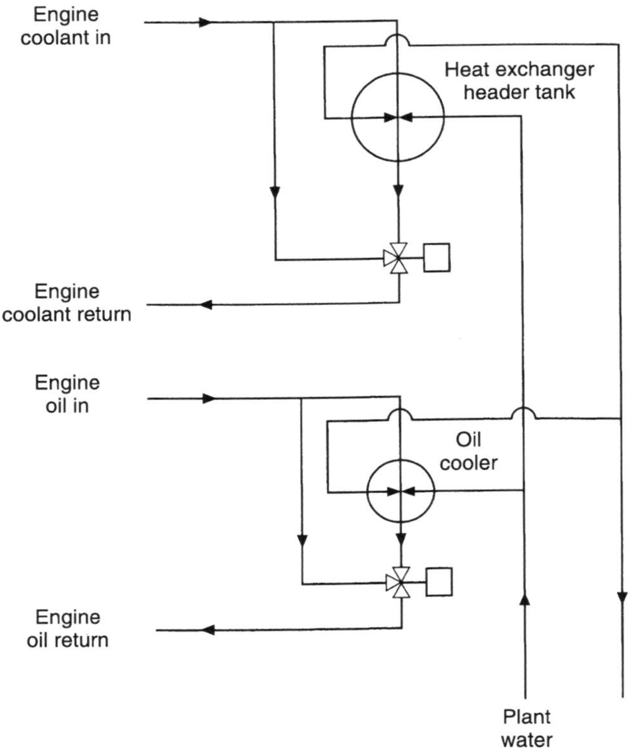


**FIGURE 6.3** The circuit diagram of the packaged coolant control unit shown in Fig. 6.4. All of the oil circuit must be below the oil level in the engine's sump.

engine loads and transient testing. Such devices, while considerably more expensive than passive types, give superior control response to rapid changes of heat load and allow a wider range of engines to be tested with the same device; also, much faster warm-up times and preheating of the engine are possible.

If engines are to be tested without their own coolant pumps, the module must be fitted with a circulating pump, commonly of the type used in central heating systems are satisfactory and easy to incorporate in the circuit (Fig. 6.6).

## Temperature control and the effects of system thermal inertia

None of the heat exchanger units described previously will operate satisfactorily if not integrated well with the engine and cell pipework.

A control time-lag is a common fault of many poorly installed coolant control systems, particularly the passive types, and has several causes. The sum of

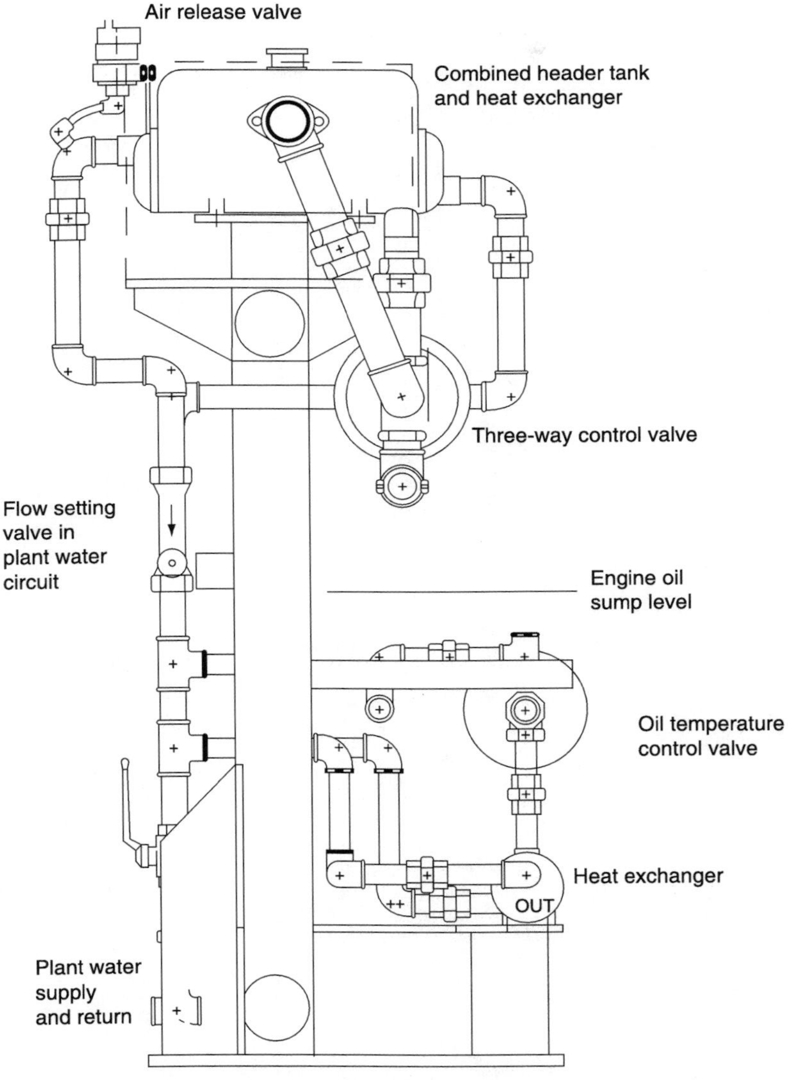


**FIGURE 6.4** Sketch of a packaged, pedestal mounted and moveable oil and coolant temperature conditioning unit. Coolant circuit uses a combined heat exchanger/header tank fitted with a radiator type pressure cap. *Note engine oil level.*

these phenomena is often referred to as the "thermal inertia" of the cooling system. To reduce thermal inertia there are three recommended strategies:

1. Reduce to the minimum possible the length of pipe and the fluid friction-head between cooler, the control valves, and engine. Reduction of friction

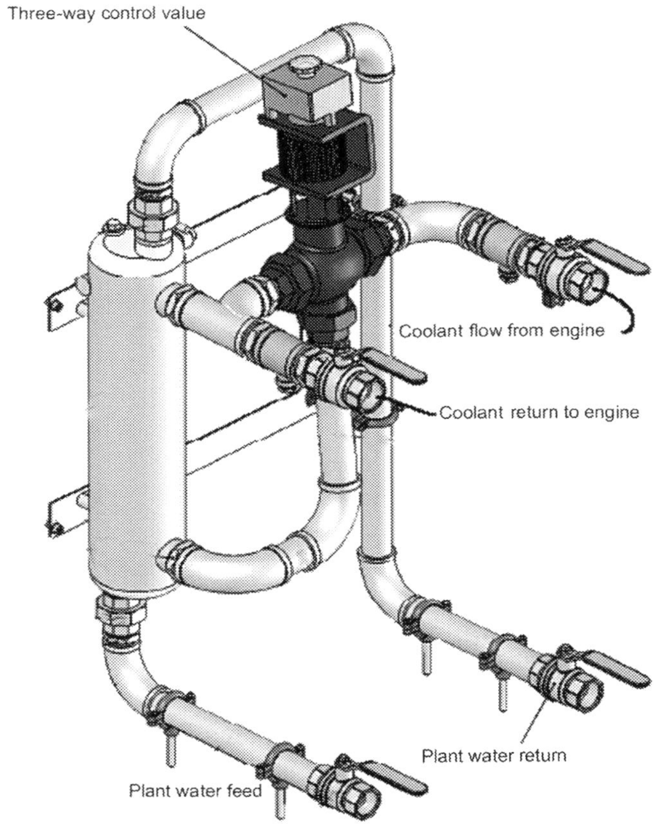


**FIGURE 6.5** Example of the same circuit shown at the top section of Fig. 6.3. and used for a simple wall-mounted engine coolant control module using a plant water cooled heat exchanger and a self-contained electrical controller operating a three-way valve.

head requires the use of long radius swept bends rather than elbows, elimination of sudden pipe diameter reductions, etc.

2. Circulate the conditioned coolant between engine and cooler/heater in a pumped circuit with an engine bypass that still maintains the required pressure to ensure flow through the engine as in type 4 units described previously and shown in Fig. 6.6.
3. Interconnecting pipes should be insulated against heat loss/gain.

Strategy 1 is best served by arranging a pallet-mounted cooling module close to the engine. In special cases such as anechoic cells, where the heat exchanger has to be remote from the engine strategy 2 is required to speed up the rate of circulation by an auxiliary pump mounted outside the cell to reduce lag and noise.

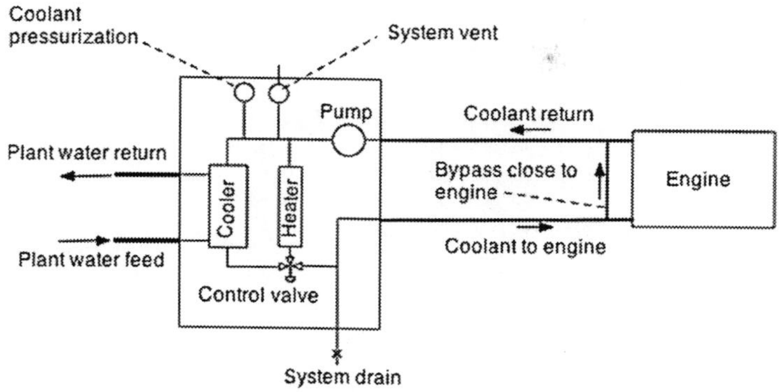


**FIGURE 6.6** Diagram of an engine coolant control system fitted with full temperature control, its own circulation pump, and a flow constricted fluid bypass close to the engine.

Three types of device are commonly adopted for the control of the secondary fluids through heat exchangers:

- independent-packaged controllers using electrically operated valves such as those produced by Honeywell
- indirect control by instruments having internal control "intelligence" and given set points from a central (engine test) control system, and
- direct control from the engine test controller via auxiliary PID software routines and pneumatically or electrically operated valves.

A proportional and integral (P and I) controller will give satisfactory results in most cases but it is important that variable-flow valves should be correctly sized. In design terms the valves have to have "authority" over the range of flows that the coolers require. A common error, which creates a circuit where valves do not have authority at low demands, is to fit valves that have a too-high-rated flow. It is very difficult to control such a system at low heat levels since the valve only has to crack open for an excess of cooling to take place and stability is never achieved.

## Chilled water circuits

In powertrain test facilities not involved in special climatic work the term "chilled water" normally covers water stored and circulated between 4°C and 8°C and may be required for the following processes:

- control room air-conditioning
- fuel temperature control
- combustion air temperature control
- thermal shock coolant conditioning module

These chilled water systems having markedly different thermal loads require designing and installing by suppliers with the relevant industrial experience.

FIGURE 6.7 Part view of a chilled water system supplying three different chilled water circuits from the chiller unit (foreground) and a vertical, insulated, stratified, buffer tank (midground aft of the chiller). Beyond the buffer tank is a combustion air treatment unit which uses a large proportion of the chilled water under certain demand conditions without affecting other connected systems. *Photo courtesy MSport Ltd.*

A common fault of badly designed chilled water systems is crosstalk, when good control of one process is lost because of a sudden change of supply pressure or temperature caused by events occurring in a separate process within a shared chilled supply. If a common chilled water system supplies several processes having differing thermal loads, each subsystem should have its own control loop and an adequately sized buffer tank should be incorporated in the system. In a closed system a thermally stratified buffer tank can be used where the system's return water enters the top of the tank and from there is drawn into the chiller, which returns the treated flow to the bottom of the tank from whence it is drawn and circulated (Fig. 6.7).

Control of liquid fuel temperature is critical to repeatability of results in engine testing; the control has a thermal rating much lower than other typical cell requirements. These small (1–2 kW) circuits are best served by having a dedicated chilled water supply; commercially available stainless-steel units of this rating, intended for the hotel industry, must be used with caution as their internal sealing may be unsuitable for liquid fuels.

## Engine thermal shock testing

To accelerate durability testing of engines and engine components, particularly cylinder-head gaskets, many manufacturers carry out thermal shock tests.

These commonly require the test system to suddenly exchange of hot coolant within a running engine for cold fluid. The term “deep thermal shock” is usually reserved for tests having a $\Delta T$ of around 100°C; such tests are also called, descriptively, “head cracking” tests. All such sudden changes in fluid temperatures cause significant differential movement between seal faces. Thermal shock tests are normally carried out in normal (not climatic) test cells and are achieved by having a source of cold fluids that can be switched directly into the engine cooling circuit via three-way valves inserted into the coolant system inlet and outlet pipes. The energy requirements for such tests clearly depend on the size of engine and the cycle time between hot stabilized running and the chilled engine condition, but they can be very considerable and often require cells with specially adapted fluid services that include a large buffer tank of chilled coolant. Mobile thermal shock systems, complete with chiller and buffer tank, are used to allow any cell fitted with suitable actuating valves in the engine coolant circuit to be used for thermal shock testing; however, their capacity is less than that of a purposely adapted cell.

Note that self-sealing couplings, commonly used to connect mobile fluid treatment modules, create a high flow resistance which can considerably add to the thermal inertia of the circuit. Globe valves and cam-locked couplings are recommended as a cheaper and a less flow-restrictive substitute.

The control for thermal shock and thermal cycling tests requires special test sequence subroutines and rapidly functioning valve actuators. There will be a need to operate the fluid control valves in the primary circuit in strict sequence in order to direct engine coolant flows to or from chilled water buffer stores. The stage “end condition” is usually defined by the attainment of a given (low) temperature in the engine cylinder head. Since this type of testing may be expected to induce a failure in some part of the engine system, it must be closely monitored by a suitably trained cell operator.

## Commissioning of cooling water circuits

Before dynamometers and any other test cell instrumentation are connected into any new primary CW system, the pipework must be pressure tested and cured of any leaks before being cleaned and the final fill of water treated appropriately. Good practice dictates that during water system commissioning the system is first fully flushed. Devices that have to be supplied with water should be temporarily bypassed by pipes that allow water to be circulated rather than to contaminate devices within the secondary circuit. If this is not possible or if large heat exchangers have to be flushed, then temporary strainers should be put in the circuit. Only when temporary strainers and line filters are not picking up debris should the strainers be removed, and the instruments connected into the circuit. In locations where water is freely available it is recommended that the first system fill of untreated mains supplied water, used for flushing, is dumped and the settling tanks cleaned. Debris will include both magnetic and nonmagnetic material such as jointing compound, PTFE tape, welding slag, and building dust, all injurious to the function of installed equipment.

It is cost and time effective to take trouble in getting the coolant up to specification at this stage rather than dealing with the consequences of valve malfunction or medium-term corrosion/erosion problems later.

After flushing the system should then be filled with clean water that has been treated to balance the hardness, pH level, and biocide level to the required specification.

## Part 2: Exhaust gas systems

Note: *The layout and special design features of exhaust systems forming part of full gaseous emission analysis systems is covered in* Chapter 17, *Engine Exhaust Emissions.*

### Safety issues with test facility exhaust systems

The gaseous emissions from ICEs are toxic to human life and must be treated as such by test facility designers and operatives. There have been many accidents caused by the poor design, lack of maintenance or careless use of these systems, some of them fatal. It is best practice that exhaust gases and any pipe or duct carrying them should, after leaving the Test Cell, exit enclosed building spaces as soon as practically possible. Particular care should be taken in the detailed design and construction of ducting, within a building space, containing diluted exhaust gases under positive pressure from a fan. This is poor practice, and wherever possible the ducting within a building space of such a system should be on the suction side of an externally mounted fan.

Explosions within exhaust systems, while on test or after final installation, are usually confined to dual-fuel engines or medium speed (non-automotive) diesel engines. In dual-fuel systems, gas leakage and build-up must be prevented by monitoring with a gas sensor. If a gas leak is detected, the gas supply valve must be immediately shut off and the engine automatically switched to diesel fuel mode. Medium speed diesel engines have one fuel pump per cylinder and explosions in the exhaust systems have occurred when one or more of the cylinders are not firing and diesel mist is pumped through the turbocharger and into the exhaust duct. Monitoring of individual cylinder exhaust temperatures and lifting the fuel pump from the cam of a nonfiring cylinder has been a marine engineer's standard method of dealing with the immediate problem. Having a guarded and chain retained explosion panel in the permanent exhaust piping of the large engine test facility is a sensible design feature.

Within the test cell the most significant risk to operation staff is contact with hot metal exhaust piping. Fixed cell sections can be permanently lagged, while the sections after the engine, particularly those running at a low level in the cell, should be guarded to avoid burns to staff and vulnerable cables. All modern, non-motorsport, automotive exhaust systems are fitted with emission control modules that require that the actual vehicle exhaust-pipe system should be employed during engine tests. This can impose problems of layout even in cells of large floor area and sometimes require the use of small profile permanent magnet motor dynamometers to allow the exhausts to run in the required configuration. If vehicle systems are to be modified it should be remembered that any change in the length of the primary pipe is particularly undesirable, since this can lead to changes in the pattern of exhaust pulses in the system. This can affect the volumetric efficiency and power output of the engine and, in the case of two-stroke engines, it may prove impossible to run the engine at all with a wrongly proportioned exhaust system. An increase in the length of pipe beyond the after-treatment devices is less critical. However, permanently fitted sections of test cell exhaust systems should always be designed to give a much lower back-pressure to the maximum gas flow than that required or specified for any test engine; control of back-pressure can then be adjusted by a permanently installed control valve. Such valves, usually of butterfly type, fitted into the oversized pipework of the cell exhaust system, need to be fitted with stainless steel internals and spindle seals that work at elevated temperatures. Automatic control and remote actuation of these valves are only fitted in cases of critically tuned systems or where engines of widely different sizes are run; otherwise, they are normally manually adjusted and locked in position after commissioning. *Turbocharged engines* in particular can have complex exhaust systems and run at such high temperatures that large areas of manifold and exhaust pipe can appear incandescent. This can represent a large heat load on the ventilating system particularly if ancillary cooling via spot-fans or air-movers are needed to prevent localized heating of near engine parts in the absence of the type of air-cooling flows that would be present in a moving vehicle.

Red-hot exhaust parts may also give false triggering of a fire system if inappropriate sensors have been fitted.

Although it can give a less cluttered cell environment, because of unseen corrosion and the fire risk caused by debris, it is generally not a good idea to run the exhaust pipes in a cell-floor duct unless the system has been specifically designed for this purpose and housekeeping is of a very high order. Such ducts have to be fitted with fuel vapor and exhaust gas detection and warning systems and be purged by the cell ventilation system (see Chapter 5: Ventilation and Air Conditioning of Automotive Test Facilities).

The practice, sometimes adopted, of discharging the engine exhaust into the cell's main ventilation extraction duct is not recommended because of several disadvantages:

- To avoid rapid acid attack and corrosion, the duct has to be made of a suitable stainless steel. Galvanized sheet ducting and particularly spiral-wound galvanized tubing will have a very limited life when used in a diluted exhaust system and is *not* recommended.

- The airflow must be increased to a level greater than that necessary for basic ventilation, to maintain an acceptable duct temperature. On a cold day this can lead to chilling in the cell.
- Soot deposits and corrosive, staining condensate in the fan and ducting are unsightly and make maintenance difficult.
- Other difficulties can arise, such as noise, variability in exhaust back-pressure.

If "tail-pipe" testers of the type used in vehicle servicing and inspection are to be employed, then easy and safe access via lockable valving to a suitable extraction tapping is necessary.

## Common test cell exhaust layouts

There are some commonly used test cell exhaust layouts as follows:

- individual cell, close coupled to ICE
- individual cell, vehicle exhaust system, scavenged duct
- multiple cells with common scavenged duct
- specially designed emission cells (see Chapter 17, Engine Exhaust Emissions)

## Individual cell, close-coupled exhaust

Such an arrangement is shown schematically in Fig. 6.8A. It may be regarded as the "standard" arrangement for a general-purpose ICE test bed and particularly for larger stationary or marine diesels. The exhaust manifold (s) is coupled to a flexible stainless-steel pipe section, of fairly large diameter, to minimize pressure waves. The gasses are led, by way of a backpressure regulating valve, to a pipe system suspended from the cell roof.

Condensate, which is highly corrosive, tends to collect in these pipes, which should be laid to slightly fall to a low point with suitable drainage arrangements. In cells designed to occasionally use smoke or particulate analysis it is usually a

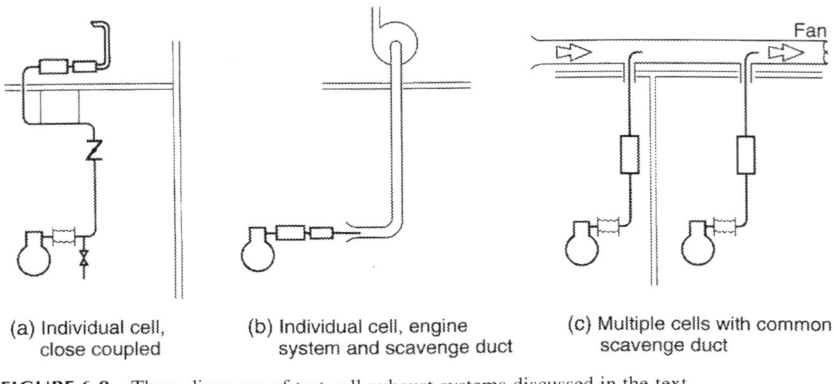


**FIGURE 6.8** Three diagrams of test cell exhaust systems discussed in the text.

requirement that the sample is taken from the exhaust pipe at particular points; some devices have quite specific requirements regarding size, position, and angle of probe insertion, and it is desirable, in order to ensure a representative sample, that there should be six diameters length of straight pipe both upstream and downstream of the probe. It is therefore good practice, when designing the exhaust system, to arrange a straight near horizontal run of exhaust pipe; such pipe sections should be easily replaced by a pipe with specific probe tappings.

## Individual cell, scavenged duct

When it is necessary to use the vehicle exhaust system, two options are available: simple and after-market systems take the pipe outside the building through a panel in the cell wall, while the more usual industrial policy is to put the tailpipe into a scavenge air system, as shown schematically in Fig. 6.8B. In this case the tailpipe is simply inserted into a bell mouth through which cell air is drawn; the flow rate should be at least twice the maximum exhaust flow, preferably more. This outflow has to be included in the calculations of cell ventilation airflow. The scavenger flow should be induced by a fan, most preferably external to the building which must be capable of handling the combined air and exhaust flow at temperatures that, if close to the cell, may reach 150°C in some circumstances.

## Multiple cells, common scavenged duct

This arrangement (Fig. 6.8C) is found in many multicell installations that carry out testing on diesel engines within the average automotive power range. It is an arrangement that is recommended only for diesel engines. In the case of spark-ignition engines, there is always the possibility that unburned fuel, from an engine that is being motored but not starting, could accumulate in the ducting and then be ignited by the exhaust from another engine. This possibility may seem remote, but "wet-start" accidents of this kind are not unknown. Note that in these layouts, silencers being part of a rigged vehicle exhaust system are used, or, in their absence, cell-specific silencers have to be fitted, near the engine, before the tailpipe runs into the dilution duct.

We can take as an example an installation of three test cells, each like that schematically illustrated in Fig. 3.3, running a 250 kW turbocharged diesel engine for which the maximum exhaust flow rate is 1365 kg/h. To ensure adequate dilution and a sufficiently low temperature in all circumstances, we need to cater for the possibility of all three engines running at full power simultaneously and a scavenging airflow rate of about 10,000 kg/h, say 2.3 $m^3$/s would be appropriate. Table 5.5 (Chapter 5: Ventilation and Air Conditioning of Automotive Test Facilities) indicates a flow velocity in the range of 15–20 m/s and hence a duct size in the region of 400 mm × 300 mm, or 400 mm diameter. As mentioned before, the scavenging fan must be suitable for handling gas temperatures of at least 150°C.

Note that, in Fig. 6.8B and C, the fan controls must be interlocked with the cell-control systems so that engines can only be run when the duct is being evacuated.

Where the engine's exhaust system is not used, the section of exhaust tubing adjacent to the engine must be flexible enough to allow the engine to move on its mountings and a stainless-steel bellows section is to be recommended. Exhaust tubing used in the area close to the UUT should be regarded as expendable and the workshop should be equipped to make up replacements. As a final point, carbon steel silencers that are much oversized for the capacity of the engine will never get really hot and can be rapidly destroyed by corrosion accelerated by condensate.

The dual use of an exhaust gas extraction duct to act as a cell purge system is not now recommended because fans that are compliant with European ATEX rules are more expensive than non-compliant units and have to be integrated with the hydrocarbon level alarm system; both are good reasons for a "stand-alone" purge system.

## Cooling of exhaust gases

There may be an operational requirement to reduce the temperature of exhaust gas exiting the cell as is the case of an engine exhaust exiting through the foam lining of an anechoic cell. This is usually achieved by using a stainless steel, water-jacketed cooler having several gas tubes that give a low resistance to gas flow. These occasionally used devices are often made specifically for the project in which they are used and there are three important design considerations to keep in mind:

- If water flow is cut off to the exhaust gas cooler, a steam explosion can be caused; therefore primary and secondary safety devices should be fitted.
- High internal stresses can be generated by the differential heating of the cooler elements; therefore some expansion of the outer casing should be built into the design.
- Since the water jacket pressure will be higher than exhaust gas pressure, tube leakage could fill the engine cylinders unless the system is regularly checked.

In installations such as production multiple "hot-test" cells, direct water spray has been used in vertical sections of exhaust systems in order to wash and cool exhaust flow into a common evacuated exhaust main. The resulting wastewater is highly corrosive and staining; therefore adequate drainage and proper disposal should be provided.

## Estimation of exhaust gas flows

The volume of raw and diluted exhaust gas flow into and out of the test cell's fixed exhaust system is highly variable, both in the rate produced at source and, to a lesser extent due to temperature change within the piping system. In order to determine, exhaust gas flow rate in mass or volume, the formula for total mass of exhaust gas is $m_a$ (mass of air) + $m_f$ (mass of fuel) the known mass may be converted to volume.

For system design purposes the calculation of maximum flow needs to be based on the fuel consumption of the largest scheduled engine at full rated power; an example of the data required and the calculation for a naturally aspirated gasoline engine of 130 kW output is shown in Table 6.4. In most

**TABLE 6.4** An example estimation of exhaust gas flow for a test cell running a 130 kW gasoline engine, with and without dilution air entrained from the test cell.

| | Value | Units |
|---|---|---|
| **Fuel** | | |
| Mass rate | 31.5 | kg/h |
| Volume rate | 42.0 | L/h |
| Specific fuel consumption | 0.242 | kg/kWh |
| **Combustion air** | | |
| Mass rate | 463 | kg/h |
| Intake density at 1 bar | 1.22 | $kg/m^3$ |
| Volume rate | 378 | $m^3/h$ |
| **Exhaust gas** | | |
| Mass rate (air + fuel) | 495 | kg/h |
| Density, after manifold 1.5 bar | 1.22 | $kg/m^3$ |
| Volume rate | 779 | $m^3/h$ |
| Density, out of cell 1.3 bar | 0.586 | $kg/m^3$ |
| Volume rate of cell | 844 | $m^3/h$ |
| **With exhaust dilution in cell** | | |
| Dilution ratio | 1.5 | |
| Mass flow at intake | 742 | kg/h |
| Density at intake (cell at −50 Pa) | 1.16 | $kg/m^3$ |
| Volume rate at intake | 637 | $m^3/h$ |
| Total mass flow (exh. + dil. air) | 1236 | kg/h |
| Mixture temperature after intake | 218 | °C |
| Mixture temperature out of cell | 60 | °C |
| Mixture density out of cell | 1.06 | $kg/m^3$ |
| Mixture volume flow rate out of cell | 1168 | $m^3/h$ |
| | 0.32 | $m^3/s$ |

cases the permanent exhaust piping in test facilities is over-sized by design to prevent flow restriction.

## Exhaust noise

The noise from test cell exhaust systems within and outside the facility building is a perennial problem for designers and operatives. It can be the subject of complaints from neighboring premises, particularly if running takes place at night or during weekends. The safety requirement to take exhaust pipes outside the building spaces as soon as possible works against the need to attenuate the noise they radiate; therefore it is necessary to reduce it within the pipe system.

Essentially, there are two types of devices for reducing the noise level in ducts: resonators and absorption mufflers. A resonator, sometimes known as a reactive muffler, is shown in Fig. 6.9. It consists of a cylindrical vessel divided by partitions into two or more compartments. The exhaust gas travels through

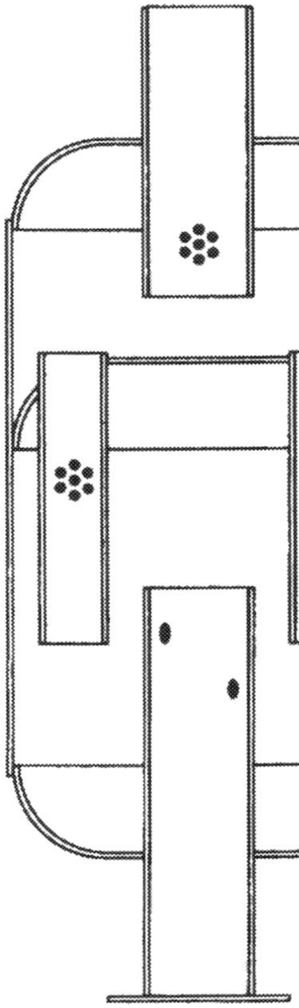

**FIGURE 6.9** A section through an exhaust muffler of the resonator or reactive type.

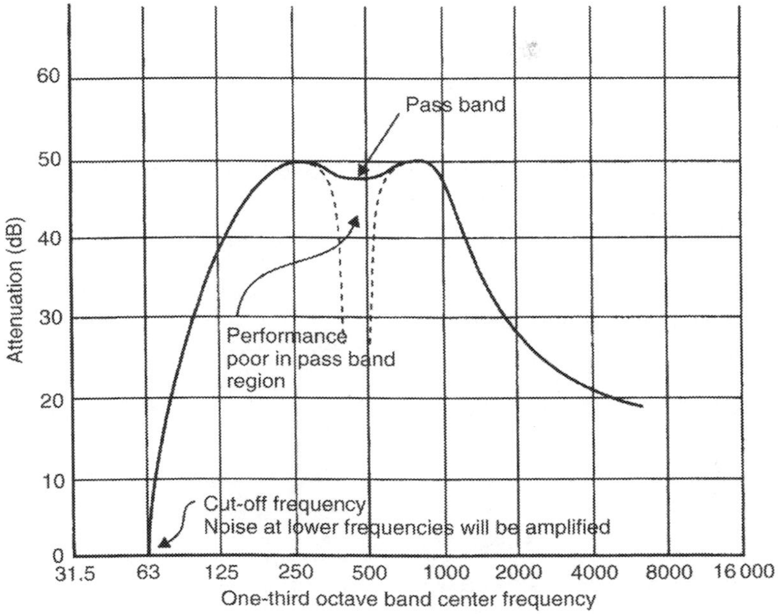


**FIGURE 6.10** Diagram of the performance of a sampled reactive muffler of the type shown in Fig. 6.9.

the resonator by way of perforated pipes, which themselves help to dissipate noise. The device is designed to give a degree of attenuation, which may reach 50 dB, over a range of frequencies (see Fig. 6.10 for a typical example).

Absorption mufflers, which are the type most commonly used in test facilities, consist essentially of a chamber lined with sound-absorbent material through which the exhaust gases are passed in a perforated pipe. Absorption mufflers give broadband damping but are less effective than resonators in the low-frequency range. However, they offer less resistance to flow. Both types of silencer are subject to corrosion if not run at a temperature above the dew point of the exhaust gas, and condensation in an absorption muffler is particularly to be avoided.

Selection of the most suitable muffler designs for a given situation is a matter for the specialist and for maximum effect needs to be optimized as closely as possible for a particular energy input range or engine; therefore the exhaust silencer may be considered as part of the engine rigging when it is engine model specific.

Even when forming part of the engine rigging it can be combined with an oversized fixed muffler which is part of the cell. Modern practice tends to use the vehicle exhaust system complete with silencers within the cell then extract the gas, mixed with a proportion of cell air, into a duct fitted, before exiting the building, via absorption attenuators. The fans used in the extraction should be rated to work up to gas/air temperatures of around 150°C and the effect of temperature on the density of the mixture must be remembered in the design process.

## Exhaust pipe termination

The noise from the final pipe section is directional and therefore often points vertically upward. However, the pipe end can, if incorrectly positioned or terminated, suffer from wind-induced pressure effects; this has led to well-studied cases over many years of difficulty in getting a correlation of results between cells. The ideal would be to terminate in a wide radius 100 degrees bend down and away from the prevailing wind.

*Remember that the condensate of exhaust gases is very corrosive, and that rain, snow, etc. should not be allowed to run into undrainable catch points or the engine.*

Fig. 6.11A shows a simple termination tube suitable for use on a single cell. This is a raw exhaust outlet that incorporates a shroud tube that acts as a plume dilution device and rain excluder.

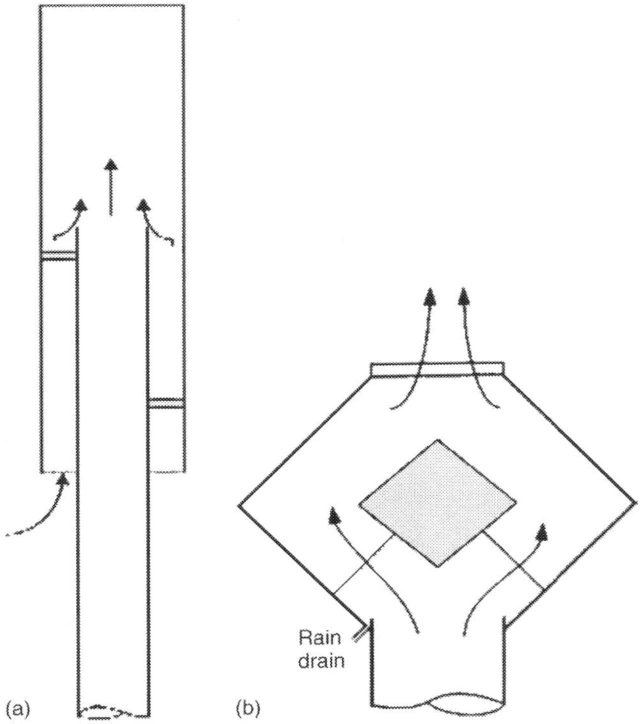


**FIGURE 6.11** (A) A simple but effective rain-excluding, roof mounted, smoke dilution tube termination fitted to a single raw exhaust outlet. (B) One of many possible designs of a building cowl for the termination of a multicell diluted exhaust system.

## Exhaust cowls on buildings

When deciding on the position of the exhaust termination outside a building, it is important to consider the possibility of recirculation of exhaust fumes into ventilation inlets. Prevention of recirculation requires the careful relative positioning of exhausts and inlet ducts in relation to each other and to the prevailing local (building affected) wind direction. The facility owner and architect may have strong opinions about the number and design of exhaust cowls. In the case of multicell facilities, the individual outlets are often consolidated into one or more chimneys, forming an architectural feature. Fig. 6.11B shows a type of rain-excluding cowl commonly used on multiple-cell diluted exhaust outlets. In all designs the exit tubes should terminate vertically to minimize horizontal noise spread.

## References

[1] BS 4959-1974, Recommendations for Corrosion and Scale Prevention in Engine Cooling Systems, British Standards Institution (BSI).

[2] W.H. McAdams, Heat Transmission, McGraw-Hill, Maidenhead, 1973. ISBN-13: 978-0070447998.

## Further reading

F. Guo, C. Zhang, Oil-Cooling Method of the Permanent Magnet Synchronous Motor for Electric Vehicle, National Engineering Laboratory for Electric Vehicles, 2019.

P.A. Vesilind, Introduction to Environmental Engineering, third ed., CL-Engineering, 2010. ISBN-13: 978-0495295853.

Chapter 7

# Energy storage

## Chapter outline



## Introduction

The transport, storage, and disposal of oils, volatile liquids and gaseous fuels worldwide are now subject to extensive and ever-developing regulation. The legislation surrounding the bulk storage of fuels is made more complicated because the safety-related risks of fire and explosion tend to be covered by one group of rules, while the risks of environmental pollution from leakage tend to be covered by another group of rules. Each set of rules may be administered by their own officials, who occasionally have policies concerning any one site that are in conflict with each other.

As the chemistry of automotive batteries together with their packaging develops so will the legislation and best practices concerning their storage

Engine Testing. DOI: https://doi.org/10.1016/B978-0-12-821226-4.00007-3

and transport. Currently (2020) for international transport lithium batteries are generally classified as Class 9 material, termed "miscellaneous dangerous goods" and are treated as such by the test industry, before they assembled within the controlled environment of a calibrated vehicle.

When planning the construction or modification of any bulk fuel or battery pack storage facility, it is absolutely vital to contact the responsible local official(s) early in the process, so that the initial design meets with concept approval.

While much of the legislation quoted in this chapter is British or European, the practices they require are valid and are recommended as good practice in most countries in the world. Much of the legislation used in handling and use of lithium ion batteries is issued from the UN. In the United Kingdom the Health and Safety Executive (HSE) is responsible for the regulation of risks to health and safety arising from work activity, while the 433 local authorities (LAs) in the United Kingdom have responsibility for the rules governing storage of petroleum and volatile fuels; each LA will have a petroleum officer.

Many of the rules and the licensing practices imposed on bulk fuel storage are designed to cover large farm (agricultural) or transport company diesel fuel systems and retail filling stations, where fuel is moved (dispensed) from the store tank into vehicles. Test facilities that dispense fuel into vehicles will fall under such rules, those that only handle fuel within a closed (reticulation) pipe system may not.

The most important reference legislations in the United Kingdom are:

- The Petroleum (Consolidation) Act 1928
- The Health and Safety at Work etc. Act 1974
- Dangerous Substances and Explosive Atmospheres Regulations in 2002
- ATEX Directive 94/9/EC
- Control of Pollution (Oil Storage) (England) Regulations 2001
- UN3480/UN3481 regulations (Lithium-Ion Batteries)

Individual countries and administrative areas within the EEC have a level of local regulation, as is the case in different administrative areas in the United States.

In the United Kingdom, if fuel, including petrol, diesel, vegetable, synthetic, or mineral oil, is stored in a container with a storage capacity of over 200 L (44 gal), then the owner may need to comply with the Control of Pollution (Oil Storage) (England) Regulations 2001.

The Explosive Atmospheres Regulations 2002 SI 2002/2776 issued to all LAs in the United Kingdom having responsibility for fuel storage sites are worth study by any responsible individual wherever based.

*It should be clear to any senior test facility planner or manager that, unless their site complies with the laws and practices that are in force in their region concerning fuel or battery storage, transport, or disposal, their site insurers will not cover the potentially ruinous costs of a fuel or battery fire or a major leakage causing environmental damage.*

## Part 1: Automotive battery storage and future testing roles

The testing of automotive battery packs is currently (2020) a rapidly developing technology although the requirements of test cells dedicated exclusively to the task are widely agreed upon and, all have a number of common features and follow a similar plan that has been briefly described in Chapter 3, Test Facility Design and Construction, and shown in Fig. 7.1:

The storage of batteries delivered to the test department needs to follow the same risk management policy adopted for any bulk fuel or energy storage:

- Storage units should be physically separate from the test laboratory building.
- The storage capacity of individual and physically separate storage units should be limited to a size that only requires limited, if any, human entry (machine accessible racks).
- The temperature within the storage space and of the packs within must be continuously monitored and connected to an alarm system; the use of thermal imaging is recommended in both store and test cell for the early discovery of "hot-cells."
- The store should take the form of a water floodable, vented box and have the required service connections to flood and drain built into the structure.

The task of the storage and testing of battery packs will expand rapidly in the future, not only because of different chemistries being used but also because of the role batteries could play beyond their first automotive life and

**FIGURE 7.1** An aerial photograph of the modern automotive battery pack test facility at Millbrook (United Kingdom). Showing the main control and data acquisition building with two battery storage chambers in front, behind run a series of individual, climatically controlled test chambers capable of full charge/discharge cycling up to 1100 V, 1400 A, and 750 kW over a temperature range of −40°C to +90°C.

in the role of they could play in the development of a nation's electrical grid, which will have to adapt to the proliferation of variable renewable energy sources. The subject of end-of-life and "second life" of automotive battery packs is relevant to the users of this book, since it will be they who will be in the lead position of having access to the skills and facilities able to safely store and test batteries, in order to judge the degree to which they may be reused or recycled. This ability comes with a new set of operational risks since the physical condition of the pack's internals will be unknown and variable.

Although no precise figure is published it is generally accepted that when a battery pack's charging capacity has fallen to 70% of original, it has come to the end of its (first) life in that below that figure the majority of drivers would judge its performance unacceptable.

Test facilities are able to use charge/discharge cycling under different and changing atmospheric conditions to learn when this "end of acceptable life" will occur (see Chapter 3, Test Facility Design and Construction).

Two factors determine the degree of residual capacity of such units:

- cycling degradation that is determined by the battery utilization in the vehicle and, to a degree, its thermal history and
- calendar degradation that is determined by not just its age but also details of its operating usage (state of charge etc.) and its environmental storage conditions.

When removed from the vehicle it is expected that they will have to undergo some form of testing (yet to be defined) to determine their residual capacity and their fate, which would be one of the following:

- *Direct recycling*: If damaged or too exhausted for second life applications.
- *Reconditioning*: Reconditioning is not expected to be widely accepted by users or original equipment manufacturers (OEMs) since newer and cheaper technologies will be available and the testing required before and after would be an onerous task.
- *Repurposing*: It is predicable that the majority of batteries considered to be suitable for second life applications would be repurposed by third-party companies or companies owned by the OEMs. The process would have to include bulk storage, partial pack disassembly, rebuilding in a different configuration, and final testing.

*The implications on national grids of, in the next decades, having GWhs of battery storage capacity are significant.*

Test facilities are currently a key part of the research into different battery chemistries, an emphasis of which is searching for new and cheaper cathode materials, particularly the reduction of expensive and scarce cobalt. Currently, lithium-ion cells have a cathode based on a number of chemistries such as those based on lithium nickel manganese cobalt oxide [(NMC-LMO), an NMC cathode blended with lithium manganese oxide], one based on lithium cobalt aluminum oxide (NCA), or based on lithium iron phosphate oxide (LFP) The search for and testing of new electrode materials and electrolytes is evolving and constantly in development.

Currently, the majority of OEMs use NMC and NMC-LMO (e.g., BMW, GM, Toyota, Mitsubishi, Daimler, Renault, and Nissan), NCA is by Tesla, and LFP by several Chinese OEM's.

## Part 2: Handling and storing volatile liquid fuels

### Safety principles

The mnemonic "VICES" is used by the HSE in the United Kingdom and is worth remembering when involved in the handling and storage of volatile liquids:

$V$ = ventilation. Adequate ventilation rapidly disperses flammable vapors.
$I$ = ignition. All ignition sources should be removed from fuel handling areas.
$C$ = containment. The fuels must be held in containers suitable for their containment with secondary devices such as trays to catch spillage and absorbent materials to hold and clean up any leakage.
$E$ = eliminate. Is it possible to eliminate or reduce some of the fuel storage containment?
$S$ = separation. Fuels should be stored in areas well separated from other storage or work areas or areas where they are exposed to accidental damage (delivery trucks, etc.).

To this list has to be added *labeling*. All fluid containers, from small sample bottles to bulk tanks, should be clearly labeled in such a way as to ensure that there is no doubt about the contents and any hazard those contents present.

*Spill kits* need to be readily available to trained operators, of all automotive test cells, so as to deal with floor spillage of oils and fuels. They should contain appropriate protective clothing, at a minimum, gloves, loose absorbent and absorbent packed in tubes to act as a dam, a plastic shovel, and thick plastic disposal bags, all the contents should be packed into a clearly labeled tub that can be used to hold the soiled materials prior to disposal. The size and contents need to be based on a risk assessment of possible spillage based on the volumes of fluids held in the cell.

### Auditing of fuel use

There have been cases where long-term fuel leakage into the environment or loss through theft has been discovered, not by management monitoring systems or facility staff, but by accountants trying to balance quantities of fuel purchased with that costed out to test contracts. Simple flowmeters measuring the flow from each tank and the accumulated flow into each cell are all that is required and are a recommended part of the quality plan for any engine test organization. For the purposes of project costing, it is important that fuel flowmeter figures for each test cell that are recorded at the start and end of each project.

## Bulk liquid and gaseous fuel storage and supply systems

For the latter half of the 20th century, it was standard practice for diesel fuel to be stored in a bunded tank above ground and for gasoline (petroleum) to be stored in a single-skinned tank buried below ground. It remains common practice, in retail filling stations, for all tanks to be buried. In the case of automotive test facilities, if built within a secure perimeter, there is an increasing preference for fuel tanks and fuel piping to be installed aboveground, where any leakage can be quickly detected. Local geology and proximity of groundwater sources or streams will play a part in determining the suitability of a fuel storage site.

The height below the datum ground level and any movement of the local water table is of critical importance. In areas where the water table is high and variable, it is not unknown for buried tanks to "float" to the surface, with consequential fracturing of connections and fuel leakage.

*Note concerning local "Water Table" level*: It is important to understand the nature and possible local variability of the water table when involved in installing subsurface fuel tanks and pipes. A good definition is: the upper level of an underground surface in which the soil or rocks are permanently saturated with groundwater in a given vicinity.

It is *not* necessarily the depth to which groundwater settles in a deep hole dug at the test location. Local geology and topography will determine the details that sometimes can vary significantly over a small area. It should also be noted that some ground materials and sands can be dangerously unstable when saturated.

The Approved Codes of Practice in the United Kingdom and in the United States require any underground fuel storage tanks (USTs) to be provided either with secondary containment or a leak detection system covering both tanks and associated pipework. Steel tanks also have to be fitted with a cathodic corrosion protection system.

The European regulations such as the ATEX Directive 94/9/EC and the EPA regulations in the United States, which in the 1980s required a major underground storage tank replacement program, encouraged many owners to convert to above ground storage. However, in many sites burial was required by local officials when the installation was considered unsightly, exposed to high solar heat load, vulnerable to vehicular collision, or vulnerable to vandalization or terrorist attack. *These criteria are important, which is why a secure perimeter and careful layout is required when planning this part of a test facility.*

When a full fuel storage license is required in the United Kingdom, the following information will usually be required (in the rest of the world attention to the same detail is recommended):

1. Site location map to a scale of 1:1250 or 1:2500 indicating all site boundaries. Two copies of site layout to scale 1:100 clearly indicating the intended layout of fuel storage and distribution system.
2. The layout plan must show the following:
   a. Location of storage tanks and tank capacities.
   b. Route taken by road delivery vehicles. Note: this must be a "drive-through" not "cul-de-sac."
   c. Position of tanker fill points and their identification.
   d. Location of pipework, including all vent pipes.
   e. Location and type description of metering pumps, dispensers, etc.
   f. All site drainage and its discharge location.
   g. Petrol interceptor location and drainage discharge point.
   h. All other buildings within the site and their use.
   i. All neighboring buildings within 6 m from the boundary.
   j. Position of liquefied petroleum gas (LPG) storage, if applicable.
   k. Location of car wash and drainage if applicable.
   l. Main electrical intake point and distribution board.
   m. Position of all firefighting appliances.

Very similar restrictions and requirements to those described previously for the United Kingdom exist in most countries in the world, and any engineers taking responsibility for anything beyond the smallest and simplest storage system should make themselves familiar with all relevant regulations of this kind, both national and local.

*Any contractor used to install or modify fluid fuel systems should be selected on the basis of proven competence and, as is usually applicable, the relevant licensing and insurance to carry out such work and the required post-installation testing.*

Both the legislative requirements and the way in which they are interpreted by local officials can vary widely, even in a single country. Where engine test installations are well established, there should be a good understanding of the requirements, whereas in localities where such systems are novel, the fire and planning officers may have no experience of the industry and can react with concern or treat the installation inappropriately as a roadside filling station; in these cases they may require tactful guidance and/or access to existing industrial installations.

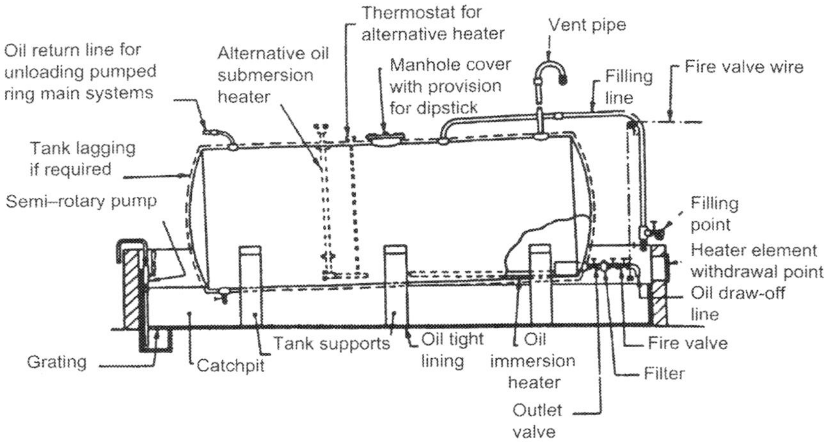


**FIGURE 7.2** Sketch of a typical aboveground, bunded, bulk fuel tank with its fittings. The semi-rotary pump is for the manual clearing of possibly polluted, storm water from the bund containment.

Fig. 7.2 shows a typical arrangement for a fuel oil or gasoline bunded storage tank according to British Standard 799.

The risk of oil being lost during tank filling or from ancillary pipework is higher than tank rupture; the UK Control of Pollution (Oil Storage) Regulations 2001 recognize this fact and require that tanks have a list of ancillary equipment such as sight tubes, taps, and valves retained within a secondary containment system. The use of double-skinned tanks is strongly recommended and increasingly required by regulation when used in underground installations [1]. Double-skinned tanks and fuel pipes should also be fitted with an interstitial leak monitoring device.

## Storage of reference fuel, drums or intermediate bulk containers

Where frequent use is made of special or reference fuels, they are usually supplied in a 200-L drum, known as a 55-gal drum in the United States and as a 44-gal drum in the United Kingdom or, much more rarely, in intermediate bulk containers (IBCs). Special provision must be made for the transportation of all such containers to and from the secure fuel store and their protection while in use.

Small bunded drum containment stores that meet legislative requirements are widely commercially available, while large sites will have specially built stores that are physically separate from the main site buildings that are serviceable by fuel deliveries in the same way as bulk tanks. Proprietary designs should allow access to a wheeled drum lifter/trolley. For medium-size facilities the commercially available packaged units, designed for use at large road building sites and built from 20-ft. ISO shipping container frames, are probably a cost-effective solution.

Fuel stores must be bunded, meaning that a total release of stored fuel will be captured and prevented from escape into the environment. ISO 9000 regulations and best practices relating to bunds dictate that they should be the following:

- Noncorrosive and resistant to oil and water.
- Able to hold 110% of the tank's capacity.
- Have an expected life span of 20 years with maintenance.
- Resistant to oil and water and have no drain-down valve or pipe.
- Every part of the tank must be contained with downward pointing valves (taps) that can be locked when not in use.

Drums are most safely handled on standard pallets and connected to cell systems when in the vertical position.

Special drum containers made from high-density plastic, designed to be transported by forklift, are available and these may be parked immediately against an outside wall of the test cell requiring the reference fuel supply. Connection and supply into the cell system can be achieved by using an automotive 12 V pump system designed to screw into the drum and deliver fuel through a connection point outside the cell fitted with identical fire isolation interlocks as used in the permanently plumbed supplies.

Finally, when planning any type of bulk fuel storage, precautions should be taken against the theft of fuel and malicious damage (vandalism). Where the risk to bulk storage tanks is considered to be high, the increased cost of burial is clearly justified.

## Decommissioning of underground fuel tanks

Sites with disused underground fuel tanks have a double problem: the land above is virtually unusable and the risk of environmental pollution increases with time. Disused tanks can either be excavated and removed or filled with inert material in situ. The old policy of pumping the tank full of a cement-based grout is not recommended because it is difficult to fill 100% and makes the tank too heavy to remove at some future date. The recommended method, after as much of the fuel as possible has been pumped out, is to have the tank filled with chemical-absorbing foam created from an amino-plastic resin, which completely fills the space, soaks up hydrocarbon residues, and does not significantly increase the mass of the tank. Specialist contractors need to carry out such decommissioning and certify the site's postdecommissioning safety.

## Test facility fuel piping

The "fuel farm" for a modern automotive test facility may be provided with a number of bulk storage tanks feeding several ring mains; these must be clearly, uniquely, and permanently labeled with their contents at all filling and draw-off points.

The expensive consequences of fuel leaks mean that modern best practice is for fuel lines to be run high above ground, where they can be seen, but not accidentally damaged, during normal site activities.

The use of standard (non-galvanized) drawn steel tubing for fuel lines is entirely satisfactory providing that they remain full of liquid fuel, but if they are likely to spend appreciable periods partially drained then the use of stainless steel is strongly recommended. The use of threaded fittings introduces some risk of leakage, although in drawn steel tube and provided there is no significant, thermally induced, pipe movement they can be satisfactory when used with modern sealants. The use of any kind of fibrous "pipe jointing" should be absolutely forbidden as fiber contamination causes "hairpin" jamming in valves and is difficult to clear.

Threading of stainless-steel tube using pipe threading machines fitted with general purpose dies often gives poor results and is not recommended. Preferably, all stainless-steel fuel lines, and certainly all underground lines, should be constructed with orbital welded joints, or with the use of compression fittings approved for the fuels concerned. Note that some fuels, such as "winter diesel" appear, from experience, to be particularly penetrating.

External fuel oil lines should be lagged and must be trace heated if temperatures are likely to fall to a level at which fuel "waxing" may take place.

Inside the test cell flexible lines may be required in short sections to allow for engine rigging or connection to fixed instrumentation. Such tubing and fittings must be specifically made and certified to handle the range of fuels being used; the generally recommended specification is for metal braided, electrically conductive, Teflon hoses. All conductive fuel lines should be linked and form part of the building's equipotential earthing scheme, as described in Chapter 4, Electrical Design Requirements of Test Facilities.

If the site has a fuel bowser for filling vehicles then, as stated elsewhere, the whole site has to comply with the local rules covering retail fuel outlets, including the type of bowser, its interconnections, color coding of its nozzles, warning signs, and electrical supply system.

## Underground fuel pipes

Whatever their type of construction it is vital that any underground fuel pipe is buried sufficiently deep and is protected from earth movement and compaction, from vehicles running over them. When running under any type of road or walkway, the route should be marked by permanent signage so that protective measures can be taken in the case of movement of construction plant, etc.

In Europe and the United States, buried fuel lines are usually required to be, and should be, of a double-walled design with the facility to check leakage from the primary tube into interstitial space (interstitial monitoring). Double-walled pipe systems are now usually made of extruded high-density plastics with the outer sleeve composition chosen for high abrasion resistance and the inner sleeve composition for very low permeability to fuels. Such pipes and fittings have to be electrofusion welded, have a maximum pressure rating for the inner containment sleeve of 10 bar, and a 30-year minimum design life.

Where single-wall steel fuel lines are used subfloor within buildings, particularly where they run under and through concrete floor slabs, they should be wrapped with water-repellent bandage (in the United Kingdom, sold as "Denzo tape") and laid on well-compacted fine gravel within a trench. The wrapping prevents the concrete from trapping the pipes and allows them to move without local stressing as they expand and contract. If and when subsurface fuel lines are run, it is good practice to lay them in a sealed trench of cast concrete sections with a load-bearing lid; the route of the trench has to be clearly marked and vehicles kept clear. Such ducts should be fitted with hydrocarbon "sniffers" connected to an alarm system to detect fuel leakage before ground contamination takes place.

*Air-locking* can be a problem in the long, quite small-bore piping used in test facility fuel systems; it can be particularly difficult to cure, so careful planning and installation is needed to avoid creation of vapor or air traps.

Pipes containing gravity-fed gasoline (from a header tank) that are run for any distance, external to the building and exposed to high solar load, are vulnerable to vapor locking and may require tall vent pipes to atmosphere to be fitted at the point at which the downward legs to cells are fitted. The termination of such vent pipes must be at a height some meters higher than the pressure head of the system under all circumstances.

## Storage and treatment of residual fuels

Many large stationary, and some medium-speed generator, diesel engines, and the majority of slow-speed marine engines before January 1, 2020 were allowed to operate on heavy residual fuels that require special treatment before use and delivery to the fuel injection system. In this fuel classification there are two traditionally used grades of fuel and one new type:

- Number 5 fuel oil, sometimes known as Bunker B.
- Number 6 fuel oil, which is a high-viscosity residual oil sometimes known as residual fuel oil or Bunker C; it may contain various impurities plus 2% water and 0.5% mineral particles (soil).
- Marine distillate and ultralow-sulfur fuel oil blends, which meets the 0.10% m/m limit imposed for 2020 onward.

The producers and particularly marine users of such fuels are under legislative great pressure to "clean up their act"; therefore large engine test facilities have new rules of which they should be aware.

The International Maritime Organization regulations to reduce sulfur oxide emissions from ships first came into force in 2005, under Annex VI of the International Convention for the Prevention of Pollution from Ships (known as the MARPOL Convention). Since then, the limits on sulfur oxides have been progressively tightened and from January 1, 2020, the limit for sulfur in fuel oil, used on board ships operating outside designated emission control areas (ECA), will be reduced to 0.50% m/m (mass by mass) from 3.50% m/m. Ships trading in designated ECA since January 1, 2015 have had to use fuel oil a sulfur content of no more than 0.10% m/m. It is important that any delivery of this type of fuel is accompanied with a bunker delivery note giving the sulfur content (% m/m) and density at 15°C ($kg/m^3$).

Any test system running engines with these fuels needs to incorporate the special features of the fuel supply and treatment systems that are typically installed in ships propelled by slow-speed diesel engines [2]. Bunker C oil needs to be preheated to between 110°C and 125°C; therefore the bulk storage tanks will require some form of heating coils within them to enable the heavy oil to be pumped into a heated settling tank. It is always necessary with fuels of this type to remove sludge and water before use in the engine. The problem here is that the density of residual fuels can approach that of water, making separation very difficult. The accepted procedure is to raise the temperature of the oil, thus reducing its density, and then to feed it through a purifier and clarifier in series. These are centrifugal devices, the first of which removes most of the water while the second completes the cleaning process. Fig. 7.3 shows a schematic arrangement. The temperature of the fuel must not be increased to more than 150°C at the engine inlet because the fuel can start to decompose.

It is also important to provide changeover arrangements in the test cell so that the engine can be started and shut down on a light fuel oil. An emergency stop resulting in a prolonged shutdown when there is heavy residual oil in the engine fuel rails creates a very unpleasant problem.

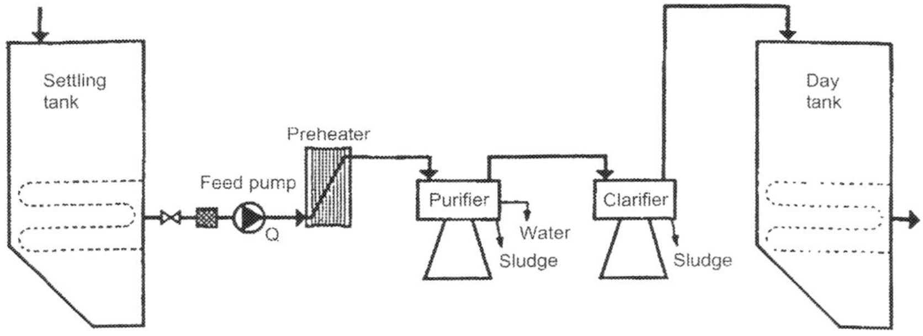


**FIGURE 7.3** Sketch schematic of fuel supply and treatment system for residual fuel oil.

## Biofuels and their storage

There are in 2020 two quite different types of biodiesel fuel commercially available, the differences of which are quite significant for test facilities storing and using both:

- Traditional biodiesel that is also known as Fatty Acid Methyl Ester or FAME is produced by esterifying vegetable oils or fats. It is defined under the standard of ASTM D6751 as "a fuel comprised of mono-alkyl esters of long-chain fatty acids derived from vegetable oils or animal fats." The use of traditional biodiesel is limited to a maximum concentration of 7% in Europe (based on EN 590 diesel standard) and sold commonly as 5% biodiesel (95% petrodiesel), which is labeled B5, and up to 20% in other parts of the world, varying from country to country.
- Renewable diesel, which is also known as hydrotreated vegetable oil, or HVO, differs significantly from the traditional FAME because of its production process during which impurities are removed from the raw materials, which are then hydrotreated at a high temperature. This means that the feedstock, unlike that of FAME, can be a wide range of waste products.

They are different products, but both are made from organic biomasses.

Test facilities involved in the preparation and mixing of traditional biofuels may have special storage problems to consider.

The cetane number of traditional biofuels is usually 50–60, and their cold resistance or waxing temperatures are higher and their shelf life are shorter than fossil diesel.

Laboratories working with the constituents of biodiesel mixtures will require storage for animal- or vegetable-derived oils in bulk or drums; these may require low-grade heating and stirring in some climatic conditions. However, any water in the fuel will tend to encourage the growth of microbe colonies in heated fuel tanks, which can form soft masses that plug filters. *In long-term storage dosing with appropriate biocide may be required.*

HVOs are straight-chain paraffinic hydrocarbons that are free of aromatics, oxygen, and sulfur and have cetane numbers in the range of 75–95 and therefore have none of the low temperature storage or microbe growth problems of the traditional fuel. It is reported to be mobile and filterable down to temperatures of −32°C.

*Ethanol/gasoline mixtures* are designated with the preface E, followed by the percentage of ethanol. E5 and E10 (commonly called "gasohol") are commercially available in various parts of Europe. E20 and E25 are standard fuel mixes in Brazil, from where no particular technical problems of storage have been reported.

*The storage and handling of 100% ethanol and methanol raises problems of security and safety; the former is highly intoxicating and the latter highly toxic, and anyone likely to drink the intoxicant probably lacks the chemical knowledge to distinguish between the two.*

## Supplying fuel to the test cell

Fluid fuel supplies to a cell may be provided by the following:

- under the static pressure head from a day tank,
- a pumped supply from a reference fuel drum, and
- by a pressurized reticulation system fed from the central fuel farm.

Day tank systems should be fuel specific to prevent cross contamination, so there may be a requirement for several. Modern safety practice usually, but not exclusively, dictates that gasoline day tanks should be kept on the outside wall rather than inside the test facility. Diesel day tanks are most frequently inside the shell building and thus protected, where necessary, from low (waxing) temperatures.

In all cases day tanks must be fitted with a dump valve for operation in the case of fire; this allows fuel contained in the vulnerable above ground tanks and pipes, to be returned to the fuel farm or a specific underground dump tank. Besides having to be vented to atmosphere, day tanks also have to be fitted with a monitored overflow and spill return system in the case of a malfunction of any part of the supply system. Fig. 7.4 shows a recommended day tank system diagrammatically.

The static head is commonly at 4.5 m or above but may need to be calculated specifically to achieve the 0.5–0.8 bar inlet pressure required by some industrial standard fuel consumption and treatment instruments.

When the tanks are exposed to ambient weather conditions, they should be shielded from direct sunlight and, in the case of diesel fuels, lagged or trace heated.

If a pumped system is used it must be remembered that, for much of the operating life, the fuel demand from the cells will be below the full rated flow of the pump; therefore the system must be able to operate under bypass, or stall, conditions without cavitation or undue heating. For this application the use of positive displacement, pneumatically operated pumps, incorporating a rubberized air/fluid diaphragm, have a number of practical advantages: they do not require an electrical supply and they are designed to be able to stall at full (regulated air) pressure, thus maintaining a constant fuel pressure supply to the test cells.

In designing systems to meet this wide range of requirements, some general principles should be remembered:

- Pumps handling gasoline must have a positive static suction head to prevent cavitation problems on the suction side.
- Fuel pipes should have low flow resistance, where smooth bends, rather than sharp elbows, are used and no sudden changes in internal cross section exist.
- Fuel lines should be pressure tested by competent staff before filling with fuel.

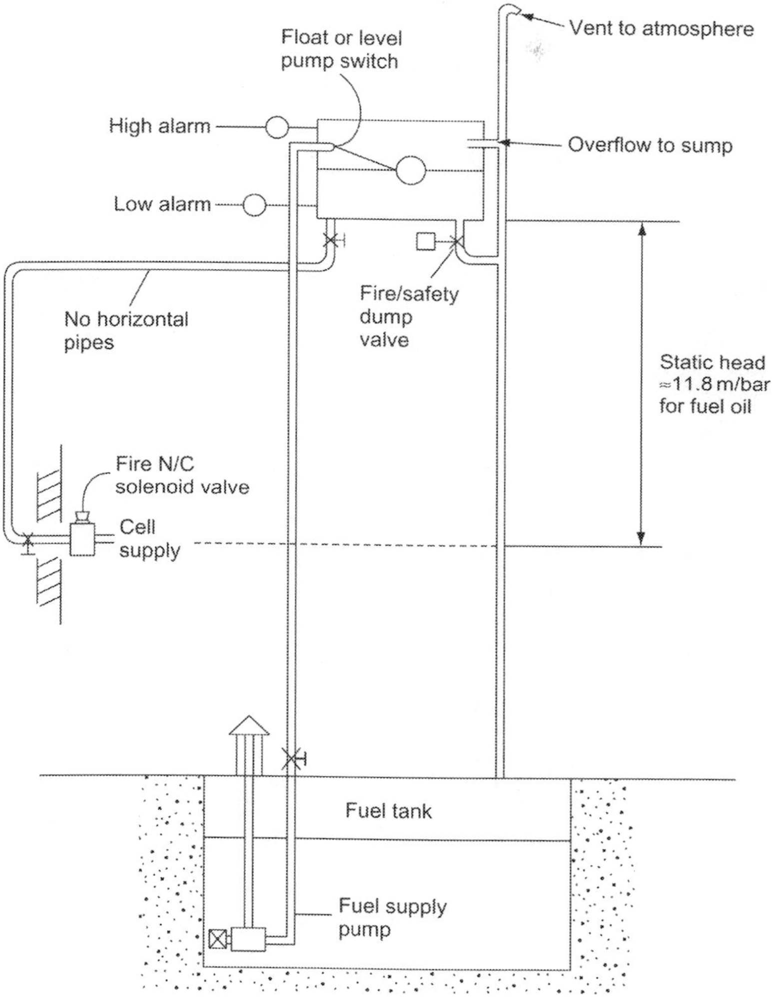


**FIGURE 7.4** A schematic sketch showing the main parts of a typical fuel day tank system combined with a subterranean bulk tank.

- Each fuel line penetrating the cell wall should be provided with a normally closed solenoid-operated valve interlocked with both the cell control system and the fire protection circuits. In some parts of the United States, local regulations require that the solenoid valves are supplemented by mechanical, normally spring-closed, valves that are held open by a soft-solder plug; in the case of a fire the plug melts and the supply is shut off.

## In-cell fluid fuel systems

*All fuel pipes containing fuels must be labeled near their point of entry and isolation.*

The fuel system in the test cell will vary widely in complexity, but the capacity of the fuel held in the cell system, including such items as filters and meters, needs to be kept to a minimum required at all times in order to reduce the potential "fire load."

In aftermarket cells dedicated to single internal combustion engines (ICE) types, the system may be limited to a single fuel line connected to the engine's fuel pump. For the occasional test, using a portable engine dynamometer system, a simple fuel tank and tubing of the type used for outboard marine engines, with capacity not more than 10 L, may be all that is necessary. These marine devices are certified as safe to use in their designated roles of containing and supplying fuel to an engine; the use of other containers is not safe and would endanger the insurance of the premises in which they are misused.

In powertrain and multipurpose development cells the requirement is likely to be for the supply of at least three different fuels plus a dedicated reference fuel system, all passing through the minimum of shared pipework but sharing fuel temperature control and fuel consumption measurement devices.

The following points should be considered in the planning of in-cell systems:

- It may be necessary to provide several separate fuel supplies to a cell. A typical provision would be three lines for diesel fuel, 95 RON (standard), and 98 RON (super) gasoline, respectively. Problems can arise from carryover and cross contamination, with consequent danger of "poisoning" exhaust catalyzers. To minimize cross contamination, it is desirable, although often difficult, to locate the common connection as close to the engine as possible. A commonly used layout provides for an inlet manifold of fuels to be fitted below the in-cell fuel conditioning and measuring systems. Selection of fuel can be achieved by way of a flexible line fitted with a self-sealing connector from the common system to the desired manifold mounted connector.
- It is a good practice to have a cumulative fuel meter in each line for general audit and for contract charging.
- Air entrainment and vapor locking can be a problem in test cell fuel systems so every care should be taken in avoiding the creation of air traps. An air-eliminating valve should be fitted at the highest point in the system, with an unrestricted vent to atmosphere external to the cell and at a height and in a position that prevents fuel escape.

- Disconnection of, or breaking into, fuel lines, with risk of consequent spillage or ingress of air, should, as far as possible, be avoided. It is sensible to mount all control components on a permanent wall-mounted panel or within a special casing, with switching by way of interlocked and *clearly marked valves*. The run of the final flexible fuel lines to the engine should not interfere with operator access. A common arrangement is to run the lines through an overhead boom. Self-sealing couplings should be used for engine and other frequently used connections.
- It is essential to fit oil traps to cell drain connections to avoid the possibility of discharging oil or fuel into the foul water drains.
- For normal automotive use aim for a flow rate during normal operation to be between 0.2 and a maximum recommended fuel line velocity: 1.0 m/s.
- Where self-sealing, quick connectors are used to switch fuels from supply lines to cell circuits, it is recommended, where possible, to use "fail-safe" connectors that physically prevent mixing circuits and the accidental poisoning of, for example, a gasoline circuit with diesel oil.

## Engine fluid fuel pressure control

There are three fuel pressure control problems that may need to be solved when designing or operating the fluid fuel supply in an engine test cell:

1. Pressure of fuel supply to the fuel conditioning system or consumption measurement instrument. Devices such as the AVL fuel balance require a maximum pressure of 0.8 bar at the instrument inlet; in the case of pumped systems this may require a pressure-reducing regulator to be fitted before the instrument.
2. Pressure of fuel supplied to the engine. Typically, systems fitted in normal automotive cells are adjustable to give pressures at the engine's own high-pressure pump system inlet of between 0.05 and 4 bar.
3. Pressure of the fuel being returned from the engine. When connected to a fuel conditioning and consumption system, the fuel return line in the test cell may create a greater back pressure than required. In the vehicle the return line pressure may be between 0 and 0.5 bar, whereas cell systems may require over 1.5 bar to force fuel through conditioning equipment and to overcome static head differences between the engine and instruments; therefore a pressure-reducing circuit, including a pump, may be required. Fig. 7.5 shows a circuit of a system that allows for the independent regulation of engine supply and return pressures. Care must be exercised in the design of such circuits to avoid the fuel pressure being taken below the point at which vapor bubbles form.

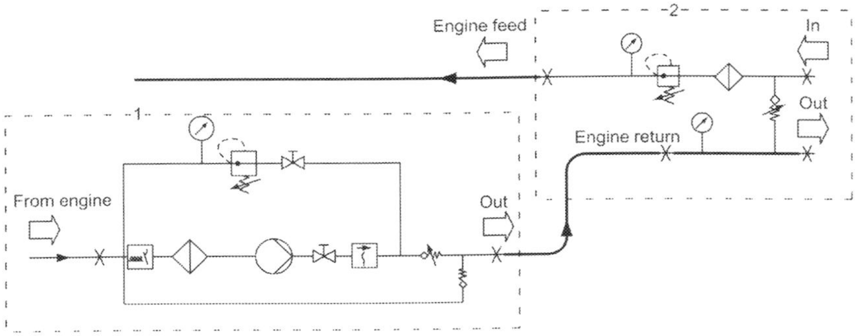


**FIGURE 7.5** Diagram of a well-proven circuit used in the regulation of fuel to and from an ICE under test. Within the dotted lines (1) is the fuel return pressure control and (2) is the fuel feed pressure control downstream of the consumption measurement station.

## Engine fluid fuel temperature control

All the materials used in the control of fuel temperature, particularly heat exchangers but also piping, and sealing materials, must be checked with the manufacturer as to their suitability for use with all specified fuels.

If fuel supplied to the engine fuel rail inlet has to be maintained at one standard temperature, as is usually the case in quality audit cells, then a relatively simple control system may use hot water circulated at controlled temperature through the water-to-fuel heat exchanger. Such an arrangement can give good temperature control for quite short and undemanding test sequences despite wide variations in fuel flow rate.

The control of the fluid fuel temperature within the engine circuit is complicated by the fuel rail and spill-back strategy adopted by the engine designer; because of this and variations in the engine rigging pipe work, no one circuit design can be recommended. Commercially available fuel conditioning devices can only specify the temperature control at the unit discharge: the system integrator has to ensure that the heat gains and losses within the connecting pipework do not invalidate experiments. It is particularly important to minimize the distance between the temperature-controlling element and the engine so that, if running is interrupted, the engine receives fuel at the control temperature with the minimum of delay.

Commercially available temperature control units typically have the following basic specifications:

- setting range of fuel temperature: 10°C–45°C;
- deviation between set value and actual value at instrument output: <1°C;
- power input: 0.4 or 2 kW (with heating); and
- cooling load: typically, 1.5 kW.

It should be noted that the stability of temperature control below ambient will be dependent on the stability of the chilled water supply, which should

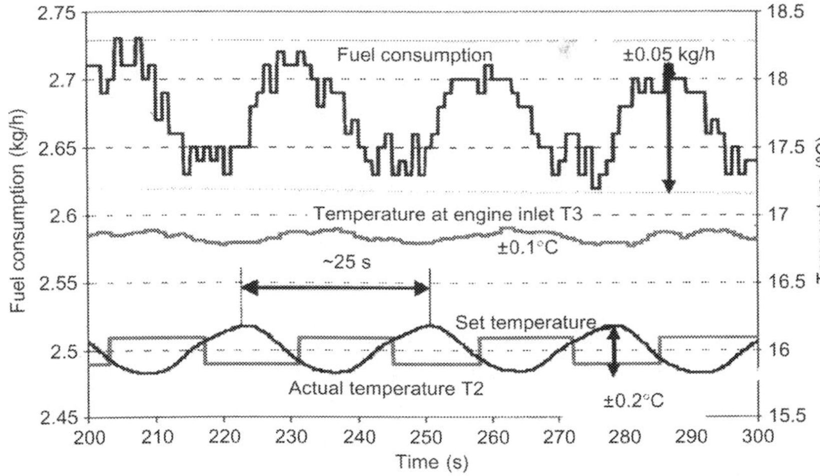


**FIGURE 7.6** Diagram showing the effects of variations in fuel temperature on consumption measurements. T2 (bottom trace) is the controlled outlet of the fuel conditioning unit superimposed on the step demand changes while T3 (middle trace) is the temperature at the engine inlet, the difference between the two is due to system damping.

have a control loop independent of other, larger, cooling loads in any shared system.

*A constant fuel temperature over the duration of an experiment is a prerequisite for accurate consumption measurement* (see Chapter 15: Measurement of Liquid Fuel, Oil, and Combustion Air Consumption).

As with all systems covered in this book, the complexity and cost of fuel temperature control will be directly proportional to the operation range and accuracy specified.

To confirm that temperature change across the fuel measurement circuit has a significant influence on fuel consumption and fuel consumption measurement, especially during low engine power periods, Fig. 7.6 should be considered. The graph lines show that making a cyclic step change in the set point of a fuel temperature controller of only $\pm 0.2°C$ within the measurement circuit produces an oscillation in fuel temperature at the engine inlet of $\pm 0.1°C$, and a variation in the measured fuel consumption value of $\pm 0.05$ kg/h (top line) is created. This equals a variation of $\pm 2\%$ of the actual fuel consumption value and shows that instrument manufacturers' claims of fuel consumption accuracy are only valid under conditions of absolutely stable fuel temperatures.

## Engine oil temperature control systems

Certain important layout requirements apply to in-cell oil cooling units. Unless the cell is operating a "dry-sump" lubrication system on the

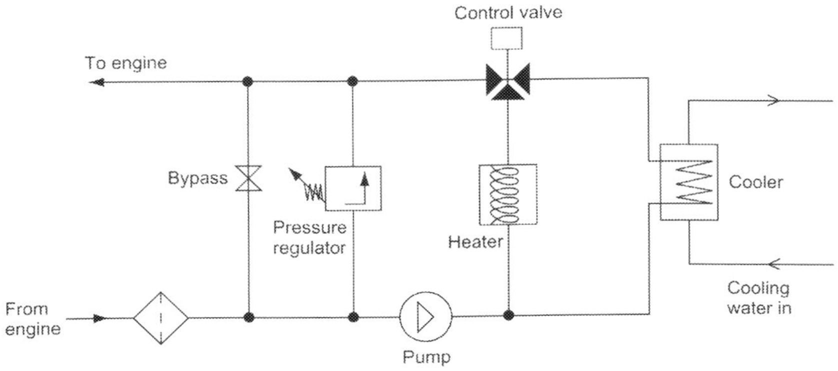


**FIGURE 7.7** Schematic diagram of a test cell oil temperature control circuit, with sensing points and control connections omitted for clarity.

engine, the entire lubrication oil temperature control circuit must lie below the normal engine oil level so that there is no risk of flooding the sump. This is particularly important when testing horizontal diesel engines where oil flooding can lead to an engine "running away." It may be necessary to provide heaters in the circuit for rapid warm-up of the engine; in this case the device must be designed or chosen to ensure that the skin temperature of the heating element cannot reach temperatures at which oil "cracking" can occur.

Fig. 7.7 shows schematically a separate lubricating oil cooling and conditioning unit. Where very accurate transient temperature control is necessary, the use of separate pallet-mounted cooling modules located close to the engine may be required; otherwise permanently located oil and cooling water modules offer the best solution for most engine testing. Once a common position for such modules was behind the dynamometer, where both units were fed from the external cooling water system and the engine connection hoses were run under the dynamometer to a connection point near the shaft end. The use of vehicle exhaust systems has made this layout less possible and transferred the modules in modern cells to one of the side walls of the engine pallet.

## Part 3, storage and handling of gaseous fuels

### Natural gas, liquefied natural gas, and compressed natural gas

Natural gas consists of about 90% methane, which has a boiling point at atmospheric pressure of −163°C. Engine test installations requiring natural gas (NG) for dual-fuel engines usually draw this from a mains supply at just above atmospheric pressure, so high-pressure storage arrangements are not necessary. In the United Kingdom the distribution of NG, to individual commercial users, is by way of a national grid and generally covered by the Gas (Third Party Access) Regulations 2004. Fire hazards are moderate when compared with LPG installations.

Engine test cells using liquefied natural gas (LNG) and NG have to comply with gas industry regulations, which may include explosion relief panels in the cell and that part of the engine exhaust system within the building. *Early contact and negotiation with the LA and supply company is vital.*

## Liquefied petroleum gas storage

LPG is often referred to as "autogas" or auto propane. In engine test facilities it will be stored under approximately 7 bar (100 psi) pressure as a liquid in an external bulk storage tank. LPG is known as one of the most "searching" of gases—that is, it can escape through gaps that would hold water and other gases. It has a density of 1.5 that of air and concentrations of 2%−10% LPG in air make a flammable mix. Test cells using LPG must be fitted with suitable hydrocarbon detectors at the lowest cell level and the ATEX precautions for the avoidance of explosive atmospheres, discussed elsewhere in this book, have to be taken. It should be clear that any LPG installation has to employ specialist contractors during the planning, installation, and maintenance stages of its life. In the United Kingdom the HSE produces a comprehensive set of guidelines and described the legislative framework under which LPG is stored and handled [3].

## Hydrogen storage

Hydrogen storage is the key technology that will enable hydrogen and fuel cell technologies in automotive powertrains to advance. While hydrogen has the highest energy per mass of any fuel, its low density at ambient temperature means it has a low energy per unit of volume.

Methods of hydrogen storage are physical containment where it is stored as:

- A compressed gas, requiring high-pressure tanks typically at 350−700 bar.
- As liquid in "dewars" or tanks stored at −253°C typically at <5 bar pressure.
- As a cold or "cryo-compressed" gas, which is a technology being developed for in-vehicle storage and like liquid storage it stores cold hydrogen at around 20K (−253°C) but in vessels able to hold "boil-off" pressures of around 350 bar.

Hydrogen is also stored within or upon solid materials (absorption or adsorption), those most investigated for vehicle applications are as follows:

- Absorbents such as MOF-5.
- Chemical hydrogen stored within compounds in either solid or liquid form as "covalently bound hydrogen."
- Interstitial storage within metal hydrides ($MH_x$).

*Bulk delivery of hydrogen* to test facilities is usually as compressed hydrogen in specially constructed multitube trailers. It can be decanted into storage on-site or, more usually and for volume users such as fuel cell test facilities, the trailer is to be left on-site and the test facility can draw directly from it.

High volume storage of hydrogen is possible using special cryogenic vessels known as "dewars," which may be rented from and maintained by the gas supplier or owned by the user. For test facilities renting the bulk hydrogen storage vessel and its support system has significant advantages particularly if the future usage is unclear and because expert advice made readily available.

### Safety considerations when working with hydrogen

The National Fire Protection Association's NFPA 704 hazard rating of hydrogen is the maximum figure of 4 on the flammability scale because it is flammable when mixed even in small amounts with air and ignition can occur at a volumetric ratio of hydrogen to air as low as 4%.

Materials used in the handling of hydrogen gas, including seals and sealants, must be carefully selected to account for their mechanical behavior in hydrogen service at the intended operating conditions.

As with all fuel lines, hydrogen piping systems must be carefully designed, supported, and installed to minimize the potential for leaks and allow for their easy detection. Since hydrogen is a lighter than air gas its detectors must be mounted at the top of enclosed building spaces and unvented roof spaces that could collect gas, particularly near lamp fittings, must be avoided.

Sources of safety guidelines concerning safety of gas cylinders and hydrogen are referenced at the end of this chapter [4,5].

## Appendix 1: Notes of the storage and properties of gasoline

***Note:*** *The EU's fuel quality improvement initiatives, up to the current Directive 2009/30/EC, have set the time line for incremental sulfur reduction and have resulted in region-wide supply of both gasoline and diesel fuel (highway and nonroad) with near-zero sulfur content.*

ICE development work is often concerned with "chasing" very small improvements in fuel consumption; these differences can easily be swamped by variations in the calorific value of the fuels. Similarly, comparisons of identical engines manufactured at different sites will be invalid if they are tested on different batches of the "same" retail fuels or fuels that have been allowed to deteriorate at differential rates.

The calorific value of automotive gasoline can be taken, for energy flow calculation in the test cell and *not* in critical test calculations, as 43.7 MJ/kg, but as sold at the retail pump in any country of the world, gasoline is variable in many details. Some variabilities are appropriate to the geographic location and season, some of which are due to the original feedstock from which the fuel was refined. The storage of gasoline in a partially filled container, in conditions where it is exposed to direct sunlight, will cause it to degrade rapidly. Reference fuels of all types should be stored in dark, cool conditions and in containers that are 90%–95% full; their shelf life should not be assumed to be more than a few months.

Octane number is the single most important gasoline specification, since it governs the onset of detonation or "knock" in the engine. Knock limits the power, compression ratio, and hence the fuel economy of an engine. Too low an octane number also causes run-on when the engine is switched off.

Three versions of the octane number are used: research octane number (RON), motor octane number (MON), and front-end octane number (R100). RON is determined in a specially designed (American) research engine: the Cooperative Fuel Research (CFR) engine. In this engine the knock susceptibility of the fuel is graded by matching the performance with a mixture of reference fuels, isooctane with an RON = 100 and *n*-heptane with an RON = 0.

The RON test conditions are now rather mild where modern engines are concerned, and the MON test not only imposes more severe conditions but also uses the CFR engine. The difference between the RON and the MON for a given fuel is a measure of its "sensitivity." This can range from about 2 to 12, depending on the nature of the crude and the distillation process. In the United Kingdom it is usual to specify RON, while in the United States the average of the RON and the (lower) MON is preferred. R100 is the RON determined for the fuel fraction boiling at below 100°C. Volatility is the next most important property and is a compromise. Low volatility leads to low evaporative losses, better hot start, less vapor lock, and in older, preinjection, engine designs, and less carburetor icing. High volatility leads to better cold starting, faster warm-up, and hence better short-trip economy, also to smoother acceleration.

## Appendix 2: Notes on the storage and properties of diesel fuels

The optimum condition for diesel storage within a bulk tank is below 20°C and above 0°C, out of direct sunlight, especially for polyethylene tanks, where surface cracking can be caused that will allow water to enter. In such conditions it is generally accepted that diesel fuel can be stored for up to 1 year without significant degradation or the need for additives. Longer than 12 months cool storage may require the fuel oil to be treated with stabilizers and biocides.

The calorific value of automotive diesel is generally taken, for energy flow calculation in the test cell, as 43,000 kJ/kg, but just as the range of size of the diesel engine is much greater than that of the spark-ignition engine, from 1–2 to 50,000 kW, the range of fuel quality is correspondingly great.

*Cetane number* is the most important diesel fuel specification. It is an indication of the extent of ignition delay: the higher the cetane number, the shorter the ignition delay, the smoother the combustion, and the cleaner the exhaust. Cold starting is also easier the higher the cetane number.

### Waxing and winter diesel

All diesel fuel contains wax because of its high cetane value. Normally the wax is a liquid in the fuel; however, when diesel fuel gets cold, around −8° C, enough the wax starts to solidify and blocks filters. In many cold climates diesel fuel is blended with kerosene or other special additives that tend to prevent the wax flakes from "clumping" together. *Viscosity* covers a wide range. BS2869 specifies two grades of vehicle engine fuels, Class Al and Class A2, having viscosities in the range 1.5–5.0 and 1.5–5.5 cSt, respectively, at 40°C. BS MA100 deals with fuels for marine engines. It specifies nine grades, Class M1, equivalent to Class A1, with increasing viscosities up to Class M9, which has a viscosity of 130 cSt maximum at 80°C.

## Appendix 3: Notes on fuel dyes and coloring

Fuel oil and gasoline are often colored by the addition of dyes at the blending stage of production. Fuels are artificially colored to enable recognition by the taxation authorities, or by the user for safety reasons to minimize the risk of using the wrong fuel type.

"Red Diesel" is sold to agricultural operators in the United Kingdom at a tax rate that is lower than that applied to the same fuel sold to the general public. Fuel "laundering" whereby such fuel is decolored and resold is highly illegal; for this reason alone, test facilities should only buy bulk fuel from fully certified suppliers. Reference fuels of the types used in critical engine

**TABLE 7.1** A list of the colored dyes used to identify fuels for the purposes of safety and prevention of tax avoidance.

| Country or group | Fuel | Dye color |
|---|---|---|
| European Union | All automotive diesel | Solvent, Yellow 124 |
| United Kingdom | Agricultural gas oil, low tax | Quinizarin, Red |
| United States | High sulfur, low tax diesel | Solvent, Red 164 |
| France | Marine diesel, low tax | Solvent, Blue 35 |
| Australia | Regular unleaded gasoline<br>Premium unleaded gasoline | Purple<br>Yellow |
| Worldwide | Aviation gasoline 100LL | Blue |
| Worldwide | Aviation gasoline 100/130 | Green |
| Worldwide | Aviation gasoline 80/87 | Red |

testing are not dyed. Table 7.1 shows examples of some national and international standards of fuel dyeing.

## References

[1] J.R. Hughes, Storage and Handling of Petroleum Liquids: Practice and Law, third revised, Hodder Arnold, 1988. ISBN-13: 978-0852642818.

[2] Winterthur Gas & Diesel AG, 'Diesel Engine Fuels' Issue 002 2018-12, Winterthur Gas & Diesel AG, Winterthur, Switzerland.

[3] <http://www.hse.gov.uk/gas/lpg/index.htm>.

[4] 'The storage of gas Cylinders', Code of Practice 44, 2016 <http://www.bcga.co.uk/assets/publications/CP44.pdf>

[5] D.K. Pritchard, M. Royle, D. Willoughby, Health and Safety Executive 2009, 'Installation permitting guidance for hydrogen and fuel cell stationary applications': UK version.

Chapter 8

# Vibration and noise

## Chapter Outline



## Introduction

The development of hybrid electric vehicles and most significantly battery electric vehicles (BEVs) has altered the work details of automotive powertrain and prime-mover testing. In addition to the vibrations and noise produced and induced by an internal combustion engine (ICE), we now have to deal with the higher frequencies of electrical drivetrains, motors, and electrical power controllers. The industry has also to consider the novel problem of a BEV being too quiet at speeds below around 20 kph (12 mph) and thus being a danger to pedestrians; there is much current (2020) research being done to define the volume and frequency of suitable warning tones. Since the electric motors used in automotive traction applications possess inherently better dynamic balance than a reciprocating piston engine, it may assumed that vibration isolation, at least at low frequencies, will be less of a problem and that research will shift to dealing with the isolation and attenuation of noise and vibration from rotors and transmission elements capable of rotating at 25,000 rpm and upward.

## Part 1: Vibration

Usually it is, or should be, the prime mover, an ICE or motor, that is the only significant source of vibration and noise in the component or powertrain

**Engine Testing. DOI: https://doi.org/10.1016/B978-0-12-821226-4.00008-5**

test cell. Secondary sources of noise and vibration within the test cell such as the ventilation system, pumps, and fluid circulation systems or the dynamometer are usually swamped by the effects of the UUT (unit under test)/ dynamometer drivetrain.

Only in the case of full powertrain test rigs may the UUT be mounted in the way that it will be in the vehicle; the "artificial" rigging arrangement of the test cell is chosen to allow specific performance measurements to be made and will usually suppress inherent vibration and therefore noise. There are several aspects to the problem of ICE, motor or transmission vibration and noise within the test environment:

- The UUT must be mounted in such a way that neither it nor connections to it can be damaged either by excessive movement or, in some cases, excessive constraint.
- Transmission of UUT-induced vibration to and via services connected to the cell structure or to adjoining buildings must be minimized.
- Noise levels in the cell should be contained as far as possible and the design of alarm systems should take into account ambient noise levels in all related work areas.

## Fundamentals: sources of vibration in the internal combustion engine

Since the majority of ICE likely to be met with are multicylinder, in-line vertical designs, we shall begin by concentrating on this configuration.

Any UUT may be regarded as having six degrees of freedom of vibration about orthogonal axes through its center of gravity: linear vibrations along each axis and rotations about each axis (Fig. 8.1). In practice, only three of these modes are usually of importance in engine testing:

- vertical oscillations on the $X$ axis due to unbalanced vertical forces,
- rotation about the $Y$ axis due to cyclic variations in torque, and
- rotation about the $Z$ axis due to unbalanced vertical forces in different transverse planes.

In general, the rotating masses are carefully balanced but periodic forces due to the reciprocating masses cannot be avoided. The crank, connecting rod, and piston assembly shown in Fig. 8.2 are subject to a periodic force in the line of action of the piston, given approximately by:

$$f = m_p \omega_c^2 r \cos\theta + \frac{m_p \omega_c^2 r \cos 2\theta}{n} \tag{8.1}$$

where $n = l/r$ in Fig. 8.2. Here $m_p$ represents the sum of the mass of the piston plus, by convention, one-third of the mass of the connecting rod (the remaining two-thirds is usually regarded as being concentrated at the crankpin center). The first term of Eq. (8.1) represents the first-order inertia force. It is equivalent to the component of centrifugal force on the line

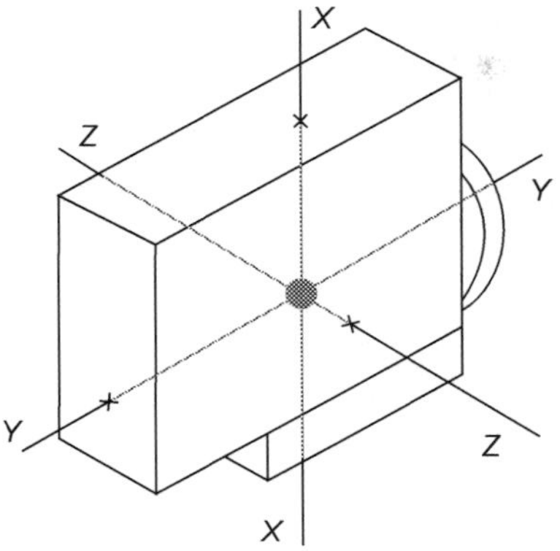


**FIGURE 8.1** Showing the principle axes and degrees of freedom of an ICE. *ICE*, Internal combustion engine.

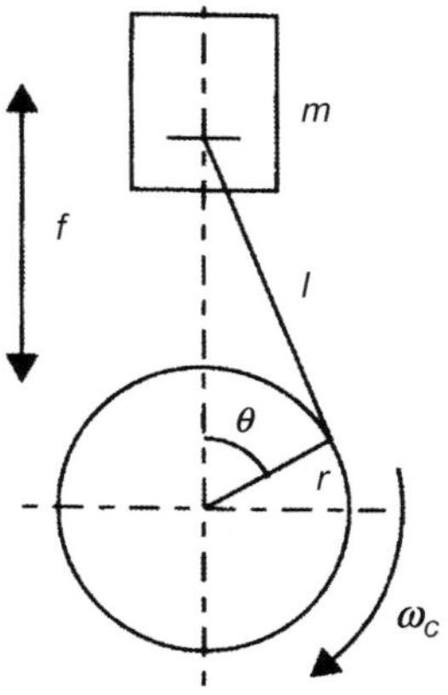


**FIGURE 8.2** Diagrammatic representation of the unbalanced forces within the connecting rod and crank mechanism.

of action generated by a mass $m_p$ concentrated at the crankpin and rotating at engine speed. The second term arises from the obliquity of the connecting rod and is equivalent to the component of force in the line of action generated by a mass $m/4n$ at the crankpin radius but rotating at twice engine speed.

Inertia forces of higher order (3 ×, 4 ×, etc., crankshaft speed) are also generated but may usually be ignored.

It is possible to balance any desired proportion of the first-order inertia force by balance weights on the crankshaft, but these then lead to an equivalent reciprocating force on the $Z$ axis, which may be even more objectionable.

Inertia forces may be represented by vectors rotating at crankshaft speed and twice crankshaft speed. Table 8.1 shows the first- and second-order vectors for engines having from one to six cylinders. Note the following features:

- In a single-cylinder engine, both first- and second-order forces are unbalanced.
- For larger numbers of cylinders, first-order forces are balanced.
- For two- and four-cylinder engines, the second-order forces are unbalanced and additive.

This last feature is an undesirable characteristic of a four-cylinder engine and in some cases has been eliminated by counterrotating weights driven at twice crankshaft speed.

The other consequence of reciprocating unbalance is the generation of rocking couples about the transverse or $Z$ axis and these are also shown in Fig. 8.1:

- There are no couples in a single-cylinder engine.
- In a two-cylinder engine, there is a first-order couple.
- In a three-cylinder engine, there are first- and second-order couples.
- Four- and six-cylinder engines are fully balanced.
- In a five-cylinder engine, there is a small first-order and a larger second-order couple.

**TABLE 8.1** Diagrams showing the first- and second-order forces in internal combustion engine of one to six cylinders.

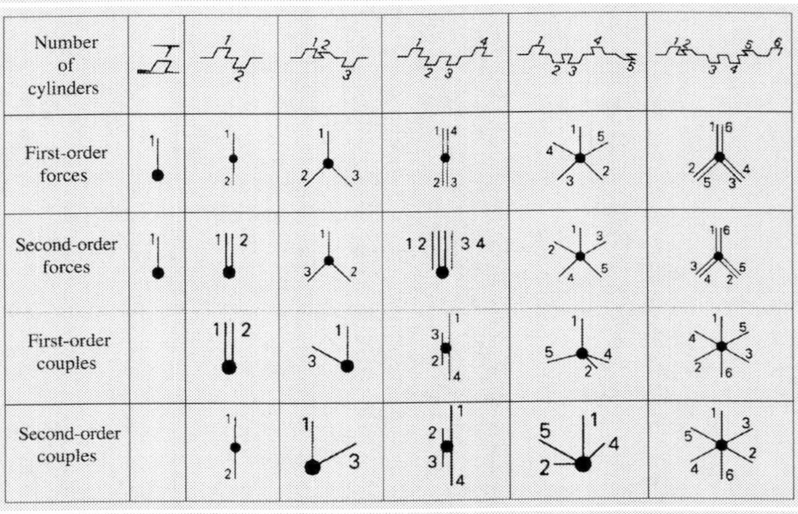

Six-cylinder engines, which are well known for smooth running, are balanced in all modes. The recent popularity of modern turbocharged, intercooled, and gasoline direct injection (GDI) fueled, three-cylinder engines, which have balanced vertical first-order forces, is in part due to the ease with which they can be packaged in the engine bay both in the mild hybrid or range extender forms.

Variations in engine turning moment are discussed in Chapter 11, Mounting and Rigging Internal Combustion Engines for Test, in conjunction with discussion on shaft selection and engine mounting. These variations of turning moment lead to equal and opposite reactions on the engine, which tend to cause rotation of the whole engine about the crankshaft axis. The order of these disturbances, or the ratio of the frequency of the disturbance to the engine speed, is a function of the engine cycle and the number of cylinders. For a four-stroke engine the lowest order is equal to half the number of cylinders: in a single cylinder there is a disturbing couple at half engine speed while in a six-cylinder engine the lowest disturbing frequency is at three times engine speed. In a two-stroke engine the lowest order is equal to the number of cylinders.

## Design of engine mountings and test-bed foundations in vibration transmission

The main problem in ICE mounting design, both in the test cell and vehicle is that of ensuring that the motions of the engine and the forces transmitted to the surroundings are kept to manageable levels. In the case of mounting an ICE in the test bed, it is sometimes the practice to make use of the same flexible mounts and the same location points as in the vehicle. This practice does not however guarantee a satisfactory solution since in the vehicle the mountings are carried on a comparatively light structure, while in the test cell they may be attached to a comparatively massive pallet and/or to a seismic block. Also, the output shaft will be different and the engine may be fitted with additional equipment and various nonvehicular service connections. All of these factors alter the dynamics of the system when compared with the situation of the engine in service and can cause fatigue failures of both the engine support brackets and those of auxiliary devices such as the alternator.

Truck diesel engines usually present much less of a problem than small automotive diesel engines, as they generally have fairly massive flywheel and are provided with well-spaced supports at the flywheel end. Stationary, nonautomotive engines will in most cases be carried on four or more flexible mountings in a single plane at or near the horizontal centerline of the crankshaft.

## Practical considerations in the design of prime-mover and test-bed mountings

Electric traction motors in their bare-shaft form, before packaging within the e-axle assembly like that shown in Fig. 3.1, which are designed for mounting within the sprung frame of the vehicle, are presented for test rigidly mounted on a very stiff steel bracket. The e-axle assemblies due to their complex casings will need to use the vehicular fixing points.

Test procedures of "bare-shaft" electric motors can use a setup that is the equivalent of the classic ICE-dynamometer arrangement, that is, the UUT coupled directly to the dynamometer with a drive shaft. Due to high rotor speeds, such rigs may be fitted with speed reducing gearboxes with torque hubs on the inlet (motor) shaft coupling hub. Since thermal management of the motor is a frequent part of the test and calibration process, the motor can be mounted onto a very substantial plate that forms the end wall of a climatic chamber, which can be moved forward to enclose the UUT (see Fig. 5.10). The mounting wall will be fitted with sealable penetrations for power, control and data cables. In its initial, precalibrated, status the combined motor and power source systems are capable of producing very high peak torques and shaft accelerations for which the rigging and dynamometer control has to be prepared.

Detailed design of mountings for the test-bed installation of an ICE and some forms of motors such as the in-wheel or hub type is a highly specialized matter. In general, the aim is to avoid "coupled" vibrations, for example, the generation of pitching forces due to unbalanced forces in the vertical direction, or the generation of rolling moments due to the torque reaction forces exerted by the UUT. These can create resonances at much higher frequencies than the simple frequency of vertical oscillation and lead to consequent trouble, particularly with the UUT-to-dynamometer connecting shaft.

## Massive foundations and air-sprung bedplates

The analysis and prediction of the extent of transmitted vibration to the surroundings is a highly specialized field. The starting point is the observation that a heavy block embedded in the earth has a natural frequency of vibration that generally lies within the range 1000–2000 c.p.m. There is thus a possibility of vibration being transmitted to the surroundings if exciting forces, generally associated with the reciprocating masses in an ICE, lie within this frequency range. An example would be a four-cylinder four-stroke engine running at 750 rev/min, we see from Table 8.1 that such an engine generates substantial second-order forces at twice engine speed, or 1500 c.p.m. Fig. 8.3 gives an indication of acceptable levels of transmitted vibration from the point of view of physical comfort.

Fig. 8.4 is a sketch of a typical small precast seismic block containing cast-in support beams and UUT mounting rails. Reinforced concrete weighs roughly 2500 kg/m$^3$ and this block would weigh about 4500 kg. Note that the surrounding tread plates must be isolated from the block, also that it is essential to electrically earth (ground) the mounting rails. The block is shown carried on four combined steel spring and rubber isolators, each

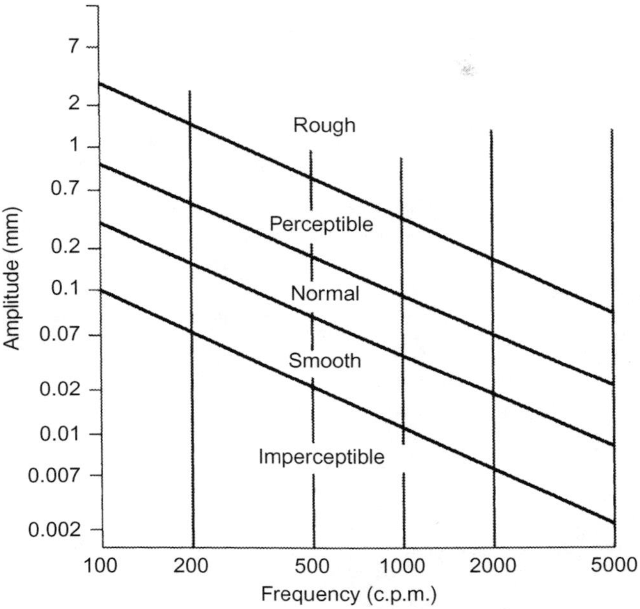


**FIGURE 8.3** The classic diagram illustrating the human perception of vibration. *Redrawn from Ker-Wilson (1959)* [1].

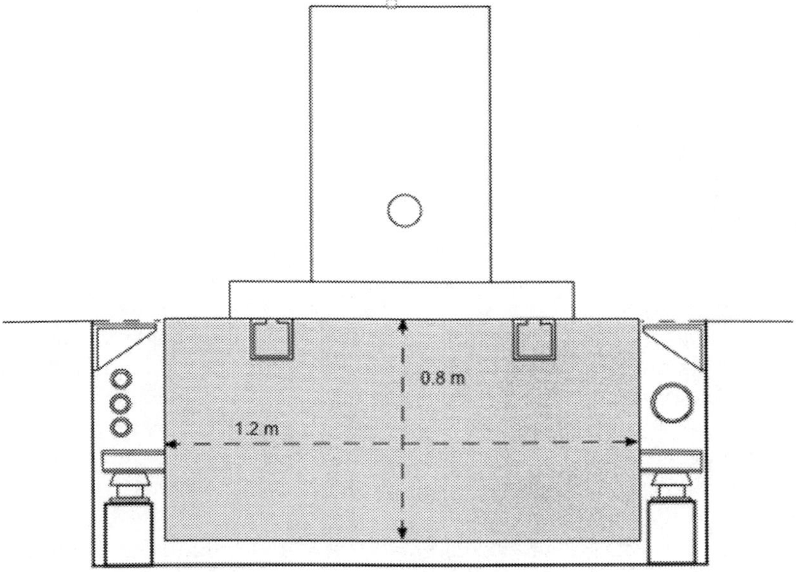


**FIGURE 8.4** A sketch from an anechoic cell of a spring-mounted seismic block with steel box section or cast tee-slotted rails cast in the top surface. Service pipes can run down each side. Pit must have a drainage outlet or pump sump and must be purged of heavier than air gasses via the ventialtion system.

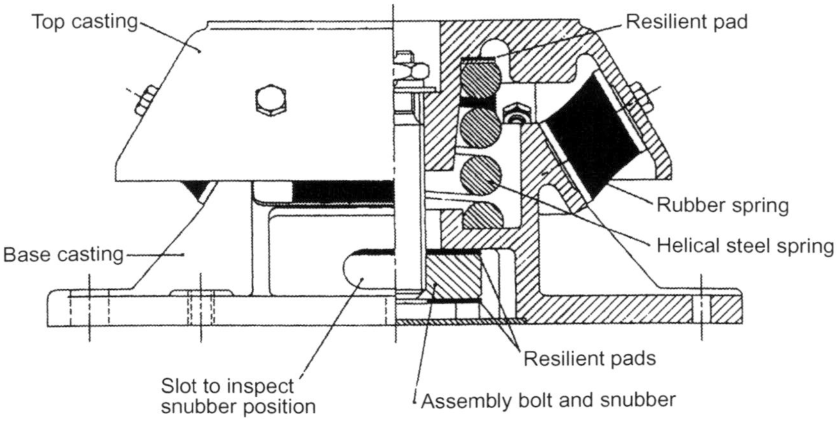


**FIGURE 8.5** Combined spring and rubber flexible machine mount.

having a stiffness of 100 kg/mm. The static deflection under the force of gravity = $mg/k$, which gives a very convenient expression for the natural frequency of vibration, shown in the following equation:

$$\eta_0 = \frac{1}{2\pi}\sqrt{\frac{g}{\text{static deflection}}} \tag{8.2}$$

From the equation the natural frequency of vertical oscillation of the bare block would be 4.70 Hz, or 282 c.p.m., so the block would be a suitable base for an ICE running at about 900 rev/min or faster. If the UUT weight were, say, 500 kg the natural frequency of block + engine would be reduced to 4.46 Hz, a negligible change. An ideal design target for the natural frequency is considered to be 3 Hz.

It is now common practice for ICE in particular to be rigged on vehicle-type engine mounts, within a trolley system, which are themselves mounted on isolation feet of the general type shown in Fig. 8.5; therefore less engine vibration is transmitted to the building floor. In these cases, an alternative to the deep seismic block is shown in Fig. 8.6A and is used where the soil conditions are suitable. Here the test bed sits on a thickened and isolated section of the floor cast in situ on the compacted native ground. The gap between the floor and block is almost filled with expanded polystyrene boards and sealed at floor level with a flexible, fuel-resistant, sealant; a dampproof membrane should be inserted under both floor and pit or block.

Where the subsoil is not suitable for the arrangement shown in Fig. 8.6A, then a pit is required, cast to support a concrete block that sits on a mat or pads of a material such as cork/nitrile rubber composite, which is resistant to fluid contamination. The latter design, shown in

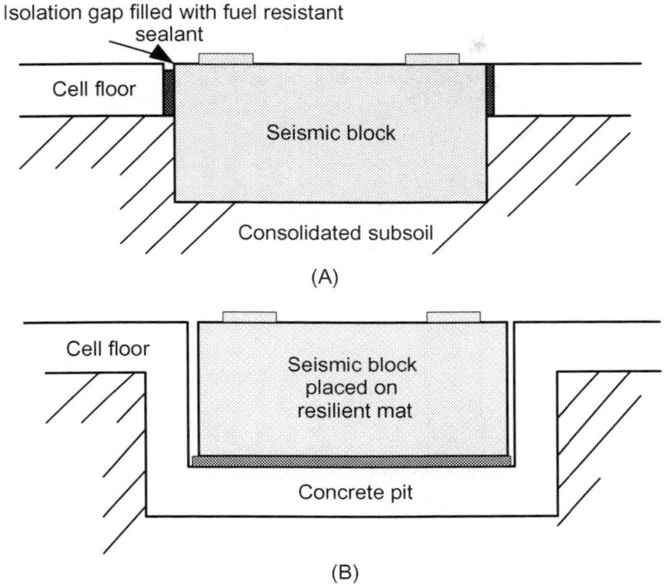


**FIGURE 8.6** (A) Isolated foundation block set into firm compacted subsoil. (B) Seismic block cast onto resilient matting in shallow concrete pit.

Fig. 8.6B while still used in some many sites, has the problem of fluids and foreign objects dropping into the narrow annular space between block and pit unless provision is made to avoid it.

Heavy concrete seismic blocks carried on a flexible membrane are expensive to construct, calling for deep excavations, complex shuttering, and elaborate arrangements, such as tee-slotted bases, for bolting down the UUT. With the wide range of different types of flexible mounting now available, it is questionable whether, except in special circumstances, such as a requirement to install test facilities in proximity to offices or in anechoic cells, their use is economically justified.

## Air-sprung bedplates

An alternative to the cast concrete seismic block is a large cast-iron bedplate supported by a self-leveling system using air springs of the type shown in Fig. 8.7.

These large multi–tee-slotted bedplates allow for the multiple of different configurations of electrical and hybrid powertrains to be laid out as part shown in Fig. 12.1; they are becoming the standard basis of modern powertrain cell design.

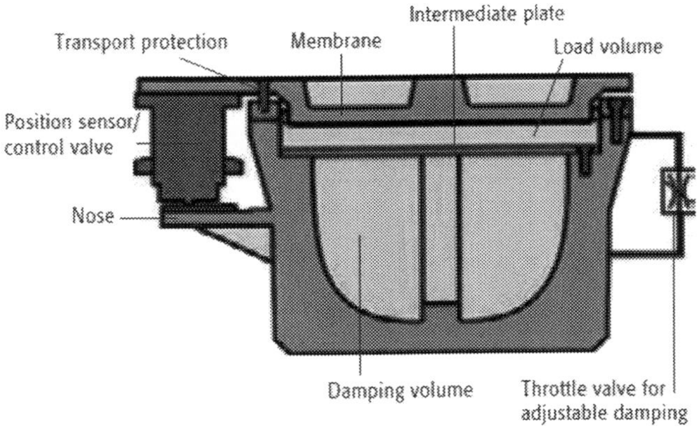


**FIGURE 8.7** A sectional view of a typical air-spring unit for supporting test system bedplates. *Courtesy: BILZ AG.*

By using air springs with automatic level control, a constant, load-independent, horizontal position of the isolated base plate can be maintained. The pneumatic spring system and high intrinsic weight of the plates give such systems an extremely low natural frequency, irrespective of the load.

The precision of surface-level maintenance such systems are capable of is unnecessary for most automotive engine test cells, but the self-leveling action is valuable when UUTs of different weights are changed on the bedplate. These systems can be tuned from "soft" where the single bedplate may be rocked by operators moving on it, to almost rigid.

Air-sprung bedplate support systems that use four or more units, such as the example shown in Fig. 8.8, connected within a dedicated compressed air circuit are now the default design in large powertrain cells. When the air supply is switched off the block or plate will settle down and rest on packers; this allows removal for maintenance. The air springs require a reliable, low-flow, condensate-free air supply that is not always easy to provide to the lowest point in the cell system. Sufficient room must be left around or within the bedplate to allow maintenance access to the air-spring units and access to hatch plates for any centrally located. The first setup of a multisection bed can take some time, particularly if the quality of the base floor is of low-quality finish and uneven.

In cases where there is no advantage in having the bedplate face at ground level (as is required for wheeled trolley systems), the bedplate can be mounted on the flat cell floor.

It is possible that plant and engines mounted on rubber viscous mounts or air-spring systems could, unintentionally, become electrically isolated from the remainder of the facility because of the rubber elements; it is vital that a common electrical grounding scheme is included

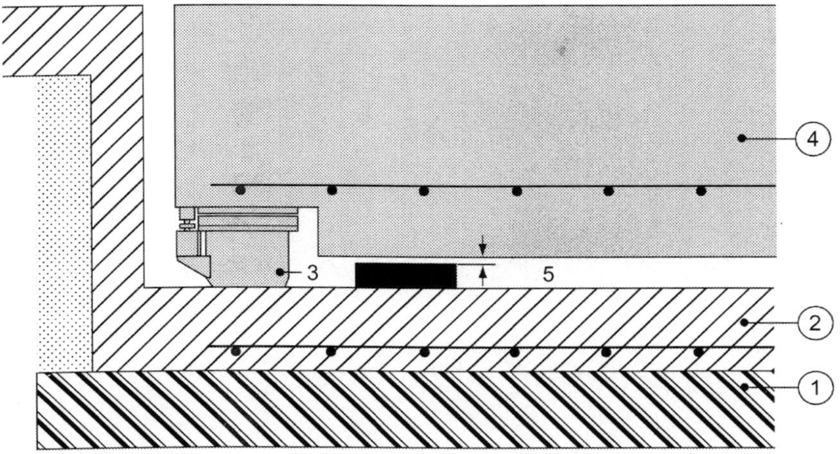


**FIGURE 8.8** Seismic blocks (4), or cast-iron bedplates (not shown) mounted on air-spring units (3) within a shallow concrete pit (2) on consolidated subsoil. (1) When the air supply is switched off the self-leveling spring units allow the block to settle on support pads; (5) the rise and fall is typically 4–6 mm. Maintenance access to the spring not shown.

in such facility designs (see Chapter 4: Electrical Design Requirements of Test Facilities).

A special application concerns the use of seismic blocks for supporting engines in anechoic cells. It is, in principle, good practice to mount engines undergoing noise testing as rigidly as possible, since this reduces noise radiated as the result of movement of the whole engine on its mountings. Lowering of the center of gravity is similarly helpful, since the engines have to be mounted with a shaft center height of at least 1 m above 'ground' level to allow for microphone placement.

Except in anechoic cells, it is common practice to mount both engine and dynamometer on a common block; if they are separated the relative movement between the two must be within the capacity of the connecting shaft and its guard.

Finally, it should be remarked that there is available on the market a bewildering array of different designs of vibration isolator or flexible mounting, based on steel springs, air springs, natural or synthetic rubber of widely differing properties used in compression or shear, and combinations of these materials. Many of the industrial suppliers provide a design and supply service and certainly, for the nonspecialist, the manufacturer's advice should be sought, with the specific test facility requirement clearly specified.

The foregoing text should be read in conjunction with Chapter 9, Dynamometers: The Measurement and Control of Torque, Speed, and Power,

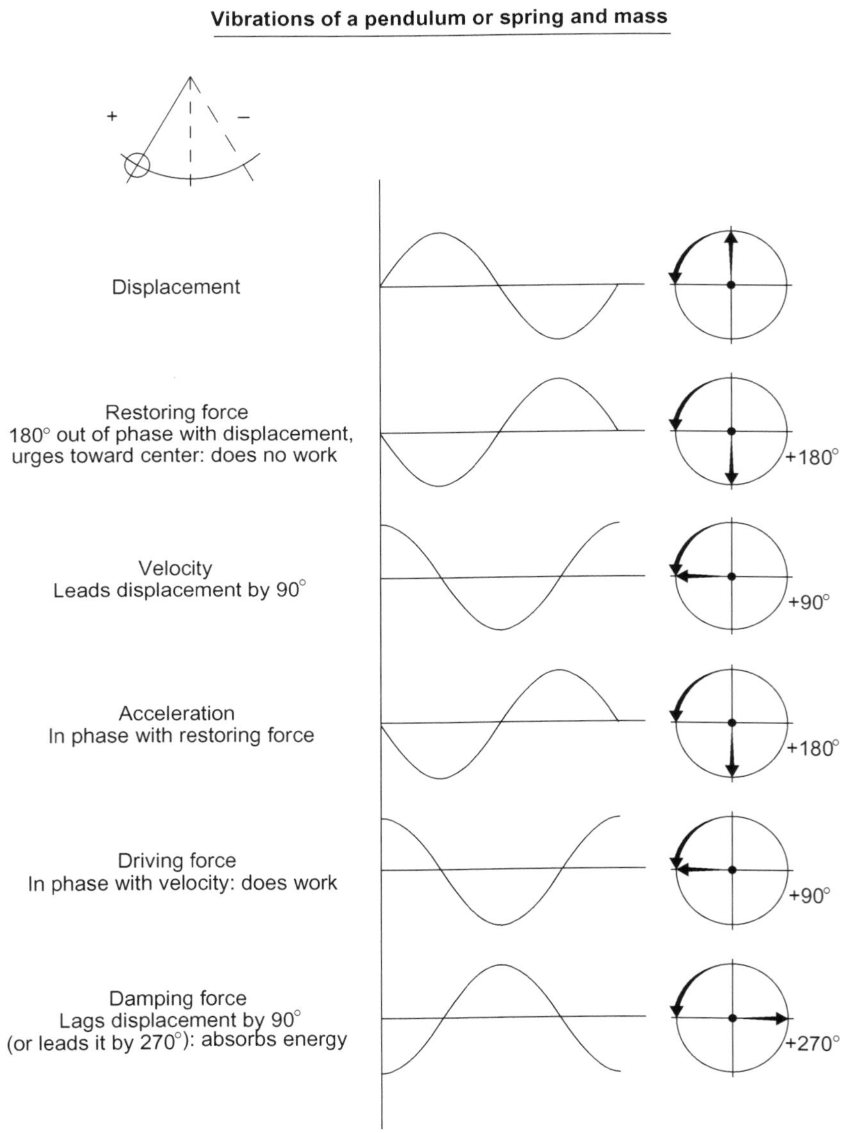


**FIGURE 8.9** Diagram of phase relationship of vibrations.

which deals with the associated problem of torsional vibrations of engine and dynamometer. The two aspects—torsional vibration and other vibrations of the engine on its mountings—cannot be considered in complete isolation. A diagrammatic summary of the vibrations of a sprung mass that may be found useful is given in Fig. 8.9.

## Part 2: Noise and its attenuation in test facilities

### Sound intensity

The starting point in the definition of the various quantitative aspects of noise measurement is the concept of sound intensity, defined as:

$$I = \frac{p^2}{(pc)} \mathrm{W/m^2}$$

where $p^2$ is the mean square value of the pressure variation due to the sound wave, that is the acoustic pressure, $p$ is the density of air and $c$ the velocity of sound in air.

Sound intensity is measured in a scale of decibels (dB):

$$\mathrm{dB} = 10 \log_{10} \frac{I}{I_0} = 20 \log_{10} \frac{\overline{p}}{p_0}$$

where $I_0$ corresponds to the average lower threshold of audibility, taken by convention as:

$$I_0 = 10^{-12}\ \mathrm{W/m^2}$$

From these definitions, it is easily shown that a doubling of the sound intensity corresponds to an increase of about 3 dB ($2 \log_{10} = 0.301$). A 10-fold increase gives an increase of 10 dB, while an increase of 30 dB corresponds to a factor of 1000 in sound intensity. It will be apparent to the reader that intensity varies through an enormous range. The value on the decibel scale is often referred to as the sound pressure level (SPL).

*In general, sound is propagated spherically from its source and the inverse square law applies.*

Doubling the distance results in a reduction in SPL of about 6 dB ($4 \log_{10} = 0.602$).

The human ear is sensitive to frequencies in the range from roughly 16 Hz to 20 kHz, but the perceived level of a sound depends heavily on its frequency structure. The well-known Fletcher–Munson curves (Fig. 8.10) were obtained by averaging the performance of a large number of subjects who were asked to decide when the apparent loudness of a pure tone was the same as that of a reference tone of frequency 1 kHz.

Loudness is measured in a scale of *phons*, which is only identical with the decibel scale at the reference frequency. The decline in the sensitivity of the ear is greatest at low frequencies. Thus, at 50 Hz, an SPL of nearly 60 dB is needed to create a sensation of loudness of 30 phon.

Acoustic data are usually specified in frequency bands one octave wide. The standard mid-band frequencies are:

31.5 62.5 125 250 500 1000 2000 4000 8000 16,000 Hz

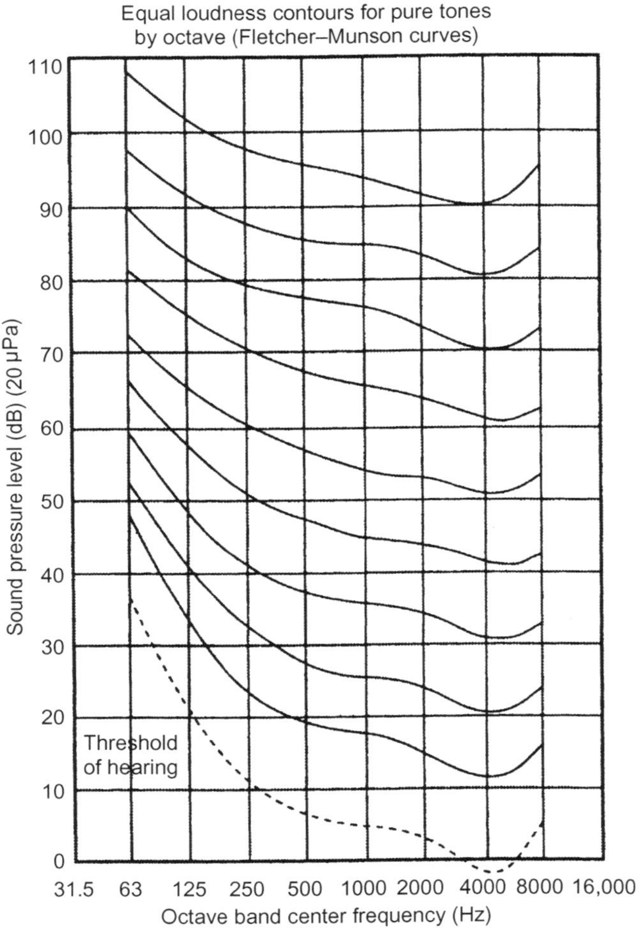


**FIGURE 8.10** Fletcher—Munson curves of perceived equal loudness.

for example, the second octave spans 44–88 Hz. The two outer octaves are rarely used in noise analysis.

## Noise measurements

Most instruments for measuring sound contain weighted networks that give a response to frequency, which approximates to the Fletcher-Munson curves. For most applications the A-weighting curve (Fig. 8.11) gives satisfactory results and the corresponding SPL readings are given in dBA. While B- and C-weightings are sometimes used for high sound levels, and a special D-tweighting is used primarily for aircraft noise measurements.

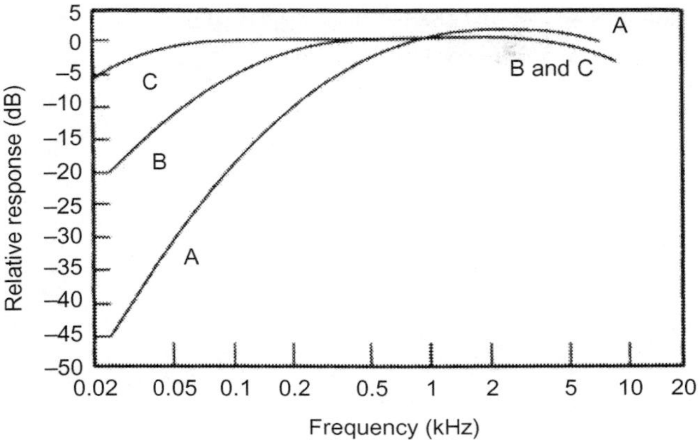


**FIGURE 8.11** Noise-weighing curves A, B, and C.

The dBA value gives a general "feel" for the intensity and discomfort level of a noise, but for analytical work the unweighted results should be used. The simplest type of sound-level meter for diagnostic work is the octave band analyzer. This instrument can provide flat or A-weighted indications of SPL for each octave in the standard range. For more detailed study of noise emissions, an instrument capable of analysis in one-third octave bands is more effective. With such an instrument it may, for example, be possible to isolate the tooth contact frequency of a particular pair of gears and thus pinpoint them as a noise source.

For serious development work on engines, transmissions, or vehicle bodies, much more detailed analysis of noise emissions is provided by the discrete Fourier transform or fast Fourier transform digital spectrum analyzer. The mathematics on which the operation of these instruments is based is somewhat complex, but fortunately they may be used effectively without a detailed understanding of the theory involved. It is well known that any periodic function, such as the cyclic variation of torque in an ICE, may be resolved into a fundamental frequency and a series of harmonics. General noise from an engine, motor unit, or transmission does not repeat in this way and it is accordingly necessary to record a sample of the noise over a finite interval of time and to process this data to give a spectrum of SPL against frequency. The Fourier transform algorithm allows this to be done.

## Permitted levels of noise in the workplace

Regulations concerning the maximum and average levels of noise to which workers are exposed exist in most countries of the world. In the United Kingdom (the Control of Noise at Work Regulations 2005) and Europe the levels of exposure to noise (peak sound pressure) of workers, averaged over a working day or week, at which an employer has to take remedial action are:

*Lower exposure action values:*

- daily or weekly exposure of 80 dB
- peak sound pressure of 135 dB.

*Upper exposure action values:*

- daily or weekly exposure of 85 dB
- peak sound pressure of 137 dB.

There are also absolute levels of noise exposure that must not be exceeded without the supply and use of personal protection equipment:

- daily or weekly exposure of 87 dB
- peak sound pressure of 140 dB.

Noise levels actually within an engine test cell nearly always exceed the levels permitted by statutes, while the control room noise level must be kept under observation and appropriate measures taken. The employer's obligations under prevailing regulations include the undertaking of an assessment when any employee is likely to be exposed to the first action level. This means that hearing protection equipment must be available to all staff having to enter a running cell, where sound levels will often exceed 85 dBA, and that control room noise levels should not exceed 80 dBA.

Internationally the noise limits for 8 hours exposure in national regulations range from the 87 dBA listed for the United Kingdom down to 85 dBA in Sweden and up to 90 dBA in the United States.

## Hearing protect devices and their use in test facilities

There is an almost universal availability of bulk dispensers of packets of sterile foam earplugs to workers in, and visitors to, industrial facilities. These are intended to meet the legal obligations of the employer and are satisfactory in the short term. However, workers who are in and out of noisy environments or who are using noisy devices should be supplied, as a minimum, with well-fitting noise-attenuating ear defenders. A problem with all noise-attenuating earphones is that they can considerably reduce spatial and situational awareness and cut off sources of verbal communication.

Operators required to go into complex powertrain test cells, or work on large engines under test, face problems in addition to noise levels that are best solved by noise-canceling headsets of the type that can isolate and enhances speech.

## Noise reverberation in the test cell environment

The measured value of SPL in an environment such as an engine test cell gives no information as to the power of the source: a noisy machine in a cell

**TABLE 8.2** A sample of the absorption coefficients for materials found in test cell construction at three frequencies.

| Wall material or covering | 125 Hz | 500 Hz | 4 kHz |
|---|---|---|---|
| Poured and unpainted concrete | 0.01 | 0.04 | 0.1 |
| Painted brick | 0.01 | 0.02 | 0.03 |
| Glass (large cell window) | 0.18 | 0.04 | 0.02 |
| Wood: 3/8" plywood panel | 0.28 | 0.17 | 0.11 |
| Perforated metal (13% open) over 50 mm fiberglass wool | 0.5 | 0.99 | 0.92 |

having good sound-absorbent surfaces may generate the same SPL as a much quieter machine surrounded by sound-reflective walls. The absorption coefficient is a measure of the sound power absorbed when a sound impinges once upon a surface and can be mathematically represented as follows:

$$\alpha = 1 - \frac{I_R}{I_I}$$

where $I_I$ is the sound intensity of the incident sound and $I_R$ is that of the reflected sound.

From this, it can be seen that the absorption coefficient of materials varies from 0 to 1.

Information on the absorption coefficients of a few materials used in test cells are given in Table 8.2. The figures indicate the highly reverberatory properties of untreated brick and concrete also the fact that it is quite strongly dependent on frequency and tends to fall as frequency falls below about 500 Hz.

It should be remembered that the degree to which sound is absorbed by its surroundings makes no difference to the intensity of the sound received directly from the UUT.

"Cross talk" between test cell and control room and other adjacent cells and "breakout" from the cell can occur through any openings in the partition walls and through air-conditioning ducts and other service pipes, where there are shared systems.

## Noise notation

**r.m.s. value of acoustic pressure** $p$ (N/m$^2$)
**density of air** $\rho$ (kg/m$^3$)
**velocity of sound in air** $c$ (m/s)
**sound intensity** $I$ (W/m$^2$)
**threshold sound intensity** $I_0$ (W/m$^2$)

## References

[1] W. Ker-Wilson, Vibration Engineering, griffin, London, 1959, ISBN-10 0852640234 (out of print in 2020).

## Further reading

Note that all BSI Standards listed below may be obtained through: https://shop.bsigroup.com/.

BS EN61260:1996, IEC 61260:1995, Specification for Octave and One-Third Octave Band-Pass Filters.

BS 3045:1981, Method of Expression of Physical and Subjective Magnitudes of Sound or Noise in Air.

BS 4198:1967, Method for Calculating Loudness.

BS 4675-2:1978, Mechanical Vibration in Rotating Machinery.

C.M. Harris (Ed.), Handbook of Acoustical Measurements and Noise Control, third ed., McGraw-Hill, Inc, New York, 1991. ISBN-13: 978-0070268685.

International Standards Related to Industrial Noise.

ISO 1999, ISO 3740, ISO 11200, IEC 60651, IEC 60804, IEC 60942, IEC 61252.

ASTM, The Experimental Setup and Dimensions for Measuring Absorption Coefficient Are According to ASTM E1050/ISO 10534-2.

W.T. Thomson, Theory of Vibration With Applications, fourth ed., Springer, 1998. ISBN-13: 978-0412546204.

Chapter 9

# Dynamometers: the measurement and control of torque, speed, and power

## Chapter Outline



## Introduction

The dynamometers used in prime-mover testing, of which there are several different types, all resist, and thus measure, the torque produced by the unit under test (UUT).

The accuracy with which a dynamometer absorbs and measures torque and controls rotational speed, by variable resistance, is fundamental to the power measurements and all other derived performance figures made in the test cell.

The most difficult question facing the engineer setting up a test facility is the choice of the most suitable type and size of dynamometer. The matching of the power- and torque-producing characteristics of the engine, motor, or turbine with the power and torque absorption characteristics of the dynamometer is vital for accuracy of data and control. Although the problem is

Engine Testing. DOI: https://doi.org/10.1016/B978-0-12-821226-4.00009-7

less critical in the days of the almost universal adoption of AC (alternating current) dynamometer systems, the problem still requires examination and careful consideration.

In order to preload powertrain development by using HiL techniques, powertrain test rigs can avoid using an internal combustion engine (ICE) as the power source by simulating its performance using modern AC dynamometer systems. Meanwhile, eddy-current dynamometers continue to find roles in automotive and fuels testing as do water-brake dynamometers in large engine and gas-turbine testing; whatever their type all modern, commercially available, dynamometers measure both torque and rotational speed.

## Units of torque measurement

The SI unit of torque is the newton meter, sometimes hyphenated as newton-meter and written symbolically as Nm or N · m. One newton meter is equal to the torque resulting from a force of one newton applied perpendicularly to a moment arm that is 1-m long:

$$T = r \times F$$

In SI units the relationship between torque and power is:

$$P_{\mathrm{kW}} = \frac{\tau_{\mathrm{Nm}} \times \omega_{\mathrm{rpm}}}{9549}$$

where $P_{\mathrm{kW}}$ is the power in kilowatts, $\tau_{\mathrm{Nm}}$ is the torque in newton meter, and $\omega_{\mathrm{rpm}}$ is the rotational velocity in revolutions per minute.

In Imperial units, still used in some parts of the English-speaking world, torque is measured in foot-pound, which is a force of 1-pound force (lbf) through a displacement (from the pivot) of 1 ft. In Imperial units the relationship between torque and power is:

$$P_{\mathrm{hp}} = \frac{\tau_{\mathrm{ft \cdot lbf}} \times \omega_{\mathrm{rpm}}}{5252}$$

The geometry of a trunnion-mounted dynamometer and the calibration devices of all dynamometers embodies the torque formula. The value $r$ is the effective length of the arm from which the force $F$ is measured, to meet the requirement of the force acting perpendicular to that arm, its centers of connection must be horizontal.

## Trunnion-mounted (cradled) dynamometers

Trunnion-mounted or cradled dynamometers are very widely used in powertrain and chassis dynamometer systems. The essential feature is that the power-absorbing element of the machine is mounted on trunnion bearings coaxial with the machine shaft. The torque generated has the tendency to turn the power-absorbing element, the "carcase," in the trunnions but it is resisted and thus measured by some kind of transducer acting tangentially at a known radius from the machine axis. Until the beginning of the 21st century the great majority of dynamometers used this method of torque measurement.

Modern trunnion-mounted machines, shown diagrammatically in Fig. 9.1, use a force transducer, almost invariably of the strain gauge type, together with an appropriate bridge circuit and amplifier.

The strain-gauge transducer or "load cell" has the advantage of being extremely stiff, so that no positional adjustment is necessary. The backlash and "stiction" free mounting of the transducer between carcase and base is absolutely critical.

The trunnion bearings are either a combination of a ball bearing (for axial location) and a roller bearing or hydrostatic plain bearing type. Hydrostatic designs, requiring a pressurized oil feed, are more complex and expensive than designs using rolling-element bearings and are sometimes used in special-purpose machines such as Anechoic Chassis Dynamometers; improvements in calibration accuracy are claimed but, when comparing machines of equally high quality, these seem to be marginal.

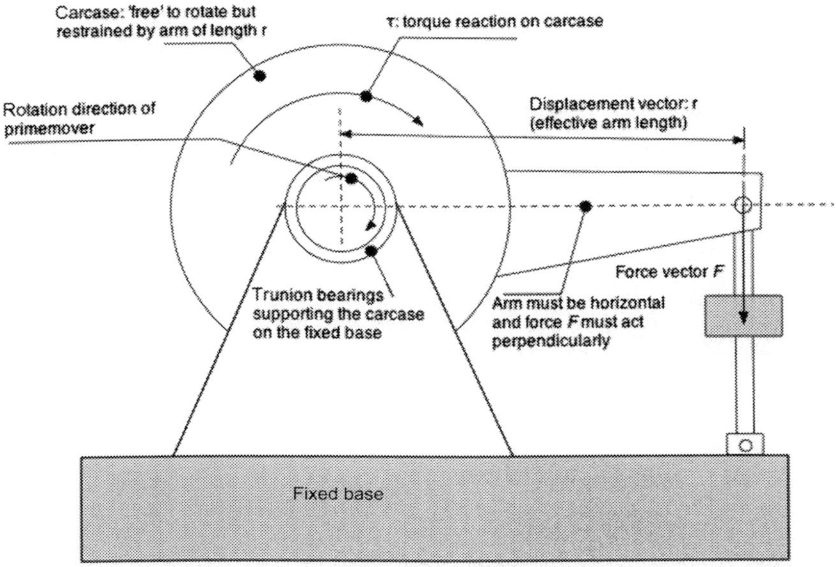


**FIGURE 9.1** Diagram of a typical trunnion-mounted dynamometer fitted with a reaction load cell. With bidirectional dynamometers torque may be calibrated and measured in either direction with the load cell in tension or compression, but compression is normally adopted for unidirectional testing.

Rolling-element trunnion bearings operate under unfavorable conditions, with no perceptible angular movement, they are consequently prone to brinelling, or local indentation of the races at the points of contact, and to fretting. This is aggravated by vibration that may be transmitted from the UUT; therefore periodical inspection and turning of the outer bearing race is recommended, in order to avoid progressively worsening calibration.

A Schenck dynamometer design replaced the trunnion bearings by two radial flexures, thus eliminating possible friction and wear, but at the expense of the introduction of torsional stiffness, of reduced capacity to withstand axial loads and of possible ambiguity regarding the true center of rotation, particularly under side loading.

## Measurement of torque using in-line shafts or torque flanges

The alternative method of measuring the torque produced in the prime mover/dynamometer system is by using an in-line *torque shaft* or *torque flange* (Fig. 9.2).

A torque shaft is mounted in the driveline between engine and dynamometer. It consists essentially of a double-flanged, pedestal bearing assembly containing shaft section fitted with strain gauges. Such systems have to obtain the signal from the rotation shaft either by using slip rings and brushes or by RF signal transmission. Fig. 9.3 is a brushless torque shaft unit intended for rigid mounting.

More common in powertrain testing is the "disk" type torque transducer, commonly known as a *torque flange* (Fig. 9.4), which is a device that is usually bolted directly to the input flange of the dynamometer. The flange has an integral, digital measurement preconditioning circuit that produces analog or digital output signals, which are transmitted, without contact, to a static antenna encircling it.

A perceived advantage of the in-line torque measurement arrangement is that it avoids the necessity, discussed next, of applying torque corrections under transient conditions of torque measurement; such corrections, using known constants, are a quite trivial problem with modern computer control systems.

However, there are potential problems with in-line torque transducer *systems* that may reduce their inherent accuracy. While in-line torque-measuring devices are widely and successfully used in many types of automotive test rigs, there can be cases where a well-designed and -maintained trunnion machine can give more consistently verifiable and accurate torque measurements than the in-line systems; the justifications for this statement can be listed as:

- The in-line transducer has often to be overrated because it has to be capable of dealing with the instantaneous torque peaks of the UUT, peaks that

**FIGURE 9.2** A modern base mounted, air-cooled, AC dynamometer fitted with a torque flange unit of the type shown diagrammatically in Fig. 9.4. Note that the oblong box immediately to the rear and above the flange contains the shaft locking mechanism used during calibration and stall-torque testing. *Photo courtesy Taylor Dynamometer/Dyne Systems.*

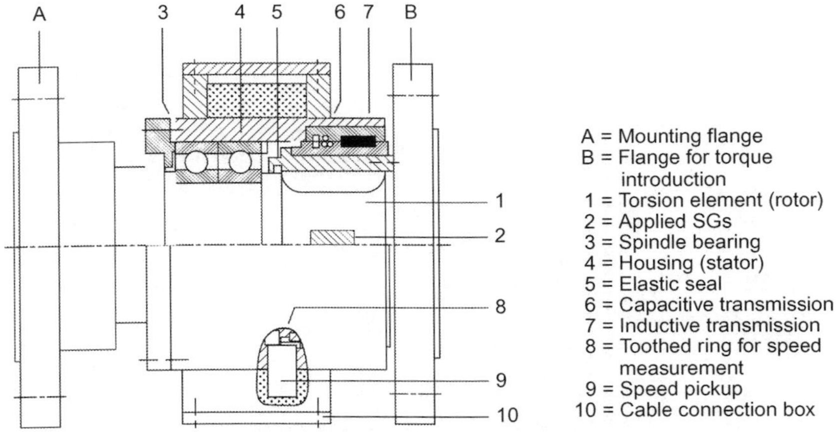


**FIGURE 9.3** Section through a pedestal-mounted brushless torque shaft unit of the type widely used in multishaft transmission and hybrid powertrain test rigs.

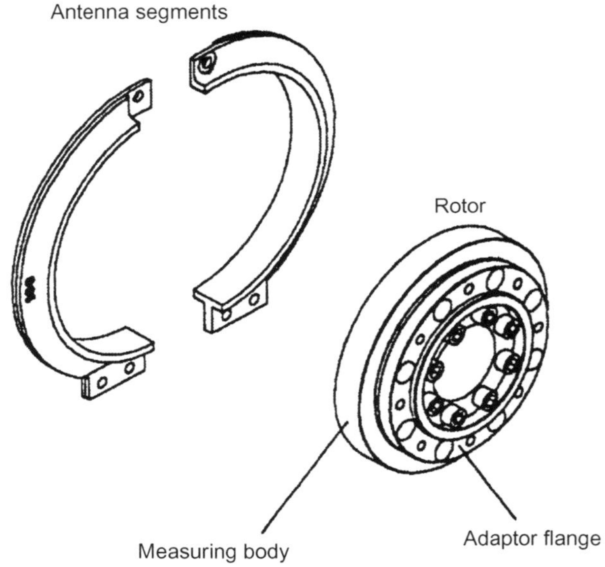


**FIGURE 9.4** Static antenna and rotational components of a torque flange unit of the type shown in Fig. 9.2 where it is shown complete with its local signal conditioning and cable plugs.

are not experienced by the load cell of a trunnion bearing machine. Being oversized the resolution of the signal is lower.

- The in-line transducer forms part of the drive shaft system and requires very careful installation to avoid the imposition of bending or axial stresses on the torsion-sensing element from other components or its own clamping device.
- The in-line device is difficult to protect from temperature fluctuations within and around the driveline.
- Calibration checking of these devices is not as easy as for a trunnion-mounted machine; it requires a means of locking the dynamometer shaft (see Fig. 9.2) in addition to the fixing of a calibration arm in a horizontal position without imposing bending stresses.
- Unlike the cradled machine and load cell, it is not possible to verify the measured torque of an in-line device during operation.

It should be noted that, in the case of modern AC dynamometer systems, the tasks of torque measurement and torque control may use different data acquisition paths. In many installations the control system of the AC machine may use its own torque calculation, while the data acquisition values are taken from an in-line transducer or reactive load cell.

*Some test facilities fit both types of torque measurement to assist in correlation of highly transient engine tests.*

## Torque calibration and the assessment of errors in torque measurement

We have seen that in a trunnion-mounted dynamometer, torque $T$ is measured as a product of torque arm radius $R$ and transducer force $F$. Calibration is invariably performed by means of a *calibration arm*, supplied by the manufacturer, which is bolted to the dynamometer carcase and carries dead weights that apply a load at a certified radius. The manufacturer certifies the distance between the axis of the weight hanger bearing and the axis of the dynamometer. There is no way, apart from building an elaborate fixture, in which the dynamometer user can check the accuracy of this dimension; therefore the arm should have its own calibration certificate and be stamped with its effective length. For R&D machines of high accuracy the arm should be stamped and matched with the specific machine.

The "dead weights" should in fact be more correctly termed "standard masses" and they should be certified by an appropriate standards authority.

Whatever the certified weight marked, it is not always remembered that their weight does not, in every geographic location, necessarily equate directly with the force they exert. While mass, measured in kilograms or pounds, is a constant throughout the world, the force they exert on the calibration arm is the product of their mass and the local value of $g$, which is not constant. The value of $g$ is usually assumed to be 9.81 $m/s^2$, but this value is only correct at sea level and a latitude of about 47°N. It increases toward the poles and falls toward the equator, with local variations.

$$F = m \cdot g,$$

where $g$ is also a variable.

As an example: in London, where $g = 9.812$ $m/s^2$ a 10 kg "weights" exert 98.12 Nm force, while in Tokyo the same weights exert 97.98 Nm, and Manila 97.84 nm, these are not entirely negligible variations if one is hoping for accuracies better than 1%.

The actual process of calibrating a dynamometer with dead weights, when treated rigorously, is not entirely straightforward. We are confronted with the facts that no transducer is perfectly linear in its response, and no linkage is perfectly frictionless.

A suitable calibration procedure for a machine using a typical strain-gauge load cell for torque measurement is as follows:

The dynamometer should be put in its normal, energized, no-load running condition (cooling water on etc.) and the calibration arm weight balanced by equal and opposite force After the whole system has been energized long enough to warm up, the load cell output is zeroed through the calibration routine of the control system. Dead weights are then added to produce approximately the rated maximum torque of the machine. The resulting torque is calculated, and the digital indicator set to this value.

The weights are removed, any error in not returning to the zero-reading noted, and weights are added again, but this time preferably in 10 equal increments, the control system display readings being noted and plotted. The weights are removed in reverse order and the incremental readings again noted and plotted.

Let us assume we apply this procedure to a machine having a nominal rating of 600 Nm torque, and that we have six equal weights, each calculated to impose a torque of 100 Nm on the calibration arm. Table 9.1 shows actual indicated torque readings for both increasing and decreasing loads, together with the calculated torques applied by the weights.

The corresponding errors, or the differences between calculated torque applied by the calibration weights and the indicated torque readings, are plotted in Figs. 9.5 and 9.6.

The machine is claimed to be accurate to within $\pm 0.25\%$ of nominal rating and in this example the machine meets the claimed limits of accuracy and may be regarded as satisfactorily calibrated.

The failure of the system to return to zero is a common problem and sometimes indicative of installation problems or wear in the linkage system. However, it is usually assumed that such hysteresis effects, manifested as differences between observed torque with rising load and with falling load, are eliminated when the machine is running, due to vibration, and it is a commonly accepted practice to knock the machine carcase lightly with a soft mallet after each load change to achieve the same result.

It is certainly not wise to assume that the ball joints invariably used in calibration arm and torque transducer links are frictionless.

**TABLE 9.1** Dynamometer calibration (example taken from an actual machine).

| Mass (kg) | Applied torque (Nm) | Reading (Nm) | Error (Nm) | Error (% reading) | Error (% full scale) |
|---|---|---|---|---|---|
| 0 | 0 | 0.0 | 0.0 | 0.0 | 0.0 |
| 10 | 100 | 99.5 | −0.5 | −0.5 | −0.083 |
| 30 | 300 | 299.0 | −1.0 | −0.33 | −0.167 |
| 50 | 500 | 500.0 | 0.0 | 0.0 | 0.0 |
| 60 | 600 | 600.0 | 0.0 | 0.0 | 0.0 |
| 40 | 400 | 400.5 | +0.5 | +0.125 | +0.083 |
| 20 | 200 | 200.0 | 0.0 | 0.0 | 0.0 |
| 0 | 0 | 0.0 | 0.0 | 0.0 | 0.0 |

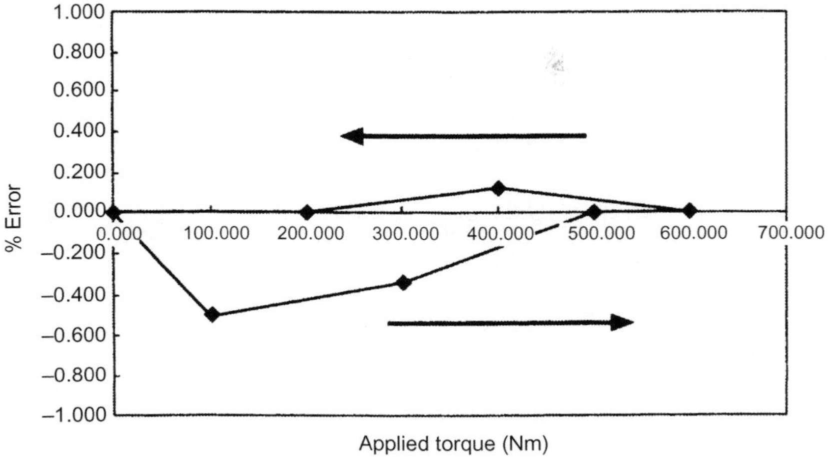


**FIGURE 9.5** A set of dynamometer calibration errors plotted as a percentage of applied torque for both incremental addition and subtraction of weights. Data taken from an actual calibration routine of a Froude eddy-current dynamometer.

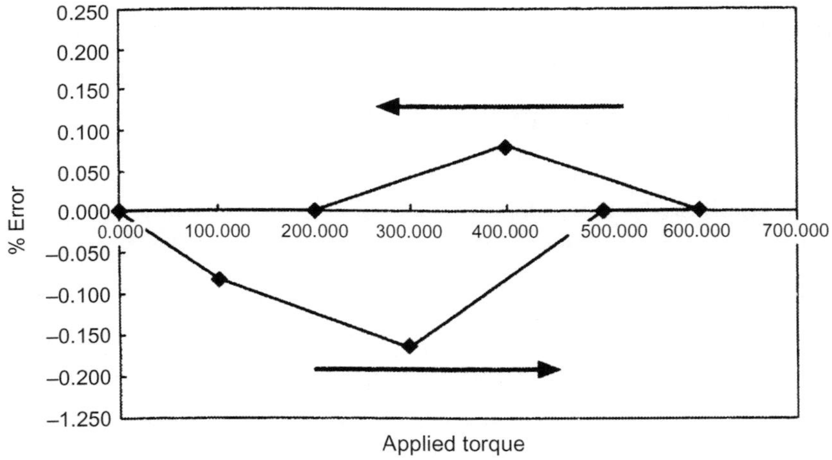


**FIGURE 9.6** A set of dynamometer calibration errors plotted as a percentage full scale (% FS) as per Fig. 9.5.

Some large water-brake dynamometers are fitted with torque multiplication levers, reducing the size of the calibration masses. Under health and safety legislation the frequent handling of multiple 20 or 25 kg weights within a restricted work area is not advisable. It is possible to carry out torque calibration by way of "master" load cells or proving rings. These devices have to be mounted in a jig attached to the dynamometer and give a verifiable measurement of the force being applied on the target load cell by means of a hydraulic actuator. Such systems produce a more complex "audit trail" used to refer the calibration back to national standards.

It is important, when calibrating any water-cooled dynamometer, that the water pressure in the casing should be at operational level, since pressure in the transfer pipes can give rise to a parasitic torque. Similarly, any disturbance to the number and route of electrical cables to the carcase must be avoided once calibration is completed.

Note that it is possible, particularly with electrical dynamometers with forced cooling, to develop small parasitic torques due to air discharged nonradially from the casing. It is an easy matter to check this by running the machine uncoupled under its own power and noting any change in indicated torque.

Experience shows that a high-grade dynamometer such as would be used for research work, after careful calibration, may be expected to give a torque indication that does not differ from the absolute value by more than about $\pm 0.1\%$ of the full load torque rating of the machine.

Systematic errors such as inaccuracy of torque arm length or wrong assumptions regarding the value of $g$ will certainly diminish as the torque is reduced, but other errors will be little affected. It is safer in most test facilities to assume a band of uncertainty of constant width; this implies, for example, that a machine rated at 400-Nm torque with an accuracy of $\pm 0.25\%$ will have an error band of $\pm 1$ N. At 10% of rated torque, this implies that the true value may lie between 39 and 41 Nm.

All strain-gauge load cells used by reputable dynamometer manufacturers will compensate for changes in temperature, though their rate of response to a change may vary. They will not, however, be able to compensate for internal temperature gradients induced, for example, by air blasts from ventilation fans or radiant heat from exhaust pipes and therefore should be protected from directional heat sources as part of the rigging routine.

Some dynamometers are suitable for "part-spanning" calibration allowing the operator to recalibrate the dyno down to half of the dynamometer's nominal torque to increase the torque measurement accuracy when running tests on an engine or motor of less than half the dynamometer's capacity.

The subject of calibration and accuracy of dynamometer torque measurement has been dealt with in some detail because it is probably the most critical measurement that the test engineer is called upon to make, and one for which a high standard of accuracy is expected but not easily achieved. Calibration and certification of the dynamometer and its associated system should be carried out at the very least once a year, and following any system change, major component replacement or new, audited, project.

Finally, it should be mentioned that the best calibration procedures will not compensate for errors introduced "downstream" of the raw data flow by software or, more rarely, hardware "corrections." The authors have been involved in several investigations into apparent test equipment inaccuracies only to find that unknown and undocumented power correction factors are running in the embedded software of the AC dynamometer's drive units or the test cell controllers. In one case an incorrectly programed temperature correction factor installed in the control system of a quality audit test bed had been the source of undetected and optimistic power figures for several years!

## Torque measurement under accelerating and decelerating conditions

With the increasing interest in transient testing, it is essential to be aware of the effect of speed changes on the "apparent" torque measured by a trunnion-mounted machine.

The basic principle is as follows:

Inertia of dynamometer rotor: $I$ kgm$^2$

Rate of increase in speed: $\omega$ rad$/s^2$

$N$ rpm/s

Input torque to dynamometer: $T_1$ Nm

Torque registered by dynamometer: $T_2$ Nm

$$T_1 - T_2 = I\omega = \frac{2\pi NI}{60}\text{Nm} = 0.1047NI\ \text{Nm}$$

To illustrate the significance of this correction, a typical eddy-current dynamometer capable of absorbing 150 kW with a maximum torque of 500 Nm has a rotor inertia of 0.11 kgm$^2$. A DC (direct current) regenerative machine of equivalent rating has a rotational inertia of 0.60 kgm$^2$.

If these machines are coupled to an engine that is accelerating at the comparatively slow rate of 100 rpm/s, the first machine will read the torque low *during the transient phase* by an amount:

$$T_1 - T_2 = 0.1047 \times 100 \times 0.11 = 1.15\ \text{Nm}$$

while the second will read low by 6.3 Nm.

If the engine is decelerating the machines will read high by the equivalent amount.

Much larger rates of speed changes are demanded in some transient test sequences and this can represent a significant variation of torque indication, particularly when using high inertia dynamometers.

*With modern computer processing data acquisition systems corrections for these and other transient effects can be made with software supplied by test plant manufacturers.*

## Measurement of rotational speed

Rotational speed of the dynamometer is measured, either by a system using an optical encoder system or a toothed wheel and a pulse sensor, the resulting signals are used by the unit's associated electronics for both control and display (Chapter 13: Test Cell Safety, Control, and Data Acquisition).

The pulse pickup system is robust and, providing the wheel to transducer gap is correctly set and maintained, reliable. It is still used entirely satisfactorily on two-quadrant dynamometers and quasi-steady-state testing. Optical encoders, which use the sensing of very fine lines etched on a small disk, need more care in mounting and operation but are a modern default standard. Since the commonly used optical encoders transmit over 1000 pulses per revolution, misalignment of its drive may show up as a sinusoidal speed change; therefore they should be mounted as part of an accurately machined assembly, physically registered to the machine housing.

It should be remembered that it is therefore necessary to measure not only speed but also direction of rotation since most automotive dynamometers are bidirectional. Encoder systems can use separate tracks of their engraved disks to sense rotational direction. *It is extremely important that the operator uses a common and clearly understood convention describing direction of rotation throughout the facility, particularly in laboratories testing electrical motors and other reversible prime movers.*

As with torque measurement, specialized instrumentation systems may use separate transducers for the measurement of engine speed and for the control of the dynamometer. In many cases engine speed is monitored separately to dynamometer speed. The control system can use these two signals to deal with shaft system twist and in order to shut down automatically in the case of a shaft failure. Measurement of crank position during rotation is discussed in Chapter 16, The Combustion Process and Combustion Analysis.

Measurement of power, which is the product of torque and speed, raises the important question of sampling time. ICEs never run totally steadily, and poor control of variable-speed electrical drives can result in speed variations over a wide range of frequencies; therefore the torque transducer and speed signals invariably fluctuate. An instantaneous reading of speed will not be identical with a longer term average. Choice of sampling time and of the number of samples to be averaged is a matter of experimental design and compromise.

## Quadrants of dynamometer operation: one, two, or four

Fig. 9.7 illustrates diagrammatically the four "quadrants" in which a dynamometer may be required to operate. Electric motors within electrical and hybrid powertrains are able to operate in all four quadrants, which is one of the reasons for the modern almost universal adoption of AC dynamometer systems.

Most ICE run counterclockwise when looking on the flywheel; therefore testing takes place in the first quadrant (looking on the dynamometer coupling).

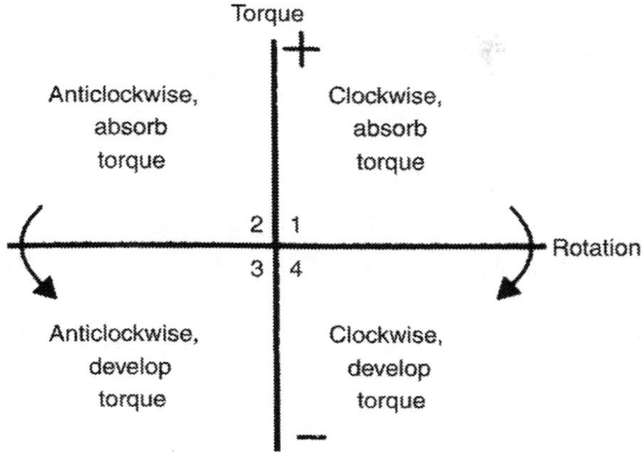


**FIGURE 9.7** The dynamometer operating quadrants drawn as if the observer was looking at the dynamometer's coupling flange.

Hydraulic dynamometers, particularly those used in the testing of large generator set or marine diesels, are usually designed for one direction of rotation, though they may be run in reverse at low fill state without damage. When designed specifically for bidirectional rotation, they are larger than a single-direction machine of equivalent power, and torque control may not be as precise as that of the unidirectional designs. The torque-measuring system must of course operate in both directions. On occasions it is necessary for a test installation, using a high-power, unidirectional, water brake, to accept engines or gas-turbines running in either direction; one solution is to mount the dynamometer on a turntable, with couplings at both ends. Most very large and some "medium-speed" marine engines are directly reversible. Eddy-current machines are inherently reversible.

When it is required to operate in the third and fourth quadrants, that is, for the dynamometer to produce power as well as to absorb it, the choice is effectively limited to DC or AC electric motor–based machines. These machines are generally reversible and, when powered by drives capable of import and export of electrical power, able to operate in all four quadrants. There are special-purpose dynamometer systems based on hydrostatic drives that operate in four quadrants, as do hybrid eddy-current/electric-motor units.

Table 9.2 summarizes the performance of machines in this respect.

## Torque/speed and power/speed characteristics of some dynamometer types

The different types of dynamometer have significantly different torque–speed and power–speed curves, and this can affect the choice made for a given application. Due to the requirement of legislative calibration test

**TABLE 9.2 The possible operating quadrants of a complete range of automotive dynamometer types.**

| Type of machine | Quadrant |
|---|---|
| Hydraulic sluice plate | 1 or 2 |
| Variable-fill hydraulic | 1 or 2 |
| "Bolt-on" variable-fill hydraulic | 1 or 2 |
| Disk-type hydraulic | 1 and 2 |
| Hydrostatic | 1, 2, 3, 4 |
| DC electrical | 1, 2, 3, 4 |
| AC electrical | 1, 2, 3, 4 |
| Eddy current | 1 and 2 |
| Friction brake | 1 and 2 |
| Air brake | 1 and 2 |
| Hybrid (absorber/motor) | 1, 2, 3, 4 |

*AC*, Alternating current; *DC*, direct current.

sequences, for many automotive powertrain engineers in the 21st century, engaged in emission homologation or engine mapping, the only viable choice of dynamometer is a four-quadrant AC unit, matched with an IBGT-based vector drive and control electronics.

Typical characteristics of the torque–speed and power–speed curves of AC machines of this type are shown in Fig. 9.8.

For engineers involved in testing prime movers in the power ranges above 1 MW, the water-brake dynamometer is probably the only viable choice.

If the reader compares the power and torque envelope shapes in Fig. 9.8 with those of hydraulic water brakes shown in Fig. 9.9, the important differences in the performance curves, within a single absorbing quadrant, between hydraulic, eddy-current (Fig. 9.10) , and AC or DC dynamometers will be clear. These differences become important when matching engines with dynamometers and give the four-quadrant machines two technical advantages:

- The AC or DC dynamometer can produce full, or near full, torque at zero speed while the hydraulic or eddy-current unit has a significant "gap" on the left-hand side of its torque curve. A common problem in matching a high-torque diesel engine with some dynamometers is shown in Fig. 9.9 where the engine's torque curve comes outside the dynamometer's torque envelope.
- The two-quadrant design dynamometers have a rising "minimum torque with speed" characteristic. This is due to rotor windage in the case of

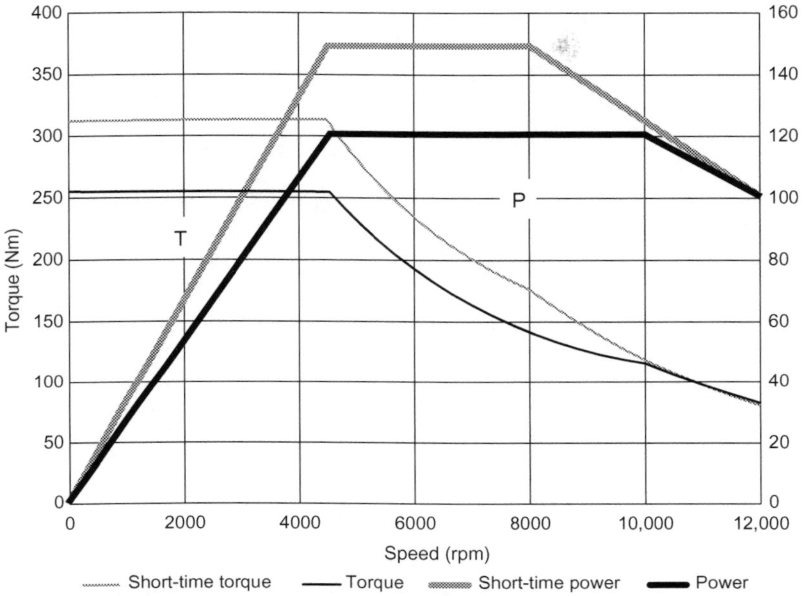


**FIGURE 9.8** Typical performance curve shape of a modern, automotive AC dynamometer. The maximum torque rating is based on the safe operating temperature of the machine rather than maximum rotor stresses; thus many machine units will have a higher, short-time, torque rating, such as +10% for 5 minutes in every hour. Monitoring of winding temperature via thermistors recommended.

eddy-current machines and torque generated by the required minimum water content in the hydraulic units; this internal resistance to rotation appears as parasitic torque in the measuring system. AC and DC machines can "motor out" such inherent internal resistance.

The key parts of the performance envelopes shown in Figs. 9.9 and 9.10 are:

1. low-speed torque corresponding to maximum permitted coil excitation;
2. performance limited by maximum permitted shaft torque;
3. performance limited by maximum permitted power, which is a function of cooling water throughput and its maximum permitted temperature rise;
4. maximum permitted speed; and
5. minimum torque corresponding to residual magnetization, windage, and friction.

In choosing a dynamometer for an engine or range of engines, it is essential to superimpose the maximum torque— and power—speed curves onto the dynamometer envelope. For best accuracy, it is desirable to choose the smallest machine that will cope with the largest engine to be tested.

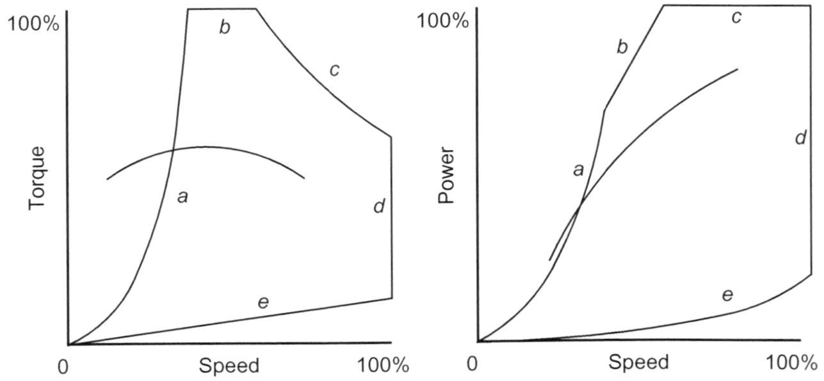


**FIGURE 9.9** An illustration of the engine-dyno mismatch problem. A typical hydraulic dynamometer torque envelope (left) and power envelope (right). The torque and power curves of an engine producing greater low-speed torque than the dynamometer can absorb have been superimposed.

## Water-brake dynamometers

"Water brakes," originally developed in England for comparing the power output of marine steam engines in 1877, were the first dynamometers to be developed and many thousands are still used worldwide. Among this family of trunnion-mounted machines are the largest dynamometers ever made, with rotors of around 5 m diameter and power absorption ratings up to 75 MW (Fig. 9.12). In addition to fixed water-brakes portable "bolt-on" designs also exist (Fig. 9.11).

In these machines the torque absorbed is varied by adjusting the quantity (mass) of water in circulation within the casing. This is achieved by a valve, usually but not exclusively fitted to the carcase at the water outlet and associated with control systems of widely varying complexity.

All such hydrokinetic machines work on similar principles. A shaft carries a cylindrical rotor that revolves in a watertight casing. Toroidal recesses formed half in the rotor and half in the casing or stator are divided into pockets by radial vanes set at an angle to the axis of the rotor. When the rotor is driven centrifugal force sets up an intensive toroidal circulation the result of which is to transfer momentum from rotor to stator and hence to develop a torque resistant to the rotation of the shaft, balanced by an equal and opposite torque reaction on the casing.

A forced vortex of toroidal form is generated as a consequence of this motion, leading to high rates of turbulent shear in the water and the dissipation of power in the form of heat to the water. The center of the vortex is vented to atmosphere by way of passages in the rotor and, within the typical automotive ICE range of rotational speeds, power is absorbed with minimal damage to the moving surfaces, either from erosion or from the effects of cavitation.

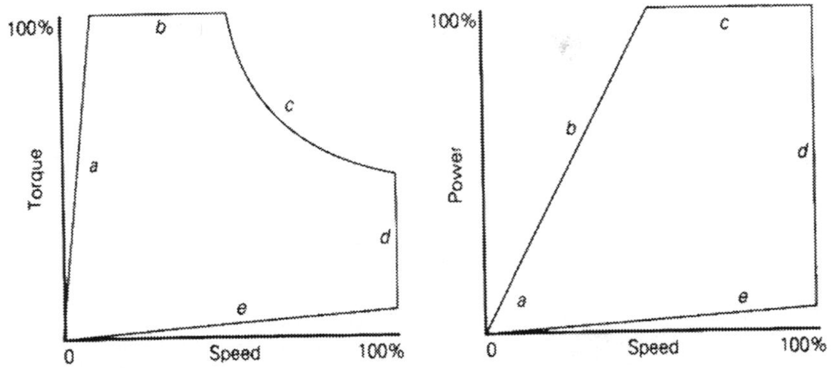


**FIGURE 9.10** Typical torque (left) and power (right) curves or performance envelopes of an eddy-current dynamometer.

**FIGURE 9.11** A "bolt-on" water-brake dynamometer rated at 1120 kW with maximum speed of 5000 rpm. Torque measurement is via a strain-gauge load cell and load is controlled through a water inlet valve. These devices may be mounted within the chassis frame of large trucks to test cooling systems, etc. *Photo courtesy Froude/Go-Power.*

As rotational speed rises the increase in water velocity and centrifugal stresses in the complex rotor becomes critical; therefore, for high-speed, lower peak torque designs, as typically used in gas-turbine testing, the rotor may be cut off at half height of the pockets; this is called a 90 degree crop. The "missing" rotor's water pocket is then formed within the stator, so the water's toroidal circulation is maintained, but with a much reduced rotor diameter capable of much higher rotation speeds and a significant change in the shape of the power envelope.

To obtain the fastest response from a water brake it is necessary to have adequate water pressure available to fill the casing rapidly; to satisfy major

**FIGURE 9.12** A Froude bidirectional low-speed water-brake 'LS' dynamometer (notice the "handing" of the vanes of the two rotors). Torque measured by four "table-leg" load cells. Power absorption capacity in MW not published as it was built for testing Naval propulsion systems but LS dynamometers were built for testing prime movers of up to 75 MW. *Photo courtesy of Froude.*

steps in power it may be necessary to fit both inlet and outlet control valves with an integrated control system.

*All these designs require a free, gravitational discharge of water so are not suitable for closed, pressurized water systems.*

## Alternating current motor–based dynamometer systems

The nomenclature and technology of electric motors has become complicated in recent years during the development of electric vehicles, with terms such as "electronically commutated motors," "brushless DC motors," and "self-synchronous AC motors" being used to describe similar devices. However, in the test facility world the demands of modern powertrain testing have dictated that dynamometers are based on asynchronous AC-induction or permanent magnet motors, combined with four-quadrant IBGT-based frequency converter control systems; these are the default choice. All these machines are designed to have appropriate, usually low, inertia to match their roles, which now include:

- As conventional motor or ICE shaft speed dynamometers either foot-mounted and fitted with inline torque transducers or trunnion-mounted and reaction load cells. Available in the ranges of torques and speeds that suit the common models of engines and motors from 1 kW up to 1000 kW.
- Use as wheel or hub dynamometers where lower than ICE speed and high torque capacity is required for and full powertrain rigs and where

inertia similar to the wheel is desirable. Power absorption figures quoted for these machines may confuse casual examination as they sometimes refer to the whole two- or four-wheel system rather than the individual hub dynamometers (see Chapter 10: Chassis Dynamometers, Rolling Roads, and Hub Dynamometers).

- As a replacement of an ICE within a gearbox or powertrain test stands where very high response capable of simulation of pulsed speed oscillations is required together with a high transient overload capacity and a mechanical inertia similar to an ICE. These test rigs are using PMM dynamometers.
- Paired systems for testing transmission systems using inline torque transducers and allowing the electrical energy to run in a loop so that only the losses of the system are taken out of the grid.
- Ultrahigh-response and physically robust units designed for electric motor or electric drive unit testing capable of speeds up to 20,000 rpm and currently covering the power range from 80 to 400 kW.
- High-response ICE test rigs where the small profile of permanent magnet motor designs is required because of restricted space within multiple axis or ICE exhaust restricted rigs.

Major international test plant suppliers, such as AVL in Europe and Horiba in Asia and Taylor Dynamometer in the United States, build AC dynamometers, control and data acquisition packages, specifically targeted at market sectors such as those listed previously while other companies, such as SAKOR in the United States, also build custom systems for customer's specific purposes.

*Note concerning the spark erosion damage of AC dynamometer bearings:*

During the development of the modern AC dynamometers, various designs suffered from premature shaft bearing failures caused by electrical arcing between the rolling elements and the races. The cause of the failures was that an electrical field is induced in the rotor creating of difference in potential between rotor and stator. This can build up to the point that arcing occurred across the bearings causing pitting of the races that led to failure. The problem has been resolved by design changes, including the use of earthing (grounding) brushes and, in some high-speed machines, the use of no-conductive bearings with ceramic rolling elements.

The physical performance of the dynamometer is dependent on being suitably matched to its electrical network and power control system. The power controls are usually large multibay devices that require to be housed in a clean and well-ventilated space, which, in the case of ICE cells, normally means that they are installed in a separate room. Due to the cost of the expensive connecting cables, the cabinet to dynamometer distance should be as short as is convenient. Exporting electrical power into a factory or national grid requires that a number of technical and legislative problems are solved (see Chapter 4: Electrical Design Requirements of Test Facilities).

There are two levels of control of the motors that form an AC dynamometer system. The primary controller is the motor drive system. The secondary controller that forms the user interface is the test cell control system (see Chapter 13: Test Cell Safety, Control, and Data Acquisition). The two systems communicate via a high-speed "intelligent" interface, modern example of which uses a "KIWI" (*k*ilohertz *i*nterface *w*ith *i*ntelligence) link or "EtherCAT." A consideration associated with integration of AC dynamometer drives is that the systems, at least in Europe, need to comply to the Machinery Directive 2006/42/EC (MD), and the associated Standard BS EN 13849-1. On sites where it is deemed that the operational risk has increased, when for instance a new AC regenerative drive and dynamometer has replaced an eddy-current unit, the requirements of the MD must be adhered to. Drive systems should be compatible to BS EN 13849-1, Performance Level d (PLd), and be supplied with a dual-channel emergency stop and Safe Torque Off (STO) signal.

*Recirculating electrical power* within powertrain test rigs is possible by using a combined power control system for two or more AC dynamometers (Fig. 9.13); it is the modern electrical equivalent of the mechanical technique of recirculating torque in "back-to-back" transmission test rigs.

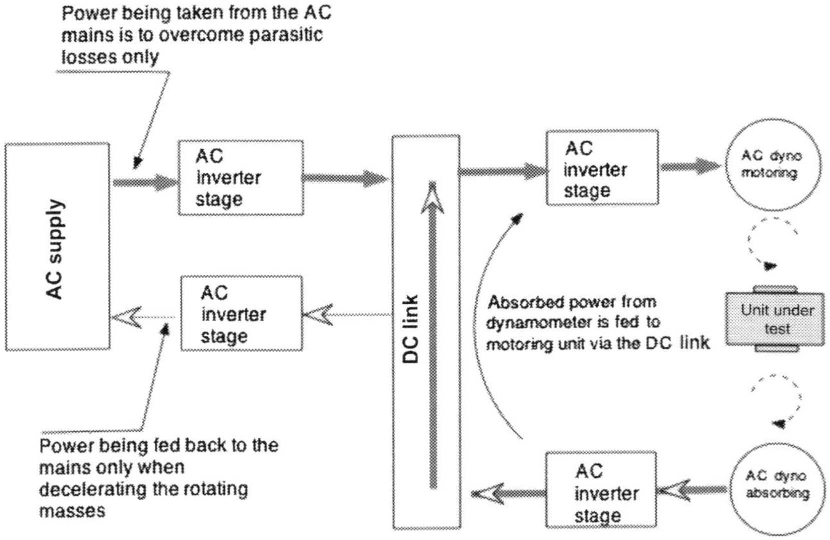


**FIGURE 9.13** Diagram showing recirculation of electrical power in a transmission test rig comprising two matched AC dynamometer systems. In this state the AC mains is required only to make up for system losses. *AC*, Alternating current.

While AC machines are, for good operational reasons, tending to be the default choice for new test facilities, their high performance does not come cheaply when compared to the cost of two-quadrant eddy-current machines. The latter are entirely suitable for much of the comparatively steady-state testing required by endurance, Fuel & Lube, or any other powertrain testing.

## Direct current dynamometers

The use and manufacture of DC dynamometers has declined markedly during the 21st century as the variable-speed AC-based technology has taken over. However, when used as engine dynamometers, they have a long pedigree particularly in the United States. They are robust, easily controlled, capable of full four-quadrant operation, and have a speed range matched to that of truck diesel engines. Their torque-measuring systems are the same as those used on AC dynamometers, although the development of torque flanges post-dates the production of many units. Speed measurement is usually via a tachogenerator or toothed wheel, or both.

The disadvantages of DC dynamometers include limited maximum speed and high inertia, which can present problems of torsional vibration (see Chapter 11: Mounting and Rigging Internal Combustion Engines for Test), and limited rates of speed change. Because they contain a commutator the normal maintenance of DC machines will be higher than those based on AC motors. Control is almost universally by means of a thyristor-based power converter. Harmonic distortion of the regenerated AC mains waveform may be a problem if isolating transformers are not used.

A modern use of DC dynamometers has been in chassis dynamometers within EMC (electromagnetic compatibility) test facilities where their absence, or low level, of radiofrequency emissions is a desirable feature although shielded AC machines are now seen in these installations.

## Packaged alternating current dynamometers for e-axle and hybrid drive testing

The multiple shaft arrangements in which AC dynamometers now have to be placed, in order to test hybrid electric vehicle and battery electric vehicle powertrain units, have required manufacturers to adopt a modular approach to their designs as illustrated in Fig. 9.14. The dynamometer unit in this packaged test stand is rated at 377 kW, 500/750 Nm, and 23,000 rpm with a rotational inertia of 0.041 $kgm^2$.

This unit is built to a fixed height and all the other dynamometer units within the packaged system are spatially adjustable. The pallet on which the UUT is very rigidly mounted, using vehicle mounting points, can be precisely adjusted to achieve the high degree of shaft alignment required.

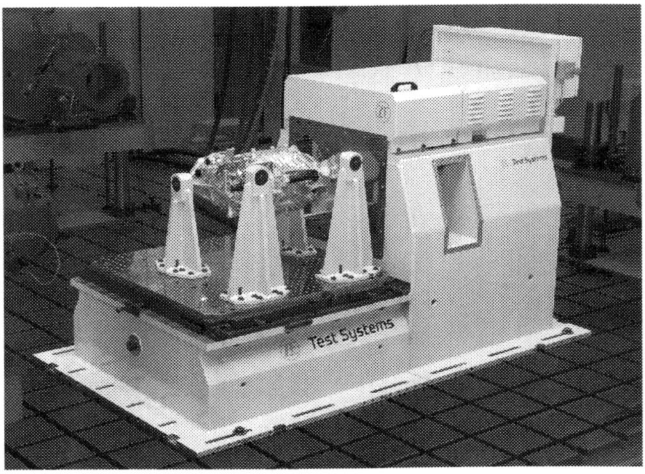


**FIGURE 9.14** An example of a packaged AC dynamometer test bench incorporating precision adjustable pallet table of the UUT. This forms the fixed node on a large bedplate whereupon various HEV and BEV shaft arrangements can be configured. *Photo courtesy ZF Passau GmbH.*

## Eddy-current dynamometers

These machines make use of the principle of magnetic induction to develop torque and dissipate power. In the most common type a toothed rotor of high magnetic permeability steel rotates in air, with a fine clearance, between water-cooled steel loss plates. A magnetic field parallel to the machine axis is generated by one or two annular coils (depending on the configuration) and motion of the rotor causes changes in the distribution of magnetic flux in the loss plates. This in turn gives rise to circulating eddy currents and the dissipation of power in the form of electrical resistive losses. Energy is transferred in the form of heat to cooling water circulating through passages in the loss plates, while some cooling is achieved by the radial flow of air in the gaps between rotor and plates.

Power is controlled by varying the current supplied to the annular exciting coils, and rapid load changes are possible. Eddy-current machines are simple and robust, providing adequate cooling water flow is maintained; the control system is simple, and they are capable of developing substantial braking torque at quite low speeds. Unlike AC or DC dynamometers, however, they are not able to develop motoring torque.

There are two forms of machine, both having air circulating in the gap between rotor and loss (cooling) plates, hence "dry gap":

1. Dry gap machines fitted with a toothed disk rotor (Fig. 9.15). These machines have lower inertia than the drum machines and a very large

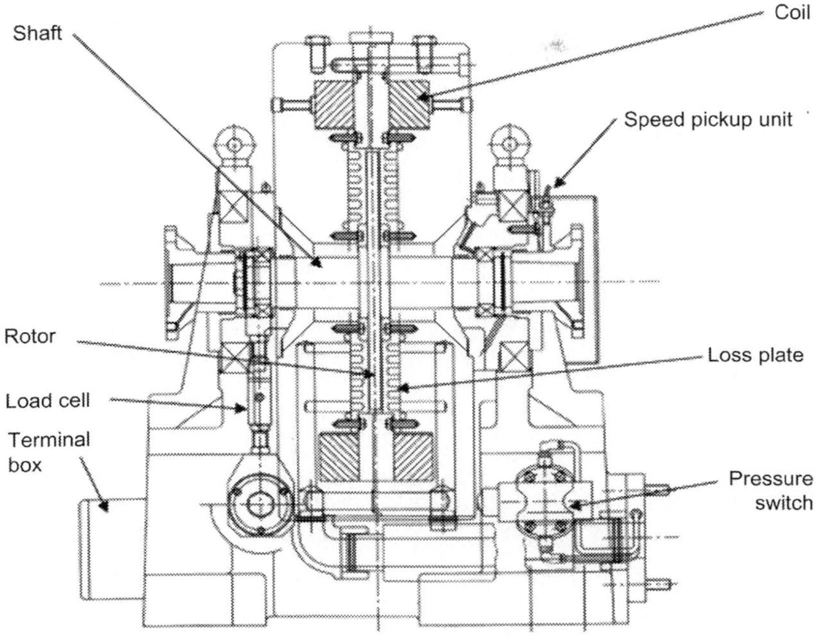


**FIGURE 9.15** Section through a trunnion bearing-mounted, dry-gap, disk-type, eddy-current dynamometer.

installed user base, around the world; however, the inherent design features of their loss plates place certain operational restrictions on their use. *It is absolutely critical to maintain the required water flow through the machines at all times; even a very short loss of cooling will cause the loss plates to distort, leading to the rotor/plate gap closing, with disastrous results.* These machines must be fitted with flow detection devices interlocked with the cell control system; pressure switches should not be used since in a closed water system it is possible to have pressure without flow. Note that the "pressure switch" shown in Fig. 9.15 is a differential pressure switch mounted across an orifice plate so is monitoring flow not pressure.

2. Dry gap machines fitted with a drum rotor are less common. These machines usually have a higher inertia than the equivalent disk machine but may be less sensitive to cooling water conditions.

## Pushing the limits of dynamometer performance

It should be understood that, in general, the limits set by the *curved lines* in any dynamometer performance envelope cannot be exceeded by test engineers because they represent the inherent characteristics of the machine design. The *straight lines* of the performance graphs, however, represent the "safe" limits set by the machine designer. In principle, therefore, provided the shaft connection has sufficient balance and strength (see Chapter 11: Mounting and Rigging Internal Combustion Engines for Test), it is possible, if not always advisable, to go beyond those (safe) performance cutoff figures to overload and overspeed most dynamometers.

The overload capacity for AC electrical dynamometer systems (the drive system/motor combination) will be quoted by the supplier and will be strictly time limited. Exceeding the overload capacity and/or the time limit, particularly in high ambient temperatures, will mean that the system protection may trip out causing an uncontrolled shutdown. Pushing DC dynamometers beyond their rated limits is not advisable as this can lead to damage to commutators.

Hydraulic dynamometers are generally robust machines and most designs are well able to deal with a moderate degree of overload and some overspeed; the critical dynamometer parameter, if running at overload for any length of time, is water exit temperature.

Froude hydraulic dynamometers, on which most designs around the world are based, were always power rated for a water flow that gave a $\Delta T$ of 28°C. The maximum safe outlet temperature is 60°C, meaning that to draw full power the inlet water must be under 33°C and, for any power overload to be sustained, the inlet water temperature must be commensurately lowered in order to maintain the 60°C outlet.

Overspeeding of water-brake machines is not advisable, particularly when running with high exit temperatures, as it may cause cavitation damage that would quickly destroy the working chambers. This destructive phenomenon can sometimes be detected by a distinctive crackling noise coming from the carcase and is caused by localized boiling and bubble collapse.

Special high-speed "cropped rotor" designs of some hydraulic dynamometers are made for testing race engines and gas turbines, etc. The power curve of this type of machine is flattened (lower torque) and shifted right (higher speed), with a much larger "hole" in the curve on the left (no torque at low speed). The Froude model designation indicates the angle of rotor crop, so the G4 unit has a full rotor while its high-speed variant has a 90 degree cropped rotor and is designated as G490.

Overloading eddy-current machines of any design is ill-advised as overheating and consequential distortion of eddy-current loss plates or coil may be caused. Like any dynamometer that converts kinetic energy to heat and removes it by flowing water, the $\Delta T$ across the machine is critical to avoid internal boiling; therefore, if maximum power or power overload is required, ensure that plenty of water is flowing through the machine with as low an inlet temperature as possible.

*Note concerning the water quantity required to absorb a given power in water brakes.*

The volume and pressure of water required by water-brake dynamometers vary according to their design, operators are advised to obey the manufacturer's instructions. There are a number of "rules of thumb" that have been used for many years, but these tend only to relate to water flow per unit time and absorbed engine power; the missing parameter in these cases is the change in temperature (delta $T$ or $\Delta T$) across the machine. The exit temperature of any water brake has to remain under the critical temperature at which internal microboiling and cavitation begins; therefore any rule for dynamometer water flow has to state the calculated $\Delta T$.

The best calculation known to the authors was produced many years ago, so has stood the test of time, and states.

*For a $\Delta T$ of 22°C the water requirement for a variable-fill hydraulic dynamometer equals 39.2 L/kWh or about 6.4 imp. gall/bhph.*

The figure of 22°C was chosen because it was considered to be the sensible limit of the ΔT achieved when using an air-blast radiator.

It is important to understand that the maximum permissible outlet temperature determines the maximum permitted inlet temperature, which in turn determines the maximum power that a given machine can absorb under those site conditions. Contractual disputes have arisen when users have failed to understand this point and then claim the dynamometer is incapable of absorbing its rated power.

## Engine cranking and starting

Cranking and thus starting an engine when it is connected to a modern AC dynamometer and control system is straightforward. Any type of two-quadrant dynamometer may present the cell designer and operator with engine starting problems and is a factor to be remembered when selecting the dynamometer and shaft system. An engine-mounted starter, if used, must be capable of cranking the combined shaft system up to a suitable speed. A dynamometer-mounted starter system must neither compromise the torsional characteristics of the driveline nor the trunnion-mounted torque measurement accuracy (see Chapter 11: Mounting and Rigging Internal Combustion Engines for Test, for the requirements of rigging or providing starting systems).

## Choice of dynamometer type

The first question concerning the choice of a dynamometer is "Does it need to be a motoring (four-quadrant) machine?". If "Yes," then an AC motor–based system may be the best, indeed only choice. Certainly, for testing E-vehicle drive motors and powertrain assemblies and in any facility doing highly dynamic ICE or transmission testing, an AC dynamometer system is the default choice. However, there are some specialist fields, such as EMC cells (see Chapter 18: Anechoic and Electromagnetic Compatibility

Testing and Test Cells), where DC or hydrostatic machines should be used because of the electromagnetic environment.

So, a question that may arise in the mind of the reader is "Why consider any other type of dynamometer than a modern AC unit?". The most obvious answer is "If full four-quadrant operation is not required why pay the considerable premium to have it?".

It must be remembered that the technology involved in variable-speed, high-power, four-quadrant AC or DC dynamometers does not come without considerable support system costs. Exporting electrical power into a factory or national grid requires that a number of technical and legislative problems are solved (see Chapter 4: Electrical Design Requirements of Test Facilities) and frequently requires the purchase of new isolating transformers and special power cables, none of which are cheap.

In summary, the final choice of dynamometer for a given application may be influenced by some of the following factors (also see Table 9.3):

1. The speed of response required by the test sequences being run. The terms steady state, transient, dynamic, or high dynamic are very subjective, but each industrial user will be aware of the speed of response required *to match the needs of their test sequences*.
2. Overloads. If significant transient and/or occasional overloading of the machine is possible, as in some motor sport tuning facilities, a hydraulic machine may be preferable, in view of its greater tolerance of such conditions. The torque-measuring system needs to have an adequate overload capacity.
3. Large and frequent changes in load. This can and does give rise to problems with eddy-current machines, due to rapid expansion and contraction of the loss plates causing distortion that leads to water-seal failure or closure of the rotor/stator air gap.
4. How are engines to be started? If a nonmotoring dynamometer is favored it may be necessary to fit a separate starter to the dynamometer shaft. This represents an additional maintenance commitment and may increase the systems' rotational inertia.
5. Wide range of engine sizes to be tested. The "turndown ratio" of any test cell system, not just the dynamometer has to be considered. When using water-brake machines it may be difficult to achieve good control and adequate accuracy when testing the smallest engines and the minimum dynamometer torque may also be inconveniently high. The greater the range of engine, the greater is the chance of hitting torsional vibration problems in any one dynamometer/shaft combinations (see Chapter 11: Mounting and Rigging Internal Combustion Engines for Test). There are many cases on record where two cells, the individual power ratings of which comfortably span half of the range of engines, produce far better results with less operational problems than a single cell catering for too wide a power range.

**TABLE 9.3 List of dynamometer types with summary of advantages and disadvantages.**

| Dynamometer type | Advantages | Disadvantages |
|---|---|---|
| *Sluice gate* Examples: Froude $DPX_n$ and $DPY_n$ models | Obsolete, but many cheap and reconditioned models in use worldwide, robust. | Slow response to change in load. Manual control that is not easy to automate. |
| *Variable-fill water brakes* Current examples: Froude "F" types, AVL "Omega" range, Horiba DT range | Units can match the most powerful prime movers built. Capable of medium-speed load change, automated control, robust, and tolerant of overload. | "Open" water system required. Can suffer from cavitation or corrosion damage. |
| *"Bolt-on" variable-fill water brakes* Current examples: units made by Piper (United Kingdom), Taylor, Go-Power (United States) | Cheap and simple installation. Available up to 1000 kW. | Lower accuracy of measurement and control than the best fixed machines. |
| *Disk-type hydraulic* | Suitable for high speeds such as required in small turbine testing. | Poor low-speed performance. |
| Hydrostatic. *Consisting of constant speed/variable "swash" power unit and variable-speed fixed "swash" dynamometer unit* | For special applications, very low inertia, provides four-quadrant performance. | Mechanically complex, noisy, and expensive. System contains large volumes of high-pressure oil. |
| *DC electrical motor* Produced by most major test equipment manufacturers in the past largely replaced by AC | Mature technology. Four-quadrant performance. Limited in automotive top-speed range. | High inertia, commutator and brushes require maintenance, harmonic distortion of supply possible. |
| *Asynchronous motor (AC)* Produced by most major test equipment manufacturers | Now a mature technology, produced in models covering all but the largest automotive prime-mover range. | Expensive. Large drive cabinet needs suitable housing. Care must be taken in connection into the facility power system. Some RF emission. |
| *Permanent magnet motor* Produced by AVL, MTS, and others | Lowest inertia, most dynamic four-quadrant performance. Small size in cell. | Very expensive. Large drive cabinet needs suitable housing. |

*(Continued)*

**TABLE 9.3 (Continued)**

| Dynamometer type | Advantages | Disadvantages |
|---|---|---|
| *Eddy current*, dry gap, water cooled | Low inertia (disk-type air gap) well adapted to computer control. Mechanically simple and comparatively cheap. | Vulnerable to poor cooling supply. Not suitable for sustained rapid changes in power (thermal cycling). |
| Friction brake | Special-purpose applications for very high torques at low speed. | Limited speed range. |
| Air brake | Cheap. Very little support services needed. | Noisy with limited control accuracy. |
| Hybrid electric/water brake | Possible cost advantage over sole electrical machine. | Complexity of construction and control. |

*AC*, Alternating current; *DC*, direct current; *RF*, radio frequency.

6. How are engines to be started? If a nonmotoring dynamometer is favored it may be necessary to fit a separate starter to the dynamometer shaft. This represents an additional maintenance commitment and may increase inertia.
7. Is there an adequate supply of cooling water of satisfactory quality? Hard water and some inappropriate water treatments that are susceptible to the high centrifuging action of water brakes will result in blocked cooling passages. The calcite deposits from hard water may break away and cause mechanical jamming of control valves. Plant water supplied at over 30°C will compromise sustained full-power use in many hydraulic and some eddy-current designs.
8. Is the pressure of the water supply subject to sudden variations? Sudden pressure changes or regular pulsations will affect the stability of control of hydraulic dynamometers. Eddy-current and indirectly cooled machines are unaffected providing inlet flow does not fall below emergency "low-flow" trip levels.
9. The electrical supply system is a major consideration when using AC and DC machines, and the reader needs to read Chapter 4, Electrical Design Requirements of Test Facilities, in full. With the exception of air brakes and manually controlled hydraulic machines, all dynamometers are affected by electrical interference and voltage changes.

**10.** Annual running time, which is not an exclusively industrial problem but has to be considered by training establishments and university facilities. From a technical viewpoint this is only a problem if the machine will spend long periods out of use; then the possibilities of corrosion must be considered, particularly in the case of hydraulic or wet gap eddy-current machines. Can the machine be drained readily? Should the use of corrosion inhibitors be considered? From a cost accounting viewpoint, it makes no sense having expensive plant that is sitting idle and not earning review, which is why comparatively cheap and robust disk-type eddy-current machines are widely used in engine test facilities with intermittent work patterns.

Chapter 10

# Chassis dynamometers, rolling roads and hub dynamometers

## Chapter Outline




**Engine Testing. DOI: https://doi.org/10.1016/B978-0-12-821226-4.00010-3**

# Part 1: Conventional chassis dynamometer or "rolling road" designs and use

## Introduction

The terms "chassis dynamometer" (or "dyno") and "rolling road" tend to be used interchangeably around the world; in this book we use the latter title to cover the smaller, packaged, aftermarket machines. The common feature of all chassis dynamometers is that the unit they test is a vehicle, fitted with its wheels that run on rollers, this differentiates them from "full powertrain" dynamometer rigs on which the vehicle, or its powertrain components in vehicle configuration, is run with the drive shafts directly connected to individual wheel or hub dynamometers.

Since the early 1970s the rise in the number of chassis dynamometers installed worldwide has been predominately due to the development of increasingly stringent vehicle emission regulations and their chassis dynamometer-based drive cycles. The majority of these machines are designed specifically for the purpose of proving compliance with such regulations and the homologation of all vehicle models sold to the general public. However, the same machines are used, sometimes within special enclosures, to optimize the performance of the "sum of the parts" that form modern vehicles. This is achieved by the vehicle driving in controlled and repeatable conditions while experiencing resistance at the wheels that produce, with appropriate accuracy, that would be experienced on the road in real life.

The advent of hybrid vehicles has complicated and multiplied the task of measuring the internal combustion engines (ICE) emissions and optimizing vehicle power sharing and control calibration; for some types of testing, it has required new dynamometer features, such as four independent wheel rollers.

The "classic" single- or double-axle chassis dynamometer may be converted, by being fitted with enhanced control and data acquisition systems, into a "Vehicle-in-the-Loop" test bed capable of a much wider range of driving style and road simulations than their original design brief. Testing the vehicle when fitted with wheel–hub torque transducers can shift the point of power measurement and remove tire-induced inaccuracies. It can also by simulation effectively turn the tires into part of the test rig rather than part of the unit under test (UUT).

Hub dynamometers and simulation systems are able to overcome the shortcomings of the unrealistic performance of the vehicle running on single-axle dynamometers where the nondrive wheels are not rotating, and their braking torque is not measured.

## Roles of the chassis dynamometer

Chassis dynamometer of single-axle, double-axle and four roller designs within automotive development and production facilities are used for the testing the whole vehicle in the following roles:

- vehicle emission testing in strict compliance with national and international regulations (see Chapter 17: Engine Exhaust Emissions);
- operation of onboard diagnostic (OBD) systems under loaded conditions and simulated faults;
- operation of hybrid vehicle control strategy, regenerative braking, and power output;
- noise, vibration, and harshness (NVH);
- drivability testing with varied driver and route profiles;
- performance under extreme climatic conditions, solar load, etc.;
- electromagnetic immunity and electromagnetic compatibility (EMC) compliance testing;
- wind tunnel performance (drag, side load stability, etc.); and
- power testing and vehicle cooling systems.

In addition to the abovementioned fact, simpler designs are used in the aftermarket for tuning and testing compliance with governmental vehicle condition tests. These varied requirements have called into existence a hierarchy of chassis dynamometers, some housed in test chambers of increasing complexity, that include—from simple to complex:

- brake testers, installed in workshops;
- in-service, small roller machines, for tuning and faultfinding installed in aftermarket or tuning workshops;
- end-of-line (EOL) production and brake test rigs, installed within the factory building; and
- mileage accumulation dynamometers installed usually in an open-sided facility where the vehicles are fitted with robotic drivers and automatic fueling systems.

## *Genesis of the "rolling-road" dynamometer*

The idea of running a complete vehicle under power while it was at rest was first conceived by railway locomotive engineers before being adopted by the road vehicle industry. As a matter of historical record, the last steam locomotives built in the United Kingdom were tested on multiple-axle units, with large eddy-current dynamometers connected to rollers with rail-line profiles under each driven axle, the tractive force being measured by a mechanical linkage and spring balance. Only the nameplate of that gigantic "rolling rail" unit, built by Froude of Worcester England, is preserved in the York Railway Museum.

Today the chassis dynamometer is used almost exclusively for road vehicles, although there are special machines designed for forklift and articulated off-road vehicles. The advantages to the vehicle designer and test engineer of having such facilities available are obvious; essentially, they allow the static observation and measurement of the performance of the complete vehicle while it is operating within its full range of power and, in most respects, in motion.

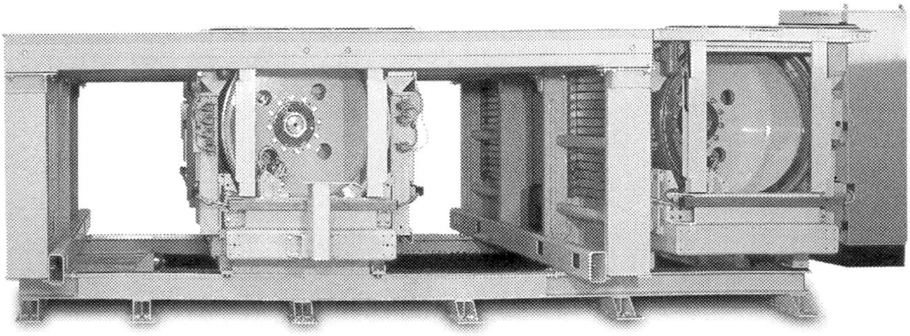

**FIGURE 10.1** A variable wheelbase light vehicle chassis dynamometer made up of two 48″ roll units mounted within a common frame. *Photograph courtesy AIP GmbH & Co. KG.*

Before about 1970 most machines were comparatively primitive rolling roads, characterized by having rollers of rather small diameter, which inadequately simulated the tire contact conditions and rolling resistance experienced by the vehicle on the road. Such machines were fitted with various fairly crude arrangements for applying and measuring the torque resistance, while a single fixed flywheel was commonly coupled to the roller to give an approximate simulation of the tested vehicle's inertial mass.

The main impetus for development came with the rapid evolution of emissions testing in the 1970s. The diameter of the rollers was increased, to give more realistic traction conditions, while trunnion-mounted direct current (DC) dynamometers with torque measurement by strain gauge load cells and more sophisticated control systems permitted more accurate simulation of road load. The machines were provided with a range of flywheels to give steps in the inertia; precise simulation of the vehicle mass was achieved by trimming the "iron inertia" by control of load contributed by the drive motor.

In recent years the development in digital control techniques and high-powered variable-speed electrical drives has meant that incremental flywheels have been dispensed with in most emission testing designs and been replaced by single (2 × 2) or double (4 × 4) roll machines reliant on electrical simulation of the vehicles' road load, route, and driver style characteristics (Fig. 10.1).

## Some limitations of the standard emissions chassis dynamometer in wider research work

A chassis dynamometer measures the tractive force generated at the tire/drum interface, which is not an accurate way to dynamically measure torque transmitted by individual half shafts because of, among other effects, the energy absorbing and damping effects of the tires.

However, because of the easy availability and convenience of a chassis dynamometer in providing a whole-vehicle support system and a means of measuring its overall performance under simulated road conditions, single-axle designs are sometimes used as the test bed for evaluating the performance of a vehicle components or control subsystem; this can for some work, prove to be poor practice.

In its original form the standard single-axle emission machine of whatever roll size was built specifically to meet the requirements of legislative emission drive cycles. Such test sequences, compared with engine dynamometer test cycles, had quite wide speed and load tolerance bands and are fit for the purpose for which they are designed, but not for precise comparison of performance while adjusting some powertrain component. Variability in test results caused by differences in ventilation airflow or in tire diameter and pressure alone can "drown out" the subtle performance changes being investigated within the powertrain; phenomena avoided within a well-designed ICE or motor test cell. If a high-fidelity test with a full powertrain configuration is required, then wheel-force measuring hubs should be considered essential.

Like all devices intended to simulate a dynamic phenomenon, the chassis dynamometer has other limitations. A vehicle, when restrained by elastic ties and delivering or receiving power through contact with a rotating drum, is not dynamically identical with the same vehicle in its normal state as a free body traversing a fixed surface. During motion at constant speed the differences are minimized and largely arise from the absence of airflow and the limitation of simulated motion to a straight line. Once acceleration and braking are involved, however, the vehicle motions in the two states are quite different. To give a simple example a vehicle on the road is subjected to braking forces on all wheels, whether driven or not, and these give rise to a couple about the center of gravity, which causes a transfer of load from the rear wheels to the front, with consequent pitching of the body. On a single-axle chassis dynamometer, however, the braking force is applied only to the driven wheels, while the forward deceleration force acting through the center of gravity is absent and replaced by tension forces in the front-end vehicle restraints. It will be clear that the patterns of forces acting on the vehicle, and its consequent motions, are quite different in the two cases. These differences make it difficult to investigate some aspects of vehicle ride and NVH testing on the chassis dynamometer.

A particular area in which the simulation differs most fundamentally from the "real-life" situation concerns aspects of driveline oscillation, with its associated judder or "shuffle."

One of the authors and the late Dr. M. Plint had occasion to study this problem in some detail and, while the analysis is too extensive to be repeated here, one or two of their conclusions may be of interest:

- The commonly adopted arrangement, whereby the roll inertia is made equal to that of the vehicle, gives a response to such disturbances as are induced by driveline oscillations, that is, judder, that are significantly removed from on-road behavior.

- For reasonably accurate simulation of phenomena such as judder, roll inertia should be at least five times vehicle inertia.
- Electronically simulated inertia is not effective in this instance; actual mass is necessary.
- The test vehicle should be anchored as lightly and flexibly as possible; this is not an insignificant requirement, since it is desirable that the natural frequency of the vehicle on the restraint should be at the lower end of the range of frequencies, typically 5–10 Hz, that are of interest.

In road simulation testing where the level of repeatability is more important that real-life simulation accuracy, there is a role for modern $4 \times 4$ chassis dynamometers in the repetition of road sections recorded during real drive emissions (RDE) testing. When unexpectedly high transient emissions are recorded at particular problematic road configurations or traffic conditions, those conditions are recorded and repeatedly re-run under laboratory conditions on a chassis dynamometer. This will certainly allow the interactions of the vehicle control systems at these "problem points" to be examined, tuned, then re-examined; the possible flaw in the strategy is that the vehicles' conditions (heat-soak etc.) prior to the emissions spike also have to be repeated, which increases the test complexity and time between tests.

## The road load equation

The behavior of a vehicle under road conditions is described by its unique road load equation (RLE). It is a fundamental requirement of a true chassis dynamometer that it has to resist the torque being produced at the vehicle drive wheels in such a way as to simulate the "real-life" resistance of the road and atmosphere to vehicle motion.

The RLE is the formula that calculates the change in torque required with any change of vehicle speed.

To simulate the real-life performance of a given vehicle, the RLE defines the traction or braking force that is called for under all *straight-line* driving conditions, which include:

- steady travel at constant speed and coasting down on a level road;
- hill climbing and descent;
- acceleration, overrun, and braking;
- transitions between any of the above; and
- effects of atmospheric resistance, load, towed load, tire pressure, etc.

The RLE for a given vehicle defines the *tractive force* or *retarding force* $F$ (Newton) that must be applied in order to achieve a specified response to these conditions. It is a function of the following parameters:

| | |
|---|---|
| Vehicle specific | |
| Mass of vehicle | $M$ (kg) |
| Components of rolling resistance | $a_0$ (N) |
| Speed-dependent resistance | $a_1 V$ (N) |
| Aerodynamic resistance | $a_2 V^2$ (N) |
| External | |
| Vehicle speed | $V$ (m/s) |
| Road slope | $\theta$ (rad). |

A common form of the RLE is:

$$F = a_0 + a_1 V + a_{(2)} V^2 + \frac{MdV}{dt} + Mg\sin\theta \quad (10.1)$$

where $MdV/dt$ is the force to accelerate/brake vehicle and $Mg \sin \theta$ is the hill climbing force.

More elaborate versions of the equation may take into account such factors as tire slip and cornering.

The practical importance of the RLE lies in its application to the simulation of vehicle performance: it forms the link between performance on the road and performance in the test department.

To give a "feel" for the magnitudes involved, the following equation relates to a typical four-door saloon of moderate performance, laden weight 1600 kg, and $C_d$ factor of 0.43:

$$F = 150 + 3V + 0.43V^2 + 1600\frac{dV}{dt} + 9.81\sin\theta$$

The various components that make up the vehicle drag are plotted in Fig. 10.2A shows the level road performance and makes clear the preponderant influence of wind resistance.

The RLE predicts a power demand at the road surface of 14.5 kW at 60 mph (96.6 kph), rising to 68.8 kW at 112 mph (180 kph) on a level road.

These demands are dwarfed by the demands made by hill climbing and acceleration. Thus, in Fig. 10.2B, to climb a 15% slope at 60 mph calls for a total power input of 14.5 + 63.2 = 77.7 kW, while in Fig. 10.2C, to accelerate from 0 to 60 mph in 10 seconds at a constant acceleration calls for a maximum power input of 14.5 + 115 = 129.5 kW.

## Calibration, coast down, and inertia simulation

Like any other dynamometer, the torque measurement system of a chassis dynamometer, in this case also used to measure tractive force, must be

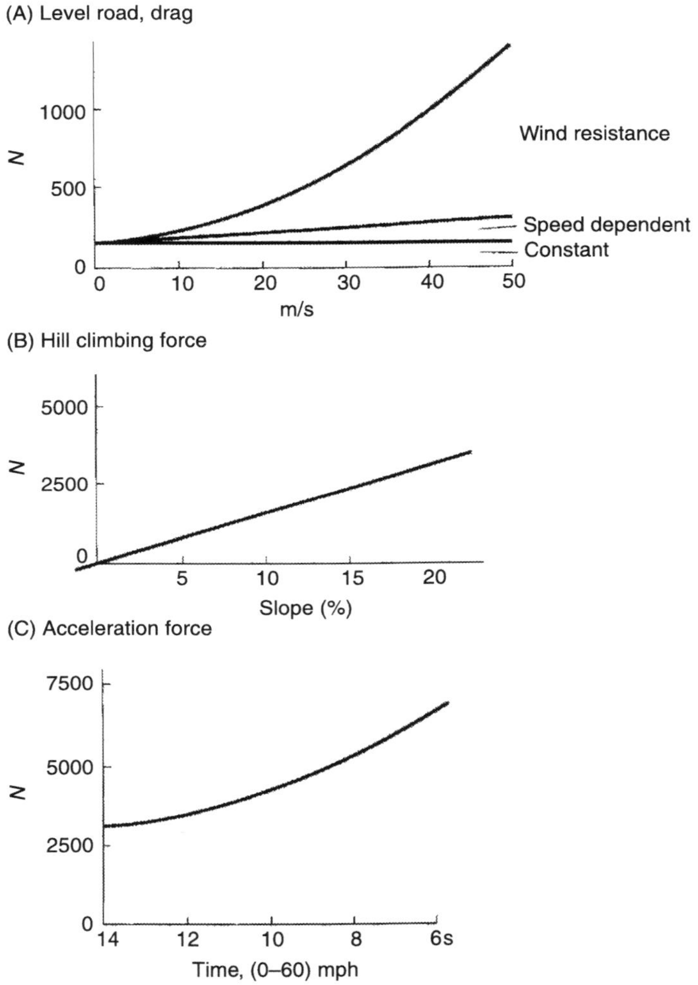


**FIGURE 10.2** Graphs visualizing drag on a vehicle of laden weight 1600 kg: (A) level road performance, (B) hill climbing, (C) acceleration force (force is in Newton).

regularly calibrated. Most machines use a strain gauge load cell that measures the torque reaction of the cradle-mounted drive motor, although a torque shaft or torque flange may also be used. These measurement systems are calibrated in a similar way to that described for engine dynamometers in Chapter 9, Dynamometers: The Measurement and Control of Torque, Speed, and Power, with the possible detailed difference that, in the case of the chassis dynamometer, less midrange, if any, calibration weights are used during *routine* checks. Since the machinery is installed below ground level, in a pit, access to the load cell is often restricted; therefore all good designs provide a calibration mechanism that allows the work to be done at cell floor level.

Speed measurement, on all but the aftermarket units, is done by an optical encoder.

Prior to any test of a vehicle, or the dynamometer itself is carried out, the machine must be run through a *warm-up routine* to thermally stabilize the whole system and reduce both parasitic losses and "stiction" in the torque measuring system. Because of the time required to fully rig a vehicle undergoing emission development tests, most machines are provided with a mechanism that can lift the (fully rigged) vehicle up, so the wheels are clear of the rolls. In this way the warm-up routine can be run without disruption to the vehicle under test except for possible slackening of the restraint straps. The control system of a chassis dynamometer in meeting the set RLE may be checked by running a number of *coast-down* tests. Using the mass and RLE for perhaps three vehicles representing the bottom, middle, and top of the range of vehicles to be run on the dynamometer, a plot of the theoretical coast-down graph of each is calculated between, typically, 130 kph (80 mph) and 16 kph (10 mph). The chassis dynamometer controller is then programed with the details of the first test vehicle, run to a rotational speed equivalent to 135 kph, without any vehicle mounted, and allowed to coast down under control; the actual speed–time graph curve of the coast down should exactly overlay that calculated.

To exactly calibrate a chassis dynamometer to a particular car, a similar technique can be adopted. This requires several actual coast downs to be carried out by the test vehicle, at the correct loaded mass, on a straight level track in near still air. The average speed–time curve of these runs is then used to check and tune the control parameters of the dynamometer controller's RLE until the good correlation is achieved.

*Note*: It is a common industrial practice to use the term "inertia" to mean the simulated or equivalent vehicle mass, rather than the moment of inertia of the rotating components. The inertia is thus quoted in kg rather than $kgm^2$.

Any differences between the mass inertia of the vehicle and that of the roller masses are compensated for through the motor control system, which must be sufficiently dynamic so that the machine's response to rapid vehicle accelerations is realistic.

The vehicle is essentially static in space when running on the rolls; therefore it has no momentum. The "missing" momentum or mass inertia is generated in the rotating roller masses. As an example, the base inertia of one particular single-axle, 125 kW, 48-in. roll dynamometer corresponds to a vehicle weighing 1369 kg (3000 lb), but, through control of its motor's torque input or absorption, it is capable of simulating the inertia of vehicles of weights between 454 kg (1000 lb) and 2722 kg (6000 lb).

It is a natural but incorrect perception when driving a vehicle in a forward direction on a chassis dynamometer, for the human driver to suppose that, if the vehicle restraint system suddenly failed, the vehicle would shoot forward. In fact, because the rolls have the momentum and they are rotating in the opposite direction to the wheels of the car, in those circumstances the car would shoot off the machine backward.

## Tire and roll diameter effects

Tires used, even for a short time, on rolling roads may be damaged by heating and distortion. Some dynamometer systems are fitted with tire cooling systems to reduce tire damage; however, all tires used for any, but short-duration tests should be specially marked and changed before the vehicle is allowed on public roads.

The majority of the rolling resistance is the consequence of hysteresis losses in the material of the tire and this causes heating of the tire. To minimize any overheating all the tires should be inflated to the same normal road maximum and to minimize differential gear load effects must be of the same condition of wear and effective diameter.

There is an obvious difference in running a tire on a flat road and running it on rollers of various diameters. A widely accepted formulation describing the effect of the relative radii of the tire and roller is:

$$F_{xr} = F_{(x)}\left(1 + \frac{r}{R}\right)^{(1/2)} \tag{10.2}$$

where $F_{xr}$ is the rolling resistance against the drum, $F_{(x)}$ is the rolling resistance on a flat road, $r$ is the radius of the tire, and $R$ is the radius of the drum.

Fig. 10.3A shows the situation diagrammatically for a "roll smaller than wheel" configuration and Fig. 10.3B the corresponding relation between the rolling resistances: this shows that rolling resistance (and hence hysteresis loss) increases linearly with the ratio $r/R$. For tire and roller of equal diameter, the rolling resistance is 1.414 × resistance on a flat road, while for a tire three times the roller diameter, easily possible on a brake tester, the resistance is doubled. A typical value for the "coefficient of friction" (rolling resistance/load) would be 1% for a flat road.

To indicate the magnitudes involved a tire bearing a load of 300 kg running at 17 m/s (40 mph) could be expected to experience a heating load of about 500 W on a flat road, increased to 1 kW when running on a roller of one-third its diameter. Clearly the heating effects associated with small-diameter rolls are not negligible and the large roller dynamometers are to be recommended. Where the effects of running on a convex surface are

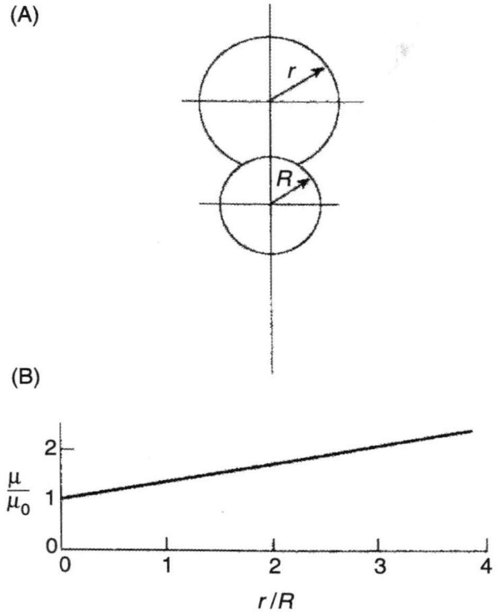


**FIGURE 10.3** (A) Diagram of the relationship between tire and a (small roll) chassis dynamometer roller diameter. (B) Linear relationship between rolling resistance and the tire/roll diameter ratio.

considered to be a significant restraint on test accuracy, flat-track or belt dynamometers may be used (see next).

## Types of chassis dynamometer

### Automotive aftermarket and end-of-line brake system testers

Service garages and government test stations are equipped with quite simple but durable machines that are used for the statutory "in-service" testing of cars and commercial vehicles. They are installed in a shallow pit and consist essentially of two pairs of rollers, typically of 170 mm diameter and often having a grit-coated surface to give a high coefficient of adhesion with the vehicle tire. The rollers are driven by geared variable-speed motors and two types of test are performed. Either the rollers are driven at a low surface speed, typically 2–5 km/h, and the relation between brake pedal effort and braking force is measured, or the vehicle brakes are fully applied and torque from the rig motor increased and measured until the wheels slip (less recommended due to likely tire damage).

The testing of a modern ABS brake system requires a separately controlled roller for each wheel and a much more complicated interaction with the vehicle's braking control system and wheel speed transducers. Modern

"EOL" brake test machines are used by vehicle manufacturers in a series of final vehicle check stations and also by major service depots; they therefore have features such as communication with the test vehicle control systems, via a "breakout box" and test loom, that are model specific.

### *Tire and brake testing dynamometers*

Tires for every type of vehicle and airplane are tested by running them on a large-diameter steel drum; the largest of which are some 5 m in diameter and the vast majority have plain machined surfaces. The tire under test is fixed on its correct vehicle wheel, which may be fitted with its complete brake assembly. The test assembly is mounted on a rigid arm, fitted with strain gauges to measure imposed and resultant force vectors in all directions. The arm is hydraulically actuated and can press the wheel onto the roller with variable force to simulate realistic vehicle loading. The arm should also be able to turn the test wheel at an angle to the drum shaft axis to simulate "skewing" of the tire to the road surface.

The drum is connected to an electrical dynamometer, the control system of which is able to simulate the vehicle mass. Aircraft tire, wheel, and brake assemblies are tested on this type of test rig, which are capable of simulating the emergency landing and braking loads of a fully laden aircraft. The most demanding aerobrake and tire test simulates aborted takeoff immediately before V1 status is reached during which the brakes and wheels of a fully loaded commercial airliner will heat up to a point that would explode the tires. The test is required to prove that destructive tire explosions are prevented by the melting of plugs built within the wheel; always an exciting routine!

However, such tire test rigs, even in the automotive sizes, can suffer from explosive tire failures and brake system fires so they require the highest possible standard of safety procedures and management relating to the exclusion of personnel from the rig area during test running.

### *"Rolling roads" for in-service testing or tuning*

For a vehicle maintenance garage the installation of a rolling road was a considerable investment, so to cater for this market a number of manufacturers produced complete packaged units, often based on nominally 8.5-in. (216 mm) diameter rollers, and requiring minimum subfloor excavation. This market was once dominated by American suppliers and grew rapidly when the US Environmental Protection Agency (EPA) called for annual emissions testing of vehicles, based on a chassis dynamometer cycle (IM 240). This test, due to the cost associated of its equipment and operation, has now been very largely abandoned and replaced by use of the vehicles own OBD 11 system, which are now judged to give a satisfactory indication of the condition of the in-service tailpipe emissions. In the United Kingdom approved

MOT test stations are fitted with the same type of modular brake test rolling roads for two-wheel drive vehicles. Four-wheel drive vehicles have a brake test carried out on the open road while fitted with an approved accelerometer.

For aftermarket facilities required to do engine tuning work beyond the basic legislative tests any road load simulation capability is usually limited to a choice of one or two flywheels covering the light and medium passenger car mass. Such units are fitted with a comparatively simple data recorder and control system capable of producing a result printout. Most of such testing is concerned with fault diagnosis, possibly using OBD readers, brake balancing, and power checks. The tests have to be of short duration to avoid overheating of engine and tires.

In-service rolling roads for testing large trucks are confined almost exclusively to the United States. They are usually based on a single large roller capable of running both single- and double-axle tractor units. The power is usually absorbed by a portable water brake, such as those manufactured by GoPower, which may also be dismounted from the roller and used for direct testing of truck engines, either between the chassis members or on an engine trolley, a useful feature in a large OEM agency, overhaul, and test facility.

## *Chassis dynamometer and rolling roads for end-of-line production testing*

EOL chassis test rigs range from simple roller sets used for first start and basic function testing of Motor cycles (Fig. 10.4) and two-wheel drive cars to multiaxle units with variable geometry of roll sets and ABS checking capability. Whatever the level of complexity all EOL rigs require some common design features and any specification for such an installation should take them into account:

- The design and construction of the machine must minimize the possibility of damage from vehicle parts falling into the mechanism. It is not unknown for small fixings, left in or on the vehicle during assembly, to shake down into the rolling road, where they can cause damage if there are narrow clearances or converging gaps between, for example, the rolls and lift-out beam, the wheelbase adjustment mechanism, or the sliding floor plate system.
- The vehicle must be able to enter and leave the rig quickly yet be safely restrained during the test. The usual configuration is for all driven vehicle wheels to run between two rollers. Between each pair of rolls there is a lift-out beam that allows the vehicle to enter and leave the rig. When the beam descends it lowers the wheels between the rolls and at the same time small "anti-climb-out" rollers swing up fore and aft of each wheel. At the end of the test, with the wheels at rest, the beam rises, the restraining rollers swing down, and the vehicle may be driven from the rig.

**FIGURE 10.4** Motorcycle EOL chassis dynamometer undergoing predelivery testing. Photograph supplied courtesy of Saj Test Plant Pvt. Ltd.

- To restrain the vehicle from slewing from side to side to a dangerous extent, specially shaped side rollers should be positioned between the rolls at the extreme width of the machine, where they will make contact with the vehicle tire to prevent further movement. These rollers must be carefully designed and adjusted so that they do not cause tire damage; they are not recommended for use with vehicles fitted with low-profile tires due to possible "pinching" damage.

Pre-1990s rigs, some still in use, that were designed with adjustable center distances between axles, had shaft, chain, or belt connections between the front and rear roller sets; this complexity has been obviated by modern electronic speed synchronization of separate drive motors.

## *Chassis dynamometers for emission testing and type homologation*

The standard emissions tests developed in the United States in the 1970s were based on a rolling-road dynamometer developed by the Clayton Company and having twin rolls of 8.625-in. (220 mm) diameter. This machine became a de facto standard despite its limitations, the most serious of which was the small roll diameter, which resulted in tire contact conditions significantly different from those on the road.

Later models, which are still in use, have pairs of rollers of 500 mm diameter, connected by a toothed belt, and between which the vehicle tire sits. The roller sets are connected by a toothed belt to a set of declutchable flywheels that simulate steps of vehicle inertia. Any adjustment of simulated vehicle inertia between the flywheel mass increments is adjusted by the control system using a small ($\leq$60 kW) four-quadrant DC dynamometer connected to the flywheel shaft. Such machines were vulnerable to abuse, notably by violent brake application that can induce drive vehicle wheel bounce and thus destructive stresses in the flywheel clutch mechanism.

When proven, high-power thyristor control systems became available, the US Environmental Protection Agency (EPA)-approved machines had a single pair of rollers of 48-in. diameter and 100% electronic inertia simulation. Most of the standard designs, now available from the major suppliers, are of this type and most are of a compact design having a double-ended alternating current (AC) drive motor with rollers mounted overhung on each stub shaft.

The performance figures quoted next are for a modern 48-in. roll chassis dynamometer designed for exhaust emission testing passenger cars to international industrial standards (in this case an AVL ROADSIM™ 48 MIM unit):

nominal tractive force in motoring mode (per axle): 5870 N at $V \leq 92$ km/h;

nominal power in motoring mode (per axle): 150 kW at $V \geq 92$ km/h and $<189$ km/h;

max. speed 200 km/h;

roller diameter 48 in. (1219.2 mm);

distance between outer roller edges 2300 mm;

inertia simulation range (2WD) 454–2500 kg;

inertia simulation range (4WD) 800–2500 kg;

max. axle load 2000 kg;

tolerance of tractive force measurement $\pm$0.1% of full scale;

tolerance of speed measurement $\leq$0.02 km/h;

tolerance of tractive force control $\leq$0.2% of full scale; and

tolerance of inertia and road load simulation $\pm$1% of calculated value (but not better than force control).

This type of standard, pretested package enables chamber installation time to be minimized.

## Four independent roll set chassis dynamometers

Differing from the two solid axle designs by having four rolls with their own motoring and/or absorbing units, these machines range from End of Line (EOL) rigs Fig. 10.4 to complex development test rigs, some having steered articulation of each (wheel) roll unit. Electronically controlled transmission systems and limited-slip differential units on vehicles can also be tested on this type of dynamometer by the simulation of individual wheels losing traction.

The production rigs permit the checking of onboard vehicle control systems, wheel speed transducers, and system wiring by simulation of differential wheel resistance and speeds of rotation.

The development of hybrid vehicles has given impetus to the development of independent four-wheel chassis dynamometers, but for much development work they suffer from the vagaries of measurement induced by the tires and the tire/roller contact and are replaced by the use of wheel or hub dynamometers (see next).

## The emission dynamometer cell environs

As with all test facilities dealing with the measurement of exhaust emissions, chassis dynamometers should be built on sites that are out of the drift zone of automotive, agricultural, and industrial airborne pollutants. All facilities have to be capable of running within the environmental conditions prescribed by current regulations which will include cold-start conditions.

Since it is more energy efficient to carry out several sub-0°C cold-start tests while the chamber is cold, the cell designers have to decide either to have the cell built large enough to cold soak multiple cars or to have a smaller dynamometer cell and a separate refrigerated cold soak room for extra vehicles, perhaps using a refrigerated ISO container.

The Worldwide Harmonised Light Vehicle Test Procedure (WLTP) requires a soak area and test bed room set point of 23°C (296K), which needs to be actively controlled on a 5-minute running average.

If both gasoline and diesel vehicles are to be tested in the same cell, then the space required for the emissions equipment can be considerable and needs careful layout. While the emission analyzers and even sample bag racks can be in a large control room, the AC or DC drive cabinets for the dynamometer must be in a separate, clean, well-ventilated room with a 15-m or less, dedicated, subterranean cable route to the motors.

The blower needs to be mounted out of working sound range of the control room on the building roof while the housing of the sample and calibration gas bottles needs to be done according to the recommendations in Chapter 17, Engine Exhaust Emissions.

## Mileage accumulation facilities

The development of the modern mileage accumulation dynamometer facility was trigged by the United states–developed EPA regulations that required that the emissions of passenger cars and light vans had to be tested after 50,000 miles, accumulated either on a track or on a chassis dynamometer using a prescribed driving cycle known as the "AMA" (40 CFR 86, Appendix IV). The cost and the physical strain of using human drivers on test tracks or public roads for driving vehicles the prescribed distances, in as little time as possible, are too high, so special chassis dynamometer systems were developed for running the specified sequences under automatic and robotic control, commonly over a period extending to 12 weeks.

In the mid-1990s the prescribed distance was doubled and resulted in the EPA allowing manufacturers to develop and submit their own strategies for proving the durability of their vehicles' emission control systems.

A number of manufacturers developed dynamometer driving cycles that were more severe than the AMA and that could reach 100,000 mi in a shorter amount of time. These are referred to as "whole-vehicle mileage accumulation" cycles.

Other manufacturers developed techniques for aging the catalytic converter and oxygen sensor on a test bench to the equivalent of the full useful life distance and then reinstalled them on the vehicle for emissions testing; this technique is called "bench aging."

Mileage accumulation dynamometers for passenger cars now tend to be of a very similar design as the emission test machines, being based on 48-in. or larger diameter, directly coupled to an AC motor in the range of 120–150 kW power rating.

In order to fully automate mileage accumulation tests, the vehicle has to be fitted with a robot driver (see section on "robot driver" later in this chapter) and a system to allow automated fueling. Because the vehicle's own cooling system is stationary, a large motorized cooling fan, facing the front grille, is essential. This fan will be fitted with a duct to give a reasonable simulation of airflow, at the vehicle's front grille, under road conditions. The fan speed is usually controlled to match apparent vehicle speed up to about 130 km/h; above that speed the noise produced and power needed become too high to be practical on most sites. The fans and their discharge ducts have to be accurately positioned and very securely anchored to the ground.

Since mileage accumulation facilities are usually housed within an open-walled structure with just a roof and run 24 hours a day, they have the potential of being "bad neighbors" and must be suitably shielded to prevent noise nuisance.

## Noise, vibration, and harshness and electromagnetic compatibility chassis dynamometers

The major features of anechoic and electromagnetic compatibility (EMC) cell design are discussed in Chapter 18, Anechoic and EMC Testing and Test Cells.

A critical requirement of an NVH facility is that the chassis dynamometer should itself create the minimum possible noise. The usual specification calls for the noise level to be measured by a microphone located 1 m above and 1 m from the centerline of the rolls. The specified sound level, when the rolls are rotating at a surface speed of 100 km/h, is usually $\leq 50$ dBA.

To reduce the contribution from the dynamometer motor and its drive system, it is usually located outside the main chamber, in its own soundproofed compartment, and connected to the rolls by way of a long shaft running through a transfer tube designed to minimize noise transfer. The design of these shafts can present problems because of their unsupported length and the nature of the couplings at each end; lightweight tubular carbon-fiber shafts are sometimes used. The dynamometer motor will inevitably require forced ventilation and the ducting will require suitable location and treatment to avoid noise being transferred into the chamber.

Hydrodynamic bearings are commonly used in preference to rolling element, but they are not without their own problems as their theoretical advantages, in terms of reduced shaft noise, may not be realized in practice unless noise from the pressurized supply oil system is sufficiently attenuated.

In a well-designed chassis dynamometer, the major source of noise will be the windage generated by the moving roll surface; this is not easily minimized at source or attenuated. Smooth surfaces and careful shrouding can reduce the noise generated by the roller end faces, but there is inevitably an inherently noisy jet of air generated where the roller surface emerges into the test chamber. If the roller has a roughened surface or is grit coated the problem is exacerbated. The noise spectrum generated by the emerging roller is influenced by the width of the gap and in some cases adjustment needs to be provided at this point.

The flooring over the dynamometer pit, usually of steel, sometimes aluminum plate, must be carefully designed and appropriately damped to avoid resonant vibrations.

A particular feature of NVH test cells is that the operators may require access to the underside of the vehicle for arranging sound recording, photographic, and lighting equipment. This is usually accessed by way of a trench at least 1.8 m deep, lying between the vehicle wheels and covered by removable floor segments.

The chassis dynamometer of an EMC facility must not emit electromagnetic noise into the cell across the frequency spectrum being used or investigated in the tests. Because of the wide spectrum of high-frequency emissions of a powerful isolated gate bipolar transistor-controlled AC dynamometer system, until very recently it was usual to specify DC motors for chassis dynamometers in EMC cells.

The subfloor layout of such complex and expensive cells has to be designed so that the thyristor-controlled DC drives and all other high-power electrical switching equipment is shielded by concrete and steel shielding, while special attention is paid to the layout of cables according to the practices recommended in Chapter 4, Electrical Design Requirements of Test Facilities. To test the EMC of systems such as ABS braking and traction control and minimize electromagnetic noise from the test equipment, independent wheel dynamometers that are based on hydrostatic motor/pump circuits have been successfully used.

**FIGURE 10.5** A large vehicle light vehicle emissions cell equipped with a 4 × 4 chassis dynamometer with variable wheelbase adjustment. The cell is stainless steel lined and with climatic control system within the cell roof. The speed following vehicle cooling fan has its final nozzle section removed. Photo by the Author.

If it is required for the vehicle running under load within an EMC chamber to be mounted on a turntable so that the running vehicle can be turned 360 degrees, in relation to the electromagnetic beam of an emitter or antenna the use of a simplified rolling-road device is to be recommended.

## Special features of chassis dynamometers in climatic cells

A typical modern 4 × 4 chassis dynamometer system built within a stainless-steel-lined climatic chamber is shown in Fig. 10.5. Such machines intended to operate over the temperature range +40°C to −10°C are built from normal materials and apart from sensible precautions to deal with condensation they are standard machines. For operation at temperatures substantially below this range and particularly those working below −25°C, certain special features and material use may be required.

Until the compact "motor between rollers" design of dynamometer became common, the dynamometer motor in climatic cells was often isolated from the cold chamber and operated at normal temperatures. With the newer designs, two quite different strategies can be adopted to prevent low temperatures causing temperature-related variability in the dynamometer system:

- The dynamometer pit can be kept at a constant temperature above or below that of the chamber by passing treated (factory) air through the pit

by way of ducts cast in the floor. This strategy submits portions of the roll's surface to differential temperatures during the static cool-down phase and it is usual to place a cover over the exposed roll portion.
- The pit can be allowed to chill down to the cell temperature but the parts of the chassis dynamometer that are crucial to accurate and consistent performance, such as bearings and load cell, are trace heated.

As with tests at ambient temperature, the chassis dynamometer in a climatic cell is always run through a "warm-up" stabilization routine before testing takes place.

Whatever the layout, components exposed to temperatures below −25°C, such as rolls and shaft, should be constructed of steel having adequate low-temperature strength, to avoid the risk of brittle fracture.

## Solar heat load testing in chassis dynamometer environmental cells

Subjecting whole vehicles and vehicle modules to simulated solar light energy and heat load, using banks of lamps that emit light within the solar wavelength spectrum, has been practiced for at least the last 25 years in order to test emissions, the efficiency of car cooling systems and material durability, color fastness, etc. Legislative emphasis on vehicle testing designed to reduce fuel consumption has meant that the power required to run vehicle air-conditioning systems has come under close scrutiny. The EPA SC03 standard defines a supplementary test procedure for passenger cars that has to be carried out in a test facility capable of providing the environment given in Table 10.1.

All legislation covering solar simulation will define standards for the spectral content and spatial uniformity; the first criterion will be the specification of the lamp units used, which should meet the EPA spectral energy distribution given in Table 10.2.

**TABLE 10.1** The required environment for The EPA SC03 standard

| Facility parameter | Parameter specification |
|---|---|
| Air temperature | 35°C (95°F) |
| Relative humidity | 40% (100 grains or 0.0648 g of water per pound of dry air) |
| Solar heat load | 850 W/m$^2$ |
| Vehicle cooling | Airflow from external vehicle cooling fan proportional to vehicle speed on chassis dynamometer (max speed 55 mph, average speed 21.55 mph) |

**TABLE 10.2 Requirement for EPA spectral energy distribution**

| | Wavelength band (nm) | | | |
|---|---|---|---|---|
| | <320 | 320–400 | 400–780 | >780 |
| Percentage of total spectrum | 0 | 0–7 | 45–55 | 35–53 |

For the EPA SC03 test the radiant energy must be uniform, within ± 15% over a 0.5 m grid at the centerline of the vehicle at the base of the window screen. Such spatial uniformity has to be achieved by the design of the lighting array and the frame in which the array is supported. The solar array frame fitted in a multipurpose chassis dynamometer cell is normally a planar design with a target area of perhaps 6 m long × 2.5 m wide; it will have to be supported from the cell roof by a mechanism that allows it to be lowered to operating height corresponding to the vehicle and similarly raised, when not in use, out of the way of normal cell operation. The lamps of commercially available solar arrays are capable of intensity variation typically of between 600 and 1100 W/m$^2$. The electrical power requirements for full vehicle arrays can be considerable, typically around 80 kW.

## *Flat-track chassis dynamometers*

There are several designs of chassis dynamometer designs that are based on an endless steel belt tensioned between, and running around, rollers horizontally disposed to each other.

The more obvious industrial requirements for such dynamometer designs are threefold:

1. Testing of snowmobiles, which are themselves propelled through an endless belt in contract with the surface on which they are traveling.
2. Testing vehicles in wind tunnels where it is required to accurately simulate the true relative motion between the underside of the vehicle and the road surface over which it is traveling. This is a situation not achieved with roller dynamometers where the cell floor under the vehicle body, unlike the real-life situation, is stationary.
3. Vehicle testing where the flat track eliminates the problems related to differences in rolling resistance between running tires on rollers and on a flat surface. In one, rather complex, design each of the four wheels sit on an individual flat belt between two rollers and a non–weight-bearing belt runs, at vehicle speed, under the full length of the vehicle to simulate the true relative movement between road and vehicle.

In all of the designs the belt supporting the vehicle or individual wheel needs itself to be supported to prevent belt distortion and sag as the tension in the belt is insufficient to maintain a flat surface when loaded with the UUT. Some small units use a rack of small support rollers to provide this midspan support, but most units use a variant of a pneumatic air bearing, feed from a high-pressure air compressor ($\sim$25 bar), mounted below the belt between the rolls, to create an air buffer underneath the steel belt in the areas of tire contact.

### *Articulated chassis dynamometers*

For quality assurance and EOL testing of articulated, off-road vehicles, such as large front-end loaders, articulated chassis dynamometers have been built. The designs use an inverted vehicle chassis frame, axles and differentials from one of the tested vehicles with rollers fitted in the place of the vehicle wheels. A two- or four-quadrant dynamometer can be fitted in the pit and connected to an extension of the (vehicle) drive shaft system. The floor plate system that is required to provide a safe floor surface, in any axle position, requires some ingenuity in the detailed design and is not a safe work area while the machine is operating.

## Restraining the vehicle on a chassis dynamometer

Whatever the type of chassis dynamometer used the test vehicle must be adequately restrained against fore and aft motion under the tractive forces generated and against slewing or sideways movements caused by the wheels being set at an angle to the rig's longitudinal axis. The restraint used must not impose loads on the vehicle that will cause body distortion or uncharacteristic tire distortion.

Restraint systems exist in three basic forms:

- restraint within the two dynamometer rollers, where the vehicle is lowered and raised by a lift-out beam positioned between the rollers;
- semi-flexible systems based on a minimum of two "tie-down" straps of the type used to fix loads on commercial vehicles attached to anchor points in the cell floor; and
- semi-rigid systems based on a floor-fixed pillar and a tie rod attached to the towing point of the vehicle.

## Restraint within two rollers

Vehicle restraint is not a great problem on the small twin-roller units as used for short-duration test work, such as the low-power EOL test rigs or the older type of twin-roller emission rigs. These rigs are fitted with a combined, pneumatically powered, mechanism consisting of a "lift-out" beam that is positioned between the pair of rollers plus, in most cases, small-diameter lateral-restraint roller mechanisms. When the vehicle drives onto the rig the lift-out beam is up and the vehicle is stopped with the drive wheels centered above them; as the beam descends, the vehicle wheels are dropped between the dynamometer's rolls and a frame comes up either side of the vehicle wheel carrying a pair of small free-running rollers (about 50 mm); these prevent the wheel from climbing out of the rolls during braking or acceleration. At the end of the test the beam comes up, lifting the vehicle, and the anti-climb-out rollers drop down, out of the vehicle path. No other restraint is required *providing the driver does not accelerate and brake violently*.

## Tie-down strap restraint

Where full restraint systems are used, the cell floor must be provided with strong anchorage points. There are many different designs of vehicle semi-flexible restraint equipment; the chain or loading strap fixed to vehicle tow points is the most common but may impose unrealistic forces to the vehicle structure and tires if overtensioned.

Three different types of strap or chain restraint system may be distinguished:

- A vehicle with rear-wheel drive is the easiest type to restrain. The front wheels, sitting on the solid cell floor, can be prevented from moving by fore and aft chocks linked across the wheel by a tie bar. The rear end may be prevented from slewing by two high-strength straps with integral tensioning devices; these should be arranged in a crossover configuration with the floor fixing points' outboard and well to the rear of the vehicle.
- Front-wheel drive vehicles need careful restraint since, with the rolls running at speed, any movement of the steering mechanism can lead to violent yawing of the vehicle. Restraints may be similar to those described for rear-wheel drives, but with the straps at the front and chocks at the rear. Human drivers require practice with the handling characteristics of the vehicle on the rolls, while for robotic operation the steering wheel should be locked; otherwise disturbances such as a burst tire could have serious consequences.
- Four-wheel drive vehicles usual rely on crossover strapping at both ends, but details depend on fixing points built into the vehicle.

When using straps, it is a good practice to tie the rear end first and then to drive the vehicle slowly ahead with the steering wheel loosely held. In this way the vehicle should find its natural position on the rolls and can be restrained in this position, giving the minimum of tire scrubbing and heating.

## Semirigid or pillar restraint

A pillar restraint, in some forms called a "sled" restraint, is attached into a vehicle's rear tow eye or a specially attachment. These restraints may take the form of a rigid, floor-fixed, steel pillar fitted with a restraint arm that is allowed to slide freely up and down as the car moves but prevents lateral and fore or aft movement. This type of restraint has found favor in powertrain development work when the drive cycles are more dynamic than those of legislative emission tests. To the human driver these semirigid restraints are generally judged to give a better, "less soggy" feel to the vehicle being driven.

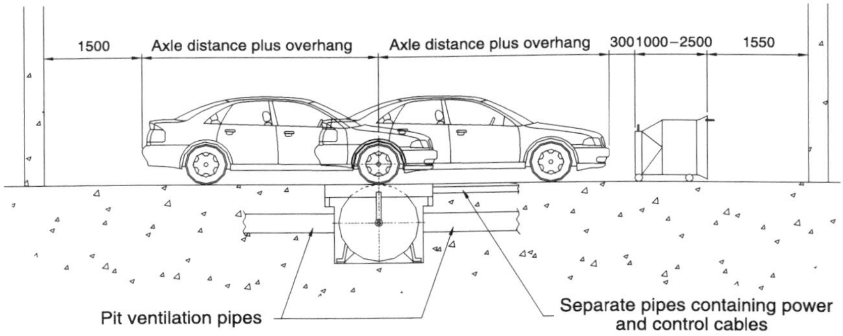


**FIGURE 10.6** Typical measurements required in laying out the *minimum* longitudinal dimensions of a cell housing a 2 × 2 chassis dynamometer and cooling fan.

### *The installation requirements of chassis dynamometers*

Whatever the type of chassis dynamometer the reader has to specify and install, there are some common planning and logistical problems that need to be taken into account.

For single-roll machines, built within a custom-built chamber, the longitudinal space requirement should to take account of front- and rear-wheel drive vehicles plus room for tie-down mechanisms and the vehicle cooling fan. A possible layout is giving the sensible minimum space required shown in Fig. 10.6.

The geometry of the chosen vehicle restraint system will determine some of the layout details.

## Chassis dynamometer cellar or pit design and construction details

Most chassis dynamometers are built on the ground floor of a building and therefore need to be installed within a pit made in that floor.

An alternative, more expensive design is to construct the chassis dynamometer's working floor at ground level but above a large cellar, with the dynamometer unit sitting on a massive plinth in the center that space. This chamber/cellar type of facility layout requires a substantial working floor (cellar ceiling) but can reduce the overall height of the facility and its ground-level footprint by providing useful floor space to house dynamometer drives and other chamber services. When of sufficient volume and with adequate air inlets, the cellar space may house the large speed following fan, the outlet air of which can be ducted up into the front of the UUT via a variable length nozzle section; this can obviate the need for a large mobile fan within the dynamometer chamber.

The classic single-axle dynamometer pit has some critical dimensions and its creation needs careful supervision and a high standard of workmanship. The exact dimensions and tolerances of the pit and point loads on the foundations within the pit should be provided by the dynamometer's manufacturer. It is usually found that the pit has to be built to a standard of accuracy rather higher than is usual in some civil engineering practices because all pit-installed chassis dynamometers have a close dimensional relationship with the building in which they are installed. The pit construction has to meet three critical dimensional standards:

1. The effective depth of the pit measured from the top surface of the foot mounting pads to the finished floor level of the test cell needs to be held to tight dimensional and level limits. Too deep is recoverable with steel packing shims, too shallow can prove disastrous to the constructor's project plan.
2. The lip of the pit has to be finished with edging steel, which has to interface accurately with the flooring plates that span the gap between the exposed rolls' surface and the building floor.
3. The centerline of the dynamometer has to be positioned and aligned, both with the building datum and the vehicle hold-down structure or rails, which are cast into the cell floor. Only very limited movement and fine alignment can be achieved by the dynamometer installers because movement on the location pads will be very restricted.

## Avoiding or dealing with pit flooding

Unless the chassis dynamometer is mounted above ground floor, its pit will form a sump into which liquids will drain. Spillage of vehicle liquids and cell washing will drain into the pit; therefore a means of removing these liquids into a foul-liquid intercept drain is necessary. All dynamometer pits should be built with their floor, other than the feet support pads, sloping slightly toward a small sump fitted with a level alarm and within easy reach of the access ladder so that pumping out of any spilt liquids can be arranged.

Flash flooding of the building due to an exceptional rainstorm, a broken water supply pipe, or the result of vehicle fire suppression have, in the past, all caused damage to chassis dynamometers, when no easily accessible method of pumping out was provided.

## Pit depth

It is normal and sensible practice to make the pit floor lower than the datum dimension and then fix into the floor precisely leveled, steel "sole plates" at the required height minus 5–10 mm to allow some upward adjustment of the machine to the exact floor level. The final leveling can be done by some

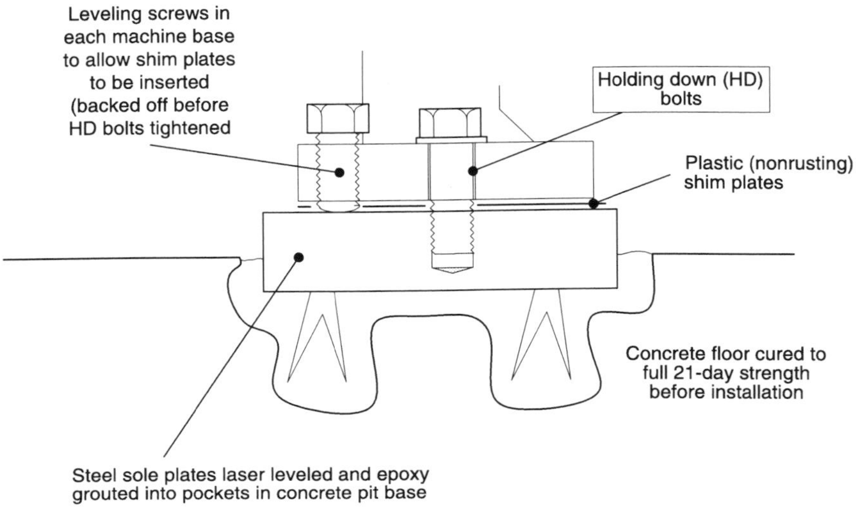


**FIGURE 10.7** A section through one of a minimum of four steel sole plates that should be precisely leveled, cast, and cured into the pit floor prior to the installation of the chassis dynamometer.

form of millwright's leveling pads, or more usually leveling screws and shim plates, as shown in Fig. 10.7.

If the site has a water table that is at or above the full pit depth the whole excavation will have to be suitably "tanked" before final concrete casting to prevent groundwater seepage.

Subfloor cable and ventilation ducts, between the control room and the pit, and the drive cabinets and the pit, require preplanning and close collaboration between the project engineer and builder; this task can complicate the job of creating a seal against groundwater leakage. When in doubt, install spare tubing capacity since it is practically impossible to add additional tubes after the facility is commissioned. Normally there will be the following subfloor duct systems, the majority of which are formed with single runs of smooth-bored plastic tubing:

- signal cable ducts between control room and dynamometer pit;
- small power cable ducts between control room and dynamometer pit;
- high-power drive cables between the drive cabinet room/space and the dynamometer pit;
- signal cable ducts between the drive cabinet room/space and the dynamometer pit; and
- pit ventilation ducts that, at a minimum, will have to include a purge duct for removal of hydrocarbon vapors (see ATEX requirements in Chapter 3: Test Facility Design and Construction).

To facilitate the installation of cables, all tubes intended to contain wiring should be installed with strong rope run through and fixed at both ends; this is used to pull through the preformed loom of cables designated for that tube.

When the installation of a "pull-through" cord has been forgotten or it has been accidentally removed, an ingenious method, twice observed by the authors, may be employed to replace it and facilitate cable installation. The animal known in England as a "Ferret," the scientific name of which is *Mustela putorius furo*, which translates into English as "stinky weasel thief," has a thin line attached to its waist and is put into one end of the tortuous, empty tube with a food lure at the other end: problem solved!

After final commissioning the open tubes, which should always slope slightly downward toward the cell and terminate about 25 mm above the cell floor to prevent spilt liquids draining out, can be plugged with fire-stopping material.

On the wall of the test cell it is usually necessary to fit a "breakout" box that allows transducers, microphones, or bus interfaces to be used for special, vehicle-related, communications between control room and test unit without compromising the integrity of the cell's enclosure.

Project managers of any major test facility building are advised to check the layout and take photographic records of all subfloor slab structures, service pipes, and steelwork before concrete is poured. Such records can be a vital aid for maintenance and in the case of facility modification as a backup to original drawings remembering that "reinforcing steel fixers," like surgeons, bury their mistakes.

*Note*: When, due to diminishing chamber access, the dynamometer has had to be located in its pit, before the building has been completed and permanent electrical services are available, it may be vulnerable to pit flooding, condensation, and roll surface damage; action must be taken by the project manager to guard against all these eventualities during this period.

## Dynamometer chamber flooring

In most large chassis dynamometer projects, the metal cell floor, above the pit, and its support structure are provided by the dynamometer manufacturer. It is usual for the flooring to be centrally supported by the dynamometer structure and to contain access hatches for maintenance; these need to be interlocked with the control system in the same manner as is recommended for engine test cell doors. It will be clear to readers that, unless the pit shape is built to a quite precise shape and size, there will be considerable difficulty in cutting floor plates to suit the pit and dynamometer frame edges. Since errors made in floor plate and edging alignment are highly visible, it is

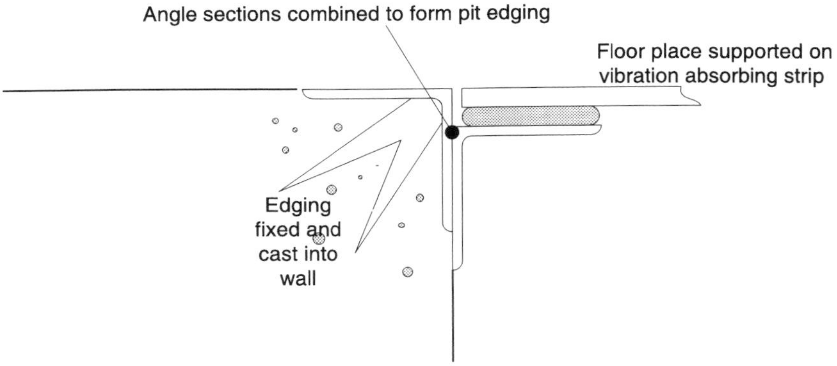


**FIGURE 10.8** Section through one form of the pit edging required to protect the concrete and support the pit flooring plates in a manner that prevents audible rattle. Floor may require separate electrical earthing (grounding).

strongly recommended that the civil contractor should supply the pit edging material and that it is cast into the pit walls under the supervision of the dynamometer supplier, before final edge grouting Fig. 10.8:

In some cases the pit edging assembly should be fixed in place using temporary jig beams to hold the rectangular shape true during grout pouring.

## Design and installation of variable-geometry (4 × 4) dynamometers

Many chassis dynamometers intended for testing four- or more wheel drive vehicles are designed so that the interaxle distance can be varied to accommodate a range of vehicles in a product family. The standard method adopted in 4 × 4 designs is for one, single-axle dynamometer module to be fixed within the common pit and for an identical second module to be mounted on rails so that it may be moved toward or away from the datum module (see Fig. 10.1). The interaxle distance for light-duty vehicles (LDVs) will be in the range of 2–3.5 m, but for commercial vehicle designs the range of the wheelbases will be greater, requiring a commensurate increase in the roll set movement. The range of movement is important because it determines the design of that section of cell flooring that is required to move with the traversing set of rolls. In the LDV designs where the total module movement is around 1.5 m, the moving plates can be based on a telescopic design with moving plates running over or under those fixed. Where the movement considerably exceeds 1.5 m, then the moving-floor designs may have to be based on a slatted sectional floor that runs down into the pit at either end, while being tensioned by counterweights, or as part of a tensioned cable loop. All these designs require good housekeeping and maintenance standards to avoid problems of jamming in operation.

The traversing roll set is normally moved by two or more, electrically powered, lead screw mechanisms, coupled to a linear position transducer so that any predetermined position can be selected from the control room. For operational and maintenance reasons it is recommended that some positional indication, vehicle-specific or actual, interaxle, centerline distance is marked on the operating floor.

LDV units are usually traversed on flat machined surface rails having lubricated footpads fitted with sweeper strips to prevent dirt entrainment; this is very similar to long-established machine tool practice and has to be installed with a similar degree of accuracy.

Large commercial vehicle rigs often use crane traversing technology, with crane rails installed on cast ledges within the pit wall design and the axle modules running within "crane beams" running on wheels, one of which on each side are connected by a common, electrically powered, shaft to prevent "crabbing."

## The housing of control and drive cabinets

The AC or DC drive cabinets and ancillary unit's drive and control cabinets are not housed within a well-designed chassis dynamometer cell, which, like that of any ICE running test cell, must be designed as a hazard containment box containing as few as possible sources of ignition or danger to occupants.

The criteria used to decide the best place for housing the large have been briefly covered in Chapter 3, Test Facility Design and Construction. In the case of chassis dynamometer facilities, it is common to provide two suitably ventilated chambers outside and alongside the main chamber, one to house the drive cabinets and the other the emissions analyzers. This leaves the main chamber an uncluttered workspace with the UUT placed center stage and the test equipment and power cables largely invisible.

The drive and control cabinets for a modern chassis dynamometer will be over 6 m long, 2.2 m high, and although only 600–800 mm deep will require space for front and sometimes rear access.

## Tire burst detectors

On all mileage accumulation rigs and others undertaking prolonged automatic (robot-driven) test sequences, there must be some form of detector that can safely shut down the whole system in the event of a tire deflating. The most common form is a limit switch mounted on a floor stand with a long probe running under the car at its longitudinal midpoint, adjusted so that any abnormal change of the vehicle's normal running body height will be detected; there needs to be one such device fitted on each side of the vehicle. These devices should be hardwired to trigger a "fast-stop" sequence.

## Loading and emergency brakes

It must be possible to lock the rolls to permit loading and unloading of the vehicle. These consist of either disk brakes fitted to the roll shafts or brake pads applied to the inside surface of one or both rollers. In normal operation these brakes have to be of only sufficient power to resist the torques associated with driving the vehicle on and off the rig rather than being an emergency braking system. Pneumatic or hydraulically powered, they are controlled both by the operational controls (vehicle loading) and are designed to be normally "on" so require active switching to be off (machine operational).

It is not good or normal practice to rely on these brakes to bring the rig to rest in the case of a test vehicle emergency: but, as normally on, they would automatically operate when there had been a major electrical power failure.

## Guarding and safety

Primary operational safety of chassis dynamometer operators is achieved by restricting human access to machinery while it is rotating. Access doors or hatches to the pit, or critical areas of the cellar, should be interlocked with the EM stop system, and latched EM stop buttons positioned in plant areas remote from the control room, such as the drive room and cell extremities.

The small exposed segment of dynamometer rolls, on which the tires rotate, is the most obvious hazard to operators and drivers when the rig is in motion and personnel guards must always be fitted. Such guards are a standard part of the dynamometer system for good safety reasons, but perhaps also because it is the only operational piece of the manufacturer's equipment that is visible to the visitor after installation and therefore will usually be fitted with the company logo. Roll cover plates, for use when no vehicle is installed, are also advisable to prevent surface damage during maintenance periods.

## Roll surface treatment

In the past, most twin-roll and many single-roll machines had a normal finish machined steel surface and no special treatment was applied. The roll surface of modern single-roll machines may now be sprayed with a fine-grained tungsten carbide coating, which gives better grip and creates a more natural tire–roller noise but causes far greater tire attrition, unless wheel alignment is of a high standard.

Brake testers and some production rigs always have a high-friction surface, which again can give rise to severe tire damage if skidding under high load occurs.

## Road shells and bump strips

For development work and particularly for NVH development, it is sometimes desirable to have a simulated road surface attached to the rolls.

Usually the required surface is a simulation of quite coarse-stoned asphalt and is achieved by use of a "road shell" enclosing the machine's steel roll. Road shell design and manufacture is difficult; they should ideally be made up in four or more segments of differing lengths and with junction gaps that are helically cut to prevent the gaps creating tire noise with a regular "beat."

Operators of acoustic chassis dynamometers now need several different surfaces for their test sequences, first to meet the different needs of customers and legislation, and second to reduce costs by transferring possibly the largest part of the test track runs to the dynamometer.

The most usual techniques for producing road surfaces for attachment to chassis dynamometer rolls are:

- Detachable cast aluminum alloy road shells made in segments that may be bolted to the outside diameter of the rolls. The surface usually consists of parallel-sided pits that give an approximation to a road surface.
- Detachable fiberglass road shells having an accurate molding of a true road surface. These shells are usually thicker than the aluminum type and in both cases the cell's floor plates must be adjustable in order to accommodate the increased roll diameter. Manufacturing techniques have developed and modern designs tend to be a sandwich construction consisting of a glass-reinforced plastic and carbon-fiber-reinforced plastic around a core of resin-impregnated foam. This produces a shell that is stronger, lighter, and less prone to delamination than the older fiberglass units.
- A permanently fixed road surface made up of actual stones bonded into a rubber belt, which is, in turn, bonded to the roll surface.

Most road shells are not capable of running at anywhere near maximum rig speed because they are difficult to fix and to dynamically balance; neither will they usually be capable of transmitting full acceleration torque. It is therefore necessary to provide safety interlocks so that the computerized speed and torque limits are set to appropriate lower levels when shells are in use. Since road shells also change the effective rolling radius and the base inertia of the dynamometer, this requires appropriate changes in the control and data processing software parameters.

Potholes in damaged road surfaces and increasing use of "traffic calming" strips at the end of high-speed road sections are sometimes simulated by fitting bump strips onto the roll surface at diametrically opposite positions; again, safety interlocks are required.

## Driver's aids

The control room needs to be in two-way audio contact with the human driver in a test vehicle on a chassis dynamometer but in addition the driver's aid (DA) display can give real-time information and operational details such as the tests "time to run."

A major function of a driver's aid (DA) screen is to give the driver instructions relating to standardized test sequences, such as production test programs or emissions test drive cycles. The early driver's displays fitted to emission homologation dynamometers took the form of an early "video game" screen where the driver had to keep the cursor, representing the car, on a line representing vehicle speed that scrolled down as the test time ran showing the speed demanded and the actual speed achieved. In the case of emissions tests the test profile must be followed within defined limits, so it is usual to include an error-checking routine in the software to avoid wasted test time.

Modern systems have evolved and can now present the driver with a full visual simulation of a route that has to be driven. The driver will normally wear a headset with microphone of the type used by rally drivers with which there is a permanently open, two-way voice channel with the control desk. The driver will also be observed by the control room operator, through CCTV, and may be sent operational instructions visually, through the external display screen of the DA.

## Fire suppression within a chassis dynamometer cell

For a more general treatment of the subject of fire suppression, see Chapter 3, Test Facility Design and Construction. The following relates specifically to the avoidance and treatment of fires in chassis dynamometer cells.

The first sensible precaution, in order to reduce the fire load in the cell, is to have the minimum required fuel in the test vehicles for the tests being run. Even with the cell's large frontal fan, the cooling airflow around a vehicle mounted on a chassis dynamometer will be different from that experienced on the open road. When running on a chassis dynamometer the underfloor exhaust system can become very hot and the fire risk from this source increases with increasing power absorption. All vehicle test facilities should be equipped with substantial in-cell mounted, handheld or hand-operated fire extinguishers, and all staff should be trained and practiced in their use.

A fixed fire suppression system is more difficult to design than in an engine test cell, because of both the larger volume of vehicle test cells and access to the seat of the fire is difficult since it may well be within the vehicle's engine bay or onboard battery.

Conventional factory water sprinkler systems are not recommended because of the high level of consequential damage if dynamometer pit flooding occurred.

Water fog suppression systems, which can be arranged to include discharge nozzles mounted beneath the vehicle and thus near the potential seat of fires, are to be recommended. Automatic gas-based systems of the type used in some engine test cells are not recommended in vehicle cells in view of cell volume and, in most cases, the need to ensure that the driver has been evacuated before they are used.

Another fire protection method, for vehicles rigged for long-duration tests, is to fit them with a system of the type designed for rally cars; this enables the driver or control room to flood the engine compartment with foam extinguishant.

There must be a clear and unimpeded escape route for any test driver. The impairment in vision, caused by steam or smoke, must be taken into account by the risk assessment, particularly in the case of anechoic cells, where the escape door positions must not be camouflaged within the cell's coned surface.

## Note concerning project management of offloading and installation of chassis dynamometer units

Even a small twin-roller chassis dynamometer machine, which is shipped and delivered as one unit, will require lifting equipment to off-load, transport, and position it in the pit. The logistics of taking delivery and first positioning of larger emission or mileage accumulation units requires careful planning and coordination with other site users.

In most cases standard 48-in. roll machines or larger units will have to be positioned in already built or, more likely, partly completed chambers. In many cases it will be necessary to provide temporary floor strengthening within the chamber to take the load of the transporting crane unit, plus the machine being moved.

The installation will often require specialist equipment and contractors to lift and maneuver the machine into a building area, of limited headroom and limited floor access, as in the case of lined climatic chambers.

It is a site manager's test piece and project phase that, particularly, if new test cell building is a construction site and when the event can be temporarily disruptive to any other site work.

## Part 2: Independent wheel (hub) dynamometers

With the development of various configurations of hybrid powertrains and the need for high accuracy and repeatability in power measurement and simulation, the age of the full powertrain dynamometer rig, using individual wheel or hub dynamometers, has come.

The physical limitations of the conventional chassis or rolling-road dynamometer, made up of one or more fixed cross-axles, have been discussed earlier in this chapter. Most of these limitations are solved by the variable-geometry full powertrain test assembly using hub dynamometers.

Such facilities divide into two general types:

- Type 1: Test areas with a flat "garage" floor using individual wheeled dynamometer units containing hydrostatic or electrical power absorption units. Such devices allow setup times to be kept short and a quick turn-around of test vehicles.
- Type 2: Custom-designed R&D cells fitted with four floor-mounted dynamometers. These facilities have all of the support services and infra-structure of a modern engine test cell and are based on a large tee-slotted air-sprung bedplate that allows for physical adaption to suit the UUT, which will range from a complete vehicle to complete or partial (simulated), powertrain assembles.

## Type 1 full vehicle powertrain test areas

The hub dynamometers of the type required for use in the flat-floor test area were first widely introduced as a viable automotive test system by the Swedish Company Rototest in the early 1990s. The first generation was based on hydrostatic motors later largely replaced by the latest AC motor units. There are now a number of manufacturers competing with AC motor-based test packages.

Such "flat-floor" facilities are finding a whole range of new test roles in the development of both hybrid and electrical two- and four-wheel drive powertrains. The dyno units are usually mounted on a wheeled frame capable of supporting its share of the vehicle mass and allowing them to be moved into position by hand and parked when not in use. As shown in Fig. 10.9 the system allows power testing with front wheel steering angle applied, which allows the influence of the constant velocity joints to studied and the units to be tested in situ.

The quick setup times of a full vehicle undergoing a simple drive cycle test, helped by the fact that all instrumentation is onboard, allows for making direct comparisons between variables such as fuels, vehicle models, engine maps, and OBD settings. Careful wheel hub mounting means that wear and tear on the test vehicle is minimized, which is ideal when dealing with expensive prototype cars.

Wheel–hub dynos, mostly of two-quadrant (eddy-current) designs have found favor with aftermarket and motorsport tuners for whom they offer improved operator safety and accuracy in power measurement. This is a market sector in which the tire slippage induced by powerful cars on small rolling roads is a major problem, only partially, and not safely, solved by getting heavy coworkers to sit over the drive wheels! For many years a major problem faced by engineers carrying out NVH testing on a chassis dynamometer is that tire noise can dominate vehicle sound measurements. One answer is to absorb the power of each drive wheel with a small

**FIGURE 10.9** Mobile (flat-floor) hub or wheel dynamometers. In this type of configuration the vehicle drive shafts can be tested through the power range at any steering angle of up to ±45 degrees. The units shown are rated at up to 500 kW and 3000 Nm of torque per wheel and equivalent road speed of up to 330 km/h. *Photograph supplied and reproduced courtesy Rototest.*

low-inertia hydrostatic or AC dynamometer, installed within a sound attenuating casing. This requires special hubs that allow the dynamometer connection to run through a dummy hub in such a way so that the vehicle can still sit on its, nonrotating, tires so giving approximately the correct vibration damping effect, while the tire contact and windage noise is eliminated.

## Type 2: full powertrain test rigs using wheel substitution (hub) dynamometers

The UUT in these rigs may be either the full vehicle, as is normally the case with Type 1, or it may be all or part of the powertrain system without any vehicle structure or body. Where modules of the powertrain are absent, their effect on the system will be simulation as part of a HiL setup. In Chapter 3, Test Facility Design and Construction, Fig. 3.14 shows most of a hybrid powertrain rigged within a Type 2 facility.

The typical performance figures of the AC dynamometers used are listed in Chapter 9, Dynamometers: the Measurement and Control of Torque, Speed, and Power, but will normally have a maximum rotational speed of 3000 rpm and range from 200 kW upward.

While the highest possible accuracy of power measurement and the range of HiL tasks able to be done in such cells is greater than that possible in Type 1 facilities, the setup task is complex and takes far longer. Each dynamometer is usually based on permanent magnet motors, the inertia of which is as close as possible to that of the wheel it is simulating. The very high response coupled with matched drive and control systems (both individual and combined) allows for dynamic ($\sim$30 Hz) simulation of a full range of real-life behavior, whether vehicle, road, or driver induced specifically:

- vehicle: road load and inertia simulation,
- road: road gradient and curvature road surface friction allowing simulation of individual wheel slip, and
- driver: driver' style simulation with manual and automatic transmissions.

Not only does such a system allow the frontloading of powertrain development modules and subsystems but the accurate rerunning of problems discovered during track or road testing. If a particular section of a route produces higher than expected emission during a RDE test (see Chapter 17: Engine Exhaust Emissions) it can be rerun, and the control variables used can changed under strict control, using such test beds. Similarly, Motorsport teams can examine and tune systems by rerunning track sections or rally stages.

## An important note concerning rigging and use of vehicle brakes when mounted in a full powertrain test bed

While the test equipment and control system of a powertrain rig can simulate the characteristics of the vehicles' RLE and the driver's inputs, it cannot directly emulate the vehicle's forward momentum when its brakes are applied; doing so may create unforeseen and erroneous results. It is also rather difficult to protect the test equipment and UUT from damage by the brake dust that results, even from mild but repeated applications. In the few occasions when brakes are absolutely required to be used on powertrain rigs, the dust has to be dealt with by drawing the contaminated air into a cyclone scrubber.

There are various test strategies to deal with these problems, some requiring significant time to rig. If braking and braking to a dead stop is required to be included in a sequence such as those testing hybrid battery regeneration and stop—start controls, the simplest solution requiring the least rigging is to insert into the sequence a "brake light on" output, which causes the dynamometer to brake the vehicle at a predetermined rate, during which the vehicle's brakes are not operated.

A more realistic simulation of braking, again applied through the dynamometers but also proportional to applied brake pressure can be achieved, but requires much more rigging effort as the vehicles brake callipers have to be removed from their disks, moved within the wheel-arch space and have a steel plate, of disk width, inserted between the pads. This arrangement allows the brakes to be applied, in the case of a robot driver with a calibrated force, and the pressure in the brake lines to be used to model the deceleration that would result and apply that braking force through the individual hub dynamometers.

## Robot drivers or gear shifters

In the years covered by the previous editions of this book the technology of robotics has advanced considerably which has meant that modern units are able to carry out the range of primary control actuation of a human driver, which includes steering. Such robotic drivers are used in, among other tests, the development of autonomous vehicles. In this chapter that concerns chassis and hub dynamometers, we concentrate on those robots that are matched with these test devices.

For a human, driving a vehicle on a chassis dynamometer for any length of time can be a tedious task and their pattern of gear changing (shifting) will vary over that time, meaning that any one test will not be exactly repeatable. To drive through a fixed emission drive cycle requires the human driver to use a visual "driver's aid" which is very similar to an early version of a video game where the driver has to match the vehicle's speed and gear with a scrolling trace on a display screen. It is widely accepted that some drivers of emission cycles have "lower $NO_x$ styles" than others, which is indicative of the variability of human driving characteristics even within a group of trained professionals.

Originally, and still primarily, a robotic driver provides a means of driving a vehicle installed on a chassis dynamometer or testing a powertrain system in a test cell in a precisely repeatable sequence and manner. These two applications have produced three different types of robotic tools, which are:

- In-vehicle, floor mounted, with driver's seat removed. This type is normally used for mileage accumulation work where the long installation time required is negligible compared to the long test period and the need for absolute reliability.
- In-vehicle, seat mounted. This type is designed to have a short installation time and no modification of the car; they are normally used in emission testing for repetitive running of legislative drive cycles.
- Test cell, floor mounted, two- or three-axis, used in powertrain testing.

However, both in-cell and in-vehicle drive robots are required to have the following common features:

- Appropriately quick setup times and a built-in routine for "learning" the gear lever positions plus, in manual transmissions, learning and readjusting for change in the clutch "bite" position. Learning is normally carried out within a programed teaching routine and involves manually shifting the gear lever with the robot attached and allowing the controller to map the $x$ and $y$ coordinates.
- Smooth and precise operation of the gear shift lever, of manual, semiautomatic, and fully automatic transmissions. Actuation in the $x$ and $y$ axes and free movement in the required arc of travel in the $z$ axis are usual. Powered operation in the $z$ axis is much more complex and usually required for engaging reverse gear so is consequentially rarer. Most

actuators are servo-motor-driven linear actuators. Pneumatic actuators give the best representation of the force applied by the human arm, since there is a smooth buildup of pressure over the synchronizer operation but are reported to be more difficult to set up.

- Operation of the clutch and throttle pedals (in manual transmissions) precisely coordinated with operation of the gear change lever.
- Removal of all load and restraint on the gear shift lever when not shifting.
- A fail—safe routine is triggered in all its own and UUT fault conditions. This will typically mean that the clutch disengaged, and the accelerator pedal released.
- Safety systems to prevent damage by overloading the lever in the case of nonengagement.
- High operational reliability, often over weeks of continuous operation.

The in-vehicle, seat-fixed designs have to overcome the problem of the compliance of their fixing and the constraints of the installation space. This means that their performance may be less capable of extended operation; they are also not capable of the most accurate force measurement or actuation in the $z$ axis, none of which are practical constraints in running emission drive cycles. The problem of wear in the gearbox actuation and in the driveline clutch has to be overcome by regularly running "relearn" routines during endurance work.

Because of their layout, most in-line, rear-wheel drive, and $4 \times 4$ vehicles have gearboxes with shift mechanisms and actuators directly connected, but cable-operated gear change mechanisms can create problems for robotic drivers. These cable designs are largely driven by NVH work that requires the decoupling of the transmission from the cabin and the use of softer engine mounts. The increased engine movement under load and cable stretch can cause the gear lever to change position during tests, which has created problems, such as the robot arm losing the gear knob position and the need for a "softer" actuation.

Chapter 11

# Mounting and rigging internal combustion engines for test

## Chapter outline



**Engine Testing. DOI: https://doi.org/10.1016/B978-0-12-821226-4.00011-5**

# Part 1: Introduction, general strategies, and rigging of engine services

## Introduction: Engine mounting strategies

The choice of how any test engine is supported and connected in the test cell all depends on the duration of time that it will remain in the cell. Some engines, as in the case of "fuels and lube" or some turbocharger testing, are themselves part of the test cell system, while others such as those used in educational facilities remain in the cell for long durations. All of which means that the engine mounting has to be robust and durable and that the time taken in rigging is not a prime concern. Mounting systems in such cases are frequently based on the type of stand shown in Fig. 11.1. Whatever system is used the techniques required to reduce vibration discussed in Chapter 3, Test Facility Design and Construction, and Chapter 8, Vibration and Noise, should be observed.

In modern internal combustion engines (ICE) test cells fast turnaround is of paramount concern; therefore palletization and prerigging has become widely developed and practised.

## Palletization of the unit under test

The high cost of a modern automotive test cell running ICEs, or hybrid ICE/electric powertrains, particularly when attached to a full emissions measurement system, is such that downtime caused by waiting for a viable unit under test (UUT) has to be minimized. In order to minimize rigging time of the UUT in the cell, advanced pallet systems have been devised, which allow the UUT to be completely prerigged in the workshop. Such systems allow all its electrical, fluid, and test transducer connections to be made and corralled into the minimum number of multipin connectors and self-sealing couplings attached to one or more pallet pillars. The pallet/UUT assembly can then be transported and precisely located on the test bed before being "plugged in," with perhaps only the shaft and exhaust pipe(s) requiring manual intervention.

The strategies that can be adopted for shaft connection will be discussed later in Part 2 of this chapter.

Such pallet systems range from simple, but robust, wheeled trollies fitted with adaptable support rails and pillars, to complex and expensive integrated support systems based on a rigid base that can be handled by a forklift, or a wheeled and steered by an electric drive unit.

All pallets should have the common feature of precision machined pads and/or dowels that allow exact positioning on matching location points on the test stand. The top surface of the pallets will form a drainable fluid catchment, with the exception of pallets used in noise, vibration, and harshness cells where they would act as a sound amplifier.

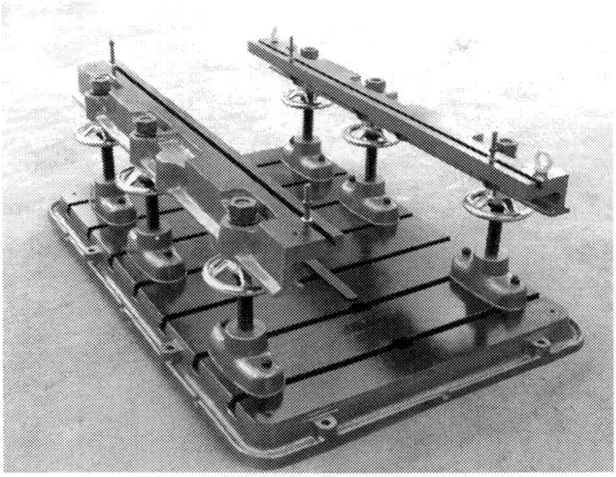

**FIGURE 11.1** A basic test cell bedplate with height-adjustable UUT mounting beams that allows for a wide range of engine mounting frames or support fittings to be used. Robust and highly adaptable such systems are used in industrial and educational facilities worldwide where speed of change of the UUT is not a high priority. *Photo courtesy SAJ Test Plant PVT.*

Palletization is adaptable to all automotive module testing, but in the case of the most complex and expensive units the number required to support a given number of cells and their disparate projects requires good workflow planning and cost–benefit analysis.

## Workshop requirements

*Every workshop engaged in ICE prerigging needs to be fitted with a dummy test cell station that presents datum connection points identical in position and detail to those in the cells, particularly the dynamometer shaft, in order for the engine to be prealigned.*

To gain maximum benefit, each cell in the facility needs to be built with critical fixed interface items in identical positions from a common datum.

Many test engineers will relate stories of days of wasted cell time because of being occupied by engines that refuse to start and that require a conference of software experts to resolve the problem. Therefore, in some large research facilities, engines are both prerigged on pallets and then started and briefly run, without loading, in a pretest shakedown area equipped with basic services, so as to check base functions before being taken to the test cell.

Powertrain pretest rigging requires highly adaptable workshop support and the provision of a wide range of suitable fittings and adaptors always needs be made available.

Making up of the engine to cell exhaust systems requires the workshop to have a welding bay.

*Electrical arc welding must never be allowed within a commissioned test cell because of the danger of damage to computers, instrumentation, and powertrain electronics by stray currents running through a common earth.*

The workshop area used for derigging should be designed to deal with the inevitable spilled fluids and engine wash activities. Floor drains should run into an oil intercept unit.

## Engine mounting methods

In test cells with lower pressure on cell "uptime" and a lower UUT throughput, a number of rigging and prerigging strategies are available.

The degree of complexity of the system adopted for transporting, installing, and removing the engine within the facility naturally depends on the frequency with which the engine is changed. In some research or lubricant test cells the engine is more or less a permanent fixture, but at the other extreme the test duration for each engine in a production cell will be measured in minutes where the time taken to change engines must be cut to an absolute minimum.

There is a corresponding variation in the handling systems:

- Simple arrangements when engine changes are infrequent and UUT occupancy of the cell is measured in weeks rather than hours. The engine is fitted with test transducers in the workshop then lifted into the cell and rigged in situ, mounted by using adapted fixings using in-service or in-vehicle mounting points and fixings, on a rigid height-adjustable pair of tee-slotted beams such as that shown in Fig. 11.1. All engine-to-cell connections are made after manual dynamometer to engine shaft alignment.
- In facilities having flat concrete floors between workshop and test cell, a manually wheeled trolley and its engine attached transducers and service connections may be used. The example shown in Fig. 11.2 is rail-mounted trolley fitted with a match plate to facilitate the automatic connection of the fluid services. Such systems require pneumatic or hydraulic trolley docking mechanisms to overcome the resistance of the self-sealing couplings on the match plate.
- For production test beds it is usual to make all engine-to-pallet connections before the combined assembly entering the cell. An automatic docking system permits all connections (including in most cases the driving shaft) to be made in seconds and the engine to be automatically filled with liquids.

In the recent past a significant cause for delays in starting testing and cell downtime has been the lack of some UUT component or connector or ECU information. In many facilities, particularly in Tier 2 and Tier 3 test cells, these problems can be frustratingly common. Missing ICE components have to be obtained and fitted as part of the rigging loom or they have to be simulated as part of an x-in-the-loop arrangement.

**FIGURE 11.2** Floor rail–mounted ICE test trolley on a turntable outside a test cell door. Trolley has engine support pillars prerigged with cable and pipes rigged into the back of self-sealing couplings allowing for fast docking of a mounted engine into the test stand. The coolant system is fitted with a standpipe having an air-release bleed valve at the high point. *Photo courtesy SAJ Test Plant PVT.*

It is still not uncommon to see rigged engines in a cell with their mounting frame festooned with modules of a vehicle loom, including the ignition key's wireless security unit.

The common causes of destructive levels of engine vibration in the test cell are discussed in some detail in Chapter 8, Vibration and Noise, but there are some general points of good practice that should be followed. It is a common practice to rig diesels for test by supporting them at the drive shaft end with a profiled steel plate bolted to the face of the flywheel housing resting on damped mounts that are fixed to the mounting frame rails on either side of the engine (shown diagrammatically in Fig. 11.3). The horizontal plane running through the mounting feet should generally be at or near that of the engine's crankshaft; this ensures that the fixings are in simple compression or tension and the connecting shaft is not subjected to lateral vibrations as the engine tries to turn about the crankshaft's centerline.

For short-duration hot and cold production tests the engine is often clamped rigidly to a common frame with the drive motor (dynamometer) using any convenient flat machined surface, or even the sump fixing bolt heads, but the whole frame will be mounted on some form of soft mounts such as those shown in Chapter 8, Vibration and Noise.

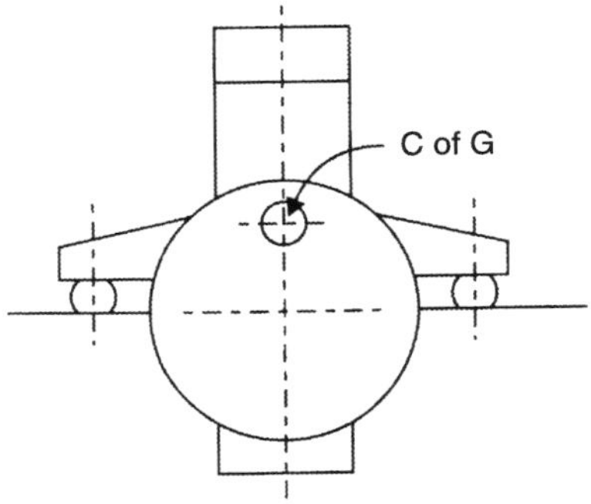


**FIGURE 11.3** Diagram of a diesel engine typically mounted on two flexible feet from a rigging frame bolted to the flywheel cover casting at the front of the engine. Rear mounts may be similar or use various ex-vehicle fixing points. Such arrangements are often based on mounting frames such as that shown in Fig. 11.1.

Most gasoline engines will be rigged for test supported from the vehicle mounting points but, as warned in Chapter 8, Vibration and Noise, this may not be satisfactory because of the differences between stiffnesses of the mounting frames and effects of other engine restraints. Failures of exhaust connections near the engine and fatigue failure of the support brackets of ancillary units, such as the alternator, are common indicators of engine mounting problems.

## Supply of electrical services to engine systems

In its simplest form a stabilized direct current (DC) power is supplied for the engine's ECU together with switched supplies and earth connections for other ancillary devices via an engine services box (ESB) mounted close to the rigged engine. The mounting in a cell of a 12-V battery to serve as the DC supply is now considered poor practice, on H&S grounds, but if adopted it must be mounted within a vented wooden box. In cells testing a variety of hybrid powertrain ICE and motors, a high-powered, grid-connected, bidirectional AC to DC power system is required. Commercially available units are designed for not only supplying stabilized DC outputs at 12, 24, and 48 V but can also act as a complete battery emulator, battery, motor, and inverter test system and can be rated up to 320 kW and 800 V DC. Such systems are physically large and are usually housed outside the cell with their cables routed to an ESB mounted on the cell wall.

## Getting the modern internal combustion engines to run in the test cell

The task of installing a modern electronically controlled ICE, which forms part of a highly integrated powertrain system, and getting it to run through its full performance envelop, in a test cell while physically isolated from some or most of the vehicle system, is not without difficulties. While there can be much glib talk about running hardware-in-the-loop (HiL) tests and substituting missing subsystems of the powertrain with software models, the task can still be time-consuming and frustrating, even for original equipment manufacturer (OEM) test facilities. The problem is even more difficult for those working in lower tier organizations external to the OEM.

In the past, the problem has been made more difficult by the lack of harmonization in the automotive industry. The complex control logic trees, resulting from interacting control devices, and proprietary signal channel labeling, together with the level of secrecy surrounding access to interface details or error codes can create an impenetrable maze for some, non-OEM, test facilities. The rapid development of hybrid and electric vehicles has compounded the complexity and increased the level and scope of confidentiality.

However, the situation improves in some areas of common interest and ISO 15765-4: 2016 is the latest version of an international standard established in order to define common requirements for vehicle diagnostic systems implemented on a controller area network (CAN) communication link. Although this standard is primarily intended for OBD systems, it also meets requirements from other CAN-based systems needing a standardized network layer protocol.

The standard is based on the Open Systems Interconnection (OSI) Basic Reference Model, in accordance with ISO/IEC 7498-1 and 10731 and their seven-level structure.

Besides the problems met in getting the engines to run correctly, some test engineers have the need to get engines to run *incorrectly*, such as when there is a need to develop OBD units or accelerate the collection of soot particles in exhaust particulate filters; these requirements can create even greater problems and require a specially a mapped ECU.

In production testing, or any test where the same model of engine is repeatedly run, a modified ECU, specifically configured for the test regime, will be used. However, unless the test is part of a full HiL experiment, the physical presence of the dedicated ECU and some transducers and actuators with its connecting loom is vital to the function of a modern engine outside the vehicle. Typically, the list will include:

- the lambda sensor mounted in the first section of exhaust ducting;
- idle running, throttle bypass valve;
- the correct flywheel with crank position transducers;
- the coolant temperature transducer;
- the throttle position transducer; and
- the inlet manifold pressure transducer.

In the case where transducers are missing, special version of the ECU and/or VCU has to be obtained or created or the signals are simulated. To cater for lower tier operations, there is a specialized market for programmable ECUs that are most often used by motorsports vehicle tuners. Such devices can be adapted to be used to run a limited range of engine tests in cells, including tests on the mechanical integrity of ancillary devices, sealants, and seals. Note that the commercially available units that "piggyback" the OEM control unit and modify the fueling and ignition signals in order to change the shape of the power curve and fuel consumption do not solve the problems faced by most engine test cell operatives.

Hardware and software tools created and supported by specialist companies such as ETAS GmbH and Accurate Technologies Inc. allow their licensees various, discrete, permitted levels of access to the ECU software code and engine maps. This is an industrial standard methodology used worldwide in the calibration laboratories of all OEM's Tier 1 suppliers and research institutions; it requires full professionally trained specialists and a high level of powertrain test instrumentation; even so, access to the lowest logic and code levels is kept commercially secret by all major ECU developers.

Whereas OEM and major research organizations have access to, or can produce verified MATLAB-based software models of powertrain modules, lower down the industrial food chain, missing transducers, or in-vehicle actuators are needed to be physically rigged in the test cell. The throttle pedal controller, which is nowadays part of a complex, closed-loop, "drive-by-wire" transducer and actuator system, is a good example and the, previously mentioned, ignition key security fob is another.

## The use of onboard diagnostic systems in the test cell

Since 1996 all cars and light trucks used on the public road in the United States have had to have ECUs that operate an onboard diagnostic system that is compliant with the SAE standard OBD-II. The EU made a version of OBD-II, called EOBD, mandatory for gasoline vehicles in 2001 and for diesel vehicles of model year 2003. The technical implementation of EOBD is essentially the same as OBD-II, with the same SAE J1962 diagnostic link connector and signal protocols being used.

The term "EOBD2" is used by some vehicle manufacturers to refer to their units that contain manufacturer-specific features or "enhancements" not part of the OBD or EOBD standard.

So whatever mix of strategies used to rig and run engines is used, the test cell operators will need a device to read the OBD-II to diagnose operational and rigging (missing signal) problems. All commercially available readers connect through a common type of 16-pin D-type connector. An OBD2 compliant vehicle can use any of the five communication protocols:

SAE J1850 PWM
SAE J1850 VPW
ISO9141-2

ISO14230-4 (KWP2000)
ISO 15765-4/SAE J2480

Each of these protocols differ electrically, (although pin designations for 12 V and GND are common), and by communication format. Clearly the chosen code reader or scan tool used must be compatible with the specific protocol of the UUT.

## Rigging engine circuits and mounted auxiliaries

The connection between cell and engine systems must not be allowed to constitute a weak point in the physical or electrical integrity of the combined system; therefore it is important to use test connectors that are at least as robust as the matching engine component. In the case of high-volume testing the cell's rigging plugs and connectors must be of a significantly higher standard than the engine components, which are not designed to sustain many cycles of connection and disconnection. In both Europe and the United States there exist specialist companies, the catalogs of which list most of the items commonly needed for engine rigging [1].

### *Air filter*

Ex-vehicular units now tend to be large plastic moldings that require supplementary support if used in the test cell; some contain a thermally operated flow device designed to speed up engine warm-up, but which will only work correctly when installed in the correct underbonnet (hood) location. When the engine is fed with treated combustion air (CA), the size and shape of the air filter will create problems and it should be replaced with a smaller, exposed element unit having the same resistance to airflow. This is a particular problem if the engine is connected to a dynamic pressure-regulating CA system, which will require a purpose-built, airtight, encasement to be made and supported in the test cell.

### *Auxiliary drives*

Many engines will have directly mounted, usually belt-driven, drives for auxiliary circuits such as air-conditioning and power steering hydraulics. The rigging of these devices rather depends on the design of the experiment and test sequences being performed. Even in tests where these devices, or the loads they impose, are of no importance, they often have to remain in place due to the configuration of the drive-belt system; in this case they may need to be either connected to a (dummy) circuit to prevent seizure or modified to only consist of the drive wheel (pulley) supported in bearings.

Note that harmonized exhaust emission legislation specifies adjustments that have to be made in the recorded power output of some engine types when auxiliary drives have been removed (see Chapter 17: Engine Exhaust Emissions).

### Charge air coolers (intercoolers)

The charge air coolers were in the past usually only mounted in turbocharged diesel vehicles but now are commonly used in gasoline and hybrid vehicles particularly those fitted with small three-cylinder Gasoline Direct Injection (GDI) engines. The diesel's intercoolers have tended to be air/air type; these present some problems in rigging in the test cell. If the ex-vehicle cooler is used as part of the engine rigging a spot fan will be required to supply the cooling airflow, but this is impossible to control with accuracy and puts a considerable amount of heat into the cell air, which may create problems when operating during high ambient temperatures.

The use of water/charge air coolers have advantages for use in the test cell as they are physically smaller and easy to rig; they are also easier to control via a closed loop within the test controller; finally, they have a larger thermal sink rate during increasing engine power. Charge coolers of the water/air type suitable for test cell use can be bought in the automotive aftermarket; they need to be of high quality, pressure tested, and not overpressurized, since the outcome of any fluid leak into the charge air is highly injurious to any ICE.

### Internal combustion engines coolant circuit

Whatever coolant control module is used (see Chapter 6: Cooling Water and Exhaust Systems), it is important to locate it as close as possible to the engine in order to reduce the system's thermal inertia and have good transient control with the minimum hysteresis. Some pallet-based systems integrate the coolant system, including heat exchangers, within the pallet or directly attached to it. Normally the connection points used are those that in the vehicle would go to the radiator. Self-sealing couplings, unless part of a pallet match plate system, are not recommended due to their high friction head and it is important not to allow the weight of overlong pipes to put undue strain on clamped pipe connections. When filling the coolant system, it is most important to bleed air out of the combined system at both the engine and control module high points; manually squeezing all accessible pipes during venting is a simple and effective technique.

### Oil cooler

ICEs, particularly those in the higher power range, may be fitted in the vehicle with auxiliary oil coolers, of which there are three major types:

- Oil coolers integrated into the engine body and circulated with coolant fluid. These normally need no rigging in the test cell.
- Coolant/oil coolers that are fitted as an external auxiliary to the engine; these can be rigged to run using cell primary cooling water or remain as part of the engine system.

- Oil coolers of the air/oil type that are, in a vehicle, a second and smaller version of the coolant circuit radiator. In most cases these are better replaced in the cell by a water/oil cooler circuit forming part of the cell's fluid control module (see Figs. 6.3 and 6.4). Some testers use the vehicle type of cooler and direct a spot cooling fan or air mover through the matrix, but this gives, at best, limited temperature regulation and puts extra heat into the cell.

## Part 2: Rigging and testing turbochargers

### Introduction

As hybrid powertrains develop, the physical differences between modern GDI engines and diesels, other than their fuel ignition processes, become less obvious and most significant to this convergent evolution, is the role of the turbocharger.

For many years turbochargers have been of a fixed internal geometry installed within a system wherein overpressurization of the engine has been prevented by the use of an internal, or external, waste gate (valve). Recent advances in materials and engine design mean that the modern turbocharger is now often fitted with a variable-geometry turbine (VGT) that is electronically controlled through the ECU. In this way system performance is optimized to deal with the wide range of operating conditions and to support the correct function of exhaust gas recirculation (EGR) and other emission control strategies. VGT systems may be servo-electric or servo-hydraulic and some do not require waste gates, but normally all applications require antisurge valves.

Antisurge or "blow-off" valves (ASV) operate as a pressure relief valve in the boost air circuit and are designed to deal with the sudden rise in air pressure that arises between the compressor and the engine when the engine throttle is suddenly closed from near, or at, wide-open throttle (WOT) operation. A complication of the operation of antisurge valves is that, to prevent incorrect fueling, the air relieved has to be returned to the compressor inlet and after the mass airflow sensor, which has already registered its passing.

The prevention of too high exhaust gas flows and therefore pressure is handled by a waste gate valve that, at a mapped point, causes a proportion of exhaust gas to bypass the turbine. Most turbochargers are fitted with internal waste gates, with a valve that is built into the turbine housing itself, but high-powered and particularly motorsport engines may be fitted with external waste gate valves that can be tuned to give adjustable boost pressures.

The moving blades and parts of the actuation mechanism of VGT units have to work in the highest temperature zone of the turbine, and because of the demands made upon the materials used, until the late 1990s, VGT designs have been confined to diesel engines, which have lower exhaust gas temperatures than gasoline engines. A problem for vehicle designers is the cost of VGT designs, which were often double that of fixed-geometry designs; therefore in truck installations, where space may not be as constrained as in light vehicles, the use of two-stage or twin-turbo designs proved to be more cost-effective. A two-stage boost system can be based on two separate turbochargers, one high pressure (HP) and the other low pressure (LP), permanently connected in series, or on two units fitted in parallel with some modulation of the gas flow to one or both units, to give optimized boost and engine power.

## Turbocharger testing

Turbocharger testing normally takes two general forms:

- Testing the turbocharger and its associated actuators and sensors, in a customer-built test stand, totally separate from any engine. In this case a combustion gas generator is required to drive the turbine.
- Testing the turbocharger mounted on an engine as an integrated system. In this case the engine provides the gas generation to drive the turbine.

The design and optimization of a turbocharger configuration relies on the availability of high-quality, empirically produced, data in the form of a turbocharger map according to standardized procedures such as SAE: J1826_19950.

The acquisition of data and the plotting of a turbocharger map is a complex process and can be time-consuming, since it is necessary to wait until thermal equilibrium is reached before map-point data can be recorded and averaged. The test process requires good understanding of the various measured parameters: pressure, temperature, airflow, gas flow, and turbine/compressor rpm. Each of these dimensions requires special measuring equipment installed in the proper locations within the test system. Installation needs to be precise and as defined by the recommendations of the standards.

The map will show, besides the measured parameters and the adiabatic efficiency "islands," three extreme working areas of the compressor: surge limit, maximum permissible rpm speed limit, and compressor choke limit.

Once a turbocharger has been mapped on a test rig it has to be fitted to an ICE and the combination has to be optimized in a system that may include waste gate and/or antisurge valves. In the motorsport and aftermarket press one may read about turbochargers running into their surge zones, more rarely into their choke zones; invariably this is due to altering the OEM's optimized ICE/turbocharger/ASV combination and inexpertly attempting to raise the charge air boost pressure.

## Turbocharger test stands

A range of test stands that are designed for testing the automotive range of turbochargers are commercially available from companies such as Kratzer Automation; these units typically use natural gas in hot gas generators, the energy output of which is variable over the full operating cycle of the intended engine installation. Although gas burners provide a virtually particulate-free gas flow, it is important for some tests to simulate operation in poor gas conditions; therefore such rigs have to be capable of simulating less than optimal gas conditions.

Nonengine-based turbocharger test rigs are cost-effectively used by manufacturers for the full range of development, quality assurance, and endurance testing, including:

- Testing over the extreme thermal range of the turbine unit and thermal shock testing.
- Testing over the full range of the pressure–volume envelope to confirm compressor surge and choke lines and the unit's efficiency "efficiency islands."
- Checking for turbine blade resonance, fatigue testing.
- Seal and bearing module performance testing and blowby tests (oil leakage into either induction or turbine side has serious consequences).
- Mapping of both turbine and compressor speed–pressure–temperature performance.
- Lubrication system and oil suitability testing, including resistance to "oil coking" and intersection heat soak following shutdown while at high turbine temperature.
- Running at overspeed and to destruction in order to check failure modes and debris retention.

### *Turbine speed and blade vibration sensing*

While many of the larger automotive turbochargers are fitted with a "once-per-revolution" speed sensing facility, during development testing it is usually desirable to measure blade passing frequency. This allows the tester to compare the actual blade arrival time at the sensor with that predicted by the rotational speed and the number of turbine blades; any difference will be caused by blade deflection and vibration (resonance). Optical laser, capacitive, and eddy-current probes can be used for the blade arrival time measurements.

## Testing of turbocharger in combination with an internal combustion engines

This must be considered as distinct from testing a turbocharged engine. An example of an instrumented test unit is shown in Fig. 11.4. In the case of testing the engine, the operation of the turbocharger is just one more variable to be mapped during engine calibration and, where appropriate, will be the subject of its own "mini-map" covering the operation of the servo-controller of the VGT vanes within the engine fueling map. For example, in engine calibration work the interaction of EGR and turbocharger geometry has to be optimized, but when testing the turbine using an engine, for speed–pressure sweeps, etc., the EGR will interfere with system control and may be switched out of circuit.

**FIGURE 11.4** A variable-geometry turbocharger (VGT) being tested while fitted to a diesel truck engine. For test, the turbine housing has been fitted with three pressure probes (wrapped transducers), three displacement transducers (two thin-threaded transducers visible, the third hidden), and thermocouples. The VG actuator is electromechanical and fitted with water cooling.

Similarly, it will be found that running the engine and dynamometer in manual control mode (position/position) will give the most repeatable results in much turbocharger testing because it eliminates the majority of engine and cell control system influences. Gas flow, from the exhaust of individual engine cylinders into the turbine, is complex and subject to pressure pulsing; therefore the pressure regime of the whole exhaust manifold and turbine has to be understood in order to optimize its design. Such test work, which is practically impossible to accurately simulate on a test bench, has to be done in an engine test cell. Such testing led to the development of twin-scroll turbine designs, which divide the gas flows into the turbine housing from pairs of cylinders in order to reduce pressure-pulse crosstalk evident only on a fully instrumented engine.

The matching of turbocharger with a particular design of engine can only truly be checked in the engine test cell, and testing includes:

- integrated turbine and compressor performance during engine power curve sequences;
- turbocharger "lag" during step changes in power demand, over the full operating range;
- testing of two-stage or twin-turbo operation;
- intercooler operation: matching performance and control;
- engine-based lubrication, cooling and seal operation, including testing on inclined engine test stands;
- effects of various exhaust pipe configurations on back pressure, or turbine, and engine performance; and

- thermal soak effects such as following sudden shutdown after WOT running, etc.

## Special turbocharger applications

*Large two-stroke marine diesel engines* have to use multiple turbochargers in order to keep the devices to manageable size; single units have now reached the point where 27,300 kW of engine power can be supported per turbocharger (MAN-B&W 8K98MC two-stroke engine).

*Motorsport applications* are usually not as restricted in their use of turbocharging by the same emission legislation as road-legal vehicles, but they do have to comply with the FIA regulations relevant to their particular specialization. World Rally Cars (WRCs) are fitted with a 1.6-L, fuel injection, turbocharged, four-cylinder engine that are fitted with an air intake restrictor mandated by regulations (2020: 36 mm) and the power output is restricted to 380 bhp. Typical boost pressure in WRC engines is 4–5 bar, compared with a maximum of up to 1 bar for a road car; this produces a high torque output having a rather flat curve from 3000 rpm to the 7500-rpm maximum. In WRC work, it is vital for the turbine to be kept running at optimum speed, whatever the driver is doing—for example, when the car is "flying" with no power transmitted through the wheels; this creates some unusual test sequences, ignition timing, and engine fueling strategies, all part of an integrated antilag system (ALS). Where emission controls are not required, throttled turbocharged engines can be controlled using air pressure rather than mass airflow, thus avoiding the ECU being confused when the antisurge valve opens, as mentioned previously. In the case of WRC engine testing, various unusual strategies are adopted as part of the ALS to keep the turbine running at its target speed.

*Motor Generator Unit—Heat (MGU-H)*: These are exhaust turbine-driven generators currently used in F1 motorsport and part of a highly complex system calibration that also incorporates a MGU-K (kinetic) system have recovers braking energy of the car. The MGU-H generator rotor can be positioned between turbine and compressor and may be calibrated so that it can be used to act as a motor to spin the rotor.

## Health and safety implications of turbocharger testing

Because of the high rotational speeds of the turbine-compressor shaft within a turbocharger and because they can be seen to glow bright red-hot in the test cell, there is often a perception that they represent an unusually high risk to cell operators and the cell fabric.

In fact, the rotating parts and the turbine blades in particular are of low mass and operate within a housing designed for their containment in the case of failure. Once again, the perception of danger is higher than the reality and in-service failures of turbochargers take the form of rotor seizure and are, almost always, due to poor lubrication. However, there are hazards associated with turbocharger testing that have to be taken into consideration, including but not exclusively:

- There is a worldwide market in aftermarket, counterfeit, turbochargers, and spare parts that are inherently dangerous to use; therefore all engine testers should ensure that the units being run are genuine OEM units or, if testing suspicious devices, full debris protection is provided.
- Bearing failures have serious implications for the unit and will be caused by impaired oil supply in the case of conventional bearings or air supply in the case of "foil" bearings.
- Seal failures can lead to contamination of engine systems and even cause a diesel engine to "runaway" (see "emergency stop" in Chapter 13: Test Cell Safety, Control, and Data Acquisition).
- Turbines and directly connected exhaust pipes may well be incandescent while the engine is running at full power; it is vital that any nearby cables and pipes are well secured and suitably shielded. Take note of the warnings concerning the use of spot cooling fans contained in Chapter 5, Ventilation and Air-Conditioning of Automotive Test Facilities, and also note that false fire alarms may be trigged by flame detectors fitted as part of an automatic fire suppression system.
- A cool down period with the engine running at idle for 5 minutes is always recommended in any test sequence involving turbocharged engines. No special guarding against debris from a catastrophic turbine failure is considered essential unless such an event is planned or anticipated during an overspeed test; the comments about the choice of cell windows given in Chapter 3, Facility Design and Construction, are relevant.

## Part 3: Connecting shafts, in context of the powertrain test facility

*Note 1: Although the ICE is inherently more difficult to connect to a dynamometer in the test cell than an electric motor, because of its inherent torsional oscillations, much of the following discussion on shaft theory and practice is relevant to electric drive systems which themselves may be subjected to drive system induced torsional vibration.*

*Note 2: A list of the notation used in calculations will be found at the end of the chapter*

The selection of suitable couplings and shafts for the connection of the engine and/or transmission to the dynamometer may appear as a simple matter but the commonly used "suck it and see" method of shaft selection can sometimes lead to spectacular failures.

Incorrect choice of or a faulty connection system design may create a number of problems:

- destructive torsional vibration causing driveline failure;

- vibration of an ICE or dynamometer leading to fatigue failure of support bracketry;
- whirling of coupling shaft ("skipping rope" mode of failure);
- damage to engine or dynamometer bearings;
- excessive wear of shaft line components such as rubber elements destroyed by internal heating;
- catastrophic failure of coupling shaft or engine flywheel; and
- difficulty in cranking and engine starting.

The whole subject, the coupling of engine and dynamometer, can cause more trouble than any other routine aspect of engine testing, but the problem of torsional vibration is not confined to the engine/dynamometer system. Automotive engineers may also have to deal with the problem in the shaft connections of hybrid vehicles; therefore a clear understanding of the many factors involved is desirable. The following text covers all the main factors to be considered, but for some problems a more extensive analysis may be necessary and appropriate references are given at the end of this chapter.

## Torsional vibration: the nature of the problem

*Note:* on background reading on torsional vibration: Since the mathematics of the subject of torsional vibration and its various manifestations is complex and not readily accessible, nor suitable for this type of book, recommendations of other authorities and a listing of the notation used are to be found at the end of this chapter.

A special feature of the problem of torsional vibration is that it must be considered afresh each time an engine or any other machine, not previously encountered, is installed within a rotative assembly in the test cell. Most importantly, it must be recognized that unsatisfactory torsional behavior is associated with the whole system, i.e., the UUT, its coupling shafts, connected devices, and a dynamometer, rather than with the individual components. All of the driveline components, which may be quite satisfactory in themselves or within an alternative system of the same nominal power may, in a particular configuration, experience severe vibration. The problem has been likened to working within an odd-shaped transparent enclosure wherein one can suddenly walk into its invisible wall without warning.

Problems arise, in the classic test cell configuration, partly because the dynamometer is seldom equivalent dynamically to the system driven by the chosen prime-mover in service.

In the case of a vehicle with rear axle drive, the driveline consists of a clutch, which may itself act as a torsional damper, followed by a gearbox, the characteristics of which are low inertia and some damping capacity. This is followed by a drive shaft and differential, itself having appreciable damping, two half-shafts and two wheels, both with substantial damping capacity and running at much slower speed than the engine, thus reducing their effective inertia. When coupled to a dynamometer this system is replaced by a single drive shaft connected, usually without a clutch, to a single rotating mass, the dynamometer, running at the same speed as the engine. Particular care is necessary where the moment of inertia of the dynamometer is more than about twice that of the prime-mover. A further consideration that must be taken seriously concerns the effect of the difference between an ICE mounting arrangement in the vehicle and on the test bed. This can lead to vibrations of the whole engine that can have consequential effects on the drive shaft.

## Effect of the overhung mass of a shaft on engine or motor and dynamometer bearings

Care must be taken when designing and assembling a shaft system that the loads imposed by the mass and unbalanced forces do not exceed the overhung weight limits of the engine bearing at one end and the dynamometer at the other. Steel adaptor plates on either end of the shaft, frequently required to adapt the bolt holes of the shaft to the dynamometer flange or engine flywheel, can increase the load on bearings significantly and compromise the operation of the system, in which case the use of high-grade aluminum is recommended. Dynamometer manufacturers produce tables showing the maximum permissible mass at a given distance from the coupling face of their machines; the equivalent details for most engines are much more difficult to obtain. The danger of bearing and shaft failure caused by cantilever overload caused by poor shaft system design should be kept in mind by all concerned as some spectacular failures have been thus caused.

## Torsional oscillations and critical speeds

In its simplest form the engine–dynamometer system may be regarded as equivalent to two rotating masses connected by a flexible shaft (Fig. 11.5).

Such a system has an inherent tendency to develop torsional oscillations. The two masses can vibrate 180 degrees out of phase about a node located at some point along the shaft between them. The oscillatory movement is superimposed on any steady rotation of the shaft. The resonant or critical frequency of torsional oscillation of this system is given by:

$$n_c = \frac{60}{2\pi}\sqrt{\frac{C_c(I_e + I_b)}{I_e I_b}} \tag{11.1}$$

If an undamped system of this kind is subjected to an exciting torque of constant amplitude $T_{ex}$ and frequency $n$, the relation between the amplitude of the induced oscillation $\theta$ and the ratio $n/n_c$ is as shown in Fig. 11.6.

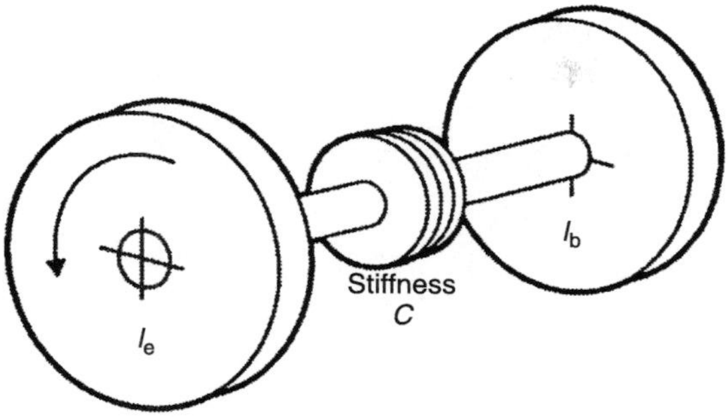


**FIGURE 11.5** A Diagrammatic representation of a classic dynamometer/UUT two-mass system. A torsional vibration demonstrator rig may be made in such a configuration by mounting a "shaft" made of piano wire, mounted in miniature bearing pedestals, and having a circular mass attached to each end.

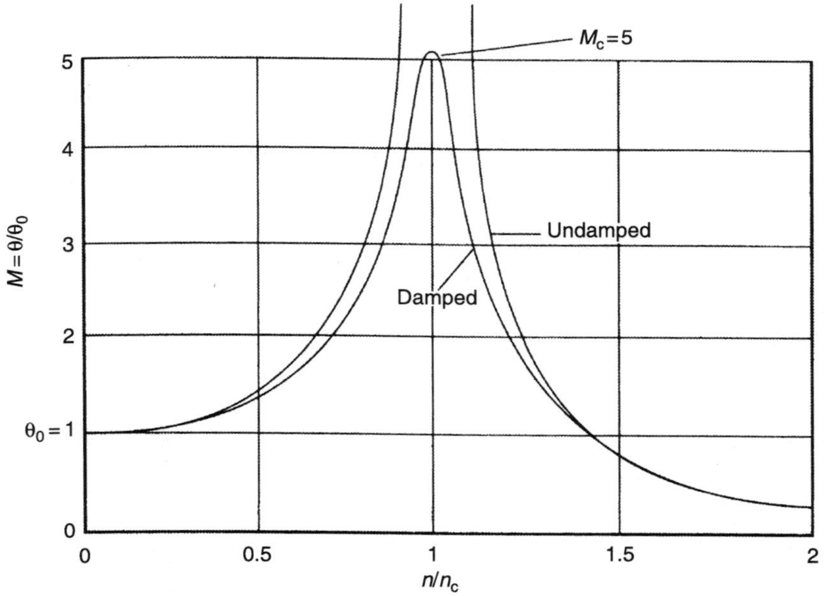


**FIGURE 11.6** Relationship between frequency ratio, amplitude, and the dynamic amplifier $M$.

At low frequencies the combined amplitude of the two masses is equal to the static deflection of the shaft under the influence of the exciting torque, $\theta_0 = T_{ex}/C_s$. As the frequency increases the amplitude rises and at $n = n_c$ it becomes theoretically infinite: the shaft may fracture, or nonlinearities and internal damping may prevent actual failure.

With further increases in frequency the amplitude falls and at $n - \sqrt{2n_c}$ it is down to the level of the static deflection. Amplitude continues to fall with increasing frequency.

The shaft connecting engine and dynamometer must be designed with a suitable stiffness $C_s$ to ensure that the critical frequency lies outside the normal operating range of the engine, and with a suitable degree of damping to ensure that the unit may be run through the critical speed without the development of a dangerous level of torsional oscillation.

Fig. 11.6 also shows the behavior of a damped system. The ratio $\theta/\theta_0$ is known as the dynamic magnifier $M$ and of particular importance is its value at the critical frequency, $M_c$. The curve of Fig. 11.6 corresponds to a value $M_c = 5$.

Torsional oscillations are excited by the variations in engine torque associated with the pressure cycles in the individual cylinders (also, though usually of less importance, by the variations associated with the movement of the reciprocating components).

Fig. 11.7 shows the variation in the case of a typical single-cylinder four-stroke gasoline engine. Any periodic curve of this kind may be synthesized from a series of *harmonic components*, each a pure sine wave of a different amplitude having a frequency corresponding to a multiple or submultiple of the engine speed, and Fig. 11.7 shows the first six components.

The *order* of the harmonic defines this multiple. Thus, a component of order $N_0 = 1/2$ occupies two revolutions of the engine, $N_0 = 1$ one revolution, and so on. In the case of a four-cylinder four-stroke engine, there are two firing strokes per revolution of the crankshaft and the turning moment curve of Fig. 11.7 is repeated at intervals of 180 degrees.

In a multicylinder engine the harmonic components of a given order for the individual cylinders are combined by representing each component by a vector, in the manner illustrated in Chapter 8, Vibration and Noise, for the inertia forces. A complete treatment of this process is beyond the scope of this book, but the most significant results may be summarized as follows.

The first major critical speed for a multicylinder in-line engine is of order:

$$N_0 = N_{cyl}/2 \text{ for a four-stroke engine} \tag{11.2a}$$

$$N_0 = N_{cyl} \text{ for a two-stroke engine} \tag{11.2b}$$

Thus, in the case of a four-cylinder four-stroke engine, the major critical speeds are of order 2, 4, 6, etc. In the case of a six-cylinder engine, they are of order 3, 6, 9, etc.

The distinction between a major and a minor critical speed is that in the case of an engine having an infinitely rigid crankshaft, it is only at the major critical speeds that torsional oscillations can be induced. This, however, by no means implies that in large engines having a large number of cylinders, the minor critical speeds may be ignored.

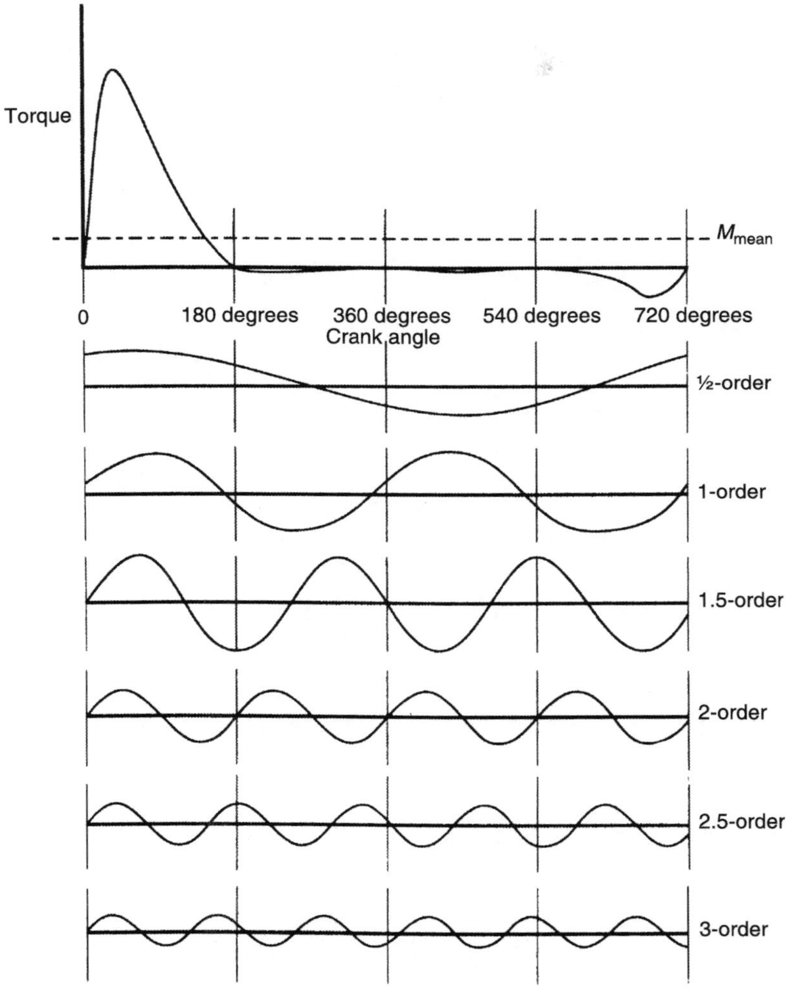


**FIGURE 11.7** Harmonic components of turning moment, single-cylinder four-stoke gasoline engine.

At the major critical speeds the exciting torques $T_{ex}$ of all the individual cylinders in one line act in phase and are thus additive (special rules apply governing the calculation of the combined excitation torques for Vee engines).

The first harmonic is generally of most significance in the excitation of torsional oscillations, and for engines of moderate size, such as passenger vehicle engines, it is generally sufficient to calculate the critical frequency from Eq. (11.1), then to calculate the corresponding engine speed from:

$$N_c = \frac{n_c}{N_0} \tag{11.3}$$

**TABLE 11.1** *p* **Factors for a four-stroke medium-speed diesel engine (Den Hartog).**

| Order | 1/2 | 1 | 1.50 | 2 | 2.50 | 3... | 8 |
|---|---|---|---|---|---|---|---|
| *p* Factor | 2.16 | 2.32 | 2.23 | 1.91 | 1.57 | 1.28 | 0.08 |

The stiffness of the connecting shaft between engine and dynamometer should be chosen so that this speed does not lie within the range in which the engine is required to develop power.

In the case of large multicylinder engines and in some powertrain configurations, the "windup" of the crankshaft as a result of torsional oscillations can be very significant and the two-mass approximation is inadequate; in particular, the critical speed may be overestimated by as much as 20% and more elaborate analysis is necessary. The subject is dealt with in several different ways in the literature but for the author Lloyd's Rulebook has been the main source of data on this subject [2].

The starting point is the value of the mean turning moment developed by the cylinder, $M_{\text{mean}}$ (Fig. 11.7). Values are given for a so-called $p$ factor, by which $M_{\text{mean}}$ is multiplied to give the amplitude of the various harmonic excitation forces. Table 11.1, reproduced from Den Hartog, shows typical figures for a four-stroke medium-speed diesel engine.

Exciting torque is:

$$T_{ex} = pM_{\text{mean}} \tag{11.4}$$

The relation between $M_{\text{mean}}$ and i.m.e.p. (*i*ndicated *m*ean *e*ffective *p*ressure) is given by (11.5a):

For a four-stroke engine:

$$M_{\text{mean}} = p_i \frac{B^2 S}{16} \times 10^{-4} \tag{11.5a}$$

For a two-stroke engine:

$$M_{\text{mean}} = p_i \frac{B^2 S}{8} \times 10^{-4} \tag{11.5b}$$

Lloyd's Rulebook, a good source of data on this subject, expresses the amplitude of the harmonic components rather differently, in terms of a "component of tangential effort," $T_m$. This is a pressure that is assumed to act upon the piston at the crank radius $S/2$. Then, exciting torque per cylinder:

$$T_{ex} = T_m \frac{\pi B^2}{4} \frac{S}{2} \times 10^{-4} \tag{11.6}$$

Lloyd's give curves of $T_m$ in terms of the indicated mean effective pressure $p_i$ and it may be shown that the values so obtained agree closely with those derived from Table 11.1.

The amplitude of the vibratory torque $T_v$ induced in the connecting shaft by the vector sum of the exciting torques for all the cylinders, $\Sigma T_{ex}$, is given by:

$$T_v = \frac{\sum T_{ex} M_c}{\left(1 + I_c / I_b\right)} \tag{11.7}$$

The complete analysis of the torsional behavior of a multicylinder engine is a substantial task, though computer programs are available and is a subject beyond the range of this edition.

It is not always possible to avoid running close to or at critical speeds, and this situation is usually dealt with by the provision of torsional vibration dampers, in which the energy fed into the system by the exciting forces is absorbed by viscous shearing. Such dampers are usually fitted at the nonflywheel end of the engine crankshaft. In some cases, it may also be necessary to consider their use as a component of engine test cell drivelines, when they are located either as close as possible to the engine flywheel or at the dynamometer. The damper must be "tuned" to be most effective at the critical frequency and the selection of a suitable damper involves equating the energy fed into the system per cycle with the energy absorbed by viscous shear in the damper. This leads to an estimate of the magnitude of the oscillatory stresses at the critical speed. For a clear treatment of the theory, see Ref. [3] or refer to damper manufacturer's design notes.

Points to remember concerning the avoidance of torsional vibration problems in the test cell are as follows:

- As a general rule, it is a good practice to avoid running the engine, or any prime mover, under power at speeds between 0.8 and 1.2 times critical speed. If it is necessary to take the system through the critical speed, this should be done off-load and as quickly as possible. With high inertia dynamometers the transient vibratory torque may well exceed the mechanical capacity of the driveline and the margin of safety of the driveline components may need to be increased.
- Problems frequently arise when the inertia of the dynamometer exceeds that of the engine: A detailed torsional analysis is desirable when this factor exceeds 2. This situation usually occurs when it is found necessary to run an engine of much smaller output than the rated capacity of the dynamometer.
- The simple "two-mass" approximation of the engine–dynamometer system is inadequate for large multicylinder ($>10$) engines and may lead to overestimation of the critical speed.

## Design calculations for coupling shafts

The maximum shear stress induced in a shaft, diameter $D$, by a torque $T$ Nm is given by:

$$\tau = \frac{16T}{\pi D^3} \text{ Pa} \tag{11.8a}$$

In the case of a tubular shaft, bore diameter $d$, this becomes:

$$\tau = \frac{16TD}{\pi(D^4 - d^4)} \text{ Pa.} \tag{11.8b}$$

For steels the shear yield stress is usually taken as equal to 0.75 × yield stress in tension [4]. A typical choice of material would be a nickel–chromium–molybdenum steel, to specification BS 817M40 (previously EN 24T) the permissible level of stress in such a shaft correctly chosen for automotive ICE or motor testing will be a small fraction of the ultimate tensile strength of the material. However, the cyclic stresses associated with torsional oscillations are an important consideration and as, even in the most carefully designed installation involving an ICE, some torsional oscillation will be present, it is wise to select a conservative value for the nominal (steady state) shear stress in the shaft.

## Stress concentrations in shafts, keyways, and keyless hub connection

For a full treatment of the very important subject of stress concentration in shafts, see Ref. [5]. There are two particularly critical locations to be considered:

- At a shoulder fillet, such as is provided at the junction with an integral flange. For a ratio fillet radius/shaft diameter = 0.1, the stress concentration factor is about 1.4, falling to 1.2 for $r/d = 0.2$.
- At the semicircular end of a typical rectangular keyway, the stress concentration factor reaches a maximum of about 2.5 × nominal shear stress at an angle of about 50 degrees from the shaft axis. The author has seen a number of shaft failures at precisely this location and angle, and for many years had a 6″ diameter example as a doorstop.

In view of the stress concentration inherent in shaft keyways and the backlash present that can develop in splined hubs, keyless hub connection systems of the type produced by the Ringfeder Corporation are now widely used. These devices are supported by comprehensive design documentation; however, the actual installation process must be exactly followed for the design performance to be ensured.

Stress concentration factors apply to the cyclic stresses rather than to the steady-state stresses. Fig. 11.8 shows diagrammatically the Goodman

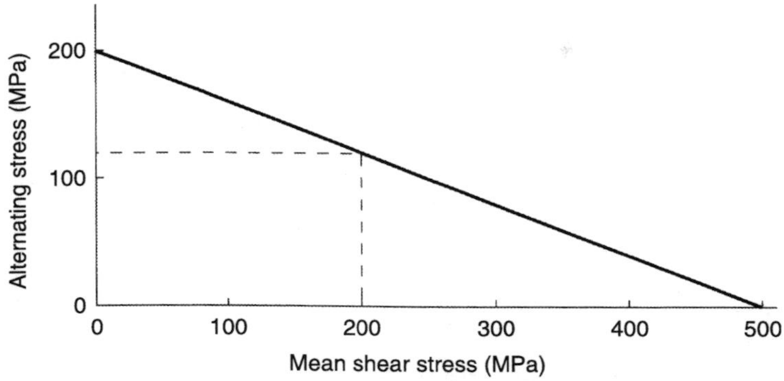


**FIGURE 11.8** A Goodman diagram: used to quantify the interaction of mean and alternating stresses on the fatigue life of a material.

diagram for EN 24T. This diagram indicates approximately the relation between the steady shear stress and the permissible oscillatory stress. The example shown indicates that, for a steady torsional stress of 200 MPa, the accompanying oscillatory stress (calculated after taking into account any stress concentration factors) should not exceed ± 120 MPa. In the absence of detailed design data, it is good practice to design shafts for use in engine test beds very conservatively, since the consequences of shaft failure can be so serious. A shear stress calculated in accordance with Eqs. (11.8a) or (11.8b) of about 100 MPa for steel with the properties listed should be safe under all but the most unfavorable conditions. To put this in perspective, a shaft of 100 mm diameter designed on this basis would imply a torque of 19,600 N m, or a power of 3100 kW at 1500 rev/min.

The torsional stiffness of a solid shaft of diameter $D$ and length $L$ is given by:

$$C_s = \frac{\pi D^4 G}{32L} \tag{11.9a}$$

For a tubular shaft, bore $d$:

$$C_s = \frac{\pi\left(D^4 - d^4\right)}{32L} \tag{11.9b}$$

## Shaft whirl

Shaft whirl is sometimes called, descriptively, the "skipping rope" mode of vibration; it may be caused by a bent shaft tube or occur within plain shaft bearings but can also occur in an engine–dynamometer system where the coupling shaft is supported at each end by a universal joint or flexible coupling. Such a shaft will "whirl" at a rotational speed $N_w$ (also at certain higher speeds in the ratio $2^2N_w$, $3^2N_w$, etc.).

The whirling speed of a solid shaft of length $L$ is given by:

$$N_w = \frac{30\pi}{L^2}\sqrt{\frac{E\pi D^4}{64W_s}} \tag{11.10}$$

It is desirable to limit the maximum engine speed to about $0.8N_w$.

*When using rubber flexible couplings, it is essential to allow for the radial flexibility of these couplings, since this can drastically reduce the whirling speed.*

It is sometimes the practice to fit self-aligning rigid steady bearings at the center of flexible couplings in high-speed applications, but these are liable to subjected to fretting problems and are not universally favored.

The whirling speed of a shaft is identical to its natural frequency of transverse oscillation. To allow for the effect of transverse coupling flexibility, the simplest procedure is to calculate the transverse critical frequency of the shaft plus two half-couplings from the equation:

$$N_t = \frac{30}{\pi}\sqrt{\frac{k}{W}} \tag{11.11a}$$

where $W$ is the mass of shaft + half-couplings and $k$ is the combined radial stiffness of the two couplings.

Then whirling speed $N$ taking this effect into account will be given by:

$$\left(\frac{1}{N}\right)^2 = \left(\frac{1}{N_w}\right)^2 + \left(\frac{1}{N_t}\right)^2 \tag{11.11b}$$

## *Shaft couplings*

The choice of the appropriate coupling for a given ICE to dynamometer application is not easy. The majority of driveline problems probably have their origin in an incorrect choice of components for the driveline and are therefore often cured by changes in this region. A complete discussion would much exceed the scope of this book, but the reader concerned with driveline design should obtain a copy of Ref. [6], which gives a comprehensive treatment together with a valuable procedure for selecting the best type of coupling for a given application. A very brief summary of the main types of coupling follows.

### Quill shaft with integral flanges and rigid couplings

This type of connection is best suited to the situation where a driven machine is permanently coupled to the source of power, when it can prove to be a simple and reliable solution. It is not well suited to test bed use, since it is intolerant of relative vibration and misalignment. However, quill shafts may be met on powertrain test beds within hybrid powertrain systems where it is essential that their designed alignment is maintained by male-female (M/F) registered machined diameters.

## Quill shaft with toothed or gear-type couplings

Splined, solid quill shafts have been successfully used as "mechanical fuses" in motorsport test-bed layouts between the engine flywheel and a bearing-mounted coupling assembly in a dummy gearbox housing. Such installations experience frequent severe power variations that such quills are designed to absorb and thus protect more valuable engine or torque transducer components that would otherwise be damaged.

In applications such as WRC powertrain testing, a 25 mm shank diameter quill shaft made from a 2.5% nickel–chromium–molybdenum high-carbon steel (S155) hardened and shot peened has been used successfully in extremely arduous torsional regimes. Such designs may find a role in similar roles during tractive motor testing where similar extremes of torque variation are experienced (see Fig. 11.9).

**FIGURE 11.9** A WRC engine test cell featuring altitude simulation through CA and exhaust pressure control. Note the large cell window, a feature of motorsport cells. The engine to dynamometer connection, capable of dealing with extremely dynamic test sequences, is via a splined quill shaft to a pedestal-mounted torque shaft in front of the AC dynamometer. *Photo courtesy MSport.*

*Hollow quill shafts and gear couplings* are very suitable for high powers and speeds and are used in steam- and gas-turbine test facilities; they can deal with vibration and some degree of misalignment, but this must be very carefully controlled to avoid problems of wear. Boundary lubrication of the coupling teeth is particularly important, as once local tooth-to-tooth seizure takes place, deterioration, even of nitrided surfaces, may be rapid and catastrophic. Such shafts are inherently torsionally stiff.

## Conventional "Cardan shaft" with universal joints

There are a number of design variations of this type of commonly used shaft, which have a type of Hooke's joint at each end as shown in Fig. 11.10. It must be remembered that a Hooke's joint is not a constant velocity joint, so to prevent a sinusoidal variation in output speed the two end joints should be in the correct relative position to each other and be set at the same "misalignment" angle.

Cardan shafts are readily available from a number of suppliers and are probably the preferred solution in the majority of automotive and "up to medium-speed" engine test assemblies. However, they can be heavy and, as warned earlier in this chapter, care should be taken with the design of the adapter plates that will be required to fit them to the dynamometer coupling. Some are fitted with torsionally soft rubber joints at one end. However, the standard ex-vehicular shafts *may not be suitable* as they will not be designed to run at engine flywheel speeds, which will be in excess of those met in vehicle applications. A small amount of misalignment between the two ends is usually recommended by manufacturers, particularly in endurance applications, to avoid fretting of the needle rollers within the joints. The shaft should be installed so that an amount of longitudinal movement in the center section is possible in either direction.

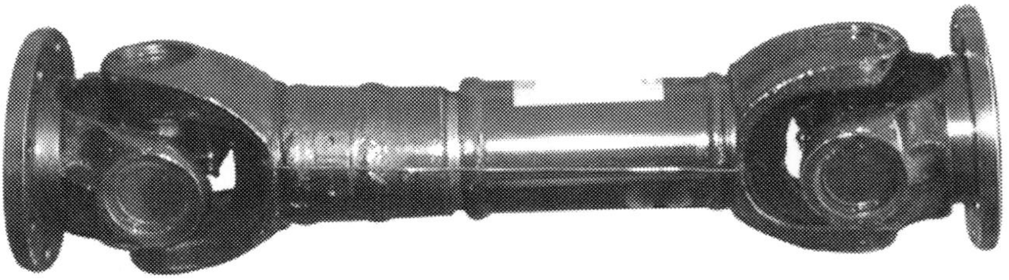

**FIGURE 11.10** A Cardan shaft shown with the splined midsection fully closed. In use it must be installed so that ± length change is possible to avoid axial loading of engine or dynamometer bearings.

## Elastomeric element couplings

Elastomeric couplings are a form of flexible coupling that uses an elastomeric polymer in one of many possible configurations between steel sections of the coupling to transmit torque

There are a vast number of different designs on the market and selection is not easy. But many manufacturers provide design and selection help and Ref. [7] may be found helpful. The great advantage of these couplings is that their torsional stiffness may be varied widely by changing the elastic elements and problems associated with torsional vibrations and critical speeds can be dealt with (see the next section).

## Torsional damping: The role of the flexible coupling

The earlier discussion leads to two main conclusions:

1. The engine–dynamometer system is susceptible to torsional oscillations and the internal combustion engine is a powerful source of forces calculated to excite such oscillations. The magnitude of these undesirable disturbances in any given system is a function of the damping capacities of the various elements: the shaft, the couplings, the dynamometer, and the engine itself.
2. The couplings are the only element of the system, the damping capacity of which may readily be changed, and for our purposes the damping capacity of the remainder of the system may be neglected.

*Note*: The term *torsional resilient* is not necessarily synonymous with *flexible* when applied to shaft couplings. Different designs are better or worse than others in absorbing radial and/or angular misalignment and the manufacturer's specification must be carefully studied before deciding on the model to use.

The dynamic magnifier $M$ (Fig. 11.6) has already been mentioned as a measure of the susceptibility of the engine–dynamometer system to torsional oscillation. Referring to Fig. 11.5, assume that there are two identical flexible couplings, of stiffness $C_c$, one at each end of the shaft, and that these are the only sources of damping. Fig. 11.11 shows a typical torsionally resilient coupling in which torque is transmitted by way of a number of shaped rubber blocks or bushes that provide torsional flexibility, damping, and some capacity to take up misalignment. The torsional characteristics of such a coupling are shown in Fig. 11.12.

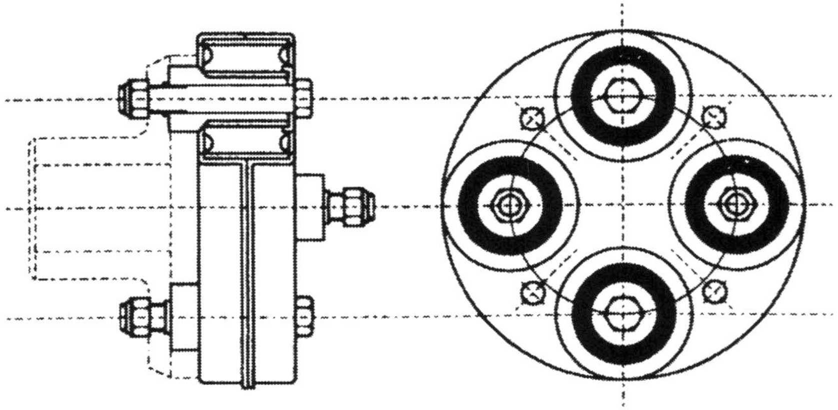

**FIGURE 11.11** A pin and rubber bush type of torsionally resilient coupling. Torsional distress within this type of coupling is characterized by a ring of rubber particles or dust found inside the shaft guard in line with the coupling joint.

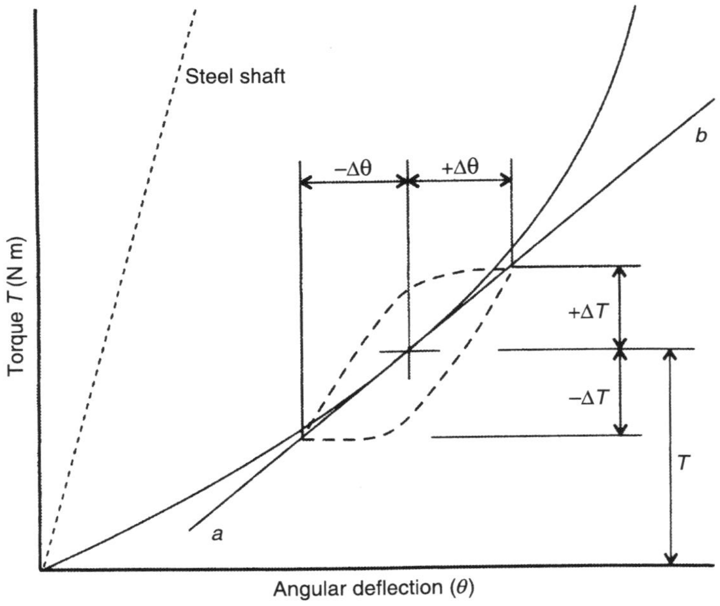


**FIGURE 11.12** Dynamic torsional characteristic of multiple-bush-type coupling of the type shown in Fig. 11.11.

These differ in three important respects from those of, say, a steel shaft in torsion:

1. The coupling does not obey Hooke's law: The stiffness or coupling rate $C_c = \Delta T/\Delta\theta$ increases with torque. This is partly an inherent property of the rubber and partly a consequence of the way it is constrained.

   The shape of the torque–deflection curve is not independent of frequency. Dynamic torsional characteristics are usually given for a cyclic frequency of 10 Hz. If the load is applied slowly, the stiffness is found to be substantially less. The following values of the ratio dynamic stiffness (at 10 Hz) to static stiffness of natural rubber of varying hardness are taken from Ref. [6].

| Shore (IHRD) hardness | 40 | 50 | 60 | 70 |
|---|---|---|---|---|
| *Dynamic stiffness* / Static stiffness | 1.5 | 1.8 | 2.1 | 2.4 |

2. If a cyclic torque $\pm\Delta T$, such as that corresponding to a torsional vibration, is superimposed on a steady torque $T$ (Fig. 11.12), the deflection follows a path similar to that shown dashed. It is this feature, the hysteresis loop, which results in the dissipation of energy, by an amount $\Delta W$ proportional to the area of the loop that is responsible for the damping characteristics of the coupling.

*Damping energy dissipated in this way appears as heat in the rubber and can, under adverse circumstances, lead to overheating and rapid destruction of the elements. The appearance of rubber dust thrown out as a ring inside the coupling guard is a warning sign.*

The damping capacity of a component such as a rubber coupling is described by the *damping energy ratio*:

$$\psi = \frac{\Delta W}{W}$$

This may be regarded as the ratio of the energy dissipated by hysteresis in a single cycle to the elastic energy corresponding to the windup of the coupling at mean deflection:

$$W = \frac{1}{2}T\theta = \frac{1}{2}T^2/C_c$$

The damping energy ratio is a property of the rubber. Some typical values are given in Table 11.2. The dynamic magnifier is a function of the damping energy ratio; as would be expected, a high damping energy ratio corresponds to a low dynamic magnifier. Some authorities give the relation:

$$\psi = \frac{2\pi}{\psi}$$

**TABLE 11.2 Dynamic magnifier (M).**

| Shore (IRHD) hardness | 50/55 | 60/65 | 70/75 | 75/80 |
|---|---|---|---|---|
| Natural rubber | 0.45 | 0.52 | 0.70 | 0.90 |
| Neoprene | | 0.79 | | |
| SBR | | 0.90 | | |

*IRHD*, International Rubber Hardness Degrees; *SBR*, Styrene–butadiene.

**TABLE 11.3 Damping energy ratio ($\psi$).**

| Shore (IHRD) hardness | 50/55 | 60/65 | 70/75 | 75/80 |
|---|---|---|---|---|
| Natural rubber | 0.45 | 0.52 | 0.70 | 0.90 |
| Neoprene | | 0.79 | | |
| SBR | | 0.90 | | |

*SBR*, Styrene–butadiene.

However, it is pointed out in Ref. [8] that for damping energy ratios typical of rubber the exact relation:

$$\psi = \left(1 - e^{-2\pi/M}\right)$$

is preferable. This leads to the values of $M$ shown in Table 11.2, which correspond to the values of $\psi$ given in Table 11.3.

It should be noted that when several components, for example, two identical rubber couplings, are used in series, the dynamic magnifier of the combination is given by:

$$\left(\frac{1}{M}\right)^2 = \left(\frac{1}{M_1}\right)^2 + \left(\frac{1}{M_2}\right)^2 + \left(\frac{1}{M_3}\right)^2 + \cdots \tag{11.12}$$

This is an empirical rule, recommended in Ref. [2].

## Selection of couplings, their torque capacity, and the role of service factors

Initial selection is based on the maximum rated torque with consideration given to the type of engine and number of cylinders, dynamometer characteristics, and inertia. Table 11.4 shows recommended service factors for a range of

**TABLE 11.4** Suggested service factors for various internal combustion engines (ICE)–dynamometer combinations. Note comments relating to severity within the text.

| Dynamometer type | Number of cylinders of and type of ICE under test | | | | | | | | | |
|---|---|---|---|---|---|---|---|---|---|---|
| | Diesel | | | | | Gasoline | | | | |
| | 1/2 | 3/4/5 | 6 | 8 | 10 + | 1/2 | 3/4/5 | 6 | 8 | 10 + |
| Hydraulic | 4.5 | 4.0 | 3.7 | 3.3 | 3.0 | 3.7 | 3.3 | 3.0 | 2.7 | 2.4 |
| Hyd. + dyno. start | 6.0 | 5.0 | 4.3 | 3.7 | 3.0 | 5.2 | 4.3 | 3.6 | 3.1 | 2.4 |
| EC | 5.0 | 4.5 | 4.0 | 3.5 | 3.0 | 4.2 | 3.8 | 3.3 | 2.9 | 2.4 |
| AC dyno and EC dyno + start | 6.5 | 5.5 | 4.5 | 4.0 | 3.0 | 5.7 | 4.8 | 3.8 | 3.4 | 2.4 |
| DC dyno. | 8.0 | 6.5 | 5.0 | 4.0 | 3.0 | 7.2 | 5.8 | 4.3 | 3.4 | 2.4 |

*AC*, Alternating current; *DC*, direct current; *EC*, Eddy current.

engine and dynamometer combinations. The rated torque multiplied by the service factor must not exceed the permitted maximum torque of the coupling.

*Note: The service factors listed in* Table 11.4 *are higher (more severe) than will be normally found in brochures issued by coupling manufacturers, this is because most design guidelines are produced for permanent installations having optimized alignment and system behavior which may not be the case in test cells. AC dynamometers designs now cover a large range of physical sizes and powers but their dynamic capabilities and applications within automotive test set-ups determine their generally high service factor.*

Different coupling manufacturers may adopt varying rating conventions, but Table 11.4 gives guidance as to the degree of severity of operation for different test situations. Thus, for example, a single-cylinder diesel engine coupled to a DC machine, with dynamometer start, calls for a margin of capacity three times as great as an eight-cylinder gasoline engine coupled to a hydraulic dynamometer.

Fig. 11.13 shows the approximate range of torsional stiffness associated with three types of flexible coupling: the annular type as illustrated in

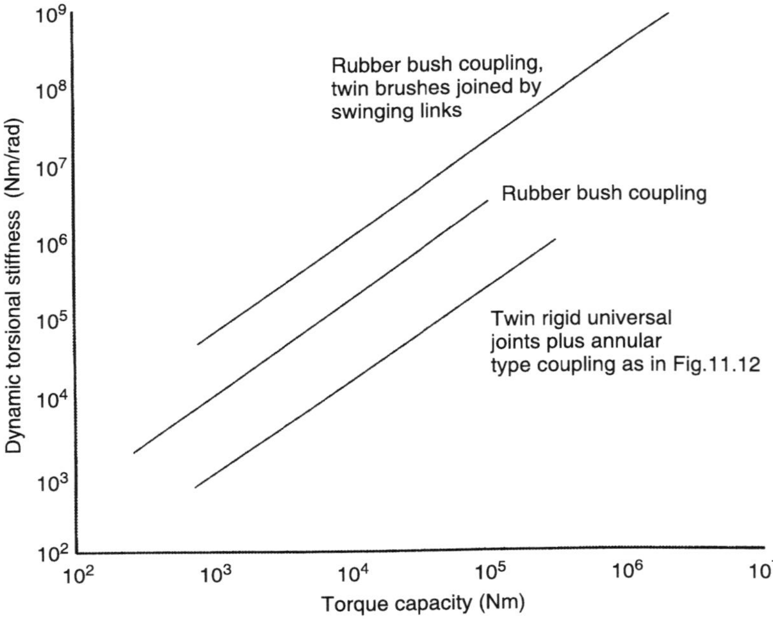


**FIGURE 11.13** Ranges of torsional stiffness for different types of rubber couplings.

Fig. 11.13, the multiple-bush design in Fig. 11.11, and a development of the multiple-bush design that permits a greater degree of misalignment and makes use of double-ended bushed links between the two halves of the coupling. The stiffness values in Fig. 11.13 refer to a single coupling rather than a double-ended shaft assembly.

## Shock loading of couplings due to cranking, irregular running, and torque reversal

Systems for starting and cranking engines are described later in this chapter, where it is emphasized that during engine starting severe transient torques can arise, which have been known to result in the failure of flexible couplings of apparently adequate torque capacity. The maximum torque that may be necessary to get a green engine over top dead center or that can be generated at irregular first firing should be estimated and checked against maximum coupling capacity.

Irregular running or imbalance between the powers generated by individual cylinders can create exciting torque harmonics of lower order than expected in a multicylinder engine and should be remembered as a possible source of rough running. Finally, there is the possibility of momentary torque reversal when the engine comes to rest on shutdown.

However, the most serious problems associated with the starting process arise when the engine first fires. Particularly when, as is common practice, the engine is motored to prime the fuel injection pump, the first firing impulses can cause severe shocks. Annular-type rubber couplings (Fig. 11.14) can fail by shearing under these conditions: in some cases, it is necessary to fit a torque limiter or slipping clutch to deal with this problem.

### *Axial shock loading*

This is a comparatively rare fault occurrence caused either by gross angular misalignment of shaft flanges (the "swash plate" effect) or thermal growth of a shaft of fixed length or having a "bottomed out" spline section. Engine test systems that incorporate automatic shaft docking systems have to provide for the axial loads on both the engine and dynamometer imposed by such a system. In some high-volume production facilities an intermediate pedestal bearing isolates the dynamometer from both the axial loads of normal docking operation and for cases when the docking splines jam "nose to nose"; in these cases the docking control system should be programed to back off the engine, spin the dynamometer, and retry the docking.

## The role of the engine clutch

Vehicle engines in service are invariably fitted with a flywheel-mounted clutch and this may or may not be retained on the test bed. The advantage of

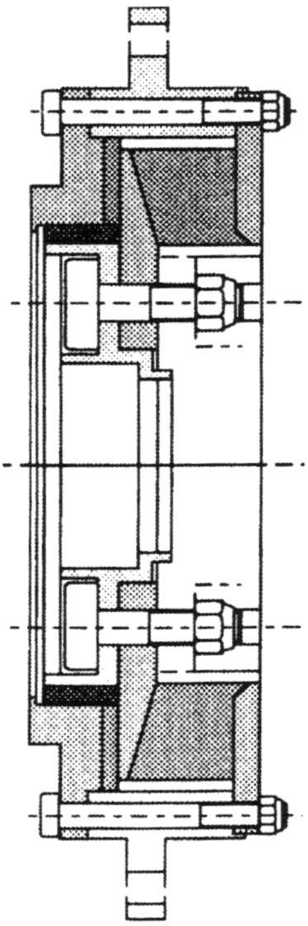

**FIGURE 11.14** A section through a typical annular-type rubber coupling.

retaining the clutch is that it acts as a torque limiter under shock or torsional vibration conditions. The great disadvantage is that it may creep, particularly when torsional vibration is present, leading to ambiguities in power measurement. When the clutch is retained a dummy (no internal gearing), gearbox is usually used to provide an outboard bearing. Clutch disk springs may have limited life under test-bed conditions.

## Dynamic balancing of driveline components

This is a matter that is often not taken sufficiently seriously and can lead to a range of troubles, including damage (which can be very puzzling) to bearings, unsatisfactory performance of such items as torque transducers, transmitted vibration to unexpected locations, and serious driveline failures. It has sometimes been found necessary to carry out in situ balancing of a composite engine driveline where *the sum of the out-of-balance forces in a particular radial relationship causes unacceptable vibration*; specialist companies with mobile plant exist to provide this service.

As already stated, conventional universal joint-type Cardan shafts are often required to run at higher speeds in test-bed applications than is usual in vehicles; when ordering, the maximum speed should be specified and, possibly, a more precise level of balancing than standard specified.

Ref. [9] gives a valuable discussion of the subject and specifies 11 balance quality grades. Driveline components should generally be balanced to grade G 6.3 or, for high speeds, to grade G 2.5. The standard gives a logarithmic plot of the permissible displacement of the center of gravity of the rotating assembly from the geometrical axis against speed. To give an idea of the magnitudes involved, G 6.3 corresponds to a displacement of 0.06 mm at 1000 rev/min, falling to 0.01 mm at 5000 rev/min.

## Alignment of unit under test and dynamometer

This appears to be simple but is a fairly complex and important subject. For a full treatment and description of alignment techniques, see Ref. [6].

Clearly it is necessary to align any two rotative machines when they are in their final running condition, that is full of fluids and therefore at running weight, with all major service connections made. When an ICE is mounted on soft mountings, it should be shaken by hand a number of times to ensure that it has come back to its natural position before being aligned.

Differential thermal growth, and the movement of the engine on its flexible mountings when on load, should be taken into account and if possible, the mean relative position should be determined. Thermal growth can be significant with some types of prime movers and often does not produce movement that is parallel with cold position, major shaft failures have occurred due to getting the alignment offset wrong so it is important to consider the subject carefully.

The laser-based alignment systems now available greatly reduce the effort and skill required to achieve satisfactory levels of accuracy. In particular, they are able to bridge large gaps between flanges without any compensation for droop and deflection of arms carrying dial indicators, a considerable problem with conventional alignment methods.

There are essentially three types of shaft having different alignment requirements to be considered:

- Rubber bush and flexible disk couplings should be aligned radially as accurately as possible as any misalignment encourages heating of the elements and fatigue damage.
- Gear-type couplings (grease lubricated) require a minimum misalignment of about 0.04 degree to encourage the maintenance of an adequate lubricant film between the teeth.
- Most manufacturers of universal joint propeller shafts recommend a small degree of misalignment to prevent brinelling of the universal joint needle

rollers. Note that it is essential, in order to avoid induced torsional oscillations, that the two yokes of the intermediate shaft joints should lie in the same plane.

Unless the shaft used has a splined center section allowing length change, the distance between end flanges will be critical, as incorrect positioning can lead to the imposition of axial loads on bearings of engine or dynamometer.

## The role and design of shaft guards

*Note: Under the latest machinery directives the shaft guard is considered as a secondary protection device, with the cell doors being the primary protection.*

The failure of a dynamometer/UUT coupling shaft is potentially dangerous as very large amounts of energy can be released in an uncontrolled and unpredictable manner. Unfortunately, such failures in ICE test beds are not particularly uncommon.

Obviously, even while running without any problem, a bare, unguarded, shaft is a clear danger to persons and anything that might get entangled with it, but a badly designed or carelessly installed shaft guard can itself represent a less obvious danger and turn a small vibration problem into a major failure.

The ideal shaft guard will contain the debris resulting from a failure of any part of the driveline and prevent accidental or casual contact with rotating parts, while being sufficiently light and adaptable not to interfere with engine rigging and alignment.

*A guard system that is very inconvenient to use will eventually be misused or neglected.*

A substantial guard, preferably made from a steel tube not less than 6 mm thick, split and hinged in the horizontal plane for shaft access, represents proved best practice. The hinged "lid" should be interlocked with the emergency stop circuit to prevent engine cranking or running while it is open. Many guard designs include shaft restraint devices, made of wood or a nonmetallic composite, that loosely fit around the tubular portion and are intended to prevent a failing shaft from whirling. These restraints should not be so close to the shaft as to be touched by it during normal starting or running, otherwise they will be the cause of shaft failure rather than its prevention.

### *Engine-to-dynamometer coupling: summary of design procedure*

1. Establish speed range and torque characteristic of the engine to be tested. Is it proposed to run the engine on load throughout this range?
2. Make a preliminary selection of a suitable drive shaft. Check that maximum permitted speed is adequate. Check drive shaft stresses and specify material. Look into possible stress raisers.

3. Check manufacturer's recommendations regarding service factor and other limitations.
4. Establish rotational inertias of engine and dynamometer, and stiffness of proposed shaft and coupling assembly. Make a preliminary calculation of torsional critical speed from Eq. (11.1) (in the case of large multicylinder engines, consider making a complete analysis of torsional behavior).
5. Modify specification of shaft components as necessary to position torsional critical speeds suitably. If necessary, consider the use of viscous torsional dampers.
6. Calculate vibratory torques at critical speeds and check against capacity of shaft components. If necessary, specify "no-go" areas for speed and load.
7. Check whirling speeds.
8. *Specify and issue alignment requirements and the assembly method of the shaft components; in the case of bolted flanges the tightening torque and the locking methods must be specified.*
9. Design shaft guards and ensure upon fitting that clearance between the guard and the shaft system is maintained over than range of possible engine movement.

## Flywheels and their role in powertrain testing shaft lines

A flywheel is a device that stores kinetic energy. The energy stored may be precisely calculated and is instantly available. The storage capacity is directly related to the mass moment of inertia, which is calculated by:

$$I = k \times M \times R^2$$

where $I$ is the moment of inertia (kgm$^2$), $k$ is the inertia constant (dependent upon shape), $M$ is the mass (kg), and $R$ is the radius of the flywheel mass.

In the case of a flywheel taking the form of a uniform disk, which is the common form found within dynamometer cells and chassis dynamometer designs:

$$I = \frac{1}{2} M R^2$$

The powertrain test engineer may expect to deal with flywheels in the following roles:

1. As part of the UUT, as in the common case of an engine flywheel, where it forms part of the engine–dynamometer shaft system and contributes to the system's inertial masses taken into account during a torsional analysis.
2. As part of a kinetic energy capture and recovery system (KERS) within a powertrain system, wherein the braking energy of the vehicle is used to store and then release energy within an encased flywheel system as in the "Flybrid KERS," which is a packaged flywheel-based M-KERS system incorporating a continuously variable transmission.

3. As part of the test equipment where one or more flywheels may be used to provide actual inertia that would, in "real life," be that of the vehicle or some part of the engine or motor-driven system.

Modern AC dynamometer systems and control software have significantly replaced the use of flywheels in chassis, engine and brake testing dynamometer systems in the automotive industry. They have been retained in very large brake test systems such as those that have to simulate and absorb the kinetic energy of the emergency braking of an express train or an aborted take-off event of a fully loaded wide-bodied jet airliner.

Any real shortcoming in the speed of response or accuracy of the simulation is usually considered, particularly in chassis dynamometers, to be of less concern than the mechanical simplicity of the electric dynamometer system and the reduction in required cell space. Some transmission test rigs still retain flywheels at the nondrive end of their wheel dynamometers where they can be easily changed or removed.

Finally, it should be remembered that, unless engine rig flywheels are able to be engaged through a clutch onto a rotating shaft, the engine starting/cranking system will have to be capable of accelerating engine, dynamometer, and flywheel mass up to engine start speed.

## Flywheel safety issues

The uncontrolled discharge of energy from any storage device is hazardous and that of flywheels can be difficult to physically contain. The classic case of a flywheel failing by bursting is now exceptionally rare and invariably due to incompetent modification rather than the 19th century problems of poor materials, poor design, or overspeeding.

In the case of engine flywheels the potential source of danger in the test cell is that the shaft fixed on the flywheel may be quite different in mass and fixing detail from its final application connection; this can, and has in recent years, cause overload leading to failure. Cases are on record where shock loading caused by connecting shafts touching the guard system due to excessive engine movement has created shock loads that have led to a cast steel engine flywheel fracturing, with severe consequential damage.

The most common hazard of test-rig-mounted flywheels is caused by bearing or driveline clutch failure, where consequential damage is exacerbated by the considerable energy that is available to damage connected devices and because of the length of time that the flywheel system may rotate before the stored energy is dissipated and movement is stopped.

It is vital that flywheels are guarded in such a manner as to prevent absolutely accidental entrainment of clothing or cables, etc. For all but the smallest automotive flywheels, it is probably not entirely practical to try and design a guard that would retain it in the case of it breaking loose, this is where the cell has to act as the hazard containment.

Test-rig flywheel sets need to be rigidly and securely mounted and balanced to the highest practical standard. Multiples of in-line flywheels forming a common system that can be engaged in different combinations, and in any radial relationship, require particular care in the design of both their base frame and individual bearing supports. Such systems can produce virtually infinite combinations of dynamic balance and require each individual mass to be as well balanced and aligned on as rigid a base as possible.

## Engine cranking and starting in a test cell

Starting an engine when it is connected to a dynamometer may present the cell operator with a number of rigging and operational problems and is a factor to be remembered when selecting the dynamometer. If the engine is fitted with a starter motor the cell system must provide the high current DC supply and associated switching; in the absence of an engine-mounted starter a complete system to start and crank the engine must be available that compromises neither the torsional characteristics of the driveline nor the torque measurement accuracy.

## Engine cranking: no starter motor on the unit under test

The cell cranking system must be capable of accelerating the engine to its normal starting speed and of disengaging when the engine fires.

An AC or DC dynamometer, suitably controlled, will be capable of starting the engine directly but the power available from any such machine will always be greater than that required; therefore *excessive starting torque must be avoided by the test controller's alarm system*, otherwise an engine locked by seizure or fluid in a cylinder will cause severe damage to the driveline.

For the many existing cells using two-quadrant machines, there are several options that are discussed next.

The preferred method of providing other types of dynamometer, most commonly eddy-current machines, with a starting system is to mount an electric motor at the nonengine end of the dynamometer shaft, driving through an overrunning or remotely engaged clutch, and generally through a speed-reducing belt drive. The clutch half containing the mechanism should be on the input side, otherwise it will be affected by the torsional vibrations usually experienced by dynamometer shafts. Any dynamometer-mounted starter system must be designed so that it imposes the minimum possible parasitic torque when disengaged, since this torque will not be sensed by the dynamometer measuring system.

To avoid this source of inaccuracy the motor may be mounted directly on the dynamometer carcase and permanently coupled to the dynamometer shaft by a belt drive and overrun clutch. This imposes an additional load on the trunnion bearings, which may lead to brinelling, and it also increases the effective moment of inertia of the dynamometer. However, this design has worked well in practice and has the advantage that, providing the combined machine is set up correctly with no residual load on the load cell, motoring and starting torque may be measured by the dynamometer system.

The sizing of the motor must take into account the maximum breakaway torque expected, usually estimated as twice the average cranking torque, while the normal running speed of the motor should correspond to the desired cranking speed. The choice of motor and associated starter must take into account the maximum number of starts per hour that may be required, both in normal use and when dealing with a faulty engine. The running regime of the starter motor can be demanding, involving repeated bursts at overload, with the intervening time at rest, therefore an independent cooling fan may be necessary.

An alternative solution is to use a standard vehicle engine starter motor in conjunction with a gear ring carried by a "dummy flywheel" incorporated in the driveline, but this may have the disadvantage of complicating the torsional behavior of the system.

## Engine-mounted starter systems

It must be understood that the standard engine-fitted starter system may be inadequate to accelerate the whole dynamometer system and the engine up to starting speed, particularly when connected to a "green" engine that may exhibit a very high breakaway torque and require prolonged cranking at high speed to prime the fuel system before it fires.

However, if the engine is rigged with an adequate starter motor, all that has to be provided is the necessary 12 or 24 V supply and an electrical relay system operated from the control desk. The traditional approach has been to locate a suitable battery as close as possible to the starter motor, with a suitable battery charger supply. This system is not ideal, as the battery needs to be in a suitably ventilated box to avoid the risk of accidental shorting and will take up valuable space. Batteries can be heavy and handling them in confined cell spaces may be considered as an H&S risk not worth taking. Modern practice is to provide the cells with an external regulated 12- or 24-V supply of which a wide range are commercially available, which include an "electrical services box" to provide additional power for ignition systems and diesel glow plugs.

## Nonelectrical internal combustion engines starting systems

Diesel engines, larger than the automotive range, are usually started by means of compressed air, which is admitted to the cylinders by way of individual starting valves. In some cases, it is necessary to move the crankshaft to the correct starting position by opening cylinder valves, to prevent compression, then either by barring using holes in the flywheel or using an engine-mounted inching motor. The test facility should include a compressor and a receiver of capacity at least as large as that recommended for the engine in service.

Compressed air or hydraulic motors are sometimes used instead of electric motors to provide cranking power but have no obvious advantages over a DC electric motor, apart from a marginally reduced fire risk in the case of a compressed air system, provided the supply is shut off automatically in the case of fire.

## Notations

| | |
|---|---|
| **Frequency of torsional oscillation** | $n$ (cycles/min) |
| **Critical frequency of torsional oscillation** | $n_c$ (cycles/min) |
| **Stiffness of coupling shaft** | $C_s$ (N m/rad) |
| **Rotational inertia of engine** | $I_e$ (kgm$^2$) |
| **Rotational inertia of dynamometer** | $I_b$ (kgm$^2$) |
| **Amplitude of exciting torque** | $T_x$ (N m) |
| **Amplitude of torsional oscillation** | $\theta$ (rad) |
| **Static deflection of shaft** | $\theta_0$ (rad) |
| **Dynamic magnifier** | $M$ |
| **Dynamic magnifier at critical frequency** | $M_c$ |
| **Order of harmonic component** | $N_0$ |
| **Number of cylinders** | $N_{cyl}$ |
| **Mean turning moment** | $M_{\text{mean}}$ (N m) |
| **Indicated mean effective pressure** | $p_i$ (bar) |
| **Cylinder bore** | $B$ (mm) |
| **Stroke** | $S$ (mm) |
| **Component of tangential effort** | $T_m$ (N m) |
| **Amplitude of vibratory torque** | $T_v$ (N m) |
| **Engine speed corresponding to $n_c$** | $N_c$ (rev/min) |
| **Maximum shear stress in shaft** | $\tau$ (N/m$^2$) |
| **Whirling speed of shaft** | $N_w$ (rev/min) |
| **Transverse critical frequency** | $N_t$ (cycles/min) |
| **Dynamic torsional stiffness of coupling** | $C_c$ (N m/rad) |
| **Damping energy ratio** | $\psi$ |
| **Modulus of elasticity of shaft material** | $E$ (Pa) |
| **Modulus of rigidity of shaft material (for steel, $E = 200 \times 10^9$ Pa, $G = 80 \times 10^9$ Pa)** | $G$ (Pa) |

## References


[1] UK Rigging Parts Catalog: Demon Tweeks, <http://www.demon-tweeks.co.uk>. RSComponents <http://uk.rs-online.com>.

[2] Rulebook, Chapter 8. Shaft Vibration and Alignment. Lloyd's Register of Shipping, London.

[3] J.P. Den Hartog, Mechanical Vibrations, McGraw-Hill, Maidenhead, 1956. ISBN-13: 978-0486647852.

[4] W.C. Young, Roark's Formulas for Stress and Strain, McGraw-Hill, New York, 1989. ISBN-13: 978-0070725423.
[5] W.D. Pilkey, Peterson's Stress Concentration Factors, Wiley, New York, 1997. ISBN-13: 978-0470048245.
[6] M.J. Neale, P. Needham, R. Horrell, Couplings and Shaft Alignment, Mechanical Engineering Publications, London, 1998. ISBN-13: 978-1860581700.
[7] BS 6613. Methods for Specifying Characteristics of Resilient Shaft Couplings.
[8] W. Ker Wilson, Practical Solution to Torsional Vibration Problems (5 vols), Chapman & Hall, London, 1963. ISBN-13: 978-0412076800.
[9] BS 5265 Parts 2 and 3. Mechanical Balancing of Rotating Bodies.

## Further reading

BS 4675 Parts 1 and 2. Mechanical Vibration in Rotating Machinery.
BS 6716. Guide to Properties and Types of Rubber.
6861 Part 1, BS 6861 Part 1. Method for Determination of Permissible Residual Unbalance.
E.J. Nestorides, A Handbook of Torsional Vibration, Cambridge University Press, 1958 ISBN-13: 978-0521203524.
W. Ker Wilson, Vibration Engineering, Griffin, London, 1959. ISBN-13: 978-0852640234.
<https://agilismeasurementsystems.com/re:turbochargerspeed>.

# Appendix

## A worked example of drive shaft design

The application of these principles is best illustrated by a worked example. Fig. 11.15 represents an engine coupled by way of twin multiple-bush-type rubber couplings and an intermediate steel shaft to an eddy-current dynamometer, with dynamometer starting.

Engine specification is as follows:

Four-cylinder four-stroke gasoline engine, swept volume 2.0 L, bore 86 mm, and stroke 86 mm.

Table 11.4 indicates a service factor of 4.8, giving a design torque of $110 \times 4.8 = 530$ N m.

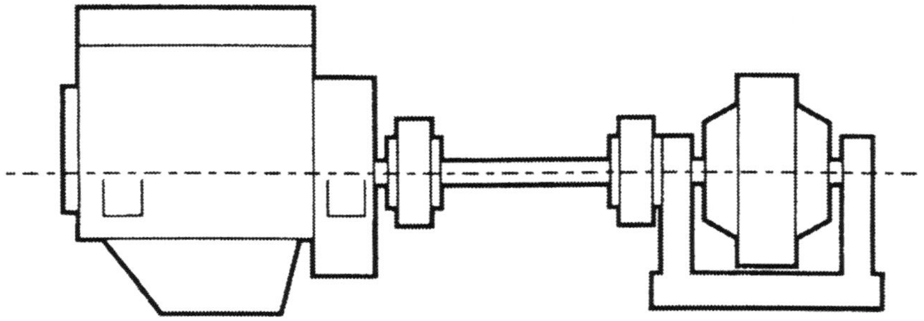

**FIGURE 11.15** Simple form of dynamometer-engine driveline used in the worked example next.

| | |
|---|---|
| Maximum torque | 110 N m at 4000 rev/min |
| Maximum speed | 6000 rev/min |
| Maximum power output | 65 kW |
| Maximum b.m.e.p. | 10.5 bar |
| Moments of inertia | $I_e = 0.25$ kgm$^2$ |
| | $I_d = 0.30$ kgm$^2$ |

It is proposed to connect the two couplings by a steel shaft of the following dimensions:

| | |
|---|---|
| Diameter | $D = 40$ mm |
| Length | $L = 500$ mm |
| Modulus of rigidity | $G = 80 \times 10^9$ Pa |

From Eq. (11.9a) and remembering the service factor of 4.8 the torsional stress $\tau = 42$ MPa, which is very conservative.

From Eq. (11.10):

$$C_s = \frac{\pi \times 0.04^4 \times 80 \times 10^9}{32 \times 0.5} = 40,200 \text{ N m/rad}$$

Consider first the case when rigid couplings are employed:

$$n_c = \frac{60}{2\pi}\sqrt{\frac{40,200 \times 0.55}{0.25 \times 0.30}} = 5185 \text{ c.p.m.}$$

For a four-cylinder four-stroke engine we have seen that the first major critical occurs at order $N_0 = 2$, corresponding to an engine speed of 2592 rev/min. This falls right in the middle of the engine speed range and is clearly unacceptable. This is a typical result to be expected if an attempt is made to dispense with flexible couplings.

The resonant speed needs to be reduced and it is a common practice to arrange for this to lie between either the cranking and idling speeds or between the idling and minimum full-load speeds. In the present case these latter speeds are 500 and 1000 rev/min, respectively. This suggests a critical speed $N_c$ of 750 rev/min and a corresponding resonant frequency $n_c = 1500$ cycles/min.

This calls for a reduction in the torsional stiffness in the ratio:

$$\left(\frac{1500}{5185}\right)^2 \quad \text{i.e. to } 3364\ \text{N m/rad.}$$

The combined torsional stiffness of several elements in series is given by (11.14):

$$\frac{1}{C} = \frac{1}{C_1} + \frac{1}{C_2} + \frac{1}{C_3} + \cdots \tag{11.14}$$

This equation indicates that the desired stiffness could be achieved with two flexible couplings, each of stiffness 7480 N m/rad. A manufacturer's catalog shows a multibush coupling having the following characteristics:

Substituting this value in Eq. (11.14) indicates a combined stiffness of 3800 N m/rad. Substituting in Eq. (11.1) gives $n_c = 1573$, corresponding to an engine speed of 786 rev/min, which is acceptable.

| | |
|---|---|
| Maximum torque | 814 N m (adequate) |
| Rated torque | 170 N m |
| Maximum continuous vibratory torque | ± 136 N m |
| Shore (IHRD) hardness | 50/55 |
| Dynamic torsional stiffness | 8400 N m/rad |

It remains to check on the probable amplitude of any torsional oscillation at the critical speed. Under no-load conditions the i.m.e.p. of the engine is likely to be around 2 bar, indicating, from Eq. (11.5a), a mean turning moment $M_{\text{mean}} = 8$ N m.

From Table 11.1, $p$ factor = 1.91, giving $T_{ex} = 15$ N m per cylinder:

$$\sum T_{ex} = 4 \times 15 = 60\ \text{N m}$$

Table 11.4 indicates a dynamic magnifier $M = 10.5$ and combined dynamic magnifier from Eq. (11.12) = 7.4. The corresponding value of the vibratory torque, from Eq. (11.7), is then:

$$T_v = \frac{60 \times 7.4}{(1 + 0.25/0.30)} = \pm 242\ \text{N m}$$

This is outside the coupling continuous rating of ± 136 N m, but multiple-bush couplings are tolerant of brief periods of quite severe overload and this solution should be acceptable provided the engine is run fairly quickly through the critical speed. An alternative would be to choose a coupling of similar stiffness using SBR bushes of 60/65 hardness. Table 11.4 shows that the dynamic magnifier is reduced from 10.5 to 2.7, with a corresponding reduction in $T_v$.

If in place of an eddy-current dynamometer we were to employ a DC machine, the inertia $I_b$ would be of the order of 1 kgm$^2$, four times greater. This has two adverse effects:

- Service factor, from Table 11.4, is increased from 4.8 to 5.8.
- The denominator in Eq. (11.7) is reduced from $(1 + 0.25/0.30) = 1.83$ to $(1 + 0.25/1.0) = 1.25$, corresponding to an increase in the vibratory torque for a given exciting torque of nearly 50%.

*This is a general rule: The greater the inertia of the dynamometer, the more severe the torsional stresses generated by a given exciting torque.*

An application of Eq. (11.1) shows that for the same critical frequency the combined stiffness must be increased from 3364 to 5400 N m/rad. We can meet this requirement by changing the bushes from shore hardness 50/55 to shore hardness 60/65, increasing the dynamic torsional stiffness of each coupling from 8400 to 14,000 N m/rad (in general, the usual range of hardness numbers, from 50/55 to 75/80, corresponds to a stiffness range of about 3:1, a useful degree of flexibility for the designer).

Eq. (11.1) shows that with this revised coupling, stiffness $n_c$ changes from 1573 to 1614 cycles/min, and this should be acceptable. The oscillatory torque generated at the critical speed is increased by the two factors mentioned previously but reduced to some extent by the lower dynamic magnifier for the harder rubber, $M = 8.6$ against $M = 10.5$. As before, prolonged running at the critical speed should be avoided.

For completeness, we should check the whirling speed from Eqs. (11.11a) and (11.11b).

The mass of the shaft per unit length is $W_s = 9.80$ kg/m:

$$N_w = \frac{30\pi}{0.50^2}\sqrt{\frac{200 \times 10^9 \times \pi \times 0.04^4}{64 \times 9.80}} = 19,100 \text{ rpm}$$

then from Eq. (11.11b), whirling speed = 12,300 rev/min, which is satisfactory.

Note, however, that if shaft length were increased from 500 to 750 mm, whirling speed would be reduced to about 7300 rev/min, which is barely acceptable. This is a common problem, usually dealt with by the use of tubular shafts, which have much greater transverse stiffness for a given mass.

*There is no safe alternative, when confronted with an engine of which the characteristics differ significantly from any other run previously on a given test bed, to following through this design procedure.*

An alternative solution to the previous worked example is to make use of a conventional propeller shaft with two universal joints with the addition of a coupling incorporating an annular rubber element in shear to give the necessary torsional flexibility. These coupling elements, shown in Fig. 11.14, are generally softer than the multiple-bush-type for a given torque capacity but less tolerant of operation near a critical speed. If it is decided to use a conventional universal joint shaft, the supplier should be informed of the maximum speed at which it is to run. This will probably be much higher than is usual in the vehicle and may call for tighter than usual limits on balance of the shaft.

Chapter 12

# Testing of electrical drive systems

## Chapter Outline



## Introduction

### The impact of electrification on automotive test facilities

At the time of writing (2020) there is a paradigm shift occurring in the automotive powertrain testing environment. The adoption of electrical propulsion and hybrid systems and the need for manufacturers provide products as close as possible to electric only propulsion in the shortest possible time is requiring the rapid adaption of existing test cells and the building of new, non-internal combustion engine (ICE), test facilities.

The outcome being a complete shift of investment away from technologies and systems for classical ICE only testing and development towards

**Engine Testing. DOI: https://doi.org/10.1016/B978-0-12-821226-4.00012-7**

environments suitable for integration of electric machine test and new powertrain concepts. One immediately visible feature of a test cell testing electric propulsion motors, or e-axles, is that the unit under test (UUT) is of small volume compared to the test equipment required.

There are a number of key areas that can be considered as basic differences, indeed simplifications, during the planning phase when designing an electric motor cell compared with ICE cells of equivalent power rating.

## Cell ventilation and safety interlocks

The cell's heat load will tend to be lower, requiring a lower flow rating of the ventilation system while retaining similar balanced inlet/outlet system (see Figs. 5.2–5.5). The ventilation system with a minimum working flow rate giving around five cell air changes per minute will need to retain all of the safety features and system interlocks discussed in Chapter 5, Ventilation and Air-Conditioning of Automotive Test Facilities, in order to retain the cell as a fire hazard containment box.

Particular attention must be made to the *appropriate electrical isolation* of all units within the cell in any given fault or alarm condition; a well-designed cell safety matrix is essential (see Chapter 4: Electrical Design Requirements of Test Facilities).

## Climatic test containment

Since the physical size of the UUT is less, and the number of service connections are fewer than those of an ICE, it is much easier to encase the UUT in an insulated box than to heat or cool the whole cell (see Fig. 5.10). A common design is based on a very rigid steel mounting faceplate, through which the dynamometer connection shaft, power and transducer cables come via a separate thermal barriered holes, forming the end wall of an insulated box that is slid over the motor to enable running in air at an elevated or reduced temperature. There are examples of such small volume electric motor test enclosures being used that have the following specifications:

- temperature range without humidity control: −55°C to +150°C
- temperature range with humidity control: +10°C to +95°C
- humidity range: 10%–95% relative humidity

## Palletization of the unit under test

Due to the lower number of service connections and their simpler format the time taken to rig and derig an electric motor unit can be much shorter than its ICE equivalent so the advantages of using palletization may be considered as less. However, prerigging of a complex electric axle drive unit, which may require connection to up to three dynamometer shafts, is clearly a good policy. To accommodate a pallet mounting system that will have to deal

**FIGURE 12.1** An e-axle unit rigged for test on a palletized fixed dynamometer test stand connected to two-wheel simulating dynamometers mounted in height-adjustable stands. *Courtesy: ZF Test Systems.*

with units of various dimensions and layouts requires a highly adaptable test cell "island" and dynamometer mountings that allow the machines to be easily moved to match the coupling face dimensions of the UUT. Due to the high rigidity and critical shaft alignment with which propulsion motors have to be mounted, it is preferable to make the pallet and prime dynamometer the fixed node of the test configuration and move other dynamometers/driven units on the common air-sprung, tee-slotted bedplate. The ZF system shown in Fig. 12.1 is in a 3E configuration and the two-wheel simulating dynamometers are mounted on height-adjustable frames to enable rapid and accurate shaft alignment. The same pallet system used in a single motor shaft configuration may be seen in Fig. 9.14.

## *Fluid services*

Cooling water may be required to serve electrical drive cabinets within the cell and chilled water will be required if refrigeration plant is installed. The compressed air supply should be treated through an air dryer.

## *Fire suppression*

While not having liquid fuels in the cell tends to reduce the cell's fire load, it does not simplify the problem of fire suppression since cable fires and fires within cabinets pose their own problems and emit toxic fumes. Water mist suppression systems are recommended in most cases when supported with appropriate handheld extinguishers (see Table 3.8 and accompanying text).

### Electrical services

Details are covered in following text and in Chapter 4, Electrical Design Requirements of Test Facilities, but the most obvious change in electrical wiring from that met by personnel used to working in purely ICE cells is the deliberately distinctive orange wiring and plugs that form the power cable harness of a battery electric vehicle (BEV) or hybrid powertrain (Fig. 12.2). All other electric services will remain as used in ICE cells fitted with AC dynamometers.

## Powertrain application—electrical power analysis overview

Electrification of the vehicle powertrain has driven the requirement and adoption of electrical measurement systems directly into the test cell environment in order to characterize the electrical propulsion system efficiencies. To do this it is necessary to measure at component level so that losses can be apportioned, and optimizations be implemented.

Traditional electrical system power analyzers can be used—and were used—when electrical motors first appeared in the automotive test cells (Formula 1 KERS implementation in 2008). However, these instruments are designed for accurate measurement on mains power networks where frequencies are not wide ranging or transient in nature. Their lack of robustness and interfacing capability to the rest of the automotive test system mean that they are not well suited to test cell environments.

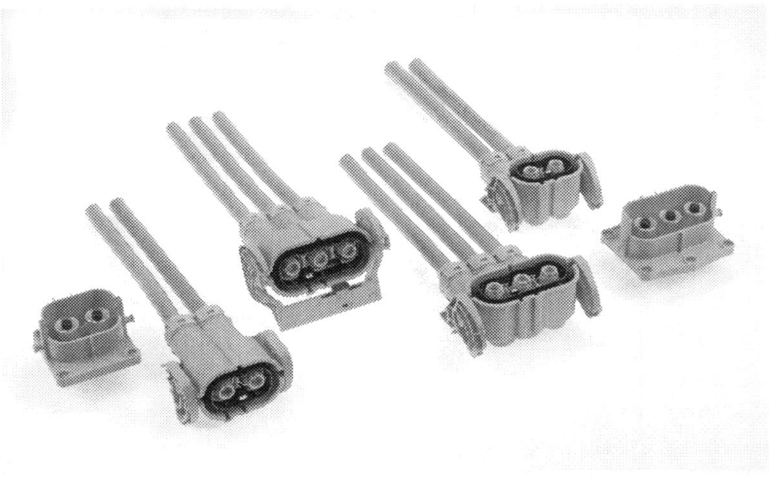

**FIGURE 12.2** A set of cables, plugs, and sockets all bright orange in color that are standard to use in electric vehicle (EV) and HEV vehicles. *HEV*, Hybrid electric vehicle. *Courtesy: Dräxlmaier Group.*

This has led to the evolution of equipment specifically to measure and characterize propulsion system electrical energy flow; with two main areas of application:

- component level measurements for motors, inverters under a full range of powers and conditions and
- holistic powertrain analysis in order to understand complete powertrain system energy flows and losses.

The test cell designer or engineer must now consider this equipment along with other powertrain instrumentation at the specification and configuration stage, in order to efficiently integrate the electrical measurements subsystems into the overall testing and data acquisition system. This allows the possibility to gain reliable, accurate time-aligned data for all of the powertrain system components.

## Powertrain electrification measurement approaches

In order to measure the required signals, access to electrical conductors is required together with appropriate signal conditioning to decouple high voltages (HVs) and to avoid rerouting of high current cabling. On a test bench, or in a test facility, this is relatively straightforward as wiring modifications can be undertaken but for measurement applications later in the development cycle, where the UUT is located within the powertrain of a vehicle, it is much more difficult to access the appropriate cables and interfaces. The wiring system used in the vehicle is normally heavily shielded to counteract noise emission and to protect against electromagnetic interference. It is also shielded and interlocked to protect against physical damage and provide personnel protection when repairing and disconnecting the HV systems. This provides a considerable challenge when trying to access individual cables for signals, even when using non-intrusive sensor technology. However, there are a wide range of voltage isolation probes, current clamps and transformers available on the market that facilitate access to the signal sources. There are also measurement and breakout boxes available that can provide a robust mechanism to modify wiring systems (Figs. 12.3–12.5), but the choice of how and with what is left to the test facility Engineer/Technician to decide.

Analog power analyzers are mostly deployed in the early stage of the development process where the electrical UUT is often characterized on a test bench. In this case analog signal capture and processing is used as the basis for acquiring relevant signals.

For vehicle-based measurements, later in the development cycle, the prospect of running or extending high-voltage cabling to measurement devices is not ideal. So, for this application, a "digital," decentralized measurement concept is more appropriate, where analog-to-digital conversion of the raw signals is undertaken by a "smart" data acquisition module, close to the point of

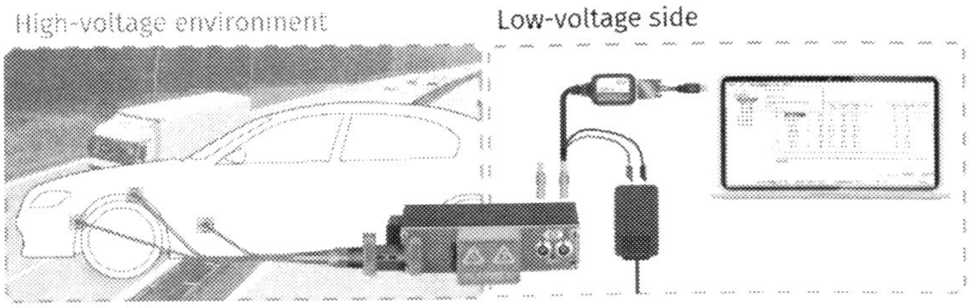


**FIGURE 12.3** Isolation-based signal conditioning is required where temperature measurements are needed within a high-voltage system (battery, e-machine, etc.). *Courtesy: CSM GmbH.*

**FIGURE 12.4** Breakout box for DC voltage and current measurement. This design has been certificated as safe and complies with relevant CE and IEC certification, appropriate IP (Ingress protection) rating is also necessary for such a module to be used safely in a prototype test environment. *Courtesy: CSM GmbH.*

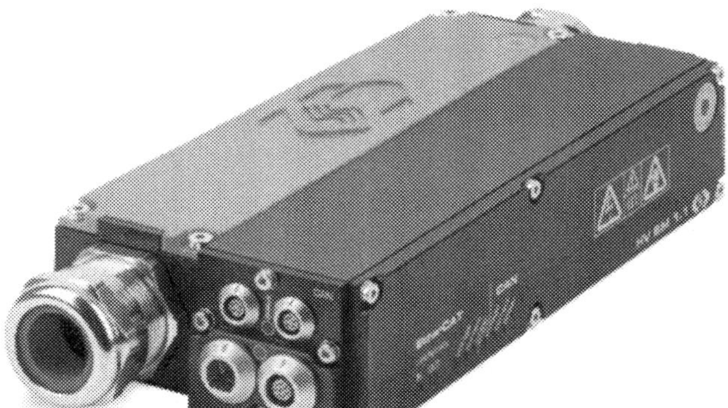

**FIGURE 12.5** Breakout boxes allow analogue to digital conversion close the to signal source. Data can then be transmitted safely in digital form using standard protocols such as CAN or EtherCAT. *Courtesy: CSM GmbH.*

measurement. With appropriate interfaces, such modules can transfer data safely (digitally) to a host where the calculations and analysis can be undertaken, this concept is ideal for holistic powertrain analysis. The modules shown in Figs. 12.3–12.5 are intelligent and suitable for such an in-vehicle measurement concept. The breakout module directly converts the analog signal into digital samples. The output of the breakout module is purely digital, acquiring and transferring data at a high sample rate ($>1$ MHz). Those samples are transferred to a host computer, equipped with software that is capable of calculating the power flows in real time. This approach for power analysis does not rely on current transducers and can incorporate safely the use of directly measuring current shunts within the module, this allows a proportionally large measurement bandwidth.

Safety and legal compliance is important point; while in a test environment there are no real restrictions on circuit modification however, if a vehicle is to be used on the public road then electrical system cable modifications and methods must be assessed and signed off by a suitably competent person, in order to comply with insurance and general safety regulations (EN 61010).

The accuracy of the complete electrical measurement chain is the most significant consideration with respect to legislation that exists, such as SAE J1711 and J1634, both of which specify AC and DC measurement accuracy of 1% or better. In general, the requirements for the test or facility engineer are as follows:

- *Voltage:* Voltage needs to be measured using an interface that can be located close to the point of connection [therefore requiring appropriate temperature and Ingress Protection (IP) ratings] with signal transmission that is galvanically or opto-isolated to ensure safe data transmission. Inputs to the signal conditioning unit should be rated at CATIII/CATIV 1000 V. Outputs must be analog for real-time power analysis and typically would be a scalable voltage.
- *Current:* Current probes, clamps and transformers are available that employ various technologies for non-intrusive current sensing (Hall effect, fluxgate, Rogowski coil). They may include integrated signal conditioning to provide voltage signals for acquisition and processing, or they may be devices that are installed permanently and require hard wiring into a test bed system. The main selection criteria will be measurement range, accuracy, and bandwidth so manufacturer data needs to be sourced in order to select the correct device and technology for the application.
- *Temperatures:* Energy storage devices and motors require temperature measurement as this is critical for efficient and safe operation. These UUT devices operate with voltage levels well above the extra-low, safe-level voltages. Therefore temperature measurement devices must be combined with appropriate signal conditioning such that the signal is galvanically isolated, so that the transmission of HV from within the UUT is absolutely not possible.

## Overview of analog power analysis system requirements

There are several categories of electrical power analyzers commonly seen for hybrid system development. As mentioned previously, commonly used devices are industrial power analyzers that are designed for the industrial and power generation market, for power quality and disturbance monitoring. However, electrical power analysis systems, specifically designed and aimed for use in powertrain testing environments, should be always specified where possible. These are based upon high-speed measurement system platforms with the appropriate real-time calculations for electrical power analysis integrated in the software. Often, they are based on existing combustion analyzer systems that give the advantage that they are capable of combined combustion and electrical power analysis, which is an ideal capability for development of complex powertrains, containing a mix of ICE and electrical propulsion systems. This system approach has the advantage that many of the required interfaces are already enabled in the hardware platform, for communication to other devices in the test environment. The electrical and combustion data can be time aligned with high precision such that highly dynamic torque monitoring can be undertaken, for the purpose of torque blending and drivability optimization.

## Signal acquisition

Most systems need a suitably ranged, analog voltage input to the measurement device inputs.

The basic requirement would be voltages and currents from each of the motor phases (typically 3 or 6 at the time of writing) on the AC side. For determining motor drive system efficiency, it would be necessary to measure DC voltage and current on the inverter supply side. While for motor efficiency tests, motor speed and torque values from a dynamometer are essential. A system overview is given in Fig. 12.6.

For in-vehicle measurement, access to the system for measuring physical speed and torque is much more difficult. These in-vehicle measurements are undertaken later in the development cycle, where the control system is calibrated such that speed and torque are derived in the controller and available on the controller bus network [Controller Area Network (CAN), etc.]. Motors are often fitted with rotational positional sensors (known as resolvers), but the signal processing for this type of encoder is not generally

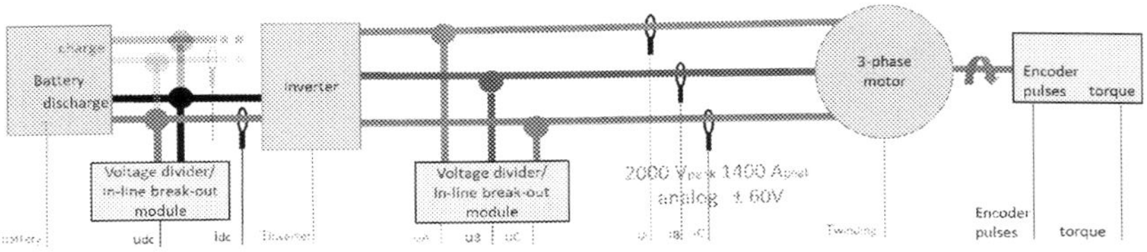


FIGURE 12.6 Overview of powertrain motor system typical measurement points.

considered in the external power analysis measurement system. More important is the ability to detect an electrical cycle period, from which result values for that given cycle (or half cycle) can be calculated. This is normally considered in the measurement software and involves quite sophisticated algorithms to detect each phase accurately—irrespective of speed or change of speed. This is the most challenging aspect and the main difference between standard power analyzers and those which are developed for automotive system measurement (where a wide range of frequencies is encountered in the motor drive system).

Other measurements of interest are temperatures—motor winding and rotor temperatures for component protection function calibration and validation of simulation models. Battery and inverter temperatures are necessary for the optimized calculation of energy balance, and simulation validation.

## *Powertrain applications—motor types*

The electric propulsion system, mechanical power output, is generated by a motor; there are many motor types available and much literature available on their characteristics. The full treatment of this topic is beyond the scope of this publication, but the motor types currently most used in vehicle powertrains are:

- synchronous, polyphase motors with permanent magnet rotors;
- synchronous, polyphase motors with DC excited rotor; and
- asynchronous, polyphase induction motors.

Synchronous motors are most widely adopted for powertrain propulsion, particularly those with permanent magnet rotors. They are ideally suited for this application due to their high efficiency, which is due to the use of permanent magnets to generate the rotor field (no parasitic power consumption for this magnetic field generation), also the absence of mechanical commutators and brushes results in low friction losses. The use of high-energy-density magnets has allowed the achievement of very high flux densities in these machines—the result of which is that high torque and power density, over a wide speed range, can be produced from a given volume of motor. Permanent magnet-based synchronous motors can be controlled easily, as the main control variables (voltage, current, frequency) are accessible and constant throughout the operation of the motor.

The disadvantages are that the rotor requires the use of rare earth materials in the magnetic construction, which is a subject of general discussion and debate with respect to supply chain risks and unpredictability of costs. In addition, rotor construction can be relatively expensive and complicated, as the magnetic material must be embedded into the rotor securely and must be able to withstand significant centrifugal force due to high rotational speeds, in the range of 20,000 revolutions/min or greater. The rare earth magnetic material is also expensive and can be demagnetized by large opposing magnetomotive force and high temperatures. There are also some safety considerations in the application. The permanent magnetism of the rotor can present a risk in the case of a short-circuit failure of the inverter drive. This is due to the constant electro motive force (EMF) that would be induced in the short-circuited windings that would generate a very large current in those windings and an accordingly large locking torque on the rotor. This torque would be uncontrolled and applied through the transmission to the road wheels, a potential fault condition that would have a significant impact on vehicle functional safety and may require the use of a decoupling safety clutch in the powertrain.

A variation of this design, to avoid the use of rare earth based, permanent magnet rotors is to use a wound field rotor that is excited by a separate DC source. This allows the ability to control the excitation field of the rotor and the possibility to rapidly demagnetize the machine in the case of short-circuit failure in the machine or inverter. The rotor winding is normally supplied by current via slip rings. However, slip rings and external circuitry limit the machine's power density and increase friction. The additional complexity also increases production costs and potentially reduces reliability.

Asynchronous motors are widely used in many industries and are a well-established technology. The basic principle is that polyphase windings generate a rotating field that induces current in a rotor that is equipped with windings (wound rotor, squirrel cage). The stator field rotates at a different speed than rotor, this produces slip (speed difference) between rotor speed and stator field rotation. The magnetic field cuts the rotor windings and produces an induced voltage in the rotor windings, since the rotor windings are short-circuited, an induced current will flow in the rotor windings that produces another magnetic field. Torque is produced by the interaction of the stator current (or stator flux) and the rotor current (or rotor flux). For a given number of rotor poles, the frequency of the AC supply determines the speed and the voltage determines the flux. Variable speed is possible through scalar control ($V/f =$ constant) or vector control (field oriented). Rotor variables are referred to the stator just as in a transformer (the induction motor may be seen as a rotating transformer) thus rotor currents are induced eddy currents—and torque is therefore only produced at asynchronous speed.

The benefit of this design for powertrain application is that three-phase induction motors are one of the most common and frequently encountered electric machines. Thus they are well established and have a simple rugged design; they have also relatively low cost and low maintenance. The main disadvantage is the heat generated in the rotor due to the circulating currents. This can be difficult to reject effectively and as such limits the power density achievable with a given motor volume.

In summary, all of the abovementioned designs have been deployed for powertrain applications (synchronous, wound field by Renault, asynchronous induction by Tesla), but by far, the most widely adopted type is the permanent magnet, synchronous design.

## *"Cogging" and torque ripple effects*

It has been assumed by the authors of this edition that fundamental design features of the permanent magnet motor and control systems will have been optimized prior to being packaged for automotive drive systems. However, mention should be made of the two inherent characteristics in the technology that can, in some circumstances and applications, cause torque variations and vibration, which are referred to as "cogging torque" or "torque ripple." While cogging torque generally refers to the phenomenon that *causes* torque variations, and torque ripple is the *effect* these variations have on motor performance. The cogging torque profile of a motor with no mechanical imbalances depends on the number of permanent magnets in the rotor and the number of teeth in the stator it can be minimized by optimizing skewing or shaping the permanent magnets to make their transition between stator teeth less step-like. In automotive applications running at high rotational speed, the motor's moment of inertia tends to filter out any effect of cogging also modern motor drives are able to compensate by techniques similar to the "antiphase" signal generation used in noise-canceling headphones.

## Power analysis

In order to characterize the performance and efficacy of the electrical propulsion system, as well as calibrate the torque delivery for optimized drivability, measurements must be undertaken to understand the electrical energy flows and losses, in conjunction with mechanical measurements for torque and speed. This is straightforward for module-based tests that can be carried out at suitable component test stands. However, at a holistic level, when a complex powertrain is assembled to be tested as a complete unit, then the requirements for electrical measurement become more application specific, which requires the use of equipment and measurement approaches, that are designed for the task.

The aim of testing is to be able to measure and characterize the performance and efficiency of each module in the electrical energy conversion chain, but as a complete system, such that the performance of the "chain" can be established beyond individual component level. For a complex propulsion system that includes an ICE, the combination of electrical and combustion measurement is important for several reasons.

- Data synchronization at high fidelity in order to characterize the individual contributions and flows between ICE, motor, and energy storage—with a suitable dynamic response to be able to measure during transient operation of the powertrain.

- Reduced complexity of measurement systems by combining power and combustion analysis in a single device capable of storage of all raw data.
- Application-specific attributes needed for powertrain electrical flow analysis in a testing environment can be accommodated in a combined system—this is different application to the traditional power analyzer function for electrical networks.
- Interfaces to other test cell equipment and hosts are already established and available in combined equipment. Industrial equipment is more likely to have fieldbus interfaces that are less common in powertrain test environments.

The most common requirement is to measure and characterize the performance of the propulsion machine and its associated drive, when installed into the complete powertrain. This involves the measurement of current and voltages between the inverter drive (Fig. 12.6) and the machine itself allowing the derivation of key results relating to power dissipation in the motor and drive system (Fig. 12.7). In addition, motor output torque can be inferred or extrapolated in order to understand torque contribution and implement effective torque blending of the e-machine torque contribution. DC current and voltage measurements on the inverter supply side allow inverter efficiency during operation to be characterized.

Total vehicle energy flows are of high interest, this involves measurement of DC energy flows but at high fidelity, so that it is possible to understand the dynamics of the DC bus during powertrain transients—particularly of

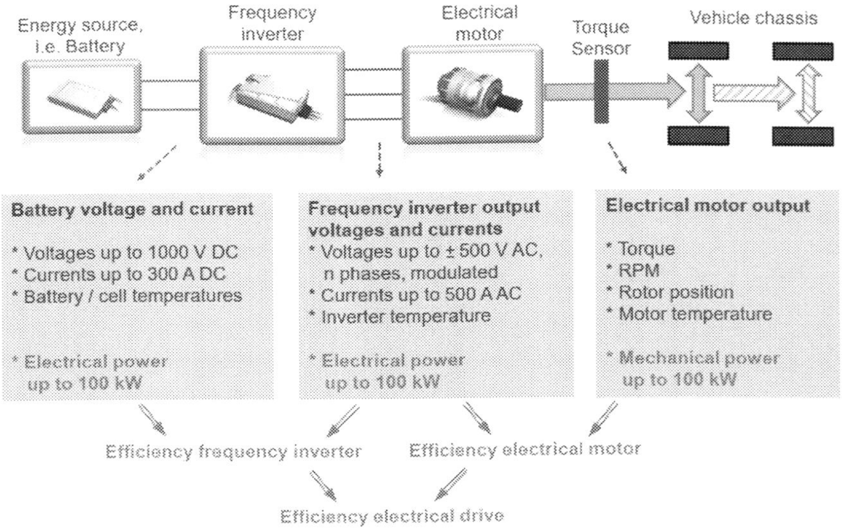


**FIGURE 12.7** Overview of powertrain electrical propulsion system typical measurements and analysis. *Courtesy: HBM GmbH.*

interest when trying to understand the contribution of highly active motors used for e-turbocharging and energy recovery, also current ripple effects when charging the battery.

## Signal processing and analysis chain for online evaluation

For motor and inverter evaluation the system needs to measure AC voltage and currents at high fidelity and resolution, storing all the raw data. However, the raw signals can be very noisy and difficult to interpret. Voltage is switched at high frequency by the inverter in order to drive the appropriate current through the motor phases and inverter switching frequencies are in the range of 16–20 kHz. In order to gain useful information from this data, real-time processing must be undertaken during runtime, by the measurement device able to:

1. Apply suitable filtering or smoothing of the signal in order to determine the approximated sine waves for voltage and current such that further processing and calculations can be performed.
2. Determine the individual cycles in real time such that cyclic results can be derived for the complete motor/inverter operation over a single, electrical cycle or half-cycle.

Following this processing the root mean square (RMS) values of current and voltage can be established for each phase and each cycle. From the instantaneous voltage and current in each phase, the instantaneous power can be derived in each phase, averaging over a cycle provides true power per phase, plus the total power by summing the phases. Cycle detection methods do vary according

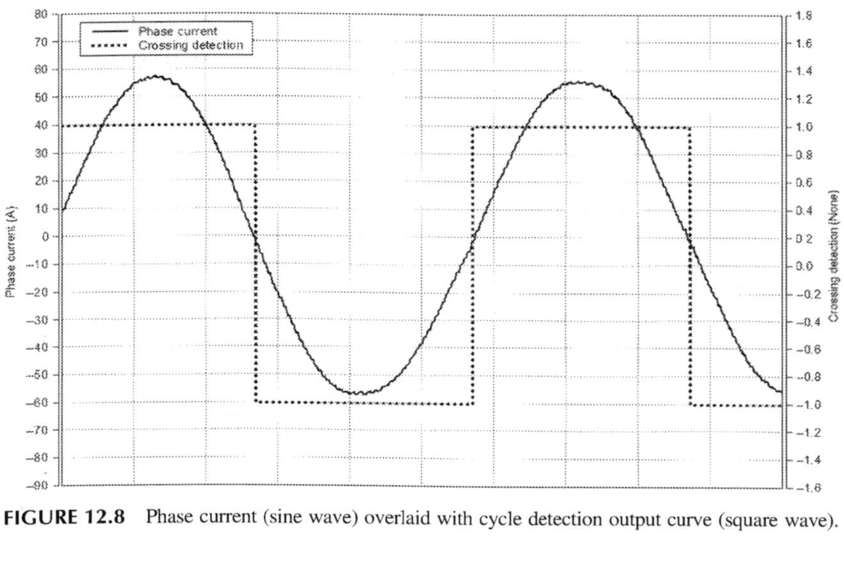


**FIGURE 12.8** Phase current (sine wave) overlaid with cycle detection output curve (square wave).

to the equipment supplier and as these are key to the performance of the equipment, the details of how they work are not stated or documented to any great detail (as this is the manufacturers' IP). The general principle can be seen in Fig. 12.8. This shows the current in a single phase, with the output of the cycle detector algorithm overlaid. This cycle detector signal can be analyzed during operation for monitoring the performance of the cycle detection algorithm during dynamic changes, in order to verify correct detection of each cycle.

Cycle detection and prefiltering are essential for further signal analysis, the speed and torque of the machine being a function of the current and frequency applied. The machine will have a transient capability and response time much faster than any ICE and in order to capture all the required detail of the motor operational dynamics, these algorithms must work from zero to maximum motor speed with high reliability. In addition, sampling rate is important, and the equipment must be capable of oversampling signals to prevent aliasing of the signal details (defined by Nyquist or Shannon theorems) and to allow more detailed analysis of high-frequency components on the signal trace, like noise or disturbances that could indicate radio frequency interference (RFI) or electromagnetic compatibility (EMC) issues.

## Electrical machine and system calculations

Once the signals are conditioned, then further analysis can be performed. Typical filtered signals are shown in Fig. 12.9.

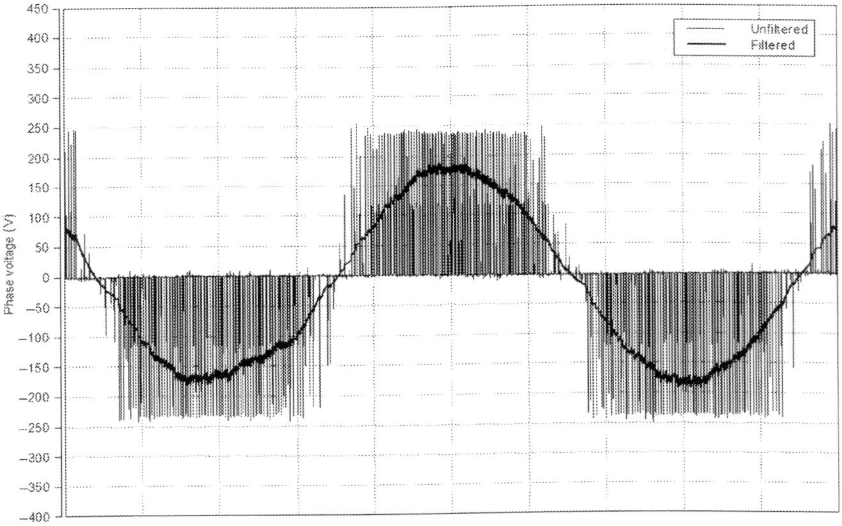


**FIGURE 12.9** Raw sampled voltage overlaid with filtered value.

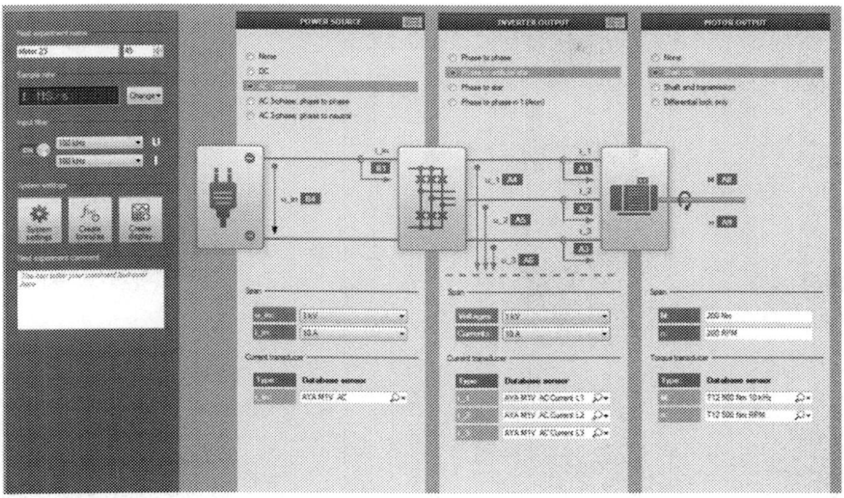

**FIGURE 12.10** User workflow-oriented interface to set up analysis of electric propulsion system. *Courtesy: HBM GmbH.*

The power calculations are based on phase voltages and currents. Therefore, before the power calculation is done, phase values must be derived for the analysis. Motors generally have star or delta windings (Fig. 12.11). In a star winding the line currents are equal to the phase currents and the line-to-star voltages are equal to the phase voltages. In a delta winding the line-to-line voltages are equal to the phase voltages, but the line currents are not equal to the phase currents. In addition, for the analysis, the user must choose the number of measured phases, otherwise known as the watt meter method.

In its simplest form the power can be derived from the product of the average voltage and current in the measured phase—and this provides the active power component dissipated in the machine windings. By establishing the phase difference between the voltage and current signals, the phase angle and power factor can be derived. The apparent power is derived from the product of the RMS values of voltage and current over a calculation cycle.

If apparent and active powers are derived, then reactive power can also be established, and this now provides all required calculations for basic three-phase power analysis. Most measurement systems are equipped with a suitable user interface, with a guided workflow to assist the user in setting the system up correctly. In addition, a formula compiler library is often provided in order that the correct calculations can be called up for the appropriate analysis of the UUT. A typical application-specific user interface is shown in Fig. 12.10.

## Line-to-phase conversion

All AC results are based on phase voltages and currents. In most cases, only line voltages are available, a transformation is necessary to derive instantaneous phase results. These transformations depend of the winding configuration of the motor. As mentioned previously, the windings for a three-phase machine can be configured as star or delta wound:

The wiring scheme can vary for three-phase power analysis depending upon whether a physical neutral point is available or not as shown in Fig. 12.11. The power calculation also depends on the number of measurements, otherwise known as the number of watt meters. For powertrain test

(A) **Delta**

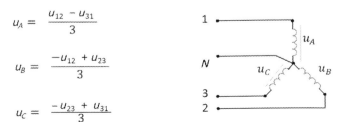

Phase voltages are equal to line voltages

$u_A = u_{31}, u_B = u_{12}, u_C = u_{23}$

$$i_A = \frac{i_3 - i_1}{3}$$

$$i_B = \frac{i_1 - i_2}{3}$$

$$i_C = \frac{i_2 - i_3}{3}$$

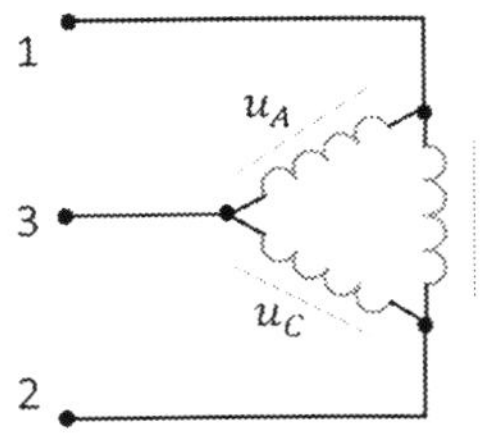


(B) **Star**

Phase currents equal to line currents

$i_A = i_{31}, i_B = i_{12}, i_C = i_{23}$

In case the user does not have access to the neutral point, the phase voltages can be virtually derived as follows:

$$u_A = \frac{u_{12} - u_{31}}{3}$$

$$u_B = \frac{-u_{12} + u_{23}}{3}$$

$$u_C = \frac{-u_{23} + u_{31}}{3}$$

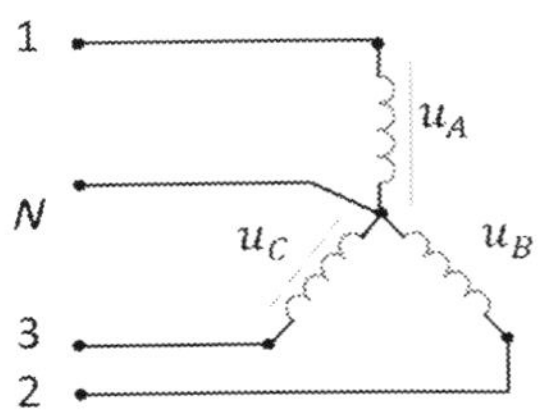


**FIGURE 12.11** Delta and star wound three-phase motor configurations, neutral point can be derived virtually if not accessible.

applications the number of measurement scenarios to consider is typically less complicated when compared to mains network power analysis. In the case of one-measurement (i.e., one phase), the power is calculated for the single phase and the total power is taken as three times that value, with the assumption of a perfectly balanced system. For most applications though a three-phase-based measurement will be used. All three phases are instrumented, the power calculations are executed for each phase individually, for example:

$$\text{Apparent power of phase one} = I1(\text{rms}) \times U1(\text{rms})$$

The total apparent power is the sum of all three phases.

## *Active and RMS power*

Most calculations involve an integral over a certain time period, which includes a number of electrical cycles. It is therefore necessary to define the calculation window by detecting these cycles—the minimum period being a single cycle. The length of a single cycle can vary considerably depending upon the drive system frequency and motor speed. The cycle detector algorithm must accurately determine the cycle length during steady state and transient conditions. From the instantaneous phase voltages and currents, analyzed over a cycle, the active power can be calculated per cycle, per phase. Eq. (12.1) reflects this:

$$P_{\text{AC, phase active}} = \frac{1}{nT} \times \sum_{t0}^{t0+nT} i_A(t) \times u_A(t)dt \tag{12.1}$$

Simple multiplication of measured voltage and current gives the instantaneous power per phase. A mean value over a cycle provides the true, active power. The phase powers can then summed to provide the active power for the machine from all phases, as shown in the following equation:

$$\begin{aligned} P_{(\text{AC, total active})} = \frac{1}{nT} \times \sum_{t0}^{t0+nT} i_A(t) \times u_A(t)dt + \sum_{t0}^{t0+nT} i_B(t) \times u_B(t)dt \\ + \sum_{t0}^{t0+nT} i_C(t) \times u_C(t)dt \end{aligned} \tag{12.2}$$

From this information the RMS values of the voltage and current, per phase and cycle (or period), can be derived as shown in the following equation:

$$I_{(A,\text{ rms})} = \sqrt{\frac{1}{nT}\sum_{t0}^{t0+nT} i_A(t)^2dt} - \text{applies to } i_A,\ i_B,\ i_C,\ u_A,\ u_B,\ u_C \tag{12.3}$$

For a three-phase, balanced system, the effective (or quadratic mean) RMS values for current and voltage can be derived as shown in the following equations:

$$I_{\text{effective, rms}} = \sqrt{\frac{{I_{A,\text{ rms}}}^2 + {I_{B,\text{ rms}}}^2 + {I_{C,\text{ rms}}}^2}{3}} \quad (12.4)$$

$$U_{\text{effective, rms}} = \sqrt{\frac{{U_{A,\text{ rms}}}^2 + {U_{B,\text{ rms}}}^2 + {U_{C,\text{ rms}}}^2}{3}} \quad (12.5)$$

### *Apparent power*

The total apparent power can then be determined from the RMS values using Eq. (12.6)

$$P_{\text{AC, effective apparent}} = I_{A,\text{ rms}} U_{A,\text{ rms}} + I_{B,\text{ rms}} U_{B,\text{ rms}} + I_{C,\text{ rms}} U_{C,\text{ rms}} \quad (12.6)$$

The apparent power is important for the power system design—the inverter rating, cabling and switchgear—as this is the power level that needs to be delivered from the energy source. Due to uncompensated phase lag in the motor windings (inductance effects), loss of power occurs and this is known as reactive power (power stored and dissipated in reactive circuit components). This can be calculated for the machine using Eq. (12.7).

### *Reactive power and power factor*

$$P_{\text{AC, reactive}} = \sqrt{{P_{\text{AC, effective apparent}}}^2 - {P_{\text{AC, total active}}}^2} \quad (12.7)$$

From the ratio of apparent and apparent power, the power factor can be derived as shown in the following equation:

$$\text{Power factor} = P_{\text{AC, total active}} / P_{\text{AC, effective apparent}} \quad (12.8)$$

The power factor serves as an evaluation of the efficiency of the conversion of energy in the system. Typical values of the power factor fall within the interval 0–1. For sinusoidal values the power factor is equal to the phase angle $\cos\varphi$. If the power factor is $\lambda = 0$, no effective power is transferred. The current that is flowing in this state simply loads the lines and other transmission equipment without doing any work. If the power factor $\lambda = 1$, only effective power is delivered to the consumer. Thus the load on electrical equipment is as low as possible for the given effective power.

## *Fundamental components*

The fundamental components of the voltage and current signals can be derived by calculating the fundamental frequency component, then reproducing a sine wave at this frequency, as shown in Eqs. (12.9) and (12.10). The same power calculations can then be executed on this fundamental data set, from this the harmonic components can be derived via subtraction, along with the total harmonic distortion, calculations as shown in Table 12.1.

**TABLE 12.1 Fundamental-based calculations.**

| Additional output signal | Default name | Unit | Description |
|---|---|---|---|
| Harmonic part of total power | $P_{AC,\ harmonic}$ | W | $P_{AC,\ total\ active} - P_{AC,\ fundamental\ active}$ |
| Fundamental reactive power | $P_{AC,\ fundamental\ reactive}$ | W | $\sqrt{P^2_{AC,\ fundamental\ eff\ app} - P^2_{AC,\ fundamental\ active}}$ |
| Effective harmonic component of $I$ | $I_{harmonic\ effective,\ rms}$ | A | $\sqrt{I^2_{effective,\ rms} - I^2_{fundamental\ effective,\ rms}}$ |
| Effective harmonic component of $U$ | $U_{harmonic\ effective,\ rms}$ | A | $\sqrt{U_{effective,\ rms}{}^2 - U_{fundamental\ effective,\ rms}{}^2}$ |
| Total harmonic distortion current | THD | % | $\frac{I_{harmonic\ effective,\ rms}}{I_{fundamental\ effective,\ rms}} \times 100$ |
| Total harmonic distortion voltage | THD | % | $\frac{U_{harmonic\ effective,\ rms}}{U_{fundamental\ effective,\ rms}} \times 100$ |

$$i_{A,\ \text{fundamental}}(t) = A_{i,\ A\ \text{fundamental}} \times \sin\left(2\pi f_{\text{fundamental}} - \varphi_{i,A}\right) \tag{12.9}$$

*Calculation of fundamental component, current in phase A shown, applicable to all phase voltages and currents*

$$I_{A,\ \text{fundamental, rms}} = \frac{A_{i,\ A\ \text{fundamental}}}{\sqrt{2}} \tag{12.10}$$

*Fundamental RMS values, analytically derived*

Based on the phase shift between the fundamental phase voltage and current, the fundamental power factor per phase and cycle can be derived:

$$\text{Phase power factor}_{\text{fundamental}} = \cos\varphi \tag{12.11}$$

*where phi is the phase angle difference between fundamental phase voltage and current.*

Note that Eq. (12.11) is true for pure sine waves only, for general, practical measured waveforms, the *true* power factor is the ratio of active and apparent power as shown in Eq. (12.8).

### Multiphase measurements

In most cases a single- or three-phase measurement will be made on a three-phase system (in the case of a single measurement, single values are multiplied by 3, balanced system assumed), three-phase motors are by far the most common. However, polyphase motors with more than three phases do exist and maybe encountered. Multiphase (typically multiple of 3: 6, 9, 12) do have some benefits in that it is possible to get more power from the low-voltage DC link. In addition, torque ripple and cogging effects can be reduced by controlling the offset between phases. For certain fault tolerant applications (non-automotive) five or seven phases can be used that provide system redundancy in case of failure of a phase. With respect to measurements, all phase-related measurements and summed components are still applicable, just applied per number of phases in the machine.

### DC component power

For propulsion system applications the energy source to the electric machine system is DC, from a battery, fuel cell, or other DC source. Therefore, in order to understand the energy conversion in the total system (not just machine and inverter), then simultaneous analysis of the DC system is also necessary. For this, voltage and currents on the supply line to the inverter should be measured with similar fidelity and precision to the AC system, although signal sampling rates can be reduced to the range of 100–500 kS/s. However, for a dynamic comparison the DC values must be sampled in the same time window or cycle window as the corresponding AC values. From this sample section an average value can be derived, the product of the mean voltage and current values provides a DC power mean value that can be compared to the inverter output measurements (12.12). The ratio of DC power at inverter in and AC power at the electric machine then provides the inverter efficiency (12.13).

$$P_{DC} = u_{DC,\ mean} \times i_{DC,\ mean} \tag{12.12}$$

*Product of DC mean values gives DC power mean at inverter input (Watts).*

$$\eta_{inverter} = \frac{P_{AC,\ total\ active}}{P_{DC}} \tag{12.13}$$

*Ratio of machine active power and DC mean power provides inverter efficiency (%).*

### Mechanical component power

Depending upon the test situation for the UUT, an additional mechanical torque and speed measurement may be required. This is very likely if the UUT is located on a dynamometer test bed as a stand-alone test object. This setup would provide the possibility to establish mechanical speed, torque, and output power of the UUT (12.14); values that can then be used to derive mechanical efficiency of the machine (12.15); and the system (12.16). Note that the speed and torque raw values must be sampled and averaged in the same windows as the electrical values so that they can be directly compared.

$$P_{\text{shaft}} = N_{\text{shaft}} \times T_{\text{shaft}} \tag{12.14}$$

*Motor shaft power, N = speed (rad/s), T = torque (Nm)*

$$\eta_{\text{motor}} = \frac{P_{\text{shaft}}}{P_{\text{AC, total active}}} \tag{12.15}$$

*Motor efficiency (%)*

$$\eta_{\text{system}} = \eta_{\text{inverter}} \times \eta_{\text{motor}} \tag{12.16}$$

*Total system efficiency from DC supply to motor shaft (%)*

### Further calculations for the system

The calculations stated provide the basis for a full, standard power analysis of the motor system during operation. However, there is further, more detailed analysis possible with the raw data once stored. This includes the following:

- Machine copper losses—Where temperature measurements in the machine are available and baseline resistance measurements have been undertaken for the windings at a known temperature.
- Machine iron losses—If electrical power, shaft power, speed-dependent mechanical losses and copper losses can be established, then iron losses can be approximated by subtraction.
- Flux linkages, electromagnetic (air gap) torque—Calculated with flux components that have been evaluated by sampling the real motor pulse-width modulation voltages and with a stator resistance, which takes into account the measured average stator temperature, this air gap torque can be defined as a reliable torque estimate.
- Detailed motor analysis—Equivalent circuit diagram, torque ripple, cogging, saturation effects.
- Detailed inverter analysis—Space vector transformation, inverter control behavior, modulation method, and switching frequency.

## Typical measurement tasks

For test stand application, mapping of the electrical machine is a standard requirement. When compared to an ICE engine mapping exercise, the execution of this test is much simpler. An electric machine is far more stable and controllable than an engine at a steady-state operating point, so that comparatively less time is required at each measurement point in order to gain statistically valid data. Therefore a typical speed and load sweep takes less time and does not require any advanced model base approaches (often used to reduce test time). A full-factorial approach can be used; a typical example may require 300 test points or more.

A typical measured torque-speed map is shown, with the input/output power map in Fig. 12.12.

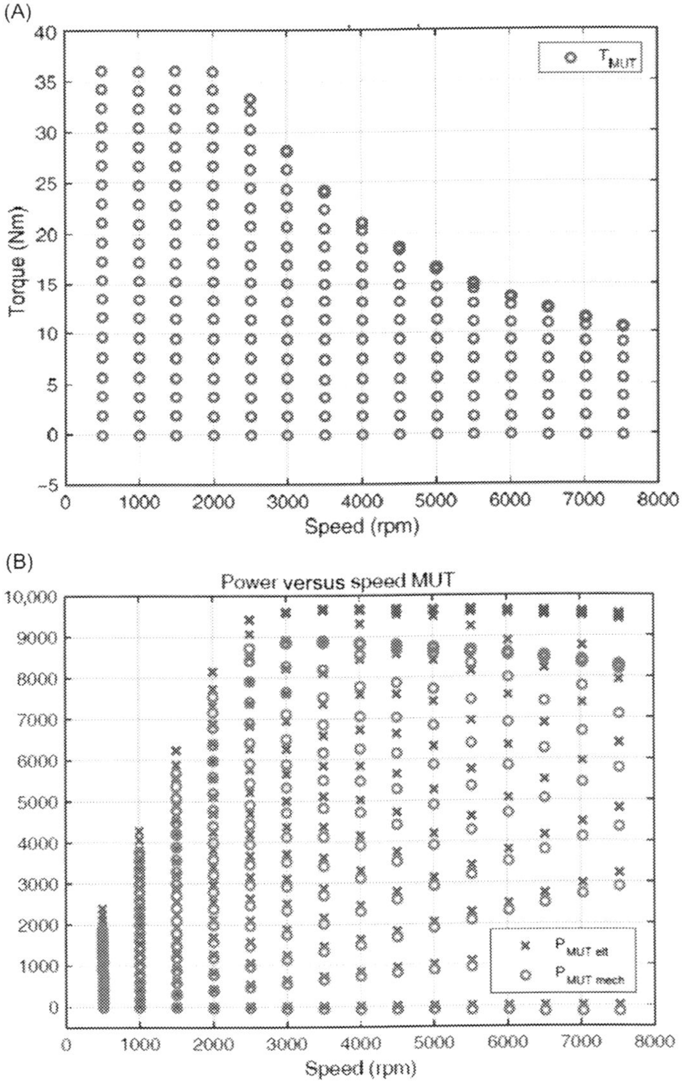


**FIGURE 12.12** Speed versus torque and speed versus power maps for a MUT (Motor under test). *Courtesy: HBM GmbH.*

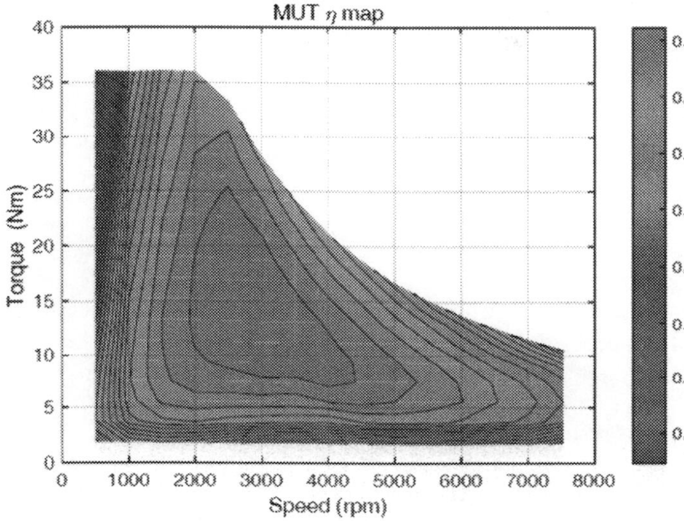


**FIGURE 12.13** Motor efficiency map. *Courtesy: HBM GmbH.*

Note that at high-speed operation, the input electrical power is almost constant, while the output mechanical power is slightly decreasing. This figure allows defining the constant power speed range, given the target output power.

A typical efficiency map of the UUT (or motor under test) is shown in Fig. 12.13. The efficiency map is extremely important to evaluate the motor efficiency for the entire torque-speed range. If required, it could be possible to combine the inverter DC input power and consequently the inverter efficiency map, and thus the drive system efficiency map could be derived.

## Integration into the test cell

It is an extremely important point to note that the chosen power analysis equipment has to be interfaced with the test control and data acquisition system and host. Many dedicated power analyzers are designed as stand-alone measurement devices, or for use in laboratory or specific industry applications; therefore they may not be equipped with an external interface, or not equipped with an interface suitable for a powertrain test environment.

The other issue is appropriate accuracy; many power analyzers are designed to be extremely accurate for signal analysis of high-power networks that typically have sinusoidal waveforms with relatively fixed frequencies and are not designed for variable frequency, complex or noisy signals like those created by a powertrain motor drive system. Neither are they designed to sample and save data over long measurement periods meaning that they can be quite unsuitable.

The current generation of combined power and combustion analyzers resolves most of these integration issues. Along with measurement systems that are also developed specifically for e-drive applications. However, it is important for the engineer involved in electrical propulsion system test equipment to consider the following when using or specifying devices:

- Measurement data capacity—Raw data is required to be stored for periods of time measured in minutes or even hours. Is a suitable data model available that can be harmonized with typical analysis tools in the powertrain testing environment, such that raw data from different systems can be aligned and analyzed together?
- Remote control—Does the system possess remote control via standard interface possibilities or technologies, to be able to start/stop measurements, transfer data, provide status information when connected to a powertrain test be automation system?
- Interfaces—Standard, fast interfaces must be available to stream cyclic calculated values to the host system for continuous monitoring and event trigger reaction. Typically, this interface does not involve remote control commands but must be capable of streaming at a sufficient rate to be able to offer closed-loop control of test conditions under dynamic operation of the UUT.
- Analysis online—Device must be capable of reducing raw data to significant, calculated components that can be viewed online and transferred via the above interface. The results must be calculated with a robust methodology so that they are reliable under operation conditions, even highly dynamic speed and load changes.

Typical examples for interfaces are mentioned elsewhere but in summary, CAN-FD and EtherCAT are appropriate for result transfer. DCOM, AK, and ASAM-GDI are suitable interfaces for remote control of an instrument by a host system.

## Powertrain system test and analysis

With increasing vehicle complexity, in combination with more rigorous and variable test procedures, it is becoming more important to fully understand the influence of real-world driving factors on the operation of a complete powertrain. This, in combination with the ability to recreate these factors in a repeatable laboratory environment, should mean that a robust and efficient powertrain development can ultimately be achieved.

Due to the increasing levels of vehicle hybridization, the robustness of $CO_2$ as indicator of powertrain efficiency decreases as those values are combined with the depleted electrical energy. Therefore the traditional approach in the assessment of on-road emissions can be problematic, such that a clear comparison between conventional and hybrid vehicles is difficult to quantify without the use of high-speed portable emission measuring system (PEMS) instrumentation and the ability to rerun the Real-Driving Emissions (RDE) test profiles in the laboratory.

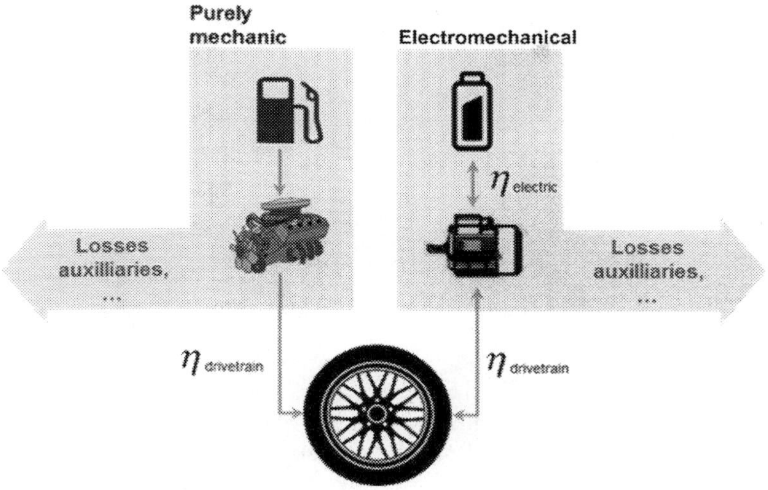


**FIGURE 12.14** Energy flow in a hybrid powertrain.

Powertrain electrification has meant many powertrain configurations have evolved. Mostly, these depend upon the level of electrical power provided to support or replace the combustion engine, but also having to be considered is the level of integration required in order to use an efficient configuration with respect to use of existing platforms plus, cost, development time, and time to market.

The energy flow map for a hybrid vehicle can be very complex. Generally, there are two different energy flow paths in an electrified powertrain—one is purely mechanical and unidirectional and the other electromechanical and bidirectional. The general flow of energy is shown in a simplified way in Fig. 12.14.

Hybrid powertrain architectures have evolved over recent years such that the simple description of series or parallel layouts is no longer sufficient. New classifications are used in order to define the level of electrification, as well as the interactions and topology of the system itself. As the system complexity increases, the degrees of freedom also increase, and this has a significant impact on the amount of effort required to characterize the chosen powertrain configuration in the application. It is then a further challenge to be able to optimize the system for the best efficiency, energy usage and, very importantly, drivability.

## Changes in legislation and associated challenges

The test requirements of the RDE procedure and instrumentation are covered in some detail in Chapter 17, Engine Exhaust Emissions.

The primary intentions of the RDE legislation are:

- to provide realistic data of real-life exhaust emissions and to impose not-to-exceed limits on defined vehicle types and models for all normal circumstances and
- to provide realistic fuel consumption, electrical energy usage and battery powered range.

The data provided by the RDE testing and particularly from high-speed analytical instrumentation incorporated in PEMS units is becoming essential for the detailed calibration of hybrid vehicles.

The calibration of the automatic stop–start system in both conventional and hybrid vehicles is a good example of the design complications that come from ever tighter legislation. The conditions required to be met for the ICE to be stopped, in different vehicle types and models, ranging from conventional ICE, through mild-hybrid (48 V) and other hybrid forms, are the result of a number of calibrated control maps, including a range of temperatures, battery state of charge (SOC), auxiliary system states, etc.

High-fidelity RDE testing is also revealing the effects of road design, particularly in built-up areas, on vehicle emissions; for example, tests have clearly revealed the concentration of $NO_x$ emissions just beyond "speed humps" due to the driver being made to slow down, with the tendency to accelerate afterwards, meaning that the gasoline engine operates momentarily lean.

Prior to the Worldwide Harmonized Light Vehicle Test Procedure, the main strategy for powertrain development activity was conducted with a strong focus on optimization defined by the New European Driving Cycle (NEDC) or similarly "benign" drive cycles. With the implementation of RDE legislation, vehicle and powertrain manufacturers are required to develop their products according to how the vehicle and the driver actually behave all types of road.

Many toolchains are being developed based on statistical techniques that aim to cover the full and broad range of operating regimes. Due to the fact that on-road testing is expensive, time-consuming, and inherently variable, most of these toolchains are aiming to create mathematical models of engine emissions under dynamic conditions and to use these models to predict emissions over a very large number of virtual RDE test cycles. However, in all cases of hybrid vehicles the calibration of the powertrain has to include the mapping of electrical energy flow, which requires a high degree of strategic optimization now based on RDE-derived data.

A common requirement within these methodologies is the accurate reproduction of on-road driving within the powertrain testing environment. This requirement brings with it a danger in that a natural need to repetitively run a "problem section" discovered on an RDE run, produces on the chassis dynamometer an increasingly unrealistic repetition of the route section's original data due to dissimilar powertrain conditions (heat soak, diesel particulate filter load, etc.) developing.

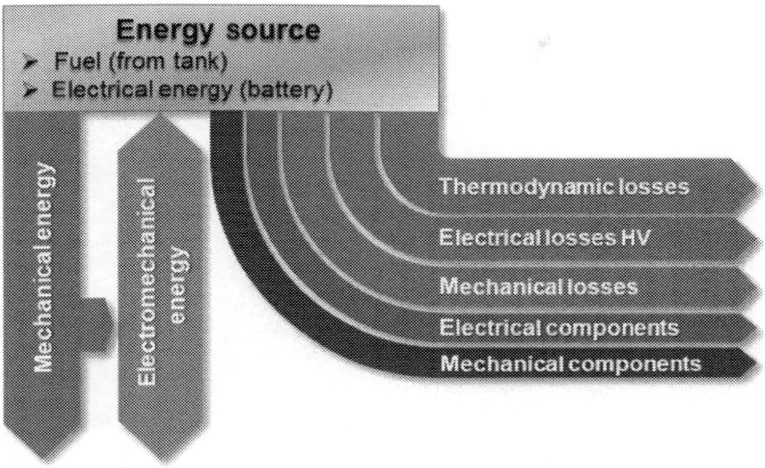


**FIGURE 12.15** Detailed knowledge of generated torque, torque path, and energy sink (road interface) is necessary for robust product development.

The requirement for compliance with RDE is also promoting vehicle manufacturers to consider applying advanced data analytical and statistical techniques (discussed in Chapter 14: Data Handling and Modeling).

This forms the requirement for holistic powertrain analysis and requires a level of data acquisition at the vehicle such that data can be gathered and used to characterize the propulsion system responses at high fidelity, under dynamic conditions. This provides a level of data precision such that replication of those conditions, in a test facility can be as accurate as possible, and detailed analysis of powertrain strategy and control system cause and effects can be fully understood by the test engineer. The Sankey diagram shown in Fig. 12.15 indicates the apportioning of losses and energy flows in a complex powertrain that combine ICE and electrical propulsion. There are two different energy sources (battery and fuel) and two energy paths for propulsion: mechanical (combustion) and electromechanical (electric machine). The electrical path is bidirectional, that is, partially reversible via recuperation. The mechanical path has a connection to the electromechanical (load point shifting, mechanical energy is used to charge the battery, etc.).

## Vehicle-based powertrain measurement and analysis

### *Typical vehicle instrumentation required*

Most of the instrumentation technology required for complete powertrain characterization already exists, the main goal is to be able to equip a single vehicle or powertrain with as much instrumentation as possible and practical. Fig. 12.16

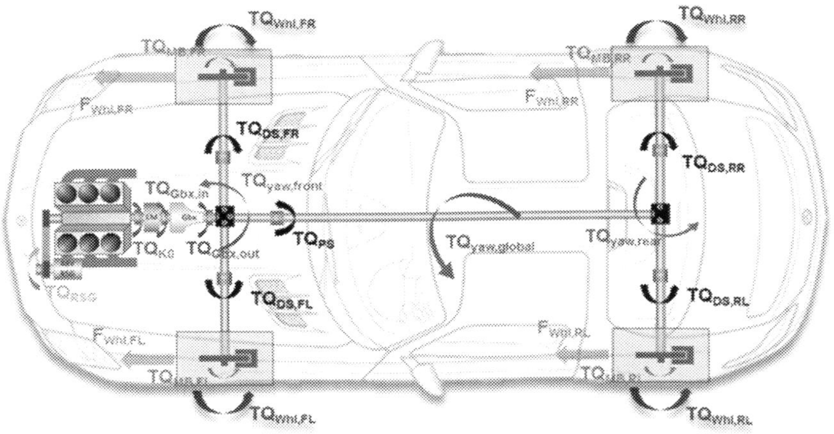


**FIGURE 12.16** Measurement locations for detailed powertrain analysis and characterization. *Courtesy: Kistler Group.*

shows an overview of the required measurement information. The more information can be gained, the higher the quality and fidelity of the analysis would be. The main goal is to gain and end-to-end characterization of the complete power flow from source to road—including all bidirectional flows in-between.

The possible measurement locations and instrumentation methods as shown in Fig. 12.16 can be listed from source to wheel in the following sections.

## Engine power

For vehicle measurements, access to engine torque can be challenging. The highest fidelity, but most difficult to engineer, is an instrumented flywheel (Fig. 12.17) or gearbox first motion shaft. This is normally a bespoke solution that can be provided by a number of companies who have experience in such instrumentation, gained from automotive and other industries. The measurement technology is normally strain gauge instrumentation with telemetry-based data transfer to a receiver unit, which can then send the signal via a wired connection to the data acquisition system.

An alternative to this is the use of engine combustion pressure measurement data to measure engine speed, derive engine torque, and calculate engine power. This can be achieved via individual cylinder indicated mean effective pressure (IMEP) values, that can be summarized to provide net torque per cycle, also to derive an instantaneous gas-based torque, which can then be extrapolated to a flywheel value in conjunction with engine friction estimation. From this in-cylinder power, can be derived in conjunction with engine speed. Also possible is to access engine torque values from the engine control unit, this value is normally derived indirectly in the controller, via a function that has been calibrated in the development process. If the control

**FIGURE 12.17** Instrumented flywheel for engine torque measurement for in-vehicle applications. *Courtesy: Greenmot.*

unit calibration is in a relatively mature state—then this torque value can be quite accurate and useable, assuming electronic control unit (ECU) or vehicle communication bus system access to that measurement label is available.

### *E-machine power*

Electric machine power, speed, and energy flows can be established quite accurately using a suitable power analyzer system, as discussed in this chapter. Efficiencies can be established easily assuming connections to DC and AC power cables/interfaces are possible in the vehicle. Note that where highly integrated mechatronic control units are deployed, these may integrate the vehicle inverter in a package that is located very close to the motor—minimizing the cabling in-between—and this can provide significant issues with integration of equipment to measure AC voltage and current. This must be considered and planned prior to any measuring campaign. It is also possible that electric machine values of interest are available as control unit labels that can be accessed with a suitable ECU or vehicle communication bus access tool. These values are derived from control unit functions that have to be calibrated. So, their accuracy and availability will be dependent on the control unit calibration state. Accessing them also depends upon having the suitable information to be able to read the data structure (an a2l or DBC file is required to understand the data message content).

In addition to electrical power measurements, mechanical power from the electric motor could be required. These values can be available from the control unit (speed, torque, etc.). It could also be possible that using analytical formula, the mechanical torque of the electric machine can be estimated from the electrical values (i.e., motor speed, air gap torque). However, it

**FIGURE 12.18** Instrumented pulley for mechanical power measurement, a concept that can be applied to a BSG unit. *Courtesy: MANNER Sensortelemetrie GmbH.*

may also be required to physically measure speed and torque in the vehicle for applications where high fidelity is required (e.g., analysis of cogging torque). This requires a more intrusive engineering approach and would require a torque measurement on the output shaft of the machine. Generally, torque flanges used at test stands are not suitable for in-vehicle integration, so an instrumented shaft is the most appropriate method—using strain gauge telemetry (as per ICE). For a belt-driven starter-generator (BSG) an instrumented pulley with strain gauge telemetry is a feasible approach, as shown in Fig. 12.18.

## Battery power flows

The battery DC power flows are easier to access in the vehicle. DC power cabling access is less challenging as these power cables will be available where they connect the battery to its auxiliaries, so the integration of measuring devices is possible along the cable length at an appropriate point. However, cabling must be adapted safely to ensure access to the correct signals with minimal disturbance of safety, EMC and RFI cabling attributes. Current clamps or transformers can be deployed on single cables, wiring interface boxes can be used to permanently modify cabling safely on prototype vehicles that must remain legally roadworthy. It is also possible to access this energy flow data that exists as ECU labels, this requires software access but no physical measuring equipment. It is also useful to measure both ECU and physical data where possible as this is useful for backup and comparison purposes (applicable everywhere where ECU labels exist and where physical measurements can also be implemented).

## *Power at driveshafts*

Propulsive mechanical power is the output of the powertrain and to establish this, torque and speed must be measured at each driven shaft. The addition of driveshaft torque measurement as a complimentary measurement to torque value at the wheels provides useful information regarding the recuperation of energy by the powertrain. This recuperation power is known by the system electrically—but shaft measurement provides the actual system input (but without the brake torque). This means that energy at the wheel, brake, and shaft can be separated and detailed information is then available for optimizing or understanding recuperation strategies.

Typically, a driveshaft is instrumented with a torque measuring mechanism (often strain gauge), encapsulating this will be a power supply and transmitter module mounted on the rotating shaft. This device can be battery or remotely powered. The signal is transferred wirelessly to a host receiver that can be hard wired into the data acquisition system. A typical system is shown in Fig. 12.19.

Once the shaft is mounted in the vehicle, the receiver unit needs to be located such that the wireless signal can be collected and converted to a

(A)

(B)

**FIGURE 12.19** Instrumented driveshaft, sensor, and transmitter module shown in detail. *Courtesy: Transmission Dynamics Ltd.*

**FIGURE 12.20** Installation of torque shaft receiver unit and aerial. Sensor and transmitter are located on the shaft surface and protected with binding, the aerial receiver can be seen as a wire "loop" around the shaft. *Courtesy: MANNER Sensortelemetrie GmbH.*

suitable signal for data acquisition. A typical complete vehicle installation is shown in Fig. 12.20.

## *Power at drive wheels*

In order to measure tractive power, instrumented wheels are required (Figs. 12.21 and 12.22), and these devices are able to measure the power and forces transmitted at the road wheel hub. Losses between the wheel hub and surface (i.e., tyre losses) can be established if the vehicle is operated and measurements undertaken on a chassis dynamometer. There are different types of measurement wheels available and they are often used to characterize the dynamic forces at the wheel during maneuvering. For this application a six-component instrumented wheel is required. For powertrain analysis a simpler instrumented wheel (single component—wheel torque only) can be used.

The measuring wheel system has three main components: wheel torque transducer, data transmission module, and onboard electronics (control unit). For data transmission from the rotating wheel to the onboard electronics, a wireless digital telemetry system can be used, or hard wiring direct to a receiver module is also possible via a suitable wiring frame. The traction torque is measured with piezoelectric quartz or strain gauge sensor technology. The signals are amplified and processed in the electronics system integrated within the wheel.

The design of certain wheel torque transducers means that they replace the middle part of the rim, thus enabling an optimum integration into the existing vehicle suspension system, that is, in the most effective position for acquiring

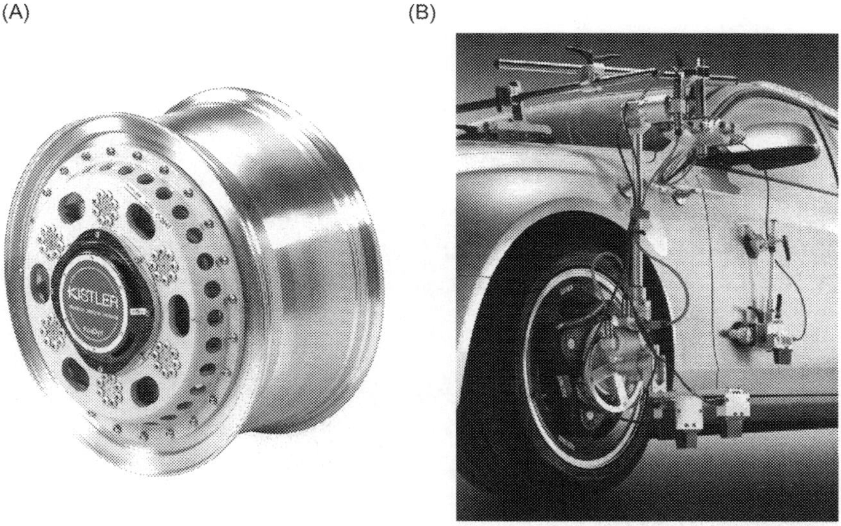


**FIGURE 12.21** (LHS) Single-component measuring wheel, (RHS) six component instrumented wheel for vehicle dynamic applications (multicomponent, torque and forces). *Courtesy: Kistler Group.*

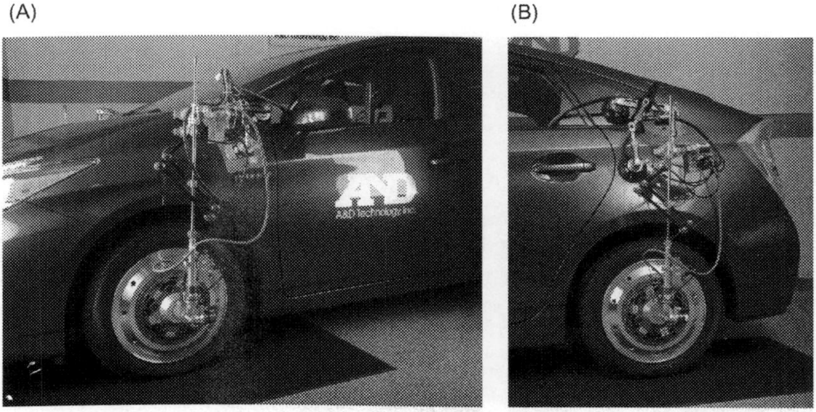


**FIGURE 12.22** Wheel measurement hub mounted on the vehicle along with associated electronics for power and signal transmission. *Courtesy: A & D Technology, Inc.*

wheel forces or torques. Mounting of the instrumented wheel torque transducer on the vehicle is comparable with changing a standard wheel. This means they can be removed and refitted quite easily, as required.

Apart from measurement of system power flows, other instrumentation and data is required to support a detailed analysis of the powertrain during operation in the vehicle. Table 12.2 shows a list of additional equipment and

**TABLE 12.2 Measurement devices for complex propulsion system analysis.**

| Device | Measuring | Purpose |
|---|---|---|
| Wheel torque flange | Wheel torques and multiaxis forces | Propulsion and braking forces, vehicle dynamic forces. |
| Instrumented driveshaft torque | Propulsion and braking torques | Propulsion and energy recuperation power. |
| Instrumented flywheel | Dynamic engine torque and speed | Engine output power and torque. |
| Combustion analyzer | Engine power and losses from in-cylinder gas work | Engine efficiency, control system, and cylinder work and stability. |
| Electrical power analyzer | AC/DC currents/voltages at e-machines | Characterization e-machine electrical performance and contribution to powertrain energy flows. |
| Electrical power flow via clamps, probes | DC currents between energy sources and consumers | Measured electrical energy flow within the vehicle power network, also to and from external charging systems. |
| Generic analog | Temperature, pressures, positional sensors | Operational status of vehicle subsystems and components, obtain ambient, vehicle and tire temperatures. |
| Vehicle bus/ network | Control variables shared between vehicle level control units | Controller variables and output of controller internal functions, vehicle status information (speeds, torques, powers). |
| Calibration tool | Measurement and calibration variables from specific control units (engine, motor, etc.) | Measurement and calibration variables, controller parameters, and lookup table values. |
| Vehicle speed over ground/ position | Distance, speed ($x$, $y$), acceleration and angular rates, pitch, and roll angle | Longitudinal and transverse acceleration, pitch and roll angle, rotation around the vertical axis of the vehicle.<br>Leveled acceleration or curve radius. Speed to any point of the vehicle, for example, center of gravity or rear axis is possible. |
| GPS position | GPS position data and time | Establishing vehicle speed, location, and acceleration. |
| Inertia | Roll, pitch, and yaw accelerations of the vehicle | Vehicle dynamic state characterization. |

measurands that can be deployed for complex powertrain analysis, along with the power flow measuring devices.

## Vehicle-based power and energy flow analysis

For holistic powertrain analysis, one of the main challenges is the data acquisition and subsequent analysis environment. So far, we have discussed the available instrumentation that can be deployed for this task. However, ideally, all the measurement systems should be acquired on a single data acquisition device with the same time stamp and resolution. This can be a challenge as it is not always the case that a single data logger is available that can interface to all the required devices. Analogue and digital signals are normally quite simple to handle with commonly used data acquisition systems, but interfaces for more complex devices, power and combustion analyzers, ECU interfaces, etc. are more challenging. This is a similar problem to test laboratory environments, and with the correct selection of equipment it is possible to create a harmonized measuring environment. If this is not possible though, data must be sampled on separate measurement devices, but a common trigger or synchronization data channels, must be available in order to time align the data. In addition, data may need resampling or cleaning up (i.e., removal of null channels) prior to data consolidation being finally executed to provide a data set suitable for processing.

A useful option is to be able to visualize all the relevant channels while the vehicle is in operation. The data logger should support this with an option to create a customized user view, along with appropriate calculations running such that the system energy balance can be viewed in real time. There are many commercial tools that could be configured to do this but, at the time of writing, and to the authors' knowledge, there is no "off-the-shelf" tool available specifically to cover this application of data viewing and analysis of energy flows.

### *Possible analysis areas*

Once the data has been prepared there are several areas of specific interest for powertrain development, apart from defining energy flows and losses. The overall vehicle and propulsion system control strategy can be evaluated and optimized. Analysis to support the functional development of the complex, hybrid powertrain is possible. A detailed understanding of the of powertrain functionalities, for example, torque vectoring and torque blending, is achievable. For vehicle controller calibration, system efficiency and component level analysis can be undertaken. Fig. 12.23 highlights the main areas of application.

The application of data science in the analysis (discussed in Chapter 14: Data Handling and Modeling) is entirely appropriate for this application. Large data sets, with many channels of continuously recorded data will be created and some appropriate method of standardized analysis, and data

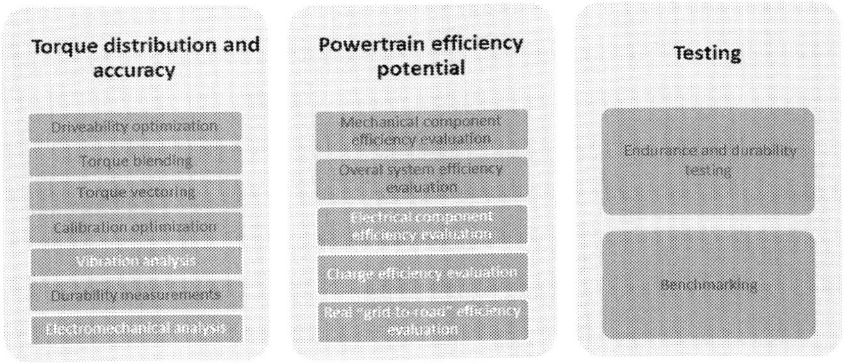


**FIGURE 12.23** Main areas of application that can be covered with a holistic powertrain analysis approach.

reduction using data mining, to search for relevant events would be a useful for understanding the data quickly and efficiently, before instrumentation is removed from a subject vehicle after a test series or program.

### *Basic analysis overview*

During the development process or benchmarking activities, knowledge of the energy flows is essential to be able to understand:

- Hybrid operating strategy.
- Efficiency of components and control strategy.
- Driveability—for example, torque vectoring makes the powertrain an active participant in vehicle dynamics.

For development engineers, the more physical measurement locations are available, the more detail can be obtained. Using the measurement instrumentation mentioned previously, based on the torque measurements at the wheels, it is possible to calculate the axle powers, as well as the total driving power using Eq. (12.17). In this equation the wheel speeds are determined from the vehicle speed and the wheel radius and the summation include:

- the two front wheel for the front axle power,
- the two rear wheels for the rear axle power, and
- all four wheels for total vehicle traction power.

$$P_{\text{axle, WT}} = \sum \tau_{\text{wheel}} \omega_{\text{wheel}} \tag{12.17}$$

Similarly, the tractive power at a dynamometer can be calculated using Eq. (12.18) based on the simulated vehicle speed and road load. This can also be resolved per axle, or combined, to give a total vehicle tractive effort.

When compared to dynamometer readings, the difference between the dynamometer power and the wheel torque power is the tire losses.

$$P_{\text{axle, DYNO}} = F_{\text{axle}} V_{\text{vehicle}} \tag{12.18}$$

The ICE power can be calculated in one of two approaches:

The first uses data from the control system relating to estimated engine torque and measured engine speed. This is calculated using Eq. (12.19).

The second approach uses the measured in-cylinder pressure data from the combustion pressure measurement system. This provides a measurement of IMEP for each cylinder, which can be used to estimate indicated torque, Eq. (12.20) and indicated engine power in Eq. (12.21).

$$P_{\text{engine, ECU}} = \tau_{\text{engine}} \omega_{\text{engine}} \tag{12.19}$$

$$\tau_{\text{engine, IND}} = \sum_{i=1}^{n} \text{IMEP}_i \times \frac{V_{cyl}}{4\pi} \tag{12.20}$$

$$P_{\text{engine, IND}} = \tau_{\text{engine, IND}} \omega_{\text{engine}} \tag{12.21}$$

Battery energy balance can be calculated using Eq. (12.22). This can be compared with the vehicle control system SOC estimation.

$$P_{\text{bat, EMS}} = U_{\text{bat, EMS}} I_{\text{bat, ECU}} \tag{12.22}$$

A *power source to road* efficiency can then be established using Eq. (12.23). This represents the ratio of driving power at the wheels to power input into the powertrain.

$$\eta_{\text{power to road}} = \frac{P_{\text{front, WT}} + P_{\text{rear, WT}}}{P_{\text{bat, ECU}} + P_{\text{engine}}} \tag{12.23}$$

Wheel torque transducers allow the high-level hybrid powertrain behavior to be identified and analyzed. In particular, they allow the breakdown of tractive power to be captured as well as mechanical power transfer between driving axles. This instrumentation approach allows a simplified analysis of energy flows to be established quite easily, with no detailed knowledge of the vehicle powertrain control system required. This facilitates identification and characterization of the different operating modes of the engine and to be able to perform an overall energy balance of the drivetrain. This approach can also be used to allow for an estimation of the efficiency of the electric elements of the drivetrain, if measurement of the power requirements of the electric auxiliaries is done synchronously.

The equipment used, along with the subsequent energy analysis, can provide a breakdown of the charging and boosting energy flows. Boosting energy can further be broken down to axle level. Charging energy can be broken down into kinetic energy recovery, front axle charging, and ICE load point shifting. An energy analysis can estimate the useable capacity of the battery based on the measured net change in SOC.

This end-to-end approach of vehicle level instrumentation along with a suitable analysis approach demonstrates the value of applying comprehensive measurement systems in order to characterize a modern vehicle propulsion system and its control strategy. Even with a limited number of standard measuring devices and techniques, these can be applied with minimal disruption and can provide valuable information regarding the operating state of a complex powertrain, gaining a second-by-second state analysis during on-road and laboratory measuring conditions.

## Example analysis, detailed powertrain characterization

Highly resolved data can be brought together to analyze specific operation conditions or phenomena in great detail. The overall measured data sets can be integrated over time to establish energy flows in the powertrain during different operating cycles, conditions, and modes. Measured data from sensors can be compared and combined with data from control units in order to characterize control system strategies or operation. Some examples of the level of detail possible are discussed below. These are from a single test vehicle, instrumented for complex powertrain analysis on the road, in real time. The vehicle had ICE and electric power with a plug-in hybrid electric vehicle (PHEV) powertrain configuration—front axle was electrically powered, rear axle ICE powered with electrical support from a BSG. Data was gathered on-road, as well as on a chassis dynamometer, driving Worldwide harmonized Light vehicles Test Cycles (WLTC) in different operating modes.

### *Boosting and tip-in, dynamic characterization*

An example of a specific condition of interest is shown in Fig. 12.24. This is a boost situation—in this measurement the powertrain was in sport mode and as such, the powertrain trimmed for maximum torque response. During steady-state driving in sport mode, the front axle is generating negative torque to pretension the powertrain, the effect of this is to store mechanical energy in the powertrain for improved tip-in response. In this case the vehicle is fitted with an ICE powering the rear axle, with an electric machine powering the front axle. The ICE axle produces propulsion torque, a proportion of which appears negatively at the front axle—effectively storing energy in torsion of the axle. During a tip-in demand the ICE is already preloaded and, if the front axle switches from charging to boosting, such that this pretensioned torque can be instantly released to accelerate the vehicle. The result is a rapid response on fast pedal transients as the preloaded ICE torque is released instantly.

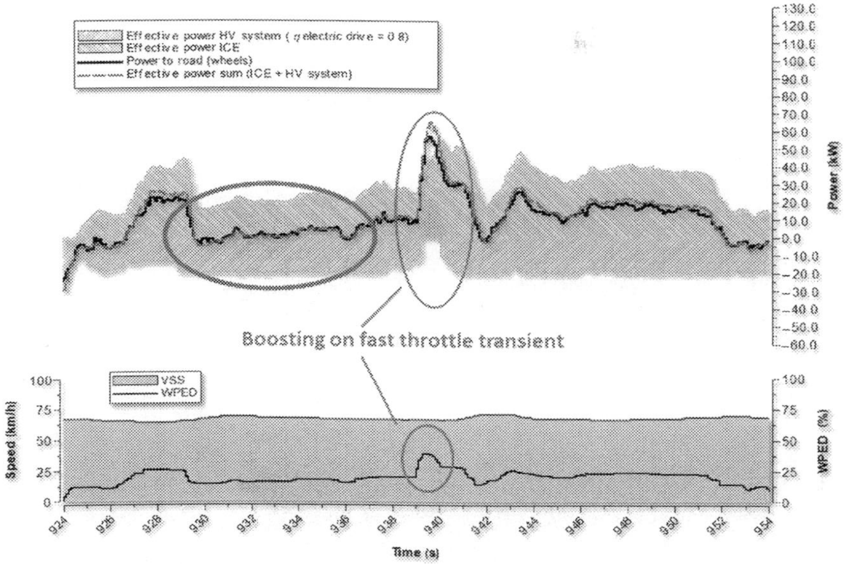


**FIGURE 12.24** Constant driving situation, ICE is generating positive torque, whereas the front axle is effectively applying a negative torque—thus tensioning the powertrain. *ICE*, Internal combustion engine.

This instant release of torque can be seen in Fig. 12.24 at 939 seconds, during a fast throttle transient. Utilizing this method effectively has two benefits:

- "Lag" of ICE (i.e., due to turbocharger) can be minimized—instant throttle response.
- Pretensioning through load point shift minimizes throttling losses of ICE.

## *Torque blending*

Fig. 12.25 shows a torque blending situation—where wheel torque exceeds estimated powertrain torque. System powers can be observed with high fidelity and are derived from torque measuring wheels. Engine and electrical drive power values can be derived from the ECU—as well as being measured via the combustion and power analyzer. A power difference can be observed between the on-road power value (black trace) in comparison to the joint ICE and battery power (dashed trace) in this data the efficiency losses were estimated. The areas during the deceleration where the on-road power is lower than the calculated power, are where the mechanical brake is participating in torque generation.

Fig. 12.26 shows the improvement in data quality when wheel torques and shaft torques are simultaneously available. In this instance the difference

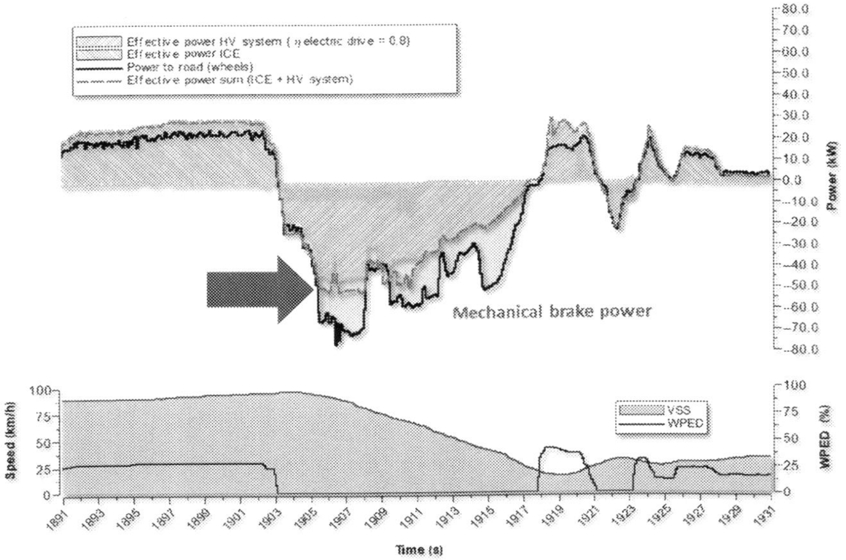


**FIGURE 12.25** Deceleration event focusing on braking power losses.

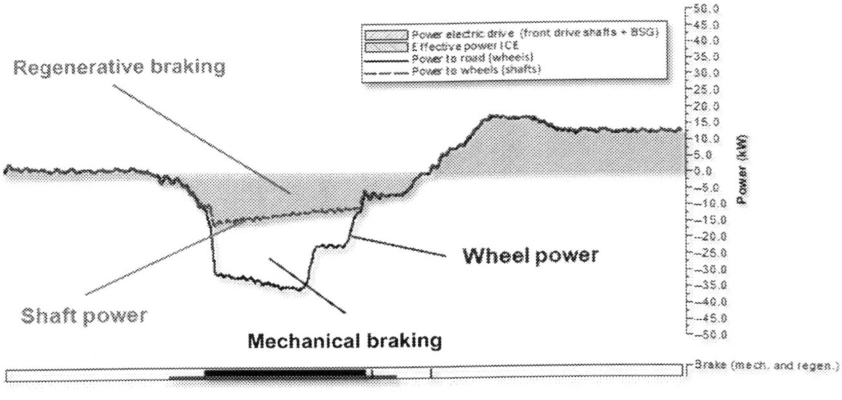


**FIGURE 12.26** Wheel torques and shaft torques during a deceleration.

between wheel and shaft torque is mechanical braking, the remainder being electrical recuperation.

## *Battery capacity estimation*

During pure electric driving the energy balance calculations, combined with the ECU derived SOC, can be used to estimate the useable capacity of the battery. This is achieved by comparing the energy balance at the battery—determined

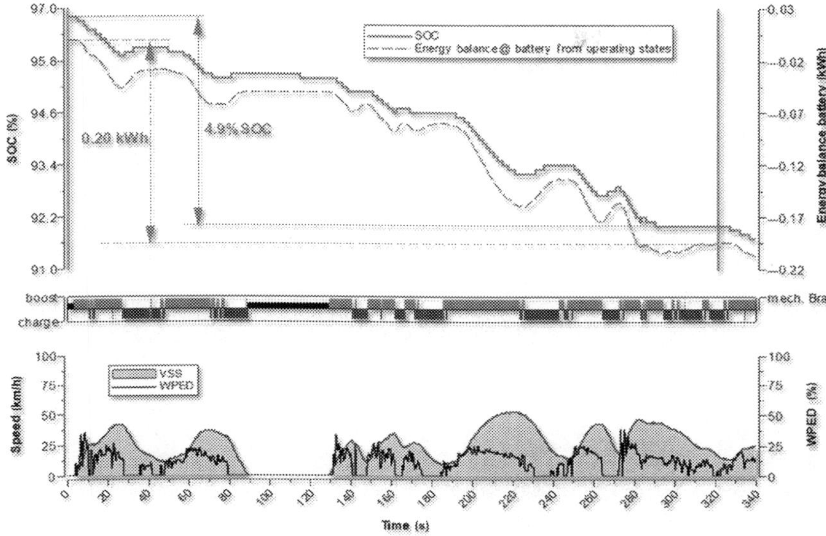


FIGURE 12.27 Battery size estimation based on energy flows for a typical WLTC.

from the operating states—with the change in SOC as shown in Fig. 12.27. Based purely on an energy balance of electric charging/boosting, front wheel torques, and ECU SOC, the useable battery capacity in this condition can be estimated.

## *Vehicle energy flows analysis*

Fig. 12.28 shows analysis of the energy flows based on the logical categorization of driving conditions for a given WLTC. In the bottom plot the logical indexes are plotted for the complete cycle. It can be seen that the test cycle is predominantly electric only for the low and medium phases: the front axle is switching between charging and boosting modes. It can be seen that the ICE is being used during certain stages of the drive cycle and in particular during the high and extra-high phases. When the ICE is active, the rear axle also switches between driving and battery charging. In this measurement series no information was available from the rear electrical machine (BSG), so the electrical state of the rear axle (whether it is "charging" or "boosting") was determined from effective rear axle power in comparison to effective combustion engine power—this is of course to be considered approximation. The top plot breaks down the front and rear axle electric power contributions used for electric propulsion. The second plot provides a breakdown of the battery charging activity during the duty cycle. The total charge energy represents the total amount of energy provided to the HV battery over the

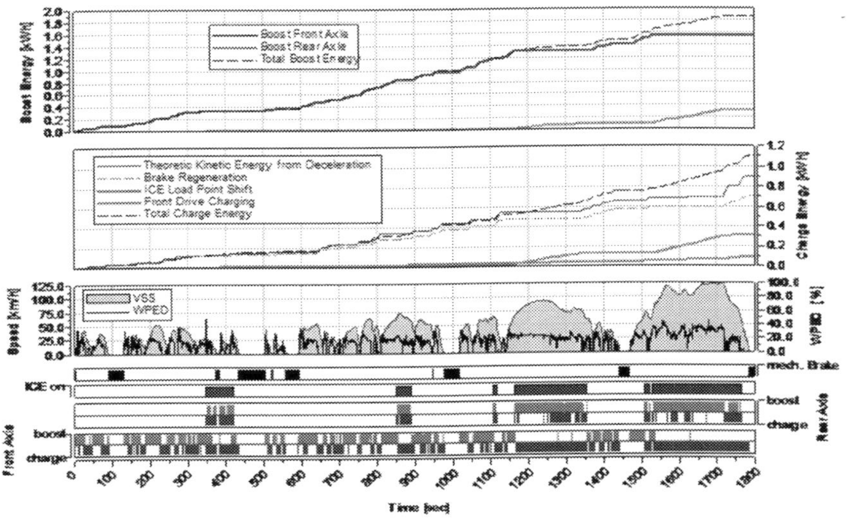


**FIGURE 12.28** Diagram giving a general impression of the energy flows in respect to operating condition for a WLTC, performed on a chassis dynamometer.

cycle. In total about 1.1 kW h is put into the battery during this cycle. It is important to note that this is not the net change in SOC over the cycle and represents only the energy provided to the battery. This energy can be broken down into three major sources:

- kinetic energy recovery during decelerations,
- front axle charging (while being driven by rear axle), and
- direct charging from engine mounted electrical machine by shifting the operating point for the ICE.

## *Summary*

Using vehicle level instrumentation for ICE and electric propulsion units, in combination with mechanical torque measurements in the powertrain, provides the ability to undertake complete end-to-end analysis of the vehicle propulsion system under dynamic operating conditions. In the examples given above the analysis of a complex hybrid powertrain was undertaken that could be applicable to a development, validation, or benchmarking type scenario. The vehicle was a complex, hybrid system using multiple axles and multiple electric machines to deploy electrical and combustion engine power. With this level of instrumentation the major energy flows can be captured in combination with basic powertrain and engine controller data—without needing information from the vehicle manufacturer. An instrumented vehicle can be tested on the road and on a chassis dynamometer to analyze these major energy flows and optimize processes for analyzing hybrid vehicle behavior. Based on this concept, the following points are valid:

- Wheel torque transducers allow the high-level hybrid powertrain behavior to be identified and analyzed. In particular, they allow the breakdown of tractive power to be captured as well as mechanical power transfer between axles.
- The instrumentation allowed a simplified analysis of energy flows to be established quite easily, with no detailed knowledge of the vehicle powertrain control system required. It allows identification and characterization of the different operating modes of the engine and to be able to perform an overall energy balance of the drivetrain.
- It is possible to allow for an estimation of the efficiency of the electric elements of the drivetrain; however, measurements of electrical auxiliaries and power units provide a greater level of fidelity and detail for the analysis.
- It is possible to derive a breakdown of the charging and boosting energy flows. Boosting energy can be broken down to axle level. Charging energy can be broken down into kinetic energy recovery, mechanical energy storage at the axle itself, and ICE load point shifting.
- The energy flow analysis can estimate the useable capacity of the battery based on the measured net change in SOC.
- On-road driving creates more extreme driving conditions than the lab-based tests. This is particularly true of braking power that is difficult to establish. However, introducing driveshaft measurements allows to one isolate mechanical work in more detail. Also, more dynamic driving behavior creates regions where the rear and front axle simultaneously produce boosting and positively contribute to vehicle propulsion. This can be identified via the combination of wheel and shaft torques.

Complex powertrain instrumentation and analysis demonstrates the value of applying end-to-end measurement systems in order to characterize a modern, vehicle propulsion system and its control strategy. This approach to powertrain characterization becomes increasingly relevant due to the changes in emission legislation, which drive characterization while driving, of gaseous and particulate pollutants, in the direction of a much more realistic measurement scenarios than synthetic cycles that have been used previously. In addition, this approach can characterize the complete powertrain, in the vehicle, in real-life operation. Although test vehicles are expensive, test facilities to replicate this level of measurement information are extremely expensive and cannot always be justified, or available. Also, these complex powertrain test environments often need on-road data to replicate—and such data must be collected on the road with an instrumented vehicle.

Of interest is the fact that the approach is scalable. A limited number of standard measuring devices and techniques, which can be applied with minimal disruption, can be used to provide valuable information regarding the operating state of a complex powertrain. However, combining more measurement systems in a synchronized system allows a greater level of precision in gaining energy flows in the powertrain, allowing a greater level of detail in a second-by-second state analysis, during on-road and laboratory measuring conditions. The main challenges for this approach, in order to undertake the required measurements, producing good-quality data for analysis, can be considered as:

- Robust equipment installation—Reliable performance of all the measuring systems is extremely important for expensive, on-road-based tests. For in-vehicle requirements the equipment and its interfaces are subjected to much more variation in the operating environment when compared to a testing facility (vibration, temperature variation, etc.).
- Power supplies—When in-vehicle, on-road, limited to batteries, with corresponding finite capacity and the potential for high currents. The amount of power drawn by all equipment needs to be established along with the capacity required. Additional instrumentation batteries can be fitted but these add to the vehicle mass, along with a parasitic load to the existing vehicle power systems.
- Data synchronization—Data will be measured by different systems that may or may not communicate to each other. This data will often have different file formats, so bringing all the data together, and time aligning it is a considerable task that should not be underestimated.
- Data access—Access to data from vehicle buses and control systems can be difficult to gain, unless you are an original equipment manufacturer (OEM). For those engineers undertaking development in an engineering service provider or benchmarking environment. The access to this data is controlled via key files (a2l, DBC, etc.). It is extremely important to clarify prior to testing what is actually available from these systems via qualified sources—reverse engineering access to information from these bus systems is difficult, time-consuming, and often unreliable.
- Measurand access—Certain physical measurements values will be required, and these may need modification of the existing vehicle—which may or may not be feasible or possible. Therefore a survey of what is available, what could be available, and what is not possible is an important study to undertake at the beginning of a measurement campaign.

## Energy storage testing and characterization

### Introduction

Powertrain testing traditionally involves physical testing, providing data for the purpose of characterization and optimization of the UUT, which in most cases is some form of rotating machine that generates positive torque (power unit) or requires a torque input (power absorber). The purpose of the test generally being to provide information such that the UUT can be optimized by the engineer, to improve efficiency of the component or system.

Electrical powertrains bring additional components to the propulsion system that are effectively static, specifically batteries and energy sources, plus the control and switching devices for electric propulsion machines. The testing of such systems is quite a different proposition for the traditional test environment, providing unfamiliar challenges. However, the overall testing goal, providing good-quality data remains the same.

## Legislative standards for the testing of energy storage devices

The testing of energy storage systems for automotive, and other applications, is a particularly sensitive issue. Highlighted in the public domain by horror stories about batteries that spontaneously ignite (thermal runaway), which cannot be extinguished using the most commonly used methods. Battery systems used in automotive applications are sophisticated devices, comprising of many individual cells that are clustered into modules, which form complete energy system packs. These units have to be robustly tested and characterized during the development process, pushed to their extreme operational limits in order to ensure safe operation under all normal and perceivable abnormal conditions.

Bearing in mind the above fact, also the complexity of the systems, there are many standards that apply for testing and developing such energy storage systems. At the highest level the standards fall under the International Electrotechnical Commission (IEC) and the International Organization for Standardization (ISO). The former concentrates on the electric components and electric supply infrastructure and the latter on the whole vehicle. The main standards relate to the following:

### *Performance*

*ISO 12405-1: Electrically propelled road vehicles—Test specification for lithium-ion traction battery packs and systems—Part 1: High-power applications*

*ISO 12405-2: Electrically propelled road vehicles—Test specification for lithium-ion traction battery packs and systems—Part 2: High-energy applications*

*IEC 62660-1: Secondary lithium-ion cells for the propulsion of electric road vehicles—Part 1: Performance testing*

*IEC 62660-2: Secondary lithium-ion cells for the propulsion of electric road vehicles—Part 2: Reliability and abuse testing*

ISO standards specify test procedures for lithium-ion battery packs and systems specifically for use in electrically propelled road vehicles. The test procedures enable the determination of the essential characteristics of performance, reliability, and abuse of lithium-ion battery packs and systems. The standard also supports the comparison the test results achieved for different battery packs or systems.

The IEC standard is similar to the ISO standard and is to specify performance testing for automobile traction lithium-ion cells and batteries—batteries that basically differ from the other batteries, including those for portable and stationary applications specified by the other IEC standards. For automotive applications, it is important to note the usage specifics; that is, the design diversity of automobile battery packs and systems, and specific requirements for cells and batteries corresponding to each of such designs. The purpose of this standard is to provide a basic test methodology with general versatility, which serves a function in common primary testing of lithium-ion cells to be used in a variety of battery systems.

### Safety

*ISO 12405-3: Electrically propelled road vehicles—Test specification for lithium-ion traction battery packs and systems—Part 3: Safety performance requirements*

*EC 62660: Secondary lithium-ion cells for the propulsion of electric road vehicles—Part 2: Reliability and abuse testing, Part 3: Safety requirements*

Specifies test procedures for lithium-ion battery packs and systems, for use in electrically propelled road vehicles to determine the essential characteristics of the safety performance of lithium-ion battery packs and systems

### Design

*ISO/PAS 16898: Electrically propelled road vehicles—Dimensions and designation of secondary lithium-ion cells*

A propulsion system battery system is a significant component of the electrified powertrain and has a huge influence on the overall powertrain and vehicle design. Depending upon the installation and package space, the battery overall volume and footprint must follow certain design constraints that are given by this standard that focuses on battery designs for road vehicles.

## Battery tests and characterizations

The tests typically carried out when characterizing batteries and their performance are presented next as an overview of all the standards.

## *Performance testing according to International Organization for Standardization 12405*

- *Energy and capacity, variable temperatures, and discharge rates*—Consisting of discharging and charging with rest phases in between. Currents, voltages, and temperatures are carefully logged and monitored during tests. Characterization of capacity under different working conditions.
- *Power and internal resistance*—To determine dynamic power capability with varying temperatures and SOC. Applies to battery packs and systems, test profile consists of an 18 seconds discharge pulse followed by a 40 seconds rest period to allow the measurement of the cell polarization resistance. After the 40 seconds rest period a 10 seconds charge pulse with 75% current rate of the discharge pulse is performed to determine the regenerative charge capabilities after the charge pulse, a rest period of the 40 seconds follows.
- *No load SOC losses*—Determines self-discharge rate when not in use. Applies to battery systems only.
- *SOC storage losses*—Capacity loss when stored. Applies to battery systems only.
- *Low-temperature cranking power*—Measuring the power capability at low temperatures. The relevant temperature is nominally −18°C with an aim to generate a data basis for a time depending on power output at low temperatures. This test applies to battery systems only.
- *High-temperature cranking power*—Intended to measure power capabilities at a high temperature of 50°C with an aim to generate a data basis for a time depending on power output at high operating temperatures. This test applies to battery systems only.
- *Efficiency*—Determines the battery system round trip efficiency by calculation from a charge balanced pulse profile. For high-power application, the energy efficiency of the used battery system has a significant influence on the overall vehicle efficiency. It affects directly the fuel consumption and emission levels of a vehicle equipped with a battery system for high-power application. This test applies to battery systems only.
- *Cycle life*—Energy throughput has a significant influence on the lifetime of a battery. For a relevant aging profile, the real conditions during driving are considered, the thermal management and monitoring of the battery system are mandatory, as well as certain rest phases are needed for equilibrium and cell balancing. This test applies to battery systems only.

### Safety testing according to International Organization for Standardization 12405 (packs) and International Electrotechnical Commission 62660 (cells)

- *Short-circuit protection*—Check operation of integrated overcurrent protection device for packs and cells via the application of a hard short circuit, for less than 1 second in 10 minutes.
- *Overcharge protection*—Charging occurs at a constant current rate, continued until the UUT interrupts the charging by an automatic disconnect of the main contactors. Test is terminated when the SOC level is above 130% or when cell temperature levels are above 55°C. Data acquisition/monitoring shall be continued for 1 hour after charging is stopped.
- *Overdischarge protection*—Testing of the functionality of the overdischarge protection. The battery management system must interrupt the overdischarge current to prevent the UUT from any further related severe events caused by an overdischarge current. The discharge test is terminated manually if 25% of the nominal voltage level or a time limit of 30 minutes after passing the normal discharge limits of the UUT has been achieved. Measurements include voltage, current, and temperature as a function of time and isolation resistance between the UUT case and the positive and negative terminals before and after the test.
- *Dewing test*—Simulates the use of the system/component under high ambient humidity. Addressing failure modes caused by electrical malfunction caused by moisture.
- *Thermal shock test*—To determine the resistance of the UUT to sudden changes in temperature. The test dictates a specified number of temperature cycles, which start at room temperature followed by high and low temperature cycling. The failure modes addressed are an electrical and mechanical malfunction(s) caused by the accelerated temperature cycling.
- *Vibration*—Test for malfunctions and failures caused by vibration—random vibration induced by rough-road driving as well as internal vibration of the powertrain. The main failures to be identified by this test are breakage and loss of electrical contact.
- *Shock*—Test is applicable to packs and systems intended to be mounted at rigid points of the body or on the frame of a vehicle. The load occurs, for example, when driving over a curb stone at high speed. The failure mode is a mechanical damage to components due to the resulting high accelerations.
- *Crush*—To characterize cell responses to external load forces that may cause packaging deformation.
- *Drop*—Simulates a mechanical load during service operation when the battery system is removed from the vehicle. During the test and for 1 hour posttest observation period, the battery system must exhibit no evidence of fire or explosion.

- *Crash test*—Simulates an inertial load which may occur during vehicle crash situation.
- *Point load contact*—Simulates a contact load which may occur during vehicle crash situation.
- *Water immersion*—Tests for robustness against water immersion scenarios, which may occur when a vehicle is flooded.
- *Thermal load*—Simulates a thermal load which may occur in a vehicle fire.
- *Cooling system*—Replicates a system failure of the thermal control/cooling of the battery pack or system.

*Note:* Crushing and penetration tests performed on battery packs have resulted in recorded thermal runaway events at test facilities in Europe, the effects of which is made more potentially hazardous when performed within an enclosed building space. Using an outdoor crash sled facility, adapted for the purpose would seem to be a sensible procedure, particularly when testing units of novel chemistry or configuration.

In most cases the document describing the standard will provide a detailed description of the test method, plus measurements to be made, as well as pass/failure requirements. There are other standards existing in addition to those above mentioned, which originate in various global regions. These are JARI (Japan), SAE (United States), and CENELEC (EU). It should be noted that all the standards are similar in terms of what performance and safety parameters are being tested. The main difference is the actual test definition and limits. Therefore it is an application-specific requirement to determine the details of the test execution. But the equipment used or needed will be broadly the same. Table 12.3 gives an overview of the relevant standards for environmental tests relating to vehicle battery systems. LV 124 is an important standard that covers tests for electric and electronic components for use in motor vehicles up to 3.5 t with a 12-V electric system. There is also standard LV 148 that covers tests for electric and electronic components in motor vehicles 48-V electrical systems. These describe one of the most important standardized environmental tests for energy storage devices for automotive applications and are particularly relevant to mild-hybrid powertrain applications. The LV 124 and LV 148 standards were put together by German OEMs but have now several variations for almost all car manufacturers around the globe. It is often the case that a battery test facility will need to cover the complete spectrum of powertrain electrification concepts from 2- to 1000-V energy storage systems—from individual cell to complete system level.

**TABLE 12.3 Overview of global standards relevant for environmental testing of vehicle battery systems.**

| Test | Test environment | Standard |
|---|---|---|
| Temperature cycles | Temperature Shock | IEC 62660, IEC 62261, ISO 12405, ISO 16750<br>UL 2580, UL 2271, UL 1973, UL 1642, UL 2504<br>SAE J2929, SAE J2464, UN 38.3<br>FreedomCAR, GB/T 31467.3, GB/T 31485<br>LV-124, Telcordia GR-3150-CORE<br>GB/T 31467.3 QC/T 743 |
| Constant, elevated temperatures | Temperature Heating | UL 2580, LV-124, BATSO 01, UL2271<br>ISO 16750, IEC 62660, IEC 62620, IEC 61960<br>ISO 12405, QC/T 743<br>Telcordia GR-3150-CORE<br>DOE-INL/EXT-15-34184 |
| Ingress Protection | Dust, spray, splash water | BATSO 02, LV-124, ISO 16750 |
| Corrosion | Corrosion in salt spray and free air | UL 2580, UL 19973, LV-124, ISO 16750<br>Telcordia GR-3150-CORE |
| Shock, vibration | Mechanical shaker | IEC 62660, IEC 62281, IEC 62133, IEC 61960, ISO 12405, IS/FDIS 6469, SAE J2929, SAE J2464, UL 2580, UL 1642, UL 2271, UL 1973, UL 2054, UN 38.3, LV-124, BATSO 01, BATSO 02, FreedomCAR, UN ECE R100<br>Telcordia GR-3150-CORE, GB/T 31467.3, QC/T 743 |
| Damp, moisture ingress | Temperature and humidity | ISO 12405, ISO 166750, GB/T 31467.3, LV-124, SAE J2929, UL 1973, Telcordia GR-3150-CORE |

*IEC*, International Electrotechnical Commission; *ISO*, International Organization for Standardization.

## The battery test facility

The requirements for electrical supply and system configuration of high-power DC systems are discussed in Chapter 3, Test Facility Design and Construction, and Chapter 4, Electrical Design Requirements of Test Facilities.

Any facility for testing batteries has to be planned with future requirements in mind; therefore a flexible and future-proof system concept must be considered in the planning stage. The test device has to provide a highly efficient test

**TABLE 12.4 Cell, module, and pack test requirements.**

| Test object | Test target | Voltage/ current | Measurements/ interfaces |
|---|---|---|---|
| Individual cell | Temperature testing and characterization | 1–6 V<br>50–400 amp | Temperature sensors required |
| Module (multiple cells) | Temperature tests<br>Climate tests | Max. 60 V<br>50–600 amp | Temperature and voltage measurement<br>Interface to module controller<br>Additional cooling system |
| Battery (multiple cells, control and management unit) | Temperature tests<br>Climate tests | Max. 800 V<br>Max. 800 amp | Temperature and voltage measurement<br>Interface to module controller and/or battery management system<br>Additional cooling system<br>Rest bus simulation |

environment with the capability of accurate state simulation, with the utmost repeatability of test conditions—as close to reality as possible. A safe test environment for the UUT, but more importantly the safety users, is of highest priority.

Facilities can be classified according to the scale and size of the UUT. An energy storage system for vehicle propulsion consists of a complete system or pack, this contains battery modules that are assembled from individual cells along with their control and monitoring systems. Therefore the current and voltage requirements for the DC loading and charging unit can vary widely. Table 12.4 provides an overview of typical requirements.

The actual tests and test profile depend upon the chosen standards for the test. Temperature management is a standard requirement, the battery test facility will need to be equipped with the capability to demand temperatures for test conditions, also to be able to cycle temperatures for test requirements. Battery systems are equipped with thermal management; therefore temperature conditioning for the UUT is also required. For maximum efficiency, all the subsystems should be slave nodes to a master control system that is also capable of interfacing to building systems, control and data acquisition systems, UUT controller, and media controller. This therefore dictates a requirement for a test automation system; however, it should be noted that many automotive test automation systems are designed for testing engines or powertrain rotating components. It is often the case that these systems are not well suited to the specific requirements for testing "static" components like energy sources or electrical drive systems. Classical test bed system software and functions are often "hierarchically" designed to consider a UUT that will rotate during test—with associated elements for control and data acquisition of such a rotating UUT. Often with these systems, functional software components are either not required, or not available when applied to the above mentioned "static" test environments. Therefore care should be taken in the planning phase to select an appropriate supplier that has a system designed for testing such components (and can prove it via reference installations). Another option is to develop a system "in-house," using components from the many generic data acquisition system component and software suppliers. This will provide a flexible system, but this approach is not suitable for complex, large-scale test environments found at automotive OEMs and is more suited to academic environments.

The required data acquisition and measurement system will consist mainly of DC currents and voltages that must be accurate to within 0.1% full scale output (FSO). These could be required at cell, module, and system level. Therefore integration of measurement devices (voltage sensors) inside a battery pack will be required (along with appropriate electrical isolation). Interface to battery system controllers is required, which may be read/write using an application system. Or read-only using a suitable bus system measurement tool (usually CAN based). Temperature measurement via embedded sensors in the UUT (thermocouples or platinum thermometry), along with acceleration sensors may be required—and these will also all require a safety concept (high-voltage isolation) to protect the user and equipment from HVs. Thermal imaging, synchronized with temperature data channels, is also useful to be able to record the temperature image and footprint of the battery package, for monitoring and early identification of critical thermal conditions.

For thermal or environmental testing, enclosures are required, these can be sized for the range of UUTs or as walk in chambers. The chamber control system will require interfacing with the overall control system. It may also be the case that the test system components are integrated within or around the chamber. This provides a comprehensive and capable test environment. However, climatic system experts and suppliers are often not experts in the areas of test object characterization—and vice versa. This then requires a talented project manager to be able to facilitate the slick integration of thermal and electrical test and load systems, during the planning, installation, and handover phase. In order to ensure that the combined facility can deliver the required combinations of thermal and electrical test conditions, this will require the following:

1. Clarity on individual systems capabilities—when combined—to meet the documented test requirements from the requirements specification.
2. Full clarity on interfaces and responsibilities—both system and personnel—during planning, installing, and commissioning.

3. Assuming above clarity is achieved between all parties. Joint agreement on the commissioning and acceptance tests and criteria, including system performance KPIs for the complete system as a whole entity.

In normal battery test and characterization environments described previously, the batteries are tested and stressed within their electrical and thermal performance limits only. For mechanical tests (drop, shock, etc.) another facility environment would be used that is more suitable for this physical abuse testing. Many new facilities that are in plan, or online, at the time of writing (2020) include a complete suite of facilities that are able to test and certify a battery for all performance aspects, to most or all of the international standards. This should be considered the best case scenario to be achieved for any planned new or upgraded facilities.

## Testing of high-power switching devices (motor inverters and drives)

The electric propulsion system consists of a motor, plus drive system (also known as an inverter) and in many cases, testing of the complete system is appropriate, particularly later on in the development cycle. In accelerated development processes, it is also the case that components may need to be tested in isolation, in order to execute parallel development workflows to support rapid system development. For these requirements, testing of inverter and drive control systems, as separate components, is required. This presents a mindset change in the test environment due to the fact that this is no longer a rotating component when tested in isolation. However, testing of high-voltage, switching devices presents its own set of challenges in the test environment. The overall need to reduce test time means that integration of a wide range of target, virtual e-motors will need to be accommodated—with varying types, performance, and tolerances. General test requirements for the inverter UUT would be:

- optimization of switching frequency
- EMI tests with different duty cycles
- torque ripple reduction
- durability
- robustness—fault insertion tests
- optimization of inverter control loop

It is possible to test the drive system with its target motor, or with a slave machine (i.e., the machine is part of the test equipment, not the UUT), but in this case, rotating machinery, cooling and mechanical systems all have to be considered, as would be the case for a normal, rotating machine test stand. In addition, using a rotating machine as an inverter test load unit can create trade-offs with respect to control dynamics, reproducibility, and safe operation at operating boundaries. Based on the above, inverter testing using e-motor emulation is a suitable alternative because:

- There is no rotating machinery required, so safety issues and costs for such a "physical" system can be avoided.
- The electric machine is represented by a model, so operating at boundary conditions is safe with no risk of hardware component damage (apart perhaps from the UUT—the motor drive system).
- Multiple load setups and configurations can be deployed as these are simple a function of a model representing the load unit connected to the UUT.
- Energy can be recirculated with greater efficiency—as it remains "electrical" energy within the test system, with no conversion to mechanical (and back again).
- Fault simulation and insertion can be integrated into test cycles—this can be executed quite simply and safely.
- Efficient and compact system setup—limited service, maintenance, and cooling requirements—installation space requirements are optimal compared to a rotating test stand.

The use of e-machine simulation allows the testing of an inverter in a workshop environment with full simulation of the vehicle powertrain concept. The test system consists of an electronic loading and drive emulation unit (similar to an AC dynamometer drive cabinet), in conjunction with a HiL capable hardware interface unit. With this test system configuration, the motor drive/inverter unit can be fully tested in normal and abnormal operating states, with automated test procedures, in compliance with the requirements for ISO 26262 (functional safety). The benefit being that the inverter system and software can be validated in combination with several target machines, even where those machines may not be available as prototypes or may only exist "virtually" as concepts. With this strategy the potential for accelerating and improving efficiency, in a test environment where significant research activities are undertaken for machine drive systems, is attractive, even if the investment required to get such a facility up and running is considerable. Note that there are other factors that the test engineer must consider, analogous to any test facility:

- Power supplies and connections—Large currents and voltages, even if recirculating within the system, need a detailed study to ensure compliance, along with a robust facility (discussed in Chapter 4: Electrical Design Requirements of Test Facilities).
- Mounting of UUT—A concept that allows safe and secure location of the UUT while under test. Must have design flexibility to cope with anticipated range of size/weights of UUTs.
- Cooling system—Inverters can be liquid cooled and require a supporting system to manage heat rejection and to create and maintain the correct

test conditions. They must be sized according to inverter power rating and efficiency within the operational envelope.

- Automation—Automation of test runs is essential for efficient operation. The system chosen must fit or align as close as possible to other systems in the working environment in order to reduce training/familiarization and to ensure data model compatibility.
- Safety—HVs and currents need a robust safety concept for the users and the facility, particularly where testing is automated or unmanned.

## Further reading


A. Lewis, 2018. Real-world to lab – robust measurement requirements for future vehicle powertrains, in: Paper Presented at Conference: 13th International AVL Symposium on Propulsion Diagnostics, 2018. Download at <https://www.avl.com/documents/1982862/8979878/Session + 7 + Real + world + to + Lab + Robust + measurement + requirements + for + future- + vehicle + powertrains.pdf>

International Electrotechnical Commission (IEC), Electrical energy storage, in: White Paper, 2011.

Cao, J., Schofield, N., Emadi, A., 2008. Battery balancing methods: a comprehensive review, in: IEEE Vehicle Power and Propulsion Conference (VPPC), 2008, doi:10.1109/VPPC.2008.4677669.

SAE J1711, 2010. Recommended Practice for Measuring the Exhaust Emissions and Fuel Economy of Hybrid-Electric Vehicles, Including Plug-in Hybrid Vehicles.

SAE J1634, 2017. Battery Electric Vehicle Energy Consumption and Range Test Procedure.

IEC 61010-1, 2010. Safety Requirements for Electrical Equipment for Measurement, Control and Laboratory Use.

IEC 61010-31, 2015. As Above but With Particular Requirements for Hand-held Probe Assemblies for Electrical Measurement and Test.

IEEE 1459-2010, Standard Definitions for the Measurement of Electric Power Quantities Under Sinusoidal, Non-sinusoidal, Balanced, or Unbalanced Conditions.

ISO 12405:2018, Electrically Propelled Road Vehicles—Test Specification for Lithium-ion Traction Battery Packs and Systems.

ISO 19453:2018, Road Vehicles – Environmental Conditions and Testing for Electrical and Electronic Equipment for Drive System of Electric Propulsion Vehicles.

IEC 62660:2018, Secondary Lithium-ion Cells for the Propulsion of Electric Road Vehicles.

ISO/PAS 16898:2012, Electrically Propelled Road Vehicles – Dimensions and Designation of Secondary Lithium-ion Cells.

ISO 26262:2011 (all parts), Road Vehicles – Functional Safety.

ISO 6469:2019 (all parts), Electrically Propelled Road Vehicles – Safety Specifications.

Chapter 13

# Test cell safety, control, and data acquisition

## Chapter Outline



Engine Testing. DOI: https://doi.org/10.1016/B978-0-12-821226-4.00013-9

## Part 1: Machinery and control system safety: stopping, starting, and control in the test cell

### Safety Legislation

The safe collection, verification, manipulation, display, storage, and transmission of data are the prime consideration in the design and operation of any test facility. The universal adoption of software programmable devices and systems in all industries means there is a commensurate rapid development in worldwide safety regulations covering their use.

There are some fundamental concepts common to almost all of the latest safety legislation, since all attempt to be relevant to the age of digital, programmable controls. These concepts are included here because they are highly relevant to anyone integrating complex control systems with programmable devices. This chapter should be read following study of Chapter 4, Electrical Design Requirements of Test Facilities, which covers the electrical power distribution and installation aspects of test facilities.

The following features apply to power train test cell systems when they are taken together to form "machinery" under the wide meaning of that word used in legislation.

Essential features of machinery design, integration, and most modern safety legislation are as follows:

- Machinery manufacturers must provide products and systems of inherently safe design and construction.

- Machinery must not start unexpectedly (e.g., keylock switches to protect remote areas).
- Manufacturers and users must take the necessary preventative measures in relation to risks that cannot be eliminated (e.g., interlocked shaft guards).
- Manufacturers and system integrators must inform users of any residual risks (e.g., discharge time of a battery simulator DC bus); therefore good documentation is vital.
- Particular to systems controlled by programmable devices, hazardous situations must not arise as the result of:
  - reasonably foreseeable human error (misuse) and
  - errors in control logic, software, or hardware.
- Hazards, their effects, and the probability of their occurrence have to be quantified by the use of factors such as:
  - component's mean time to failure,
  - identification of common cause (of) failure,
  - severity of injury that might result from the hazard,
  - frequency of exposure to the hazard, and
  - (ease of) possibility of avoiding the hazard.

While the manufacturer is responsible for delivering the safety components and the system integrator is responsible for the system, it must be the responsibility of the user to select the safe-operating parameters when integrating the unit under test (UUT). The degree to which the user can be "reasonably" foreseen to misuse a system, before rendering it hazardous, will probably be decided in the future by lawyers equipped with perfect hindsight.

## European Machinery Safety Regulations

There are two major pieces of safety legislation covering the European Economic Area:

- The Machinery Directive 2006/42/EC, which covers many qualitative aspects of safety, some appropriate to specific industries, such as ergonomics, hygiene, and environment.
- The "New Machinery Directive" (EN ISO 13849-1), which tends to concentrate on the quantitative categorization of hazards, with particular concentration on machinery under the control of programmable control systems. Depending on the performance levels (PL) identified, the directive leads to the appropriate choice of control system components and architecture.

From the start of 2011, there was a lot of discussion about the detailed implementation of EN ISO 13849-1, which replaced the machinery directive EN 954-1 under which test cells had been designed and built up to and during 2010. There was some confusion concerning the status, under legislation, of levels of modifications carried out to test facilities that were built under older legislation, which have been running for some years and under established procedures; these problems have largely been resolved. However, it is still the case that if a test facility built to pre 2011 standards is the subject of a substantial modification or equipment update, it should be brought up to the latest standards in every respect.

Much of the detail of the directive affects the test plant system manufacturer, supplier, and system integrator rather than the test engineers who use the plant, although the latter bear the heavy responsibility of understanding the safe operation of the system and are required not to make unauthorized modifications to it that materially affect its (safe) operation.

The five PL ratings (PLr) a, b, c, d, and e, defined in ISO 13849, together with the equivalent safety integrity level (SIL) defined in IEC 62061, are shown in Table 13.1. The major test plant suppliers in Europe are generally working to PL(d) and SIL 2, but it is the responsibility of the systems integrator to ensure that the control system(s) are compliant with the European harmonized EN ISO 13849-1 safety standard.

The new legislation is tending to move designers toward the use of standardized system and subsystem architecture using components that meet or exceed the PL levels required. Component catalogs are now listing such features as mean time to dangerous failure (MTTFd) for items used in automated (test) systems in a form such as:

*[Device name]: MTTFd: 50 years, DC: medium, Cat 3, PLe*

As a final note, it should be remembered that, according to recent reports, 50% of industrial accidents are caused by operator error or "human behavior"; therefore training in the operation and maintenance of power train test equipment is vital, even after safety legislation has rendered it as safely designed and constructed as is reasonably possible.

**TABLE 13.1** Legislative safety levels to the two different European Standards.

| PL ISO 13849-1 | Average probability of failure to danger level per hour ISO 13849-1 | EN/IEC 62061 safety integrity level (SIL) |
|---|---|---|
| A | $\geq 10^{-5}$ to $<10^{-4}$ | |
| B | $\geq 3 \times 10^{-6}$ to $<10^{-5}$ | 1 |
| C | $\geq 10^{-6}$ to $<3 \times 10^{-6}$ | 1 |
| D | $\geq 10^{-7}$ to $10^{-6}$ | 2 |
| E | $\geq 10^{-8}$ to $10^{-7}$ | 3 |

*PL*, Performance level.

## US Machinery Safety Regulations

In the United States the Occupational Safety and Health Administration (OSHA) is the federal organization that oversees industrial safety. However, any test facility project manager will also need to understand the local requirements of the National Fire Protection Association (NFPA) standards, and most major automotive original equipment manufacturers (OEMs) have their own regulations specific to power train testing.

The OSHA publishes regulations in "Title 29 of the code of Federal Regulations" (29 CFR)1 and standards covering industrial machinery are in Part 1910. Standards such as 29CFR 1910.147, Subpart J, which covers the rules of locking out machines during maintenance, or Sub part O, which covers guarding, are mandatory. Many OSHA standards have equivalent ANSI standards, while some standards are voluntary, or advisory, local fire chiefs or responsible officials may insist that they be met. Once again, early contact with local officialdom is the best way of avoiding problems later in a modification or build project.

## Emergency stop function

### *Do not start something you cannot stop*

The emergency stop (*EM stop*) function is, in older hardwired systems, sometimes quite separate from any automatic alarm monitoring system embedded into the cell control system.

Most EM stop systems may be visualized as a chain of switches and relays in series, all of which need to be closed before engine ignition or some other significant processes can be initiated. When the chain is opened, normally by use of manually operated press buttons or relays linked to an oil pressure switch, the process is shut down, and the operated button is latched open until manually reset. This feature allows staff working on any part of the test cell complex to use a safety procedure based on using an EM stop station as a self-protection device by pressing in an EM stop button and hanging a "do not start" notice on it. The notice has to bear the time of application and name of the user; in normal circumstances, only the person named may remove it.

In modern systems built to the New Machinery Directive, the EM stop chain will probably be operated through a specially constructed safety PLC connected to double-pole switches and relays, all built to Cat 3, PLe standards, and the wiring scheme will follow an architecture that meets the same PL.

EN ISO 13849-1 requires that all field-mounted devices in the EM stop chain, including push buttons, shaft guard limit switches, and cell door limit switches, need to be double-pole (Cat 3) devices.

In general, the EM stop system is initiated by humans rather than devices; the exceptions are safety-critical systems such as the fire alarm system, which often have a hardwired safety relay connection in the EM stop circuit. This particular "fire alarm" feature is not always available or desirable because, provided the cell is fully staffed, the damage to a valuable prototype UUT, through the rapid sequence of events that would result from the automatic, and possibly spurious, triggering of a fire alarm-initiated shutdown, can be avoided by the more appropriate actions of trained operators.

The traditional result of triggering the (non fire) EM stop circuit was to shut off electrical power from all devices directly associated with the running of the test cell. The major perceived risk was seen as electrical shock. In this model, dynamometers are switched to a "de-energized" condition, fuel supply systems are shut off, ventilation fans are shut down, but services shared with other cells are not interrupted and fire dampers are not closed.

Modern practice, particularly for cells fitted with electrical four-quadrant dynamometers, is for the EM stop function to shut off engine fueling and ignition; shaft rotation is then stopped through the controlled application of load, using a pre programed "fast stop" sequence. In an electric or hybrid cell the same policy is followed with electrical feed to the UUT cut and shaft rotation stopped by application of dynamometer load. In both cases, power to the dynamometer(s) is cut when the zero-speed signal is recorded.

BS EN 60204-1: 2006 (Safety of Machinery) recognizes the following categories of EM stop:

- Uncontrolled stop (removal of power): Category 0. This is the traditional function commonly used in engine cells fitted with manual control systems and two-quadrant dynamometers.
- Controlled shutdown (fast stop with power removed at rotation stop or after timed period): Category 1. This is the normal procedure for engine cells fitted with four-quadrant dynamometers where, under some circumstances, shutting down the engine fueling may not stop rotation.
- Controlled stop (fast stop with power retained at rotation stop): Category 2. Normally operated as a "fast stop" button rather than an "EM stop" button in modern test cell practice (see next).

The control desk area may be fitted with the following "emergency action" buttons:

- large red EM stop buttons (latched)—one will be desk mounted, but there may be at least two repeater buttons mounted on the cell wall and on any auxiliary control box;
- fast stop button (hinged cover to prevent accidental operation)—mounted only on the control desk; and
- fire suppression release button—wall mounted and possibly break-glass protected.

It should be clear to all readers that operational training is required to enable operators to distinguish between situations and to use the most appropriate emergency control device.

A rare and sometimes confusing emergency that can occur when testing diesel engines is when the engine "runs away." In this uncontrolled state the engine will not respond to shutting of the fuel rack as it is being fueled from its own lubrication oil. This can happen if the sump has been overfilled or seals in the turbocharger fail; engines with worn bores and those built in a horizontal configuration can be particularly susceptible. When dealing with a runaway engine, hitting a Category 0 EM stop is completely the wrong action; cutting the fuel supply will have little effect, and removing load from the dynamometer will allow the engine to accelerate to a destructive overspeed. The engine must be stalled by applying load with an automated fast stop function or under manual control of the dynamometer. To recognize the risk of such events and avoiding them mean that the training of staff, particularly those running with worn or rebuilt diesel engines, is important.

It is important to use the safety interaction matrix, introduced in Chapter 3, Test Facility Design and Construction, to record what equipment will be switched, to what state, in the event of any of the many alarm or shutdown states being monitored. There can be no one rule that fits all cases; the risk assessment process that determines the safety actions matrix for a test facility must differentiate between the situations of, for example, a remotely monitored transmission endurance cell and an F1 power train cell manned by several highly trained engineers.

Everything possible should be done to avoid spurious or casual operation of the EM stop system or any other system that initiates cutting off power to cell systems. Careful consideration has to be given before allowing the shutting down of plant that requires long stabilization times such as humidity control for combustion air or oil temperature control heaters. It is also important not to exceed the numbers of starts per hour of motors driving large fans, etc. that would cause overheating of windings and the operation of thermal protection devices.

## Hardwired instrumentation in the test facility

Hardwired instrumentation and safety systems, once widely required, are becoming increasingly rare with the availability of digital devices that meet the requirement of safety legislation. Only devices that are part of gas detection and fire suppression systems are now required to be hardwired, in the latter case using cables compliant with BS 5839.

Power and signal cabling to the several motors built into a test facility need to be marshaled within a motor control panel (MCP) that, for multicell facilities, will be large and complex; therefore it is usually positioned on the services floor with ventilation equipment, motors of which, among others, are connected to it. The status of the services operated from the MCP (motorized dampers, fan motors, cooling water pumps, etc.) needs to be visible to the test cell operator(s) either by a simple indicator lamp board or through a mimic panel being part of a building management system (BMS).

## Computerized monitoring of test cell and unit under test alarm signals

All computer-based test cell control systems allow the safe-operating envelope for both test cell equipment and the UUT to be defined and protected by user-defined alarms. Alarms can be programed in many ways. Their setting and modification may be protected by a hierarchy of passwords, but the default form is usually based on the following alarm levels:

- high-level warning—no automatically initiated action,
- high-level test shutdown or other automatic remedial action,
- low-level warning—no automatically initiated action, and
- low-level test shutdown or other automatic remedial action.

Where alarm channels are inappropriate to safe operation (e.g., low coolant temperature = shutdown), these alarm levels can be switched off or assigned values outside the operating range.

Primary alarms, often high-level shutdowns, which are protected by the highest password level access, should refer to limits of performance of the installed test plant, rather than the UUT, the performance limits of which should be lower. These test plant limits, if exceeded, would endanger the integrity of facility equipment and therefore personnel.

Levels set to protect the UUT may be global, meaning they are active throughout the whole test or test stage specific. The global alarms usually protect the integrity of the whole test system, while the stage alarms may protect either the UUT or the validity of the test data. Thus it might be necessary during a particular test sequence to hold the engine's coolant temperature between close limits while a series of secondary measurements is made; by setting the high and low warning alarm limits to those limits, the validity of the test conditions is audited.

Some test sequence editors will allow alarm actions to be set by Boolean logic statements; they may also allow the sensitivity (time lag) to be changed to avoid false alarms caused by short-duration peaks.

Most engine test software contains a rolling buffer that allows examination of the values of all monitored channels for a selectable time before any alarm shutdown. Known as "historic" or "dying seconds" logging, this buffer will also record the order in which the alarm channels have triggered, information of help in tracing the location of the prime cause of a failure.

## Security of electrical supply

The possible consequences of a failure of the electricity supply, or such events as individual subcircuit fuse failures, must be carefully thought out and recorded in the safety matrix. The safe shutdown of rotating equipment in the case of a mains electrical supply failure needs to be considered carefully, since the dynamometers will have lost the ability to absorb torque and will have effectively become flywheels.

Equally, an automatic or unsafe restart after power is restored following a mains supply failure is particularly dangerous and must be inhibited by system design, the supply switching, and operator training.

It is usual to design safety systems as normally de-energized (energize for operation), but all aspects of cell operation must be considered. The best design practice of cell control logic and now legislation absolutely requires that equipment, such as solenoid-operated fuel valves at the cell entry, must be reset after mains failure through an orderly, operator initiated, restart procedure.

The installation of an uninterrupted power supply (UPS) for critical control and data acquisition devices is to be recommended. This relatively inexpensive piece of equipment needs to provide power for the time required to save or transfer computer data.

## Building management systems and services status displays

Test cell function is usually totally dependent on the operation of a number of building and site services, remote and hidden from the cell operator. A failure of one of these services may directly or indirectly (through alarm state caused by secondary effects) result in a cell shutdown. The common and time-wasting alternative to the display of the service's running status is for operators to tour the facility, checking the status of individual pieces of plant at start-up, after a fault condition and shutdown. As stated elsewhere, a centralized control and indication panel for services is of great value; however, such system status displays must indicate the actual status of each unit, rather than its control demand status (A key lesson learnt from the Three Mile Island accident).

A modern BMS can have an interactive mimic display showing not only the running state of services but also the status of fire dampers, valves, and the temperature and pressure of fluids. These PLC-based systems may have the complete start, stop, and alarm state sequence logic programed into them, but cell designers must carefully consider the interaction of the BMS and the cell controller. All alarm states triggered by one system must have the correct (safe) response within the other. It is particularly important to avoid test shutdowns being triggered by trivial or unrelated fault conditions detected by the BMS. It is equally important that the building services local to a cell are controlled appropriately to any fault in the cell; for example, a high hydrocarbon alarm has to switch the affected cell ventilation to maximum purge, rather than shutting ventilation down.

## Test cell start-up and shutdown procedures

Failure of any test cell service, or failure to bring the services on stream in the correct order, can have serious consequences, and it is essential to devise and operate a logical system for start-up and shutdown. In all but highly automated production tests, a visual inspection of the UUT must be made before each test run. Long-established test procedures can create complacency and lead to accidents. When investigating cell accidents, the authors have noted the frequency with which the phrase "I assumed that . . ." arises, as in the real-life example "I assumed that the night shift had reconnected the drive shaft."

*Visual inspection beats assumption every time.*

## Checks before start-up

On the basis that routine calibration procedures have been followed, and that instrument calibrations are correct and valid, then the following type of pre-run checks should be made, or status confirmed:

- No "work in progress" or "lockout" labels are attached to any system switches.
- UUT/dynamometer alignment within set limits and each shaft bolt has been tightened to the correct torque and the fact recorded and signed for in the control desk diary or appropriate record.
- Shaft guard in place and centered so that no contact with shaft is possible (if appropriate, it is a good practice to rock the engine on its mounts by hand to see that the rigging system, including exhaust tubing, is secure and flexes correctly).
- All loose tools, bolts, etc., removed from the test bed.
- UUT support system tightened down.
- Fuel system connected and leak proof.
- Engine oil at the correct level.
- The valves for all fluid services are in the correct position for running.
- Fire suppression system is indicating a "healthy" status.
- Ventilation system available and switched on to purge cell of flammable vapor before start-up of an engine.
- Check that any cell access doors, remote from the control desk, have warnings set to deter casual entry during test.

The engine or motor start process is then initiated.

## Checks immediately after start-up

- Oil pressure is above trip setting. In the case of manually operated cells the alarm override is released when pressure is achieved.

- If stable idle is possible and access around the engine is safe, a quick in-cell inspection should be made. In particular, look for engine fuel or oil leaks, cables or pipes chafing or being blown against the exhaust system and listen for abnormal noises.
- In the case of a run scheduled for prolonged unmanned running, test the emergency shutdown system by operating any one of the stations.
- Restart and run the test.
- Note: If safe entry to a newly started cell can be managed by a trained operator, in addition to a quick visual inspection, their hearing and sense of smell can often pick up insipient faults before the control system.

### Checks immediately after shutdown

- Allow a cooling period with services and ventilation left on.
- Shut off fuel system as dictated by site regulations (draining down of day tanks, etc.).
- Carry out data saving, transmission, and backup procedures.

## Part 2: Open- and closed-loop control of engine and dynamometer

True open-loop control of a modern ICE is an arguable concept because they run under the control of their electronic control unit (ECU), which, for most of the time, is a closed-loop controller.

There are many motorsport test cells that run under manual control, which is not necessarily open-loop, using hardwired or even mechanical linkage controls for throttle position and dynamometer load.

Open-loop control of the rotational speed and load applied to transmission test rigs is possible in manual control mode, during which primary alarms should still be active. If the automatic control of speed or load is poor, the best strategy in investigating instabilities is to eliminate as much of the control system as possible by switching to open-loop control where possible. In this way the source of the disturbing influence or the interaction of two systems may be identified.

The control software programed in a modern engine ECU and electric dynamometers makes wide use of ROM-based lookup tables rather than (sometimes additional to) complex control algorithms. However, the controllers of many single- and two-quadrant dynamometers and auxiliary devices, such as coolant temperature controllers, use proportional–integral–derivative (PID) feedback loop-based control (Fig. 13.1).

There are many web-based tools and books on the subject of PID tuning and no space here to go into detail; it is sufficient to record that the roles of each of the terms expressed mathematically in Fig. 13.1 are as follows:

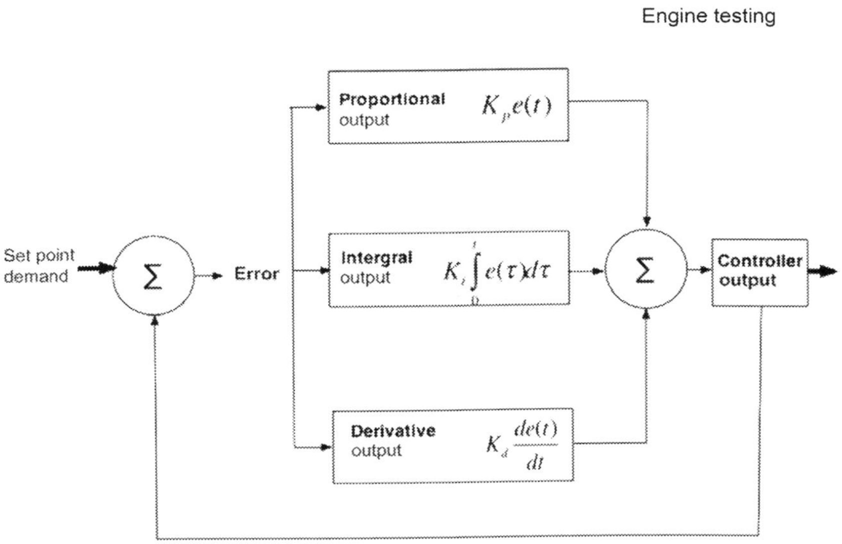


**FIGURE 13.1** PID feedback control diagram. Tuning parameters are $K_d$, derivative gain; $K_i$, integral gain; $K_p$, proportional gain; SP − CV, set point control variable minus the measured value; $t$, time; $S$, error; and $s$, dummy integration variable.

- Proportional gain effect—the controller output is proportional to the error or a change in measurement.
- Integral gain effect—the controller output is proportional to the amount of time the error is present. Integral action tends to eliminate offset from the set point.
- Derivative gain effect—the controller output is proportional to the rate of change of the measurement or error.

Up to the early 1980s, analog PID controllers were based on electronic hardware containing one miniaturized, variable resistor "pot" per gain term. The tuning of these devices was somewhat of a "black art" because there was no visual or numerical indication of what value of gain was being used, other than the observed effects on the rotating machinery when the pots were adjusted (see Table 13.2). Nowadays, PI and PID controllers are almost universally based on digital computers and PLCs that present to the operator a number, albeit a rather arbitrary number, relating to the channel gain, and thus the size of changes in value being applied may be judged.

## Problems in achieving control

Some engines are inherently difficult to control because of the shape of their torque−speed characteristics. Highly turbocharged engines may have abrupt changes in the slope of the power−speed curve, others may have sluggish

**TABLE 13.2 The effects, on the control of steady-state and step demand performance, of increasing any one proportional–integral–derivative parameter independently.**

| Parameter | Stability | Overshoot | Rise time | Settling time | Steady-state error |
|---|---|---|---|---|---|
| $K_p$ | Degrades | Increases | Decreases | Little effect | Decreases |
| $K_i$ | Degrades | Increases | Decreases | Increases | Large decrease |
| $K_d$ | Improves | Small decrease | Small decrease | Small decrease | No effect |

response to step demand changes due, for example, to the time taken for speed changes in the turbocharger; such characteristics can make the optimization of a PID-based control system that has to operate over the full performance envelope, very difficult.

Inexperienced attempts to adjust a full three-term PID controller can lead to problems; it is well worth recording the settings before starting to make changes so that one can at least return to the starting point. It is good practice to make sure that elementary sources of trouble, such as mechanical backlash and "stick-slip" in control linkages, are eliminated before delving into control variables in software.

Sinusoidal speed signals caused by poor encoder installation or intermittent speed signals caused by the wrong pickup clearance at the engine flywheel are frequently the cause of poor engine/dynamometer control. A speed encoder must be accurately aligned and fitted with a coupling that does not impose lateral loads on the instrument's shaft.

While the development of complex, adaptive, and model-based controllers will continue, the fundamental requirements such as good mechanical and noise-free electrical connections still have to be observed.

## The test sequence: modes of prime mover/dynamometer control

A test program for an engine coupled to a dynamometer is, primarily, a sequence of desired values of engine torque and speed. This sequence is achieved by manipulating only two controls: the engine's power output, usually by way of its throttle system, and the dynamometer torque absorption setting.

For any given setting of the throttle the engine has its own inherent torque–speed characteristic, and similarly, each dynamometer has its own torque–speed curve for a given control mode setting. The interaction of these two characteristics determines the inherent stability, or otherwise, of the engine–dynamometer combination.

## Note on mode nomenclature

In this book the naming convention of modes of control used in the United Kingdom and widely throughout the English-speaking world has been adopted. The alternative nomenclature and word order used in many major German speaking and mainland European organizations are shown in Table 13.3.

The engine or throttle control may be manipulated in three different ways:

- to maintain a constant throttle opening (position mode),
- to maintain a constant speed (speed mode), and
- to maintain a constant torque (torque mode).

The dynamometer control may be manipulated:

- to maintain a constant control setting (position mode),
- to maintain a constant speed (speed mode),
- to maintain a constant torque (torque mode), and
- to reproduce a particular torque–speed characteristic (power law mode).

Various combined modes are possible, and in this book, they are described in terms of engine mode/dynamometer mode.

## Position/position mode

This describes the classic manually operated engine test. With the engine running the throttle is set to any fixed position from that giving idle to wide-open throttle (WOT), the dynamometer control similarly set in such a position to absorb the required torque, and the system settles down, hopefully in a stable state. There is no feedback: this is an open-loop system. Fig. 13.2A shows a typical combination, in which the engine has a fairly flat torque–speed characteristic at fixed throttle opening, while the dynamometer torque rises with speed. The two characteristics meet at an angle approaching 90 degrees, and operation is quite stable. Certain

**TABLE 13.3** Control mode nomenclatures.

| English (first term will refer to Engine) | German (first term may refer to the dynamometer) |
|---|---|
| Speed | "n" |
| Position | "Alpha" |
| Torque | Torque |

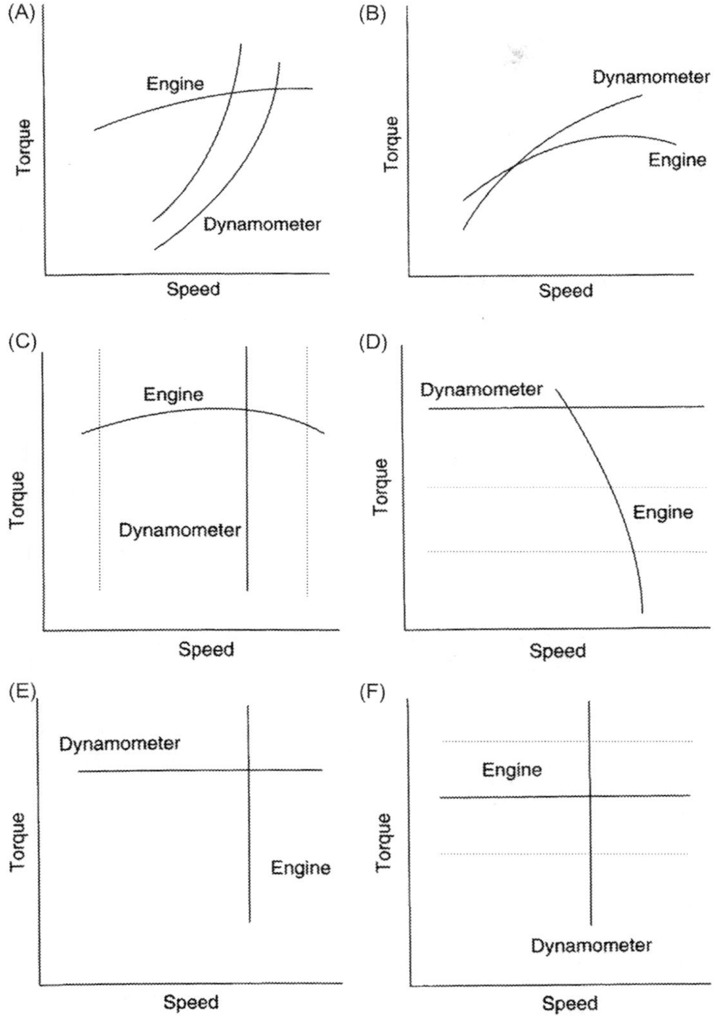


**FIGURE 13.2** (A) Position/position mode with stable control of hydraulic or eddy-current dynamometer. (B) Position/position mode with unstable control of hydraulic dynamometer. (C) Position/speed mode point. (D) Position/torque mode testing of a speed governed engine (speed droop characteristic exaggerated). (E) Speed/torque mode. (F) Torque/speed mode. Speed is held constant by the dynamometer and engine throttle controls torque at that speed. When the two characteristics meet at an acute angle, as shown in (B), good control will not be achieved. Position/position mode is useful in system faultfinding in that the inherent stability of the engine or dynamometer, independent of control system influence, may be checked.

designs of variable fill water brake dynamometer, with simple outlet valve control, may become unstable at light loads; this can lead to hunting, or to the engine running away.

## Position and power law mode

This is a control mode that can be visualized as the operator driving a boat; as the throttle is opened, the dynamometer torque rises with a curve speed characteristic of the form:

$$\text{Torque} = \text{constant} \times \text{speed}^n$$

When $n = 2$, this approximates to the torque characteristic of a marine propeller, and the mode is thus useful when testing marine engines. It is also a safe mode, helping to prevent the engine from running away if the throttle is opened manually by an operator in the cell.

## Position and speed mode

In this mode the throttle position is set manually, but the dynamometer adjusts the torque absorbed to maintain the engine speed constant whatever the throttle position and power output, as shown in Fig. 13.2C. As the speed setting line of the dynamometer moves back and forth, the intersection with the engine torque curve gives the operating point. This is a very stable mode, generally used for plotting engine power curves at full and part throttle openings.

## Position and torque mode (governed engines)

Governed engines have a built-in torque–speed characteristic, usually slightly "drooping" (speed falling up to 3% as torque increases). They are therefore not suited for coupling to a dynamometer in speed mode; they can, however, be run with a dynamometer in torque mode. In this mode the automatic controller on the dynamometer adjusts the torque absorbed by the machine to a desired value (Fig. 13.2D). Control is quite stable. Care must be taken not to set the dynamometer controller to a torque that may stall the engine.

## Speed and torque mode

This is a useful mode for running in a new engine, when it is essential not to apply too much load. As the internal friction of the engine decreases, it tends to develop more power and since the torque is held constant by the dynamometer, the tendency is for the speed to increase; this is sensed by the engine speed controller that acts to close the throttle, shown in Fig. 13.2E.

## Torque and speed mode

This is an approximate simulation of the performance of a vehicle engine when climbing a hill. The dynamometer holds the speed constant while the engine controller progressively opens the throttle to increase torque at the set speed (Fig. 13.2F).

## Precautions in modes of the four-quadrant dynamometer

Where the dynamometer is also capable of generating torque, it will be necessary to take precautions in applying some of the abovementioned modes of control. Such precautions are built into all reputable commercial systems but need to be understood by those assembling their own test plant or writing control software. When a four-quadrant dynamometer is in speed mode, it is capable of maintaining the set speed even if the UUT has failed in some way and has ceased to deliver power. Therefore when in speed mode, such dynamometers testing engines should be working in "absorb only" rather than "motor and absorb" control mode.

### *Throttle actuation*

Except in cases when an engine throttle lever is directly cable controlled, with a marine type of "tee handle" system at the control desk, prime-movers require the test cell to be equipped with some form of remote and precise throttle, or throttle-pedal unit, actuation, for both manual and automatic control. The actuators may be based on rotary servomotors, printed circuit motors, or linear actuators and fitted with some form of positional feedback to allow closed-loop control.

Most throttle actuators need to be fitted with some form of stroke adjustment to allow for 0% (shut) to 100% (WOT) positions to be determined for different engines and linkage configurations. Setting up this precise relationship between the controller's datum of "closed" and WOT with the physical position of the engine's fueling device is an important, safety-critical, rigging task. It must be arranged in such a way that UUT movement on its mountings does not induce throttle movement.

Some units will have a built-in force transducer that can be used as a force-limiting device or a friction clutch can provide the same protection to the physical system. Typical specification ranges of proprietary devices are as follows:

| | |
|---|---|
| Constant force applied at Bowden cable connection | 100–400 N |
| Maximum stroke length | 100–160 mm |
| Cable connector speed of travel | 0.5–1.5 m/s |
| Travel resolution and positional repeat accuracy | ± 0.05 mm |
| Maximum shifting force | 20 N (push and pull) |

It must be remembered that the forces applied by the actuator to the engine lever will depend on the *mechanical leverage* within the engine rigging.

The specification of positional repeat accuracy (above) requires that all cables and linkages used must be free from backlash or good closed-loop control will not be achieved. An important safety feature of a throttle actuator is that it should "fail safe," that is, automatically move the engine control lever to the stop position if there is a power failure.

While entirely possible, it is not common nowadays to drive the electronic throttle controller (ETC) directly as a discrete "analog out" voltage because the ECU requires a number of feedback signals from the ETC in order to function correctly, which is why the complete throttle pedal assembly is often used.

## Part 3: Test control software, choosing supplier, and test sequence creation

### Choosing a test system supplier

The world's engine test industry, up until very recently, was served by a relatively small number of companies, the flagship product of which is a control system and comprehensive suite of software designed specifically for automating engine, power train, transmission, and vehicle tests, acquiring test data and displaying results. The recent proliferation of manufacturers producing electric and hybrid vehicle subsystems has meant that there is an increasing number of non proprietary systems being used which, unless based on a depth of appropriate experience, may bring with them some risk to operators and data.

Several of these companies also make the major modules of instrumentation required for testing all types of prime movers, such as dynamometers, engine fluid temperature controllers, fuel meters, and exhaust emission equipment. However, most facilities are made up of modules of plant coming from several of these specialist suppliers, along with locally made building services, all of which have to be integrated into one optimally functioning whole.

The software suites sold by major test equipment suppliers are the outcome of many man-years of work; it is not cost-effective or very sensible for even a highly specialized user, however, experienced, to undertake the task of producing their own test control hardware and software from scratch.

Prospective buyers of automated test systems should, if possible, examine the competitive systems at the sites of other users doing similar work. As with any complex system, the following general questions should be considered, in no order of importance:

- Does the system support ASAM (Association for Standardization of Automation and Measuring Systems) standard interfaces for communication with third-party devices? For engine calibration and many research tasks, the communication interface with an engine ECU is vital—and for data exchange a modern cell control system must meet the requirements of the ASAM ODS (Open Data Services) standard that deals with data models, interfaces, physical storage, and data exchange.
- How intuitive and supportive to the workflow does the software appear to be? Or does the software provider understand the user's world in which the software has to function?

- Purchase cost is clearly important but also consider the cost of ownership, including technical support, upgrades, and additional licenses.
- Can the postprocessing and reporting required be carried out directly by the new software or can data be transferred to other packages in use on-site?
- How good is the training and service support in the user's geographical location?
- In the case of international companies, does the software support screen display of the languages required?
- Does the proposed system run on hardware and software platforms that are supported by the company's IT policy? This is a subject that can generate considerable internal heat in organizations. The author has witnessed in the past cases where compliance with a corporate IT diktat, based on policies for use of computers in business systems rather than industrial control, has compromised the choice and increased the price of test cell computer systems.
- Has the system been used successfully in the user's industrial sector?
- Can existing dynamometers, instrumentation, or new third-party instrumentation be integrated with the new system without performance degradation?
- Is data security robust and can the postprocessing, display, and archiving required be accommodated?
- Can the data generated by the new system be stored in the existing data format? This "heritage support" may be both important and difficult to achieve.

## Main requirements and description of the automation system software environment

The complexity of the task of efficiently using any type automotive test bed to develop and calibrate power trains and their component parts has increased exponentially of the last decade. A modern hybrid vehicle will contain, in addition its ECU, many other control units and "intelligent devices" all of which have to be included in the vehicle calibration process. The number of individual parameters, their functions, and their manufacturer's naming conventions adds to the complexity of designing any sort performance-optimizing test sequence.

This complexity puts great pressure, not only on the development of software tools but also on the training staff capable of handling them and understanding the data they produce. The stage by stage development of test bed control and data acquisition software tools and the differentiation of test engineer's roles have tended to produce suites of software packages that can be both integrated vertically and be used sequentially where the results of one can be passed seamlessly to the next. A vital feature of such software tools is that they will embed and continue to develop industrial best practice for the process they control and where mutually advantageous will be based on low-level, industrial standard tools and communication protocols.

A modern test facility will have to use software tools that will perform the following functions:

1. A test bed controller (TBC) controls not only the function of the UUT and the devices connected to it but also the test environment and services. Those services may be directly controlled by one or more intelligent controls working according to the TBC demands.
2. An "add-on" tool stores and allows the integration of real-time application models for the simulation of a device physically missing from the configuration of the UUT, typically such tools deploy MATLAB and Simulink for modeling of engine and power train components and systems.
3. A range of HiL (Hardware-in-the-Loop) add-on tools allow for the simulation of absent modules, a driver, vehicle road load equation (RLE), and road characteristics and full vehicle simulation capability.
4. A tool that disciplines the performance of, and the data from, a test sequence so that it complies with the requirement of calibration certification.
5. A host computer tool provides for central data correlation from a wide range of automation, simulation, and measurement systems. This is the package that allows all of the test data collected to be presented in formats that allow for technical management to make informed decisions.
6. Any facility carrying out combustion analysis (see Chapter 16: The Combustion Process and Combustion Analysis) will need to have the software tool to post-process and report the results of test sequences.
7. A vehicle-mounted data acquisition and reporting device can take real-life vehicle performance data in a form that can be seamlessly integrated into the TBCs sequencing editor for rerunning in the test cell.
8. The software tool allows Calibration engineers to interface with an ECU in order to produce, edit, and run calibration test sequences through the TBC. This tool is also required to gather data from the vehicle bus networks during runtime.

## Main requirements of the automation system hardware environment

There are many suppliers of systems for test bed control and data acquisition. There are also many suppliers of more general data acquisition and control systems as these are used in other industrial applications where safe and efficient operation of plant systems is required. These are often known as SCADA (supervisory control and data acquisition) systems. Therefore it is important to understand the specific requirements of a power train test bed

(PTTB) system in order to be able to specify, or even understand fully, the requirements of such an application. An overview of the main function blocks and areas is given below, but this list should not be considered exhaustive:

- UUT/dynamometer controller—this is a subsystem that controls the status of the UUT and the dynamometer—implementing the demands received from the execution of the test run. It is also ensuring that the test cell, UUT, and dynamometer all operate within safe, globally defined limits. The system also responds to critical safety situations to ensure a safe response to protect operators, test subjects, and environments. It is a real-time, deterministic system with hardwired safety loops and basic monitoring. In addition, the UUT/dynamometer PID controllers run within this system, in real time, to ensure good response and stabilization to external demand values. This device interfaces to the test automation system or can be fully integrated into it, running on a PC-based real-time platform. Interfaces to the dynamometer and UUT are also integrated into this subsystem.
- Data acquisition—a hardware and software system that allows the connection of various types of sensors or transducers, located at the UUT, in order to measure and save data to characterize the states of the UUT at any stage of the test and also to provide monitoring inputs—to protect the UUT for extreme operating conditions. Most test systems use distributed acquisition hardware models that have the capability to connect to various sensing technologies, converting the raw signal to a physically correct representation of the measured value—correctly calibrated and linearized, with a named channel and units. This value is digitized and transferred to the automation system for acquisition and storage. The approach of distributed modules allows full scalability of the system and reduces cable length between the sensor and the point of signal digitization, as these modules can be located in the test cell close to the engine, connected with a daisy chain topology.
- Device interfaces—in addition to sensors and transducers, the test system needs to acquire data from "special" measurement devices. The application of these devices varies (and is discussed later), but in general, they possess local intelligence—meaning that they are measurement subsystem that connects to the host for control and data transfer. The UUT controllers (ECUs and also now MCUs - Motor Control units) can also be considered as special measurement devices as data from the ECU or the UUT communication bus systems will also need to be incorporated into the test system data model.
- Data storage/access—the test system will contain one or a number of PCs as a test system hardware platform. Data storage and data model are key topics. The hardware must support a mechanism to store the data—often

a database—that must be made secure and safe; that is copied and backed up. Also required is the capability to transfer, store, and archive data in a transparent and accessible format, for any stakeholder who needs it. Test systems only produce data; the key to efficiency is facilitating easy access to that data for anyone with permission, at anytime, with the greatest security and minimal transfer delay. Test system hardware must also be robust—integration of backup power supplies (UPS) with power supply monitoring is essential in a modern test environment.

- User interface—this is an important topic as it is the main touch point between the user and the system, it is the human–machine interface for the whole test environment. Therefore the layout of this environment is critical to system efficiency and ease-of-use. Multiple displays (screens) are required for the software and control system tools, and many keyboards and mouses need to be organized. A hardware-based interface for manual operation needs to be easy to use and access—often physical interfaces are preferred by test bed users, as opposed to virtual control elements on a computer screen (i.e., control wheels, buttons). Space for administrative tasks and documentation is also required. The importance of this factor multiplies in the largest test environments, with multiple installations, where operators and engineers move around the test facility according to their projects and tasks.

## Power train electrification and its effect on the test system environment

The trend toward power train electrification has a significant impact on the design and application of the industry standard test environment and the design target that this environment is created to achieve. The major shift in application is driven by emissions regulations (as is so often the case)—but the paradigm shift is due to the need to provide a test environment that can replicate some real driving conditions—as opposed to a synthetic, legislated test cycle. There are areas listed below that can be considered as the main areas of change:

- Dynamometers—much more focus is now directed at complete power train characterization, not just the prime mover. As discussed in Chapter 9, Dynamometers: The Measurement and Control of Torque, Speed, and Power, and Chapter 10, Chassis Dynamometers, Rolling Roads, and Hub Dynamometers, the current generation of power train test cells requires dynamometers positioned at wheel—with higher torque and lower speed capability.
- Integration of battery emulation—a complex power train now involves an electric machine and inverter that requires a DC supply. For power train out-of-vehicle testing, a battery emulator provides a much higher degree

of freedom and security for testing the power train at boundary operating conditions (see Chapter 12: Rigging and Running of Electrical Drive Systems).

- Integration of runtime co-simulation—as mentioned earlier, simulation and testing are now coupled. Most test automation systems have some level of simulation capability, but in order to replicate real driving, complex driver, vehicle and environment models must be executed in real time to create the dynamic test sequence required. This normally requires an X-in-the-loop hardware system to run concurrently with the TBC (Test Bed Controller).
- Test sequences—steady-state characterization is no longer relevant for a complex power train; most testing is undertaken as transient profiles with continuous data recording required from all test devices, synchronously.
- Data processing—large data sets are created from transient tests where simple plotting of data is no longer sufficient, tools are required to reduce and make sense of large test data sets. The latest developments in data science are required in order to support efficient data processing, and this creates a significant shift in the mentality of test environment users as automated data processing brings data analysis closer to the test environment. While the developments in data science tend to take post-test analysis tasks to specialists, who are often far removed from the physical testing, tools have to be provided for the power train engineer to use and "mine" such ever-increasing data masses. This topic is covered in depth within Chapter 14, Data Handling and Modeling.
- Data acquisition—state-of-the-art instruments and special measurement devices, capable of continuous and synchronous recording, are required and available for ICE-based measurement requirements. However, the integration of electrical propulsion elements requires new measurement systems to characterize integrated electric motors and overall system performance during test runtime. Recording and visualization of energy flow and measurement of efficiency for the complete power train are of vital importance.
- Safety is discussed in some detail in Chapter 2, Quality and Health and Safety Management, and Chapter 3, Test Facility Design and Construction; however, the integration of vehicle systems with voltages above extra low level (and thus classified as greater than low risk according to IEC 60038) in a propulsion system is a new consideration.
- Knowledge—power trains can be considered complicated mechatronic systems, fundamentally mechanical systems with electronic control. Current power trains now incorporate a new domain of electrical power systems, and this is unfamiliar technology to many of those working in power train development environments. Thus changes in development processes are required, along with reskilling or recruitment of staff with relevant knowledge. However, the critical path is still to integrate teams,

personnel, and systems for holistic development, as opposed to developing the systems in isolation and then uncovering integration issues late in the development process.

## Test automation tools integration

It is often the case, when procuring and equipping a facility, that avoiding "near" single sourcing can provide benefits in terms of choice and freedom to select the best equipment based on experience or industry knowledge. The downside of this approach is the "integration" or "interface" topic. This normally raises an issue during commissioning of a system or when a system is in use and a problem occurs. In theory, connectivity of systems is provided via known interfaces, using documented protocols. However, in practice, there are always variations and deviations from the standards by each supplier. The consequence of which is failure in part or full of an interface; then the problem is finding a supplier or person with the right skill and knowledge and the desire to take ownership and solve a problem.

A specific example would be an interface between an ECU calibration tool and a test system automation tool. Both domains require different suppliers as there is no crossover in solutions or knowledge. The application system supplier is an embedded software tool expert, and the test system supplier is an automation/test expert, so the domains touch but there is often limited or no overlap. The challenge then is being to get the stakeholders talking to resolve an issue on your behalf, which is easily said not easily done.

A similar problem can exist with respect to test data file and database formats, file conversion, loss of fidelity during conversion; these are topics that can absorb significant time and resource discussing and resolving with suppliers. These problems can justify the single source trajectory. In practice, the risk of a mixed-supplier test system can be reduced by effective project management and the identification of a contractually responsible systems integrator in the installation and commissioning phase. Appointing a lead project manager, experienced in test systems and on-site installation, can considerably reduce the suffering all round where complex systems are to be integrated.

## Editing and control of complex engine and motor test sequences

An engine, motor, or transmission test program can, in its simplest form, consist of a succession of stages of speed and torque settings, together with the demand values sent to other devices in the test cell system, the whole format looking rather like a spreadsheet. Modern, highly dynamic test sequences still have to contain the same fundamental speed, torque, and analog and digital signals, both in and out, but they more resemble, or can actually be constructed from, a Simulink or other proprietary software's block diagram, or model-based development environment.

Software suites designed to semi-automate ICE calibration work are capable of calculating and generating the test set points within a test sequence designed to find boundary conditions, such as engine knock.

Most power train test profiles or sequences, with their precisely ramped changes in values, are beyond the capacity of a human operator to manually run accurately and repeatably, a fact borne out by the variability seen in the results from human-driven, vehicle emission, and drive cycles.

Although the terminology used by test equipment suppliers may vary, the principles involved in building up test profiles will be based on the same essential component instructions.

Test sequence "editors" range from those that require the operator to have knowledge of software code, through those that present the operator with an interactive "form fill" screen, to those where blocks of, proven and ever more complex, subsequences are assembled in the required order of running.

The nature and content of the questions shown on the screen will depend on the underlying logic that determines how the answers are interpreted. It is important that the user should understand the test cell logic; otherwise, there is a risk of calling up combinations of control that the particular system cannot run, though it may not be able to indicate exactly where the error lies. Using a simple example, if the control mode is set for throttle as "position" and dynamometer as "speed," the editor will require instructions in terms of these parameters; for example, it will not be able to accept a throttle control instruction in terms of speed because it requires a percentage figure.

## Basic test sequence elements

In those test sequence, editors, which are ideal for producing "nontransient" test sequences, are based on a simple "form fill" layout; the following elements will be included:

- Mode of control (covered earlier in the chapter). As a historical note, most analog-based control systems required engine speed and load to be brought down to some minimum value between any change of test mode, while modern digital controllers are able to make a "bumpless transfer," a term still used.
- Engine speed and torque or throttle position. The way in which these parameters are set will depend on the control mode. A good sequence editor will present only viable options.
- Ramp rate. This is the acceleration rate required or the time specified for transition from one state to the next.
- Duration or "end condition" of stage. The duration of the stage may be defined in several ways:

  - at a fixed time after the beginning of the stage;
  - on a chosen parameter reaching a specified value, for example, coolant temperature reaches 80°C as in a "warm-up" stage; and
  - on reaching a specified logic condition, for example, on completion of a fuel consumption measurement.
- Choice of next stage. At the completion of each stage (stage x), the editor will "choose" the next stage. Typical instructions governing the choice are to:
  - run stage x + 1,
  - rerun stage x a total of y times (possible combinations of "looped" stages may be quite complex and include "nested" loops), and
  - choose the next stage on the basis of a particular analog or digital state being registered (conditional stage).
- Events to take place during a stage. Examples are the triggering of ancillary events such as fuel consumption measurement or smoke density measurement. These may be programed in the same way as duration or end condition mentioned earlier.
- Nominated alarm table. This is a set of alarm channels that are activated during the stage in which they are "called." The software and wired logic must prevent stage alarms from modifying and overriding primary "global" safety alarms.

Such sequence generation remains perfectly adequate for a wide range of prime mover testing on test beds using two-quadrant dynamometers such as those engaged in fuel and lubrication or endurance work.

## The "road-to-rig" testing concept

The ultimate form of power train testing may be envisaged as a PTTB that is able to continuously expose the UUT and its interactive component parts to precisely the same inputs and to cause them to produce precisely the same outputs, as they would while driving in the "real world." It is to be able to reliably reproduce, during a test run of the same time duration, the same fuel consumption and emissions as was measured on the road in a vehicle of precisely the same configuration, driven by a driver having consistent characteristics.

Having achieved and recorded such a proven, high-fidelity, test sequence, it would then be valuable to be able to rerun it, with one or more of the vehicle parameters or driver's characteristics altered so as the determine cause and effect of such changes and as a means of performance improvement.

Driven by RDE (real drive emissions) test legislation and aided by the research it has required, we now have extensively instrumented vehicles so that, aided by a collection of sometimes disparate, software suites, we are able to supplement and convert that data into test bed control sequences. All this means that the automotive test industry is working ever nearer to achieving true in-cell simulation of the road, that is, "road to rig" (R2R).

On the path to the ultimate target, there will be several, easier to achieve and useful stages. The present state of R2R test development has already produced and is using the concept of converting road data in two-dimensional (2D) and three-dimensional (3D) PTTB test sequences. The 2D models use only road gradient and vehicle speed to recalculate, during the sequence, the classic RLE (see Chapter 10: Chassis Dynamometers, Rolling Roads, and Hub Dynamometers). The 3D models use data relating to vehicle dynamics such as lateral acceleration from road curvature to modify the RLE variables such as effective mass and rolling resistance and altitude to modify the ECU's A/F ratio map.

No packaged software toolchain currently exists that will seamlessly take RDE test vehicle–derived data and convert it into a 3D test bed control sequence. Therefore those final control sequences have to be assembled from the road data, by human intervention, in stages, using a patchwork of existing software suites, many from the test industry but some of which, such as those using GPS-derived road information, that have no test industry heritage. The R2R task is not yet a one "button pressing" exercise but may become so in the lifetime of this edition.

A test bed can only cause and influence the UUT's performance through:

- the acceleration and deacceleration of its dynamometers,
- the control of the UUT services and cell climate, and
- the commands of its engine control system.

It will be clear that every system within the test facility system, whether for prime mover, full power train or full vehicle (chassis dynamometer), has to be capable of meeting the highly dynamic simulation requirements imposed by R2R testing. Its own friction and windage losses must be compensated for and ambient cell air temperature and pressure held within close limits.

As the physical content of any power train UUT decreases and is replaced by HiL using virtual models, the veracity of those models is absolutely crucial. It is also crucial that those same reference models used to calibrate the control units of the UUT and also to calculate and modify the instantaneous RLE used by the test bed are as accurate as possible. The proven maturity of the calibration of all power train control devices is an essential part of the path toward repeatability of results and true R2R realization.

The neutral gear coast-down model, used and "gamed" if not abused for many years in emission homologation testing, is one such model, now vital in defining the friction losses of the vehicle through its internal friction added to wind/air resistance, tire losses, etc.

It has to be recognized that there are realistic limits to the simulation accuracy of some variables. The most variable performance will always be that of the driver with variability increasing directly with the increasing detail of his or her movements, as exemplified by throttle pedal dither or "busyness" and manual gearchange actions.

Of course, current industrial PTTB practice can and does use characterized and standardized driver models, from "aggressive youth" to "aged weekly shopper" in order to run repeatable, reference, (non-R2R) test drive simulations.

The varied quality of the transmission model and the ECU reaction to gear change transients (shifting) compounds the driver-induced variability of the test results and their reproducibility. A continuous feedback and model verification loop is vital.

As the development process of power trains experiences the expected decline in numbers of UUT that include an ICE, the task of R2R testing will simplify as the drive-by-wire electric-powered light vehicles reduce the number of prime mover variables and bring effective discipline those of the driver inputs.

## Part 4: Data acquisition and the transducer chain

The electronically encoded information, defining the state of any measured parameter, travels through a cable from the transducer and then through signal conditioning, usually involving amplification and/or analog-to-digital conversion (ADC), to a storage device and one or more displays. During this journey, it is subject to a number of possible sources of error and corruption, many of which are covered in more detail in Chapter 4, Electrical Design Requirements of Test Facilities, and Chapter 19, The Pursuit and Definition of Accuracy.

### Data channel names

*It is vital for test data to have a consistent and coherent naming convention attached to each data channel.* If data is to be shared between cells or sites or is rerun at a later date, an identical naming convention must be used. The process of attaching a name to a measurement channel, from transducer through terminal number, to display and finally to the calibration certificate, is done at the time of system calibration and is sometimes called the system's parameterization. It should be obvious to the reader that without such discipline, there is no realistic chance of anything but the crudest comparison of test results and all chances of true correlation are effectively lost.

## Calibration of instrumentation

All professional engine test software suites will contain calibration routines for all commonly used transducers and instrument types and, in cases of industrial standard equipment, the exact models. Often these routines follow a "form fill" format that requires the operator to type in confirmation of test signals in the correct order, permitting zero, span, intermediate, and ambient values to be inserted in the calibration calculations. Such software should lead the operator through the calibration routines and also store and print out the results in such a way as to meet the company's ISO 9001 quality control and certification procedures. The appropriate linearization and conversion procedures required to turn transducer signals into the correct engineering units will be built into the software.

For the most demanding calibration the complete chain from transducer to final data store and indication needs to be included, but for normal post-commissioning work, it is not practical, nor cost/time-effective. Common practice for temperature channels is to use certified transducers and a calibrated instrument called a resistive decade box, which is injected into the measuring system at the point into which the temperature transducer is plugged. Several measuring points throughout the transducer's range are simulated, checked, and adjusted against the display. Similarly, pressure channels are calibrated by attaching a portable pressure calibration device, made by companies such as Druck or Fluke, to the transducer housed in a transducer box.

## Transducer boxes and distributed I/O

To measure the various temperatures and pressures of the UUT (engine, motor, or vehicle), transducer probes have to be attached and their signal cables connected into the data acquisition system. In many cases the transducer probes will be intrusive to the UUT and fitted into test points sealed with compression fittings. Along with the single probe cables, there will be cables from other instruments such as optical encoders requiring connection within the test cell in order for the signals, in raw form, to be transmitted to the signal conditioning device. This network of short transducer cables needs to be marshaled and collected at a *transducer box* near the engine.

The best modern practice is to reduce the vulnerability of signal corruption by conditioning the signal as near as possible to the transducer and transmit them to the computer through a high-speed communication link; such implementations have greatly reduced the complexity of test cell installation.

Older installations will have a loom of discrete cables in a loom running from the transducer box through the cell wall to a connection strip in the control cabinet where signal conditioning takes place.

Transducer boxes will contain some or all of the following features:

- external, numbered or labeled, thermocouple or PRT (platinum resistance thermometer) sockets;
- pressure transducers of ranges suitable for the UUT, possibly, including two with special ranges: inlet manifold pressure and ambient barometric;
- external labeled sockets for dedicated channels or instruments; and
- regulated power supply sockets for engine-mounted devices such as an ECU.

Transducer boxes may have upward of 50 cables or pipes connected to them from all points of the engine under test; this needs good housekeeping to avoid entanglement and contact with hot exhaust components.

The transducer box can be positioned as follows:

1. On a swinging boom positioned above the engine, which gives short cable lengths and an easy path for cables between it and the control room. However, unless positioned with care, it will be vulnerable to local overheating by radiation or convection from the engine; in these cases, forced ventilation of the enclosed boom duct and transducer box should be provided.
2. On a pedestal alongside the engine, but this may give problems in running cable back to the control room.
3. On the test cell wall with transducer cables taken from the engine through a lightweight boom.

## Choice of instruments and transducers

It is not the intention to attempt a critical study of the vast range of instrumentation that may be of service in engine testing. The purpose is rather to draw the attention of the reader to the range of choices available and to set down some of the factors that should be taken into account in making a choice.

Table 13.4 lists the various measurements. Within each category the methods are listed in approximately ascending order of cost.

The subject of instrument accuracy is treated at more length in Chapter 19, The Pursuit and Definition of Accuracy, but notes that:

- Many of the simpler instruments, for example, spring balances, manometers, Bourdon gauges, liquid-in-glass thermometers, cannot be integrated with data logging systems but may still have a place, particularly during commissioning or faultfinding of complex distributed systems.
- Accuracy costs money. Most transducer manufacturers supply instruments to several levels of accuracy. In a well-integrated system a common level is required; over- or underspecification of individual parts compared with the target standard either wastes money or compromises the overall performance.

**TABLE 13.4 Common instrumentation and transducers for frequently required measurements.**

| Measurement | Principle application | Method or transducer |
|---|---|---|
| Time interval | Rotational speed | Tachometer<br>Single impulse trigger<br>Starter ring gear and pickup<br>Optical encoder |
| Force, quasistatic | Dynamometer torque | Dead weights and spring gauge<br>Hydraulic load cell<br>Strain gauge load cell |
| Force, cyclic | Stress and bearing load investigations | Attached strain gauge plus wireless transmitters<br>Stain gauge transducer<br>Piezoelectric transducer |
| Pressure | Fluid flow systems, liquid or gas | Liquid manometer<br>Bourdon gauge<br>Strain gauge pressure transducer |
| Pressure, cyclic | In-cylinder, inlet and exhaust events, fuel rail | Toughened strain gauge transducer<br>Capacitive transducer<br>Piezoelectric transducer |
| Position | Throttle and other actuators | Linear variable displacement transducer<br>Rotary optical transducer |
| Displacement, cyclic | Needle lift<br>Crank TDC | Eddy-current transducer<br>Inductive transducer<br>Capacitive transducer |
| Acceleration | NVH investigation<br>Shaft balancing | Strain gauge accelerometer<br>Piezoelectric accelerometer |
| Temperature | Fluids, gases, mechanical components, in-cylinder | Liquid in glass<br>thermocouples (various)<br>PRT<br>Thermistor<br>Electrical resistance<br>Optical pyrometer |

*NVH*, Noise, vibration, and harshness; *PRT*, platinum resistance thermometer; *TDC*, top dead center.

- It is cheaper to buy a stock item rather than to specify a special measurement range.
- Some transducers, particularly force and pressure, can quickly be destroyed by overload. Ensure adequate capacity or overload protection.
- Always read the maker's catalog with care, taking particular note of accuracies, overload capacity, and fatigue life.
- Some transducers and instruments cannot be calibrated at site but have to be returned to the manufacturer; support contracts covering such devices may be time and cost-efficient.

## Measurement of time intervals and speed

Time can be measured with great accuracy. However, for linked events, which occur very close together in time and are sensed in different ways, with different instruments, it can be very difficult to establish the exact order of events or to detect simultaneity. This is of prime importance in combustion analysis and engine calibration when highly transient events such as the pressure pulses in fuel rails need to be taken into account.

A single impulse trigger, such as a hole in the rim of the flywheel, is satisfactory for providing a position and timing datum for counting purposes but is not ideal for locating top dead center (TDC) (see Chapter 16: The Combustion Process and Combustion Analysis, for detailed coverage of TDC measurement). Multiple impulses picked up from a starter ring gear or flywheel drillings used by the engine's ECU may be acceptable.

For precise indications of instantaneous speed and speed crank angle down to tenths of a degree, an optical encoder is necessary. However, mounting arrangements of an encoder on the engine may be difficult in ICE testing as if, as is usually the case, the drive is taken from the non-flywheel end of the crankshaft, torsional effects in the crankshaft may lead to errors that may differ from cylinder to cylinder. When encoders are fitted to the free ends of both dynamometer and the UUT, particularly an engine, there will be a lack of observed synchronization that is due to torsional windup and release in the shaft system, so care must be exercised when deciding the source of signals for different purposes.

## Force, quasistatic

The strain gauge transducer has become the almost universal method of measuring forces and pressures. The technology is very familiar, and there are many sources of supply. There are advantages in having the associated amplifier integral with the transducer as the transmitted signals are less liable to corruption, but the operating temperature may be limited.

Hydraulic load cells are simple and robust devices still fairly widely used in portable dynamometers.

## Measurement of cyclic force

The strain gauge transducer has a limited fatigue life, and this renders it unsuitable for the measurement of forces that have a high degree of cyclic variation, though in general, there will be no difficulty in coping with moderate variations such as result from torsional oscillations in drive systems.

Piezoelectric transducers are immune from fatigue effects but suffer from the limitation that the piezoelectric crystal, which forms the sensing element, produces an electrical charge that is proportional to pressure change and thus requires integration by a charge amplifier. Such transducers are unsuitable for steady-state measurements but are the universal choice for in-cylinder and fuel injection pressure measurements. Piezoelectric transducers and signal conditioning instrumentation are in general more expensive than the corresponding strain gauge devices.

## Measurement of pressure

To avoid ambiguity, it is important to specify the mode of measurement when referring to a pressure value. The three modes are:

- absolute,
- relative, and
- differential.

Pressures measured on a scale that uses a complete vacuum (zero pressure, as in a theoretical vessel containing no molecules) as the reference point are said to be *absolute* pressures.

Atmospheric pressure at sea level varies but is approximately $10^5$ Pa (1000 mbar) and is an *absolute* pressure because it is expressed with respect to zero pressure. Most engine test instrumentation is designed to measure pressure values that are expressed with respect to atmospheric pressure; thus they indicate zero when their measurement is recording atmospheric pressure. In common speech, relative pressures are often referred to as *gauge* measurements. Because of the difference between an absolute pressure value and a gauge pressure value, the actual value of atmospheric pressure in power train test facilities needs to be recorded daily as a reference datum for all calculations.

Note that the barometric pressure measured in an engine test cell, and experienced by the engine, will often be some 50 Pa lower than the value taken elsewhere in the facility due to the effect of a pressure-balanced ventilation system. Engine inlet manifold pressure will vary between pressures above and (generally) below the "reference" barometric pressure; this is another form of relative pressure, sometimes called "negative gauge pressure."

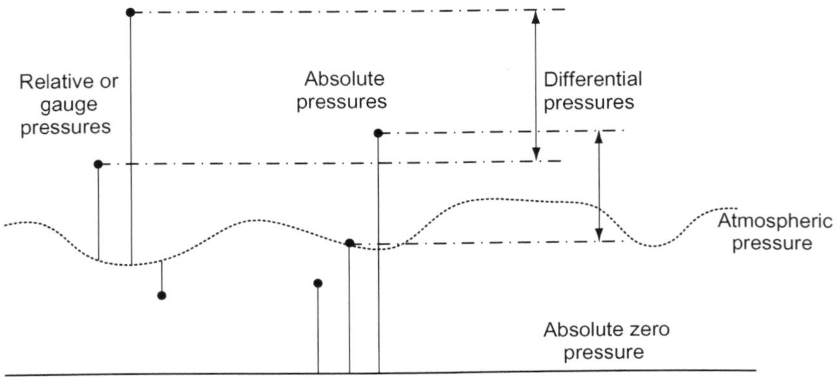


**FIGURE 13.3** Diagram showing relationship between absolute, relative, and differential pressures.

In applications such as the measurement of flow through an orifice, when it is necessary to measure the difference in pressure between two places, the reference pressure may not necessarily be either zero or atmospheric pressure but one of the measured values; such measurements are of *differential* pressures (see Fig. 13.3).

## *Electronic pressure transducers*

There are large ranges of strain gauge and piezoelectric pressure transducers made for use in engine testing, most having the following characteristics:

- Pressure transducers normally consist of a metal cylinder made of a stainless steel, with a treaded connection at the sensing end and a signal cable attached at the other.
- The transducer body will contain some form of device able to convert the pressure sensed at the device inlet into an electrical output. The signal may be analog millivolt, 4–20 mA, or a digital output connected to a serial communications interface.
- The output cable and connective circuitry will depend on the transducer type, which may require four-, three-, or two-wire connection, one of which will be a stabilized electrical supply.
- Each transducer will have an operating range that has to be carefully chosen for the individual pressure channel; they will often be destroyed by over pressurization and be ineffective at measuring pressures below that range.
- Each transducer will be capable of dealing with specific pressurized media; therefore use with fuels or special gases must be checked at the time of specification.

Capacitive transducers, in which the deflection of a diaphragm under pressure is sensed as change in capacity of an electrical condenser, may be the appropriate choice for low pressures, as an alternative to strain gauge or piezoelectric sensors.

Calibration of the complete pressure channel may be achieved by the use of certified, portable calibrators. Some special pressure transducers such as barometric sensors and differential pressure transducers may need annual off-site certification.

Pressure sensors used in the engine testing environment, because of their expense and vulnerability to damage, are normally mounted within the transducer box with flexible tubes connecting them by self-sealing couplings to their measuring point. Their vulnerability to incorrect pressure or fluid makes it imperative that the connectors are clearly marked at their connection points on the transducer box.

### Other methods of pressure measurement

The liquid manometer has a good deal to recommend it as a device for indicating low pressures. It is cheap, effectively self-calibrating and can give a good indication of the degree of unsteadiness arising from such factors as turbulence in a gas flow. It is recommended that manometers are mounted in the test cell with the indicating column fixed against the side of the control room window.

The traditional Bourdon gauge with analog indicator is the automatic choice for many fluid service sensing points such as compressed air receivers.

## Displacement

Inductive transducers are used for "noncontact" situations, in which displacement is measured as a function of the variable impedance between a sensor coil and a moving conductive target.

The hall effect transducer is another non-contact device in which the movement of a permanent magnet creates an induced voltage in a sensor made from gold foil. It has the advantage of very small size and is used particularly as an injector needle lift indicator.

Capacitive transducers are useful for measuring liquid levels, very small clearances, or changes due to wear.

## Acceleration/vibration

There are three basic types of transducer for measuring vibration:

- *Accelerometers*. These are based on piezoelectric sensors that generally have the widest frequency range and typically an output of 100 mV/g. They are extremely sensitive to the method of attachment, and it is vital to follow the manufacturer's instructions. Accelerometers are normally used for high-frequency vibration readings, and the output is integrated electronically to velocity (mm/s). Piezoelectric sensors should not be used for frequencies of less than about 3 Hz but tend to be more robust than strain gauge units.

- *Velocity transducers.* These usually work by the physical movement of an internal part within a magnetic field; they are therefore more prone to damage than accelerometers, and they have a lower frequency range.
- *Proximity transducers.* Sometimes called "eddy probes," these are used as noncontract vibration transducers. They tend to be rather delicate until correctly installed but can measure a very wide range of frequencies and are particularly good at frequencies below the range of the other types of transducer. The normal air gap between transducer face and the vibration surface is usually between 2.5 and 5 mm.

## Temperature measurement

### *Thermocouples*

Most of the temperatures measured during engine testing do not vary significantly at a high rate of change nor is the highest degree of accuracy cost-effective. The types of thermocouple commonly used in engine testing are listed next.

For the majority of temperature channels the most commonly used transducer is the type K thermocouple (see Table 13.5), having a stainless steel–grounded probe and fitted with its own length of special thermocouple cable and standard plug. To produce a calibrated temperature reading, all thermocouples need to be fitted within a system having a "cold junction" of known temperature with which their own output is compared. Thermocouples of any type may be bought with a variety of probe and cable specifications depending on the nature of their installation; they should be considered as consumable items and spares kept in stock.

### *Platinum resistance thermometers*

When accuracy and long-term consistency in temperature measurement higher than that of thermocouples, or for measuring temperatures below 0°C, is required, then PRTs are to be recommended. The PRT works on the

**TABLE 13.5 Commonly used thermocouple types.**

| Type | Internal materials | Temperature range (°C) |
|---|---|---|
| Type T | Cu–Ni/Cu | 0 to +350 |
| Type J | Fe–Ni/Cu | 0 to +800 |
| Type K | Ni/Cr–Ni/Al/Mn | −40 to +1200 |

principle of resistance through a fine platinum wire as a function of temperature. PRTs are used over the temperature range −200°C to 750°C.

Because of their greater durability or accuracy over time, it is common practice to use PRTs in preference to thermocouples for within-device control circuits. They require different signal conditioning from thermocouples and, for engine-rigged channels, are fitted with dedicated plugs and sockets within the transducer box.

### *Thermistors*

Thermistors (thermally sensitive resistors) have the characteristic that, dependent on the nature of their sensing "bead," they exhibit a large change in resistance, either increasing or decreasing, over a narrow range of temperature increase. Because they are small and can be incorporated within motor windings, they are particularly useful as safety devices. Thermistors are generally available in bead or surface mounting form and in the range from −50°C to 200°C.

### *Pyrometers*

Optical or radiation pyrometers, of which various types exist, are noncontact temperature measuring devices used for such purposes as flame temperature measurement and for specialized research purposes such as the measurement of piston-ring surface temperatures by sighting through a hole in the cylinder wall. They are effectively the only means of measuring very high temperatures.

Finally, suction pyrometers, which usually incorporate a thermocouple as temperature-sensing device, are the most accurate available means of measuring exhaust gas temperatures.

### *Other temperature-sensing devices*

Once a standard tool in engine testing, liquid-in-glass thermometers are now very rarely seen; however, they are cheap, simple, and easily portable. On the other hand, they are fragile, not easy to read, and have a relatively large heat capacity and a slow response rate. When set within a magnetic casing, they retain some practical use during commissioning of process plant distributed over a wide location. The interface between the body, solid, liquid, or gas, of which the temperature is to be measured and the thermometer bulb needs careful consideration.

Vapor pressure and liquid-in-steel thermometers present the same interface problems and are not as accurate as high-grade liquid-in-glass instruments but are more suitable for the test cell environment and are more easily read.

Mobile, handheld thermal imaging devices are more useful for fault-finding and safety work than for collection of test data. They are used for on-site commissioning work on large stationary engine systems.

## Note on analog-to-digital conversion accuracy

Whatever the level of computerization in the test facility, the signals from all transducers having analog outputs that have to be digitized, conditioned, and linearized, both for display and storing in the chosen format. The converted values are usually stored electronically in binary form, so the resolution is usually expressed in bits, and proprietary systems are usually available with ADC of 8 to 24 bit resolution. The signal range and resolution should be selected as appropriate to the accuracy and the significance of the signals to be processed.

It tends to give an illusion of accuracy (see Chapter 19: The Pursuit and Definition of Accuracy) to use a 16-bit or higher resolution of ADC for measurement channels having an inherent inaccuracy an order of magnitude greater. For example, a thermocouple measuring exhaust gas temperature may have a range of 0°C–1000°C and an accuracy of ± 3°C.

## Smart instrumentation and communication links

The test cell control and data acquisition computer will, along with taking in individual transducer signals, also have to switch on and off and acquire data from complex modules of instrumentation, a list of that includes:

- fuel consumption devices—gravimetric or mass flow,
- oil consumption,
- engine "blowby,"
- exhaust gas emission analysis equipment,
- ICE combustion pressure analysis instruments,
- e-machine and inverter electrical power analysis,
- interfaces to control modules for calibration and measurement, and
- interfaces to power train bus systems for measurement.

The integration of these devices within the control system will require special software drivers to allow for communication covering basic control, data acquisition, and calibration routines.

The interface between device and test cell computer may be by way of conventional analog I/O, 0–10 V, digital I/O, or by serial interface such as RS232.

Combustion analyzers and other complex instruments are increasingly equipped with high-speed outputs based on CAN bus (controller area network) (ISO 11898 compatible) with baud rates, which are dependent on cable length, of about 1 Mb/s; 15 m is usually the supplied length, with 25 m as the recommended maximum.

## Control for endurance testing and "unmanned" running

Due to the need to obtain a return on the very high investment costs, most full power train test rigs built in the second decade of the 21st century are capable and used for 24 hour running, of which at least 8 hours are unmanned or remotely monitored.

Unmanned running, in the context of power train testing, means that no operator is present within the control room outside set working hours, and during that time the test cell runs under automatic control. Strict procedures have to be in place to cover the event of a shutdown alarm occurring; these include safe automatic shutdown, remote alarm monitoring, competent technician support, and backup being "on call." A detailed historical log with an adequate rolling buffer needs to be part of the logging system so that unmanned shutdowns can be analyzed.

Techniques such as key life testing, aimed at accelerating the attrition experienced by power train components during real life, attempt to reduce the duration and therefore the cost of endurance tests. However, there are several, fairly standard, engine and transmission test sequences that are of 400 hours or more duration, which are very expensive to run and call for a level of performance and reliability in the test equipment that is an order of magnitude higher than that of the UUT.

Cells running endurance and unmanned tests should not be vulnerable to spurious shutdowns caused by faults within indirectly connected systems. For the same reason, shared systems should be minimized so far as is possible. For endurance work with low levels of human supervision, duplication of transducers and careful running of cables to prevent deterioration by wear or heat over long periods are good practices.

## Control for dynamic engine drive cycles

In the past, there was much debate on the exact definition of dynamic or transient testing; much depended on the type of UUT and the technology being used. What was considered a highly dynamic test sequence in one industrial sector was considered positively pedestrian in another; therefore the authors in the early editions of this book, and others in the industry, classified the titles being used in terms of four characteristic timescales, where the times quoted represented the time in which the control instruction is sent and the change in force at an AC dynamometer air gap is sensed.

- $< 2$ ms = high dynamic, true zero inertia/torque simulation, power train dynamic simulation up to 40 Hz.
- $< 10$ ms = dynamic, driver, road load and gradient simulation, power train dynamic simulation up to 8 Hz.
- $< 50$ ms = transient, test defined in speed/torque pairs of less than 1 second.

- $>100$ ms is in the region where steady-state conditions begin to exist, in automotive work.

It is perhaps worth pointing out that before the arrival of modern dynamometer controllers during what may be described as "classical" steady-state testing, the transition ramp rates between successive test conditions were highly variable and were rarely used to trigger data collection. In modern drive cycles the transient states are the test sequence stages.

Exhaust emission legislation continues to be based on engine and vehicle testing incorporating transient drive cycles, and because of legislative pressure and the development of gas analyzers with faster response times applied to RDE testing, the benign drive cycles of the past are being replaced worldwide.

In the transient test frequency band range of 0.2 seconds, we are concerned essentially with torsional vibrations in the engine–transmission–road wheel complex, ranging from two-mass engine–vehicle judder, with a frequency typically in the range 5–10 Hz, to much higher frequency oscillations involving the various components of the power train.

This kind of power train investigation, however, involves the ICE to a secondary degree; in fact, for such work the engine may now be simulated by a low inertia, permanent magnet, electric motor–based dynamometer system.

In the transient test timescale, work is characteristic of gearshifts; correct behavior in this range is critical in a system called upon to simulate these events. This is a particularly demanding area, as the profile of gearshifts is extremely variable. At one extreme, we have the fast electrohydraulic shifting of a race car gearbox, at the other gear changes in the older commercial vehicle, which may take more than 2 seconds, and in between the whole range of individual driver characteristics, from aggressive to timid.

Whatever the required profile, the dynamometer must impose zero torque on the engine during the period of clutch disengagement, and this requires precise following of the "free" engine speed. The accelerating/decelerating torque required is proportional to dynamometer inertia and the rate of change of speed:

$$T = \frac{I\mathrm{d}\omega}{\mathrm{d}t}$$

where $T$ is the torque required (Nm), $I$ is the dynamometer inertia (kgm$^2$), and $\mathrm{d}\omega/\mathrm{d}t$ is the rate of speed change in rad/s.

A gear change involves throttle closing, an engine speed change, up or down, of perhaps 2000 rev/min, and the reapplication of power. Ideally, at every instant during this process the rate of supply of fuel and air to the engine cylinders and the injection and ignition timings should be identical with those corresponding to the optimized steady-state values for the engine load and speed at that instant; it will be clear that this optimizing process makes immense demands on the mapping of the power train management system.

A special problem of ICE control in this area concerns the response of a turbocharged engine to a sudden demand for more power. An analogous problem has been familiar for many years to process control engineers dealing with the management of boilers. If there is an increase in demand, the control system increases the air supply before increasing the fuel input; on a fall in demand the fuel input is reduced before the air. The purpose of this "air first up, fuel first down" system is to ensure that there is never a deficiency of air for proper combustion. In the case of a turbocharged engine, only specialized means, such as variable turbine geometry, can increase the airflow in advance of the fuel flow. A reduction in value from the (optimized) steady-state value, with consequently increased emissions, is an inevitable concomitant to an increase in demand. The control of maximum fuel supply during acceleration is a difficult compromise between performance and drivability on the one hand and a clean exhaust on the other.

## Data display

The manner in which data is displayed on the operators' screens is usually highly configurable by the user, perhaps dangerously so.

Individual operators, if allowed, will adapt the data display according to their own abilities in using the software and to suit their own immediate requirements and their esthetic tastes; the result may not produce a clear representation of data according to its operational importance or be readily understandable by colleagues. Most large test facilities impose the use of one or more, task-appropriate, data display templates on test cell operatives as part of their processes under ISO 9001.

These comments and the example shown in Fig. 13.4 only cover the main "engine driver's" screen, the main function of which is to ensure that those parameters of major significance to the test and those covered by primary alarms, such as engine oil pressure and engine coolant, are able to be monitored within a clear uncluttered display. There may be several other screens displaying data from other cell instrumentation and the cell services, all of which can be considered of secondary importance to the safe running of the in-cell machinery and the UUT.

The appropriate use of analog or digital representation is briefly discussed in Chapter 19, The Pursuit and Definition of Accuracy, but it will be found informative to visit facilities testing widely different types of prime movers in order to see the best practices found to be appropriate to each specialization. As an example, in those test cells testing gas turbines and some medium-speed diesel engines, it is found that there is a wide use of multiple vertical bar displays; this is because it is the equal balance of multiple temperatures or cylinder pressures that has to be immediately apparent to the operator, rather than the exact measurement.

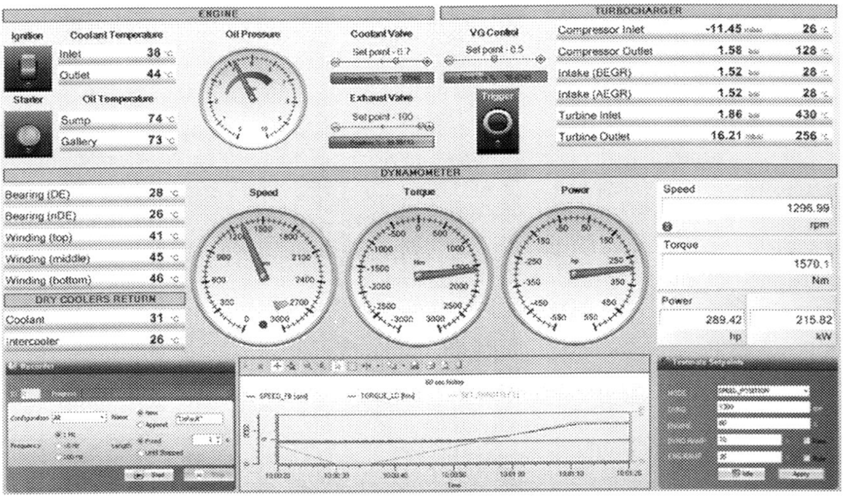


**FIGURE 13.4** An example of a typical test cell operator's display screen. The analog representation of speed, torque, power, and oil pressure follows good practice. The test sequence was concerned with turbocharger operation; therefore there is a display section covering that item. The bottom graph represents a scrolling 60-s log of speed, torque, and other chosen parameters. Source: *The University of Bradford HyperC.*

One great advantage of the digital display of an analog (dial) format, as shown in the engine speed display in Fig. 13.4, is that the dial can be easily scaled appropriately, from zero to maximum, to the limits of operation of the UUT.

## Part 5: Data collection and transmission

It is perhaps the ever-increasing speed of the recording, processing, transmission and storage of data, and the availability of virtual models of vehicle components that are responsible for the most revolutionary changes in the practice of vehicle and power train testing in the last 15 years.

It should be appreciated that, if given lack of care in their collection and storage, computerized data collection and processing systems are inherently no more accurate, nor traceable, than old paper systems. All data should have an "audit trail," back to the calibration standards of the instruments that produced them, to prove that the information is accurate, to the level required by the user. It is a prime responsibility of the test cell manager to ensure that the installation, calibration, recording, and postprocessing methods associated with the various transducers and instruments are in accordance with the maker's instructions and the user's requirements.

It is necessary to repeat a statement previously made:

*It is vital for test data to have a consistent and coherent naming convention attached to each data channel; if data is to be shared between cells and sites it is vital that an identical naming convention is shared.*

Bad data are still bad data, however, rapidly recorded and skillfully presented.

## Traditional paper-based test data collection

It should not be assumed that traditional methods of data collection should be rejected; they may still represent the best method of monitoring a device or complex widely dispersed system under test. This is particularly true in cases where it is necessary to ensure an operator regularly patrols the test field, as is the case with a very large marine engine installation or industrial process plant. It is also the most cost-effective solution in certain cases where budgets and facilities are very limited.

Fig. 13.5 shows a traditional "test sheet." The example records the necessary information for plotting curves of specific fuel consumption, BMEP (brake mean effective pressure), power output, and air/fuel ratio of a gasoline engine at full throttle and constant speed, with a total of 13 test points. All the necessary information to permit a complete and unambiguous understanding of the test, long after the test engineer has forgotten all about it, is included.

Such simple test procedures as the production of a power–speed curve for a rebuilt diesel engine can perfectly well be carried out with recording of the test data by hand on a pro forma test sheet. In spite of the low cost of computer hardware, there may be little justification for the development and installation of a computerized control and data acquisition system in the engine rebuild shop of, for example, a small city bus company, handling perhaps less than 20 engines per month. If an impressive "birth certificate" for the rebuilt engine is required, the test sheet may easily be transferred to a computer spreadsheet and accompanied by a computer-generated performance curve.

In the data flow shown in Fig. 13.6 the only postprocessing of the test data is the conversion of test data into an engine "birth certificate" that may be simply a reduced and tidy copy of the original data. The archived original copy may be legally required and be available for use for technical and commercial statistical analysis.

## Chart recorder format of data

Chart recorders represent a primitive stage in the process of automated data acquisition. Multipen recorders, having up to 12 channels giving continuous analog records, usually of temperatures, in various colors, still have a useful role, particularly in the testing of dispersed systems such as installed marine

TEST SHEET– I.C. ENGINES

UNIT NO.

| | | | | | |
|---|---|---|---|---|---|
| DATE | CUSTOMER | | | | WORKS ORDER NO. |
| DYNAMOMETER TORQUE ARM 265 mm | | | | | |
| ENGINE VARIABLE COMP | BORE 85 mm | STROKE 825 mm | CYLINDERS 1 | SWEPT VOL. 468 cc | FUEL SHELL 4-STAR; OIL SHELL 20W30 |
| BAROMETER 755 mmHg | AIR TEMP. 21°C | AIR BOX SIZE 150 liter | | ORIFICE DIA. 18.14 mm | FLOWMETER NO.; FUEL GAUGE 'O' |
| POWER kW = $\frac{F_N n}{36,000}$ | b.m.e.p = 7123 $F_N$ kN/m$^2$ | | FUEL liter/Hour = $\frac{180}{1}$ | | NOTE: COMPRESSION RATIO 7.1 |

| | TACHO | COUNTER | | | | | | | FUEL | | | AIR CONSUMPTION MEASUREMENT | | | | | | | | REMARKS |
|---|---|---|---|---|---|---|---|---|---|---|---|---|---|---|---|---|---|---|---|---|
| | rev/min | rev | sec | rev/min | BRAKE LOAD | POWER | b.m.e.p | 1sec 50/cc | liter/hour | liter/kW hr | EXH °C | HEAD $cmH_2O$ | TEMP °C | VOL F/R l/s | EFFY. η VOL | MASS F/R: g/s | a/l RATIO | | | |
| | | 2x | | n | $F_N$ | kW | kN/m$^2$ | | | | | $h_o$ | $T_A$ | $V_a$ | | $m_a$ | | | | |
| | | | | | | | | | | | | | | | | | | | | |
| RICH | 2000 | 780 | | 2012 | 50.5 | 2.022 | 360 | 47.0 | 3.830 | 1.357 | 620 | 6.15 | 21 | 4.02 | 0.827 | 6.89 | 7.38 | | | |
| ↑ | | 834 | | 2002 | 67.0 | 3.726 | 477 | 50.0 | 3.600 | 0.966 | | 6.15 | | 4.92 | 0.630 | 5.89 | 7.85 | | | |
| | | 891 | | 1022 | 68.0 | 3.630 | 484 | 52.5 | 3.429 | 0.945 | 500 | 6.13 | | 4.91 | 0.655 | 5.88 | 8.23 | | | |
| | | 962 | | 2025 | 71.0 | 3.993 | 506 | 57.0 | 3.158 | 0.791 | | 6.10 | | 4.90 | 0.620 | 5.87 | 8.92 | | | |
| | | 1044 | | 1989 | 74.0 | 4.089 | 527 | 63.0 | 2.857 | 0.699 | | 6.05 | 21.5 | 4.88 | 0.629 | 5.84 | 9.81 | | | |
| | | 1156 | | 2010 | 74.5 | 4.160 | 530 | 69.0 | 2.609 | 0.627 | | 6.00 | | 4.86 | 0.620 | 5.82 | 10.71 | | | |
| | | 1304 | | 2019 | 74.5 | 4.178 | 530 | 77.5 | 2.323 | 0.556 | | 5.97 | | 4.85 | 0.616 | 5.80 | 11.99 | | | |
| | | 1355 | | 1995 | 74.5 | 4.129 | 530 | 81.5 | 2.209 | 0.535 | 675 | 5.93 | | 4.83 | 0.621 | 5.79 | 12.58 | | | |
| | | 1560 | | 2035 | 69.0 | 3.900 | 491 | 92.0 | 1.957 | 0.502 | 685 | 5.93 | 19 | 4.83 | 0.609 | 5.79 | 14.20 | | | |
| | | 1615 | | 1988 | 65.2 | 3.601 | 464 | 97.5 | 1.846 | 0.513 | | 5.87 | | 4.81 | 0.620 | 5.76 | 14.98 | | | |
| | | 1698 | | 2008 | 60.0 | 3.347 | 427 | 101.5 | 1.773 | 0.530 | 673 | 5.95 | | 4.84 | 0.618 | 5.80 | 15.70 | | | |
| ↓ | | 1792 | | 2010 | 52.0 | 2.903 | 370 | 107.0 | 1.682 | 0.580 | | 6.03 | | 4.88 | 0.623 | 5.84 | 16.66 | | | |
| WEAK | | 1908 | | 2000 | 35.0 | 1.944 | 249 | 114.5 | 1.572 | 0.809 | 682 | 6.03 | 19 | 4.88 | 0.626 | 5.84 | 17.83 | | | |
| | | | | | | | | | | | | | | | | | | | | |
| | | DENSITY OF FUEL 0.75 kg/liter | | | | | | | | | | | | | | | | | | |

**FIGURE 13.5** A traditional test sheet.

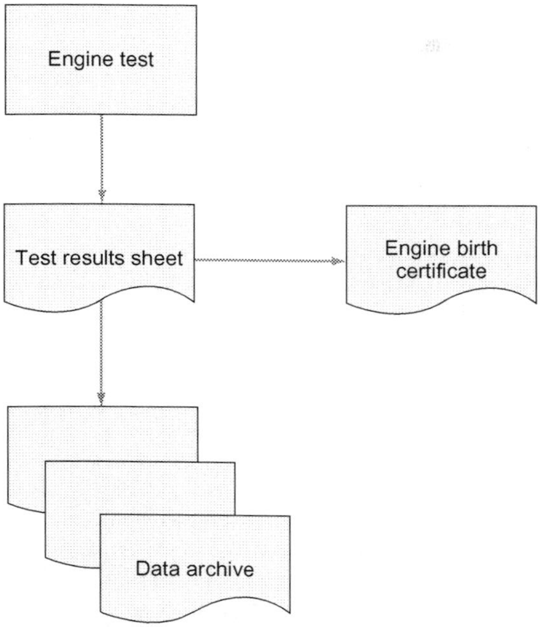


**FIGURE 13.6** Data flow based on precomputerized systems.

installations or of large engines "in the field" where robustness and the ease with which the record may be annotated are of value.

It is the manner in which the data is displayed as a colored graph of values against time, now emulated by computerized displays, which is the real value to test engineers who need the clear and immediate indication of trends, rates of change, and the interaction of data channels, in a paper form that can be notated contemporaneously with events.

## Data recording with manual control

At all levels of test complexity, the appropriate degree of sophistication in the management of the test and the recording of the data calls for careful consideration.

As the number of channels of information increases and there is a need to calculate and record derived quantities such as power and fuel consumption, the task of recording data by hand becomes tedious and prone to error. There are a number of simple PC-based data acquisition packages on the market, particularly in the United States, that do not have any control function and are designed for the aftermarket engine tuning and testing facility. These systems, at their most basic, take time-stamped "snapshots" of all connected data channels on the key command from the operator.

## Fully automated test cell control and data-handling systems

Most of the following text will be based on these fully computerized systems now found in good technical universities as frequently as they are in the facilities of OEMs. A small danger for students having their first, perhaps only, experience of engine testing on these modern systems is that the majority of that experience is vicarious, through watching a screen displaying highly processed numbers.

## Data transfer systems and interface protocols

The electronically controlled power trains of today, installed within electronically controlled vehicles, are tested within electronically monitored and controlled test facilities, so to cope with the two-way flow of this data traffic, the development of types of high-speed data transmission systems has had to take place. A computer "bus" is a subsystem that allows the orderly and prioritized transfer of data between components inside a computer and between computers and connected equipment; the test engineer will meet a number of different types in interfacing with both test objects and test equipment.

Computer buses can use both parallel and bit-serial connections and can be wired in either a multidrop or daisy chain topology or connected by switched hubs. The main test bed computer may be physically linked, through CAN bus or another standard communication bus, to both the UUT and major modules of microprocessor-controlled instrumentation. A common example of such an interface is that of exhaust emission analyzing equipment. Linking the test bed control computer with such an instrument is rarely a trivial task, and it is essential, when choosing a main control system, to check on the availability of the correct version of software, device drivers, and compatibility of data transfer protocols. The descriptions in the following sections give an overview of the most common data buses seen at test objects and systems in power train testing environments.

## Automotive vehicle bus system

### *Controller area network*

A very widely used protocol in vehicles and power trains (as well as migrating to other domains, for example, industry, medicine, aerospace, rail) is the *controller area network or CAN bus*. This is a multimaster, broadcast serial bus that has become a standard in automotive engineering and is one of the protocols used in OBD-11 vehicle diagnostics. The CAN bus runs over either a shielded twisted pair (STP), unshielded twisted pair (UTP), or ribbon cable, and each node uses a male nine-pin D connector. The most common physical media for a CAN bus is a twisted pair, 5 V differential signal, which allows

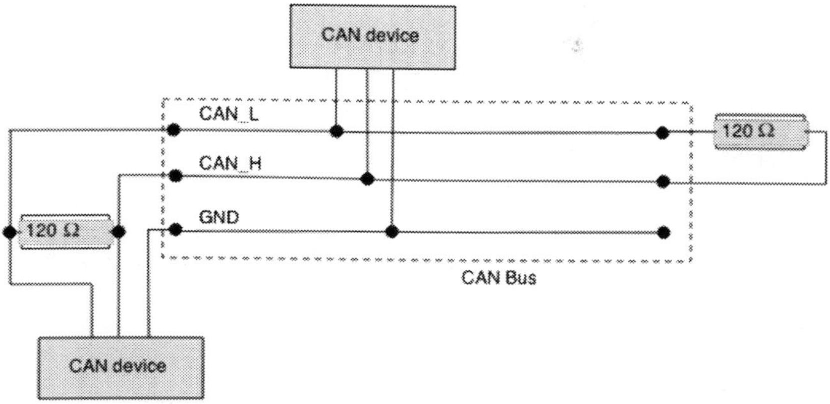


**FIGURE 13.7** The common physical media of the CAN bus (ISO 11898-2), terminated at both ends with 120 Ωwhen running at high speeds to prevent "reflections" and to unload the open collector transceiver drivers.

operation in the high-noise environments of some automotive installations (see Fig. 13.7).

The nodal devices are typically sensors, actuators, and other vehicle control devices such as the ABS (antiskid braking) system; these devices are connected through a CAN controller. To ensure that CAN bus devices from different suppliers interoperate and that a good standard of EMC and RFI tolerance is maintained, the ISO 11898-2 (cars) and SAE J1939 (heavy vehicles) families of standards have been developed. The high level of inbuilt error checking also gives the protocol the high level of reliability required in some of the safety-critical automotive systems.

Pressure on the data transfer rates beyond the capabilities of the original CAN bus have meant that several other protocols and standards (some optically based) have been developed; many are manufacturer or manufacturer consortium specific, and it is not clear at the time of writing which will become a future standard. Examples are given in the following subheadings.

## CAN-FD (Controller Area Network Flexible Data-Rate)

This development of CAN was released as a direct response to the need to increase data throughput on bus systems used in vehicle for power train system control and communication. Also as a response to the development of FlexRay and the associated threat to the dominance of the use of CAN. In simple terms the data payload of each CAN message was increased along with further optimization of the message structure that provides an increase in data rate suggested to be 30 times faster than classic CAN. CAN FD can function on the same network hardware and is backward compatible. It is therefore ideal for vehicle manufacturers as the increased performance requires minimal changes to the vehicle architecture and is therefore very cost-effective.

### FlexRay

FlexRay was devised in order to address the high data throughput needs of mission or safety-critical vehicle—particularly the needs of X-by-wire control systems. Data throughput is in the range of 10–20 Mb/s with deterministic performance. The disadvantage of FlexRay is increased cost and disruption when implementing on a given vehicle architecture (that has to be adapted and redesigned). Nevertheless, FlexRay has been employed on several key vehicle platforms and is adopted now as a standard for high-performance vehicle bus systems.

### Local Interconnect Network

For connecting a variety of simple peripherals to vehicle networks, where lower cost and performance are tolerable, LIN (Local Interconnect Network) bus is widely adopted. LIN is a simple serial communication system based on a single-wire bus, and it is compatible with the CAN multimaster network via a LIN master/slave gateway. It is normally used for body control applications and not applicable to power train control applications.

### Media Oriented Systems Transport

The MOST (Media Oriented Systems Transport) bus was developed for the networking of multimedia and infotainment systems and the transmission of audio and video data in the vehicle. It is an optical bus and therefore highly immune to electromagnetic effects as it does not cause any interference radiation and it is insensitive to electromagnetic interference. It is not relevant for power train development applications.

### Automotive Ethernet

The adoption of Ethernet technology for in-vehicle communication, measurement, and diagnostics—as well as for communication between electric vehicles and charging stations—is an emerging trend. In addition, the developments of advanced driver assistance systems functions for automated and autonomous driving as well as infotainment system adoption also create a use case for Ethernet as a data transfer backbone in the vehicle, allowing easy connection of the different communication domains of a vehicle electrical/electronic architecture.

At the time of writing, there is limited uptake for power train testing applications—although this will change if ECU access for calibration is made widely available via automotive Ethernet.

## Personal computer—industrial buses/protocols

### *RS232*

A simple, well-established serial interface—but now very dated—such that modern PCs are generally not equipped with this interface port. However, there are still many pieces of industrial equipment that use this interface, and it is still encountered in use on many instruments and measurement systems, often for debugging and diagnostics.

### *USB*

It is very widely used interface that is available on all PC equipment and allows plug-and-play connections to external devices as well as hot plugging. Currently, at V3.0 with a 4 GBits per second data throughput capability, it is a suitable bus for computers and peripherals but is less common for the connection of mission-critical components.

### *AK (Arbeitskreis)*

AK is a software protocol that is quite widely used at automotive test beds for special measurement devices (particularly emission and flow measuring devices). Note that AK is the software layer, with some standardized commands but the protocol can run on different hardware layers, for example, RS232 or TCPIP. The general command set is not well documented in the public domain and is also now quite old. The manufacturer-specific and device-specific documentation must be sought for any particular application instance, in order to locate the correct commands for the application task on that device. The basic message structure forms a telegram with a minimum of 10 bytes. Request and response telegrams are used to communicate bidirectionally.

### *Ethernet*

It is the most widely used data network technology for computer systems and networks and also used as a communication layer between measurement devices and systems at the test bed where an average level of performance is acceptable (i.e., it is not deterministic).

### *Profibus/Profinet*

ProfiBus (Process Field Bus) is a very widely used fieldbus system in Europe for automation in industrial environments. It contains a family of protocols aimed at factory automation tasks. ProfiNet (Process Field Net) is an industry technical standard for data communication over industrial Ethernet, designed for collecting data from, and controlling equipment in industrial systems, it can deliver data at high throughput (1 ms or less). These bus systems are normally used where industrial machines are connected to controllers or networks. Most commonly in power train environments, they are used for the connection of electrical machine drive systems and controllers that are used as dynamometers. Both technologies are administered by the international umbrella association PROFIBUS and PROFINET International (PI).

### *EtherCAT*

A more recently developed interface that was conceived to apply Ethernet technology for automation applications that require short data update times (cycle times; $\leq 100\ \mu s$) along with low jitter (for precise synchronization $\leq 1\ \mu s$). EtherCAT is easy to integrate and deploy in a given environment as the master bus requires no additional extension board as it is easily implemented on any Ethernet adapter. Many test and simulation system providers are now providing EtherCAT support for their systems, and it is likely that this fieldbus will be widely adopted in power train test environments.

This above mentioned list is not exhaustive, and there are many other legacy bus systems for communication between digitally based measurement devices and systems. There is also continuous development of many new systems, some of which are successful, some become obscure. There is much documentation and information freely available on all these systems, as well as manufacturer-specific documentation. It is not often the case that a test system engineer or user must choose the bus or network to be used. It is more usual that it is just necessary to familiarize with the tools and technology for a specific bus system when it is encountered or relevant.

## Association for Standardization of Automation and Measuring Systems standards

ASAM is an important, nonprofit organization that develops and promotes standardization for tools and protocols used in the development of automotive systems and associated test environments (see further reading section for web link). The organization is formed by representatives from international automotive OEMs, tool suppliers and vendors, engineering service providers, and research institutes—all involved in the automotive industry. The standards are conceptualized and developed in working groups that are composed of experts from the member companies and bodies. This promotes and enables platforms for the exchange of data or information within a toolchain or workflow. The ASAM group nominates itself as the legal "owner" of these standards and is responsible for their distribution and marketing.

The standards are clustered into logical groups.

### *Measurement and calibration*

These standards relate to control unit calibration tools and interfaces between the control unit and tool, as well as the tool and host system. In this group of standards read–write access to the data in ECU memory, meta description of the data, storing the data in files, and describing the calibration process are included. Examples are CCP (CAN Calibration Protocol—ECU interface), MDF (Measure Data Format—File storage format), and CDF (Calibration Data Format), as well as MCD-1 (Communication protocol), MCD-2 (Data model), and MCD-3 [application programming interface (API) specification for automation system interface].

### *Diagnostics*

These standards define the data model and application interface for the diagnostic server functions within control units. They facilitate the description of diagnostics capabilities of ECUs needed throughout the life cycle of a vehicle from development, testing, production to after-sales, and service and also exchange of diagnostic information between partners in the development process. Examples are MCD-2/3D.

### *Controller networks*

These standards are used for the design, configuration, monitoring, and simulation of communication on automotive networks. The standard supports autogeneration of software code for ECUs and the configuration of test tools for simple testing of ECUs. The standard consists of a generic interface description and technology-specific extensions for the most common automotive BUS interfaces. Technology-specific properties are described for each network port, and the interface description contains a list of sent and received signals for each ECU. The standard is known as MCD-2 NET and is also known as FIBEX—Field Bus Exchange Format.

### *Software development*

These standards support the control unit embedded software development process. They include a description and documentation of control unit software, also a description of change requests, and block sets for model-based software engineering. Current development processes for control strategies rely upon simulation, rapid control prototyping, and XiL technologies. They often use block diagram, so-called model-based specifications in functional models to express their designs. The use of standard block sets, as defined in this standard, ensures that intellectual property is not locked into a proprietary tool environment and can be easily transferred between compliant tools. The standard also covers data exchange formats for this environment. Examples are MDX that describes ECU functions and allows integration of these functions without having to access ECU source code. Also, MBFS (model-based function specification), which defines a block library for model-based design that contains typical functions, needed in automotive control algorithm specifications.

### Test automation

This is probably the most significant areas of the ASAM standards for those engineers involved in the specification and configuration of power train test cells because these are the standards for test and measurement system interfaces, including APIs for programmatic access to sensor and actuator devices, access to measurement and calibration systems, access to HiL systems variables, and access to design of experiment (DoE) tools with their related formats for test descriptions.

Of particular note are as follows:

- ACI—interface between test automation and ECU application tool.
- ASAP-3—RS232 communication protocol between a test automation system and a measurement and calibration system connected to an ECU.
- ASAP-3 MCD—calibration and measurement purposes in development, testing, and production of ECUs.

*Note that ASAM MCD-3 MC currently coexists with the older ASAM ASAP3 standard, which is dependent on specific interfaces (RS232, Ethernet) but is still widely used.*

- GDI (Generic Device Interface)—developed for providing an independent integration interface between measurement and control devices and test bed automation systems.
- iLinkRT—uses a protocol-based and event-triggered communication method between ECU and host system, which is more efficient than other data communication methods. Using this protocol ECU labels can be delivered to the host system at a rate of kHz, thus allowing the implementation of highly dynamic test procedures that rely on information from ECU labels in the engine control loop cycle—such as those used for ICE air path calibration routines.
- XiL—API definition for the communication between test automation tools and XiL test benches. The standard supports test benches at all stages of the development and testing process—most prominently model-in-the-loop (MiL), software-in-the-loop (SiL), and hardware-in-the-loop (HiL). The notation "XiL" indicates that the standard can be used for all "in-the-loop" systems.

### Data management and analysis

These standards are for storing, retrieving, and analyzing mass data captured during simulation, testing, production, and the operation of vehicles.

There are many test data postprocessing applications and tools on the market, which have either a proprietary plug-in architecture or no plug-in capabilities at all. Customized solutions for such applications such as importers, algorithms, or graphic elements cannot be easily reused in other applications and thus would require significant porting efforts. The CEA (Component for Evaluation and Analysis) standard defines an application framework and functional components for the evaluation and analysis of test measurement data and overcomes these issues.

ODS is the standard relating to databases for storing of test bed data. The standard is primarily used to set up a test data management system on top of test systems that produce measured or calculated data from testing activities. Tool components of a complex testing system can store data or retrieve data as needed for proper operation of tests or for test data postprocessing and evaluation. Centralization and management of data and parameters, across large test fields, is an integral component that this standard covers.

### *Environment simulation*

This standard set provides a description of road networks for driving and traffic simulation, including the specification of driving maneuvres and test scenarios. As the complexity of automated driving functions rapidly increases, the requirements for test and development methods are growing.

The OpenDRIVE file format provides the precise analytical description of road networks, including the exact road geometry descriptions, including surface properties, markings, signposting, and logical properties such as lane types and directions. Other relevant standards relate to interfaces (open simulation interface) and the description of complex, synchronized maneuvres that involve multiple entities such as vehicles, pedestrians, and other traffic participants (OpenSCENARIO).

### *Model interfaces in virtual test environments*

The evolving trend of cosimulation during test procedures requires that model based, and physical test objects must be able to run in parallel, with deterministic levels of communication performance, in order to exchange data and variables between the physical and virtual test domains. This has promoted the development of integration platforms in testing environments to facilitate this concept. An essential requirement is that a multicomponent, multiphysics-based test environment can be created that includes all components—real and simulated, irrespective of their domain—physical or virtual, mechanical or electrical, and hydraulic or pneumatic. This allows the creation of a so-called "virtual" vehicle or power train that then forms the UUT as a total entity, able to perform virtual test drives. A test environment as described above appears to create the ultimate facility—but such environments are now existing and are required to be able to deal with the complexity of a modern power train, in conjunction with requirement to fulfill current emissions legislation with high degrees of freedom on the test cycle. The goal of such an environment is to be able to combine physical elements with models from a variety of standard tools and environments. In order to support this, the Functional Mock-Up Interface (FMI) standard (see further reading) was developed and is now actively used in desktop simulation environments, also in virtual testing environments such as those described earlier. This standard allows the integration of models from various domains that can be encapsulated as Functional Mock-up Units (FMUs), with interfaces to integration platforms—thus facilitating full interconnectivity of models and physical/simulation test environments

The development and promotion of the FMI standard was driven by the need to simplify the creation, storage, exchange, and use/reuse of dynamic system models, from different simulation environments, for MiL/SiL/HiL simulation of cyber-physical systems, as well as other applications. The FMI specifications are published under the CC-BY-SA (Creative Common Attribution-Share Alike 4.0 International) license (see further reading) (i.e., the license used by Wikipedia). The standard is in two parts covering:

- *Model Exchange*—the modeling environment can generate a C code representation of a dynamic system model that can be utilized by other modeling and simulation environments. Models are described by differential, algebraic, and discrete equations with time-, state- and step-events. If the C code describes a continuous system, this system is solved with the integrators of the environment where it is used, that is, the solver that exists within the tool.
- *Co-simulation*—to provide an interface standard for coupling of simulation tools in a cosimulation environment. The data exchange between subsystems is restricted to discrete communication points. In the time between two communication points, the subsystems are solved independently from each other by own solver within the FMU.

The two interface standards have many parts in common. In particular, it is possible to utilize several instances of a model and/or a co-simulation tool and to connect them together. The interfaces are independent of the target environment. This allows generating one dynamic link library that can be utilized in any environment on the same platform. A model, a co-simulation slave or the coupling part of a tool, is distributed as an FMU that contains all required XML files, binaries, and C code zipped in a single file. The standard has been widely adopted in the community of tool and software suppliers, also by OEMs and automotive industry suppliers. At the time of writing (2020), it is claimed that over 100 tools and environments support the standard.

## An overview of power train calibration and optimization

The term "calibration" is used by test engineers to describe the process by which the operational and control characteristics of a vehicle and its power train system is optimized, through manipulation of all its control mechanisms, to suit a specific geographical/vehicle/driver combination while remaining within the limits of environmental and safety legislation.

As power trains increase in the complexity of every area of technology, in order to meet legislative standards, the task of power train calibration is being made more difficult by the increase in the number of parameters with which it has to deal. This "curse of dimensionality" is driving the calibration process beyond the capabilities of individuals and their available timescales and is driving the development of special tools, techniques, and software, often through the collaboration of industry and academia, worldwide.

The optimized control parameters, when found, are stored within the "map," essentially a multi axis lookup table or set of tables, held in the memory of the engine ECU, the power train, and vehicle control units. Thus characterizing an engine, outside its intended vehicle, requires that it be run on a test bed capable of simulating not only the vehicle's road-load characteristics but also the full range of the vehicle's performance envelope, including the dynamic transitions between states.

To reduce the testing time required to produce viable ECU maps, and to keep the complexity of the operator's role in the mapping process to sensible levels, software tools designed to semiautomate the tasks are used.

During the characterization process, in order to discover boundary conditions, such as "knock," an ICE will be run through test sequences that seek and confirm, by variation of control parameters of which ignition timing and fueling are only two, the map position of these boundaries at a wide range of vehicle operating conditions will be found. While running these tests to find the limits of safe operation, conditions of excessive temperatures, cylinder pressure, and "knock" require fast detection and correct remedial action that may be beyond the capabilities of a human operator. Experimental, online model-based tools contain test sequence control algorithms that allow some degree of automation for the location of these boundary conditions.

A typical sequence for a gasoline engine is that called "spark sweep" wherein, at a given speed/torque operating point, the spark timing is varied until a knock condition is approached, the timing is then backed off and the knock point approached with finer increments to confirm its map position. The test sequence then sweeps the ignition timing so that a fault condition, such as high exhaust temperature, is located at the other end of the sweep. Given sufficient data, the software can then calculate and extrapolate the boundary conditions for a wide operating envelope without having to run all of it.

Advanced DoE tools allow the test engineer to check the plausibility of data, the generation of data models and obtain visualization of system behavior while the test is in progress.

As implied in the previous paragraphs, the ultimate target of the calibration engineer is to be able to set the operating profile of the test engine and then set the sequence running, only to return when the task of producing a viable engine map has been automatically completed.

## Part 6: Virtual testing

Virtual testing where part of the UUT system is physically "missing," is a topic that was treated as a novel concept in the previous edition of this book; in the meantime, it has gained considerable significance. The market demand to create new products and innovations faster has forced power train development processes to evolve considerably to allow the use of "frontloading" techniques that enable more of the workload to take place earlier in the development cycle. In most case the context of these activities centers around the vehicle control system calibration (populating lookup tables and parametrizing control system functions in order that the power train fulfills the required engineering targets).

Hardware-in-the-loop (HiL) testing is a well-established approach that has been complimented by pure virtual test solutions in order to perform calibration tasks in an office environment on a PC. Given verifiable models, the use of HiL simulators as a complete virtual test bed the limits of the design space can be explored, characterized, and calibrated before any real testing of physical units is done. Bringing the virtual and real test bed closer has been a development of modeling techniques and integration of systems such that a power train calibration engineer can use the same environment (automation, acquisition, and analysis tool base) right through the development process from virtual verification to final validation in the vehicle test environment.

### Virtual testing domains and clarification of terminology

It is important to understand the terminology encountered in virtual testing—a HiL test is a widely used term—but the "hardware" could now refer to several topologies—with the risk of subsequent confusion. Therefore for the sake of clarity:

- Hardware-in-the-loop (HiL)—this term originally derives from the simulation systems used to test embedded control units. A powerful computer is used to execute real-time model of the "missing" hardware, the system has an I/O system that is physically connected to the control unit inputs and outputs. The simulation system is capable of providing a virtual environment both electrically and dynamically. The real and simulated subsystems interact continuously over time, in a closed loop. The time step for each loop cycle defines the fidelity of the simulation that is achievable.

- Model-in-the-loop (MiL)—this is normally an early development stage activity. A "model" of the control system is developed in a suitable tool environment that allows characterization of a system controller function along with a suitable plant model and simulation test in order to test the function and performance of a controller in a completely virtual setting. The use case is to verify or test whether the controller or function is feasible for implementation in the target controller.
- Software-in-the-loop (SiL)—derived from the previous model environment or separately coded. In this instance the control function is contained in deployable code that can be tested on a virtual ECU that has a virtual processor that matches the physical, intended target. This means the execution of the *actual* code can be tested, not just the function of the controller. A current trend is a complete virtual vehicle system with simulated ECUs and bus connections such that the complete vehicle architecture and functionality can be verified early in the development process.
- Processor-in-the-loop (PiL)—in this case the deployable code, verified in the SiL development stage, is deployed on the actual processor hardware but still retaining the plant model simulation environment. The simulation system must include an evaluation board that contains the target processor for deployment of the test code.
- Function-in-the-loop (FiL)—this approach establishes a direct, software-based connection between the simulation system and the target ECU control functions by means of a plant bypass. It is similar to PiL but the target is the *actual target ECU processor*. This allows testing of ECU functions without the need for signal conditioning hardware (as PiL). A test case could be an ECU where new functions have been developed but where the I/O requirements are unchanged thus no requirement to test using HiL where physical–electrical interfaces are required.

Note that the main difference between PiL and FiL is the test hardware. PiL typically deploys a prototyping board as the target. FiL uses the target ECU along with bypass technology in order to directly stimulate the processor—bypassing the ECU I/O interface.

These approaches have their own benefits and disadvantages, but in general, they are used in a combined workflow in order to enable accelerated development of software and functions and verify them early in the process, thus allowing a development process that tests the software incrementally, in environments that progressively become closer to reality but staying virtual. Virtual test environment allows testing of extreme scenarios with reduced risk as there is no real hardware (i.e., a vehicle or engine or personnel) to damage. Software and functions can be developed and tested very early in the development process, before a physical ECU is available. In the case of

MiL and SiL the speed of the simulation is not limited to real-time execution and can be executed as fast as the processor running it can achieve. This means test cycles can run faster than real time, and more tests can be executed in a given time frame, in conjunction with test automation, test can be run order of magnitude faster—gaining precious time early in the development process.

## Virtual testing and simulations in the physical test environment

Virtual ECU development is a broad topic beyond the scope of this book. However, the classical HiL test environment and workflow has evolved and been adapted to complement the physical power train test environment. This basically involves using a combination of advanced simulation at the same time as physical testing, with adapted workflows—in order to increase efficiency in the development process. The main applications, at the time of writing, are as follows:

- Deploying a HiL-based system as a mechanism to provide advanced simulation in order to be able to recreate complex RDE drive cycles as a test profile in a physical test environment. High-fidelity driver, vehicle, and environment models interact on the HiL hardware and create a real-time demand profile in time steps according to the processing capability of the simulation hardware. This is communicated to the test automation system controller as speed/load demands such that the physical test environment can recreate the real world test experience.
- Using a HiL system in combination with a physical ECU, along with a test bed control and data acquisition system creates a virtual test bed for the purpose of calibrating the ECU early in the calibration development phase—but without the need for a physical UUT. The test automation system is used to control the test and deliver the test plan, often using a model-based approach to the calibration task in hand. The HiL system interacts with the ECU electrically and dynamically as would normally be the case—but under control of the test automation system. The main difference is that a highly developed engine or motor system model, running on the HiL system, is used in place of a physical test subject and dynamometer. This model combines physical and empirical modeling techniques and is prevalidated as a virtual representation of the target UUT. The complete environment interacts with the user using the same tools and user interface as a physical test bed. It is suggested that virtual test beds could replace more than 30% of their physical counterparts and can reduce to load on remaining physical test beds but up to 80%.
- General co-simulation tasks are being adopted in order to accelerate the overall power train development process in order to parallelize testing, using a combination of physical and virtual test objects. There are many

combinations of hardware, tools, and modeling approaches available, and there is no single solution that can be defined. However, complex the power train configuration, it can be simulated and virtual component simulation can run in parallel to a physical test object.

The key to successful execution of combined testing and simulation activity relies heavily on several critical points that a test bed or test environment engineer must consider:

- Interfaces—this is particularly important between the simulation and test systems. However, there are no special protocols required and, in most cases, standard well-established technology will be used. If this cannot provide the required level of performance, then alternative approaches will need to be used (i.e., simulation platform in the test bed automation environment). The level for simulation performance needs to be clear before this topic can be considered (see Goals next)
- Models—using or choosing the correct type of model is dictated by what is available and what is possible. The classic approach of physical modeling provides the possibility to characterize any system, but the level of accuracy, parameterization effort, and execution time limits the possibilities of deployment. Empirical models are ideal for characterizing complex interactions and can be executed quickly but need training, can be difficult to reuse, and are limited to staying within design space (no extrapolation possible).
- Goals—it is important to have a clear mindset on what is required and what is achievable with the tools and models available. Using physical or numerical simulation, it is possible to simulate the impact of the engine design on pollutant emissions and fuel consumption, but the level of accuracy is not sufficient for ECU calibration. Verified model-based approaches allow the development of robust empirical models and enable the modeling and prediction of pollutant emissions and fuel consumption on the whole engine map, allowing the consideration steady-state operating points, as well as the air path and injection parameters with a level of accuracy suitable for calibration tasks (e.g., $5\%-10\%$ on $NO_x$). Current and future legislation does demand more dynamic test cycles where air system control error has a greater impact—producing a loss of accuracy of up to 25% error $NO_x$ per cycle. Therefore if the goal is clear, for the given development phase, then the appropriate modeling approach to combine with the physical test can be chosen and provide good results
- Platforms—sharing of models and model parameters are a general subject for product lifecycle management tools where the goal is to be able to reuse simulations across a development process. Concepts exist for managing models across a CAE (computer-aided engineering)-driven development environment, and this concept is known as FMI (Functional Mock-up Interface, discussed earlier in this chapter). This technology has

migrated to power train testing environment, and several of the tools used for power train development can support import and export into this FMI platform, which can provide a mechanism for sharing models for co-simulation and exchange. Note that such a systematic mechanism can be very effective; but it can also be quite disruptive to implement. It is up to the user and the environment to decide if such a platform will bring any benefit or efficiency gain.

## Virtual driver models in power train calibration and drivability testing

*All models are wrong, but some are useful (George E. P. Box).*

The pressure on "time to market" and the development programs associated with engines and power train units do not allow the testing of embedded control systems to wait for a vehicle or engine prototype to be available. All new development schedules require that hardware-in-loop simulation will be used in parallel with the development of the mechanical and electrical modules, and such simulation requires virtual models of the missing modules.

While modeling of the mechanical and electric functions and interactions is becoming demonstrably ever more useful, feedback from the vehicle aftermarket reveals that some of the models used in power train optimization are not yet fully reflecting real-life use or the expectations of drivers. The problem may be seen more as a marketing problem than a calibration engineer's problem because the non-specialist driver's expectation of the essential parts of driving experience has remained regionally static while the vehicle technology, and therefore function, continues to change significantly. A modern vehicle with a continuously variable transmission feels and sounds disturbingly different from, for example, the five-speed manual transmission–equipped car for which any European driver with 30 years' experience has as a mental model of a vehicle's "normal" behavior. The author had occasion to "rescue" a terrified driver from a powerful car, the DSG power train and vehicle stability system (VSC) of which became "undrivable" in moderately deep snow. The driver had no idea how to switch the gearbox to manual or to switch off the VSC and was determined to sell the car at the first opportunity. Perhaps, a control system and driver display model that was more sympathetic to the average, technically unaware, driver could have been used in the power train development? There is a discussion to be had concerning the need to, and advantages of, "recalibrating" inappropriately experienced drivers in order that they derive optimum benefit from modern hybrid vehicles, but this book is not the place.

This all raises the questions of how drivability is defined and how it is included in the model. Presently, the calibration engineer's proxy for drivability is the rate of change of actuators; no doubt this area of modeling will develop.

Since HiL test techniques are only as good as the models on which they are based, a heavy responsibility rests on the quality control, verification, and correct application of models. However, as the total store of performance data, including previously unmodeled fault combination events, builds during their life, the models should improve. Model improvement requires that the in-service feedback loop works efficiently and quickly, rather than working through the motoring press and, that modern indicator that HiL has failed, the model recall process for a control software upgrade.

HiL testing, as shown diagrammatically in Fig. 13.8, is used by engineers to carry out two key tasks of ECU calibration:

- vehicle performance optimization during the creation of control parameter maps and
- testing of the diagnostic performance of the ECU under fault conditions as required by OBD legislation and proprietary service support.

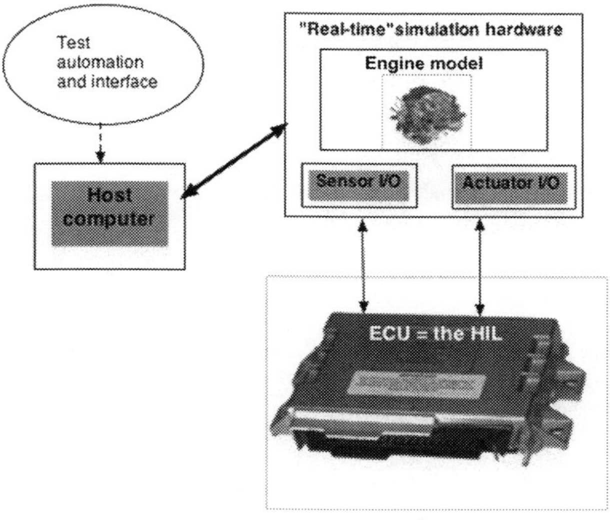


**FIGURE 13.8** The engine or complete power train is simulated in a model programed into high-speed electronic hardware together with models of sensors and actuators. The model simulation has to be an order of magnitude faster in operation than the ECU with which it is connected. The ECU is real, it is the hardware in the (simulation) loop. The host computer provides the test automation and the user interface.

A key benefit of HiL testing is that the ECU can be pretested in its full range of operation and in all primary fault conditions without endangering the hardware it will subsequently control. The multitude of vehicle and engine system faults, with their intended, and unintended, interactions can be examined in this virtual test environment, to an extent that would be simply impossible in the traditional engine or vehicle test bed. Along with the practicality of testing operation and fault conditions, the HiL test sequences give the great benefit of repeatability and the ability to rerun transient conditions to determine cause and effect.

As experience builds, the models used become more verifiable, but test engineers have to decide on the operational constraints that need to be imposed on the system. These constraints are determined both by the boundary conditions required to ensure that, within all "normal" real-life conditions:

- The engine runs in safe conditions of combustion (knock avoidance, etc.)
- Emissions in defined test conditions are within legislative limits.
- Good drivability within the limits of the range of drivers targeted by the marketing of the final vehicle model.

While boundary conditions and abnormal operation conditions can be safely and "cheaply" explored during HiL testing and it is therefore ideal in system development and university teaching, it can remove the tester from the reality of their test object; this decoupling of an increasingly important part of the test process needs to be guarded against. Driving the famous Nordschleife of the Nürburgring on a computer gaming machine may be convincingly realistic to those used to screen-based simulations, but (once again in the experience of the authors) it does not adequately prepare one for the real thing.

## Further reading

Association of Standardization of Automation and Measuring Systems (ASAM), <www.asam.net/>.

Benedict R.P., Fundamentals of Temperature, Pressure and Flow Measurements, John Wiley, 1984. ISBN-13: 978-0471893837.

BS 6174, Specification for Differential Pressure Transmitters With Electrical Outputs.

Creative Common Attribution-Share Alike 4.0 International <https://creativecommons.org/licenses/by-sa/4.0/>.

Functional Mock-Up Interface standard <https://fmi-standard.org/>.

National Physical Laboratory, Good Practice Guides. <http://www.npl.co.uk/>.

Nise N.S., Control Systems Engineering, John Wiley, 2011. ISBN-13: 978-0470646120.

Chapter 14

# Data handling and modeling

## Chapter outline



## Introduction

Data science is now commonly manifesting its presence in our daily lives through social media analytics, targeted advertising, number-plate recognition, etc. It is a term that is often used to capture the breadth of the widely discussed subject of artificial intelligence. Much of the work of data

Engine Testing. DOI: https://doi.org/10.1016/B978-0-12-821226-4.00014-0

scientists is far beyond the scope of this book, owing to a technical library couched in foundational statistics and an entire universe of software jargon that provides a significant barrier to entry for those outside the community. Some of its mathematics is fairly esoteric and made more difficult for engineers to completely understand because of the unfamiliar use of terms coopted from classic physics, such as entropy. In the context of data science, entropy is a measure of uncertainty.

Data scientists utilize their core skillset to create models from which inferences and predictions are made; these may be deployed within a business or customer facing process to, either directly or indirectly, improve decision-making. In our field of powertrain testing, data science can be used to derive the maximum benefit from a vast range of data generated, not only from the units' test processes but also data from the production process prior to testing and the units' service life post-testing. Large companies in the automotive industry now see the test process data as a subset in a much larger data set, that covers their whole product life—a "cradle to grave" data set from which cause and effects far outside, but possibly seen within, the test cell may be determined. These companies are employing data science teams in order to use and reuse their generated data to the maximum benefit of their processes and customers.

The growth in significance of data mining and machine learning (ML) is partially driven by the reduction in cost of computation, coupled with significant public interest and the exponential growth in data available in many domains, for example a connected vehicle can produce up to 4 GB of data per second! This abundance of resource has generated immense enthusiasm among the established engineering community, which has led to an entirely new demographic of newcomers to the data science space.

Often, data science activities are deployed to extract information from large data sets that exist before the explicit purpose of the data collection exercise was conceived. This contrasts with engineering data collection that, normally, should always consider the use to which the data will be put once it has been collected. The distinction is important since many are incorrect in believing that value can be extracted from any relevant data set or mass. Within engineering and physical sciences, many hours are often spent on planning experiments and considering the requirements in order to ensure that the resulting data will meet these. One such process is known as design of experiments (DoEs). One of the first experimental design processes was conceived and exemplified by Ronald Fischer as a tea tasting experiment (see further reading) and is largely based on frequentist statistics to explicitly design an experiment that will yield data of the quality required to support the task at hand. It may be possible with some good fortune to win the "data lottery" and happen upon something useful; however, for practitioners from an engineering background, a more considered, more certain approach is required.

The following chapters will discuss data arising from ICE testing and how it is subsequently deployed within engineering development processes while making overtures to data analytic tools that often exist outside of common engineering practice. This includes foundational ML, which is presented for clarification. The following discourse is as generic as is allowed in a text of this nature and attempts to expound the topic for interested, nonexpert readers; there is a wide range of literature to which the reader is directed, for example Refs. [1–5], for more information.

As a general principle, this chapter is structured around the data life cycle and the main stages that it is subjected to once collected, with some specific applications considered. It is assumed that the reader is familiar with the experimental process, the subject of much of the remainder of the book. Good quality data is a prerequisite to the various stages of the data life cycle as is proper experimental design, without which everything that follows is of questionable value.

Considered in overview, the data life cycle is separated into several stages:

- *Data acquisition and storage*—How is data to be obtained (a non-trivial task in powertrain testing), quality assessed at point of collection, and made available for users of the data.
- *Data preprocessing*—The processes associated with converting data from a raw form to a format useable by statistical techniques and stripping out erroneous or unusable observations from a large set (also referred to as data cleaning in the data science community).
- *Data utilization*—The end goal of "performing data science" by transforming the data or using it to train statistical models, upon which predictions and decisions are made at the design/development level. Note that this area of the chapter will cover many conventional "data-driven techniques" that already exist in engineering practice and typically follow a more conventional system's identification approach. It is wise to consider these existing processes in parallel with state-of-the-art techniques, to understand how they may also be improved by modern advances.

Modern powertrain development and related test facilities generate significant amounts of data with volumes increasing daily, necessary due to exponential growth in complexity, as degrees of freedom (DOF) are added to the powertrain system to achieve legislated emissions levels and market expectations.

Powertrain test data is generated as a result of established processes and the purpose of the data collection exercise should, therefore, be known as a priority factor. This is important since it imposes requirements on the data; accuracy, coverage, and certainty that need to be met for the exercise to be a success. This contrasts to common practice in applied data science approaches that often give little to no regard to data quality with all data treated as equal. The reality is that all data is *not* equal, and the underlying quality of the data is a key consideration since it informs everything that follows.

For any given testing exercise, a targeted problem must be well defined and repeatable. This is to establish a baseline against which robust improvement can be measured. Only then may changes be made (in the form of a design change, calibration effort, controller redesign, etc.) and validation data collected. If the initial problem or solution is incorrectly defined, often additional rounds of experimentation and analysis may be wastefully executed. For example, using the above framework, for any number of repeats $n$, of the targeted testing procedure, there may typically be up to $2^n$ experiments performed.

Consider also that typical powertrain testing routinely incorporates untargeted or loosely targeted experiments, for the purposes of mechanical validation and technical objective verification. Further investigations and targeted testing may also result from observations made in these standardized test routines. The assertion that powertrain development comprises strictly data-driven decisions is reasonable, based upon the sheer volume of data collection that is mandated. Data is collected to highlight problems, characterize them, and validate their solutions at every turn.

The implementation of data science techniques has its own challenges for application in engineering test environments:

1. There is a dominating principle of robustness in engineering practice. Every observation must be repeatable. A data-driven approach to decision-making derives utility from large data and general trends but often lacks the specificity of observations from targeted test data. Robustness practice is partially driven not only by legislative scrutiny but also by the strictly empirical mindset of engineers when it comes to knowledge acquisition:

   *I'll believe it when I see it, and then recreate it.*

2. Propulsion systems are complex. For any given development program, the corresponding system-level model will be characterized by many parameters that require tuning. Over the course of the program, designs are changed, control strategies are improved, and parameters are tuned concurrently and frequently through the process. Fast-moving development presents a unique data challenge. The system being characterized develops rapidly in small, incremental steps, such that data sets can be quickly rendered obsolete.
3. The above mentioned complexity is also reflected in organizational structure and the way it impacts data management. Powertrain design often takes a systems engineering approach, under which individual teams are responsible for the development of a subsystem (aftertreatment, coolant paths, air system, etc.). Collaboration of sub-system teams will exist where technical objectives compete or overlap, or where boundaries exist between subsystems. To the extent that each team has its own priorities and testing, data

acquisition and storage are generally not unified. Often the lack of a common information architecture for engineering data means that data analysis exercises occur primarily at the subsystem level, not the top-level system.

For these and other reasons, there are issues intrinsic to engineering data analytics in practice. This demonstrates that often, there is a gap in the requirements between engineering data and the more conventional data science process. In order to fully integrate the benefits of data science directly into engineering processes, this gap must be bridged and is an area in which many innovations will likely be made over the coming years.

## Part 1: Traditional data processes

### Acquisition, storage, and quality

The following section presents the landscape of tools, processes, and standards upon which powertrain data analysis is currently performed. The reader will note that topics such as statistics and data management are broached. Often this results in an overlap in technical knowledge between the engineering and data science communities, but with differing vocabulary and application domains. In reality, a great deal of the knowledge is transferrable.

Modern, high-speed data acquisition systems are capable of acquiring vast amounts of data that may be totally irrelevant to the purposes of the test or stored in a way that makes it virtually impossible to correlate with other tests. The first basic rule is to keep a consistent naming scheme for all test channels, without which it will not be possible to carry out postprocessing or comparison of multiple tests. Since, for practical purposes, all vehicle test work is relative, rather than absolute, this ability of comparison and statistical analysis of results is vital. Mere acquisition of the data is the "easy" part of the operation. The real skill is required in the postprocessing of the information, the distillation of the significant results, and the presentation of these at the right time, to the right people, and in a form they will understand. This calls for an adequate management system for acceptance or rejection, archiving, statistical analysis, and presentation of the information.

*Data is as valuable as it is accessible by the people that need to use it.*

The information technology (IT) manager, in charge of the storage and distribution network of the data, is important in the test facility team and should be local, accessible, and involved. One vital part of the management job is to devise and administer a disaster recovery procedure and backup strategy. Normally the IT manager will have under his or her direct control the host computer system into which, and out of which, all the test bed level and postprocessing stations are networked. Host computers can play an important role in a test facility, providing they are properly integrated and secured. The relationship between the host computer, the subordinate computers in postprocessing offices, and the test cells must be carefully planned and disciplined. The division of work between cell computer and host may vary widely. Individual cells should be able to function and store data locally if the host computer is "offline," but the amount of postprocessing done at test bed level varies according to the organizational structures and staff profiles. In production testing the host may play a direct supervisory role, in which case it should be able to repeat on its screen the display from any of the cells connected to it and to display a summary of the status of all cells. With modern communication networks, it is possible for the host computer to be geographically remote; some suppliers will offer networking as part of a "site support" contract. This ability to communicate with a remote computer raises problems of security of data.

## An overview of post-acquisition data processing, statistics, and data mining

Much of this book is directly and indirectly concerned with the ways in which the accuracy of raw data is as high as practically possible and corruption of the raw values is avoided during its primary processing and storage. The possibilities of data corruption and misrepresentation by postprocessing are endless and the audit trail may be torturous; therefore a prerequisite of any test data handling system is to maintain an inviolate copy of the original data, proof against accidental or malicious alteration. The first level of postprocessing, whereby individual tests are converted into a report for the use of a wider group than those conducting the test, such as production or quality assurance, is very formulaic and optimized for its specific purpose and statistical analysis. In research and development work, the test data may be subjected to later analysis not directly related to the initial purpose of the test and used statistically in a wider investigation. The processes whereby data becomes statistics contain terms and concepts that are frequently misused.

It is vital that the basic, mathematically correct, nomenclature of data analysis is understood by both those at its source and its end use. In particular there is common confusion in the use of the terms "data" and "statistic," as in data analysis and statistical analysis. An illustrative example follows:

- The torque produced by a particular engine at 3000 rpm is not a statistic; it is data, a fact (accurate within the restraints of the experiment). To a statistician it is a specimen.
- The torque produced by 10% of the same engine type at 3000 rpm is also data. To a statistician it is a sample.
- The torque produced by all the engines of the same type at 3000 rpm is also data. To a statistician it is a population.

- The average of the population's (all the engines) torque at 3000 rpm is not a statistic; it is data. But the average of the torque value produced by taking the average of the sample is a statistic.

*A statistic is an estimate of a parameter based on data collected from a sample of a population.*

In the case of averaging, commonly used in simple processing of powertrain data, qualification of the simple average has to be made because we need to know if the spread of torque figures was over a narrow or wide range of values, and whether there were numerically more engines above or below the average value. The first qualification is standard deviation and the second is called the skew of the population. The processing of such data sets can be carried out by any analysis software; the presentation of the results requires understanding of the underlying data and the test processes. Since test acquired data is expensive in time and money, any business will need to make the widest possible use of it. A very small powertrain or vehicle test operation usually stores test data in a serially dated, flat form, database. Most importantly they also store the data as the developing and applied experience of their operative(s). Thus when test-related problems or phenomena are met more than once, the system relies on human memory to find previous occurrences so that resolution can be based on previous data or methodology.

Large organizations, having vast quantities of data and a turnover of many employees, must develop the same learning ability by structuring data storage, so it can be efficiently processed in the future and so that it adds to the corporate wisdom. Data mining is a practice carried out by staff trained in the subject and usually far removed from the test cells (discussed later in this chapter); these, like the design and use of a relational database, are highly complex subjects and are served by their own specialist literature. All this means that the test engineer must be provided with tools containing proven data handling and presentation templates that use audited formulae. In addition, security of data needs to be built into any tool so that accidental or deliberate changes to data can be seen or prevented. The difference between statistical analysis and data mining is the subject of specialist debate. For readers interested in data produced from testing propulsion systems ICEs and transmissions, the following has been found useful:

- Statistical analysis uses a model to characterize a pattern in the data.
- Data mining uses the pattern in the data to build a model.

In other words, during statistical analysis you know what pattern in your data you are looking for, while in data mining (which is usually carried out on very large data sets) you are looking for patterns, but you may, before starting, have little idea of their form or use. Two recommendations to test and development engineers, based on experience and observation of data analysis, can be summed up by:

- Restraining the impulse to extrapolate beyond available data.
- Being warned by the exclamation attributed to a geologist:

  *If I hadn't believed it, I would never have seen it!*

The presentation of data in graphical format has rather less literature than statistics but is increasingly capable of being misused with the aid of powerful graphic software. This is one more reason for the test engineer to be provided with proven (data plotting) tools containing appropriate presentation templates that use audited formulae.

## Data analysis tools for the test engineer

Proprietary data analysis and presentation tools tend to be specific to areas of research such as population trends or medical research, but a few have been adapted and designed for the engineering test industry. Some tools are specifically designed to deal with the different types of data from test bed, notably combustion analysis and emission equipment, and provide most of the mathematical, statistical, and graphical tools useful to automotive development engineers.

If the "upstream" users of the test facility, that is, the engineers requesting test work to be carried out, are not aware of the capabilities of the data acquisition system, they will not make optimum use of it. It is, therefore, essential that they should be given proper training in the capabilities of the facility and its data in order to use it to maximum advantage. This statement will appear to be a truism yet, in the experience of the authors, training in the capabilities of modern test equipment and data analysis is often underfunded by purchasers after the initial investment, which leads to the facility's skills and use leveling out well below best possible practice.

## Physical security of data

The proliferation and increased reliability of devices used for the storage and processing of electronic data has made the risk of its security more of a problem of misappropriation than of accidental loss. The privately owned computer system on which this book is being written automatically and wirelessly backs up all changed files every 30 minutes to a separate, secure storage device. The danger of catastrophic loss through hardware failure is greatly reduced from the level of that same risk that existed when the first edition was written in 1992, but the danger of human data management failings, of misfiling or overwriting, is undiminished; it is all about management.

In the 21st century, data security relies on management procedures: the appropriate distribution, access to and archiving of data, and the use of only authorized postprocessing tools and templates.

The control system software will have its own security system to guard against unauthorized copying and use. It will probably be sold with a user license that will, through embedded code capable of shutting down the system, define and restrict the licensee to use the program on a single, identified machine.

Security of data systems from malicious programs is a priority of the IT department and the imposition of rules concerning internet access on the company computer system. Some test facilities, such as those involved with military or motor sport development, can be under constant and real threat to the confidentiality of their data. Physical restriction of access of personnel is built into the design of such test facilities. However, modern miniaturized data storage devices and phone cameras seem to conspire to make security of data increasingly difficult; it is a management role to balance the threat with sensible working practices. Meanwhile, differential access to computer network systems must be imposed by password or personal "swipe card." In most large powertrain test facilities, there are three levels of password-enforced security:

- *Operator*—Typically, at this level tests may be started, stopped, and run only when the "header information," which includes engine number, operator name, and other key information, has been entered. Operator code may be verified. No alarms can be altered and no test schedules edited, but there will be a text "notepad" for operator comments, to be stored with the other data.
- *Supervisor*—On entry of password this level will allow alteration of alarm levels and test schedules, also activation of calibration routines, in each case by way of a form fill routine and menu entry. It will also give access to fault-finding routines such as the display of signal state tables.
- *Engineer*—On entry of the password, this level is allowed access to calibration and debugging routines.

There will certainly be a fourth and lowest level of access available only to the system supplier, which can be called "commissioning"; this allows system setup and will not generally be available to the user.

## Data repeatability and reproducibility

ISO 21748:2004 defines repeatability as "the precision of measurement under repeatable conditions," that is, conditions that remain identical from test to test, for example, unit under test (UUT), operator, test cell, and equipment. Reproducibility is defined as precision of measurement under reproducibility conditions; this differs in that while the UUT and test method remains the same, the test may be undertaken at a different laboratory with a different operator and using different equipment.

Several factors can reduce the quality of data from the test preparation through test execution and postprocessing of data. Many data quality issues arise as a result of misunderstanding of the task that may be down to miscommunication or poor planning in general. Several solutions to data quality may be adopted; standardization, qualification, and automation (see further reading). Standardization can improve process efficiency as there is less need to cater for individual setups.

As an example of typical differences in data repeatability and reproducibility, results are presented from several emissions measurement studies below in Table 14.1. Most notable are the difference in repeatability between particulate emissions and the others considered, a consequence of influences from the vehicle, equipment, and environment.

*Note:* It may be value to read the section of Chapter Two entitled "Cell-to-Cell Correlation of ICE performance" when considering reproducibility of test results.

Variations are inherent in measured data and observed in every test environment. To effectively manage variation, there needs to be knowledge of the system under test and the test environment. Sources of variation can then be identified at each stage of the process to provide an approximation of the precision expected. The magnitude of variation is dependent on the factors that influence the measurement being made, the repeatability of particulate

**TABLE 14.1 Comparison of uncertainty defined by ISO 21748:2004 as two multiplied by the standard deviation divided by the mean $2\sigma/\mu$.**

| | Uncertainty ($2\sigma/\mu$) (%) | | | | | |
|---|---|---|---|---|---|---|
| | CO | HC | $NO_x$ | $CO_2$ | PM | PN |
| Reproducibility | 25.5 [6][a] | 27.5 [6][a] | 10.55 [6][a] | 5.65 [6][a] | 102.90 [7], 150 [6][a] | 171 [8], 112.5 [6] |
| Repeatability | 28 [6][a] | 23.5 [6][a] | 6.9 [6][a] 107.2 [9][DISI] 29.2 [6][PFI] | 1.15 [6][a] | 69.84 [7]93 [6][a] | 183 [8] 156.5 [6][a] |

[a]*Average obtained from two different vehicles.*

emissions, for example, are generally low as there are numerous noise factors that have a large effect relative to the measurement signal. The performance of an ICE is expected to change over time as components age; some components may initially show improvement, such as with metal-to-metal seals as the surface wears in. The performance of other components is expected to reduce, such as fuel injectors as carbon deposits in the nozzles and reduces the atomization of fuel.

As previously mentioned, the use to which data is eventually put determines the requirements on it; as a consequence of this, the collection method and indeed the UUT operating points at which it is collected should be a matter for careful consideration.

One possible use for data in the process of ICE development is that collected for the purpose of engine or electronic control unit (ECU) calibration, that is, the set of data stored in the nonvolatile memory of the ECU which ultimately determines together with inputs from sensors, the position of actuators as the ICE is operated. The intent is that the position of these actuators is optimal to some objective(s)—for example, best driveability, highest efficiency, and/or lowest emissions. The values that are stored on the ECU are normally ascribed labels hinting at their role within the control strategy and may take the form of individual values, arrays, or tables of data. They are collectively known as calibration parameters.

Establishing the optimal values (to achieve desired ICE performance) of the calibration parameters is a significant undertaking that relies on the collection of volumes of data that describe the behavior or output of the ICE relative to known actuator settings, for example, by varying Spark Angle in a gasoline ICE, assuming all other actuators remain fixed, one can determine the relationship between the Spark Angle and the torque produced by the ICE, the volume of hydrocarbon emissions, efficiency, etc. This data is typically used to identify static models that allow an engineer to interpolate between the collected data points and thus understand over the continuity of the operating space the ICE behavior. The models are used within an optimization process that determines, given what is understood of the behavior of the ICE, the best position to set actuators such that the calibration objectives (fuel consumption, driveability, etc.) are met. Finally, the results of the optimization are used to determine the calibration parameter values.

In order for the calibration objectives to be achieved, the underlying data (information about the engine's behavior) has to meet certain requirements; it must through analysis provide enough information over the range of possible actuator settings to fully and conclusively understand how outputs change with changing inputs; it must be of high enough accuracy to be representative of the "truth" and it must be of sufficient precision for the engineer to be confident that for given actuator setting, the outputs observed once will be observed again, that is, repeated.

## Accuracy and precision

*Note:* The reader should also refer to Chapter 19 that covers the general points of accuracy and precision.

The statistical language that describes our understanding of measurement quality/error is simple and captured in the concepts of accuracy and precision. Both should be maximized to give a true and reliable estimate of the system under investigation. Accurate measurements are low in systematic error, and their mean value (or other central population parameter) describes well the true value of a system, subject to some variance across multiple measurements. Inaccurate measurements (i.e., those that reliably deviate from the true value) are a strong indication that the collection methodology is biased, perhaps due to poor measuring equipment calibration or a fault in the experimental hardware setup.

Precise measurements are low in variability or random error. While a precise system can be greatly inaccurate and fail to describe physical reality, its estimate can at least be described as reliable, or "tightly grouped" as visualized in the right-hand side of Fig. 19.1 (refer to Chapter 19: The Pursuit and Definition of Accuracy). Poor precision is unlikely to be driven by human or experimental factors. It can be accounted for by low-quality measuring equipment or "noise factors" such as electrical noise or even external inputs of a mechanical kind (acoustic inputs). For imprecise systems, confidence must be improved by the acquisition of more measurements.

Sources of inaccuracy and variability are often independent; therefore it is possible to have highly accurate but imprecise measurements, or highly precise measurements with poor accuracy. It is, therefore, imperative that both factors are accounted for when assessing measurement quality.

It should be noted that the performance of prediction models may also be viewed through an accuracy and precision lens. In this case, it is the mean and variance of prediction error relative to a measured data point that is assessed. In the world of statistics and data-driven modeling the negative analogues of accuracy and precision are referred to as bias and variance. Their treatment in data science and ML will be explored later in the chapter.

## Design of experiments

Data collection in powertrain development is a cost-intensive exercise. This is due to the expense of the equipment, the skilled manpower required to operate it, and the high dimensionality of modern power unit control systems. The term dimensionality here refers to the number of combinations it is possible to set the output actuators at. For example, a powertrain controller with only two actuators, say with three possible settings each, would require nine experiments to collect data at every possible combination of actuator, one with 10 actuators with 10 settings each would require $10^{10}$, that is, 10,000,000,000 experiments! It is not uncommon to have more than 10 actuators on a modern ICE control system, each with many more than 10 settings.

In order to make the problem tractable, DoEs methods are used to reduce the number of test points required. Methods for DoEs typically fall in one of the three categories; classical, space filling, and optimal designs. The choice of which is selected depends on the level of prior knowledge the engineer has about the system. Typically, where there is little knowledge, space filling designs are used; where knowledge is high optimal, designs are chosen with classical designs fitting somewhere in the middle.

Space filling experimental designs seek to fill what can be thought of as a multidimensional volume that represents all possible combinations of actuators under investigation, with a finite number of test points. The aim is to get sufficient coverage of this volume to typically fit a model, meeting some requirements, that interpolates between points. In deciding the number of points to use the time available is often the deciding limitation. Several methods are available for distributing points within the volume, but common approaches are Euclidean distance maximization, that is, maximizing the distance between all adjacent points or a pseudo random generator a method known as "Latin Hypercube Sampling" shown in Fig. 14.1.

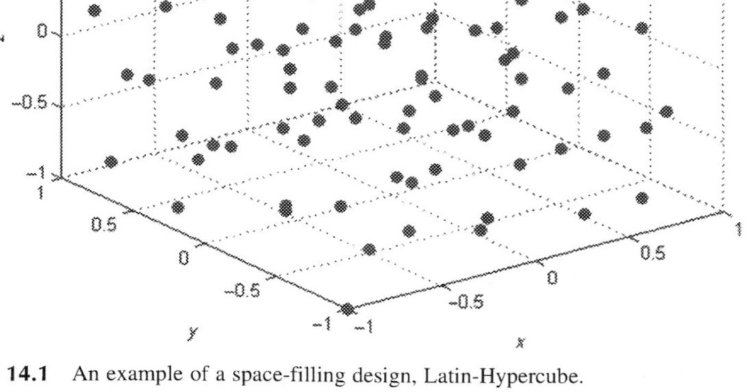

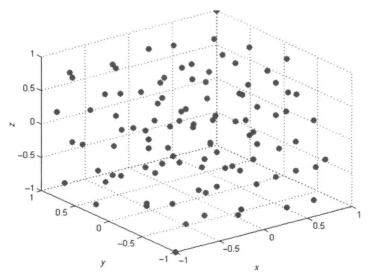


**FIGURE 14.1** An example of a space-filling design, Latin-Hypercube.

Classical designs offer a further reduction on number of test points through some rigor and principles derived from foundational statistics. The prior knowledge requirements for a classical approach are:

- A strong causal understanding of the system; identification of a response variable and its contributing "factors" (explanatory variables).
- A basic knowledge of boundaries or required areas of the design space to investigate. This sets the number of "levels" tested for each factor.
- The ability to replicate test points multiple times to quantify stochasticity in the result.
- The ability to randomize test sequencing to negate bias in test procedure and limit the effect of confounding influences on the experiment.

It is these sets of limitations that define how many test points must be run, as opposed to the space filling approach wherein a maximum number of test points are used to achieve confidence. An example scenario for a classical approach may be the optimization of an ICEs cylinder geometry relative to a performance characteristic such as brake-specific fuel consumption.

In this case the factors are easily defined based on the DOF in the designer's control (such as injector spray angle and height), the design space to be explored is controlled by size and budget constraints, as well as intuitions on the part of the design engineer as to what is reasonable (i.e., spray angle is limited by the conditions of wall wetting, height is limited by piston head collision). Finally, replicability and randomization are achievable because the design permutations are likely to be performed in simulation.

Beyond these basic principles there exist an array of design options. For example, systems characterized by curved surfaces will require the use of quadratic or higher order models for which the Box–Behnken and central composite designs are most appropriate.

Finally, there may be systems for which a high degree of prior knowledge is available (down to the required model structure) and for which the experiment is constrained. The constraints may either be the following:

- The number of runs achievable in a time or budget frame is extremely limited.
- The design space to be explored is small, but rigorous characterization is still required.

For such problems the use of optimal designs is warranted. Provided the underlying model structure for the system is known, it is possible to optimally design the experiment according to some objective, typically related to obtaining the maximum information from the experiments to fit the model, these are known as design optimality criteria. In D-optimal design (Fig. 14.2) the quality of a proposed experiment is a function of how narrow the confidence intervals are for each parameter estimate. The optimization in

| Design | | | | | |
|---|---|---|---|---|---|
| A | | B | | C | |
| 0.75 | 0.25 | 1.0 | 0.0 | 1.0 | 0.0 |
| 0.75 | 0.25 | 1.0 | 0.0 | 1.0 | 0.0 |
| 0.50 | 0.50 | 0.5 | 0.5 | 1.0 | 0.0 |
| 0.50 | 0.50 | 0.5 | 0.5 | 0.0 | 1.0 |
| 0.25 | 0.75 | 0.0 | 1.0 | 0.0 | 1.0 |
| 0.25 | 0.75 | 0.0 | 1.0 | 0.0 | 1.0 |
| $X_A$ | | $X_B$ | | $X_C$ | |

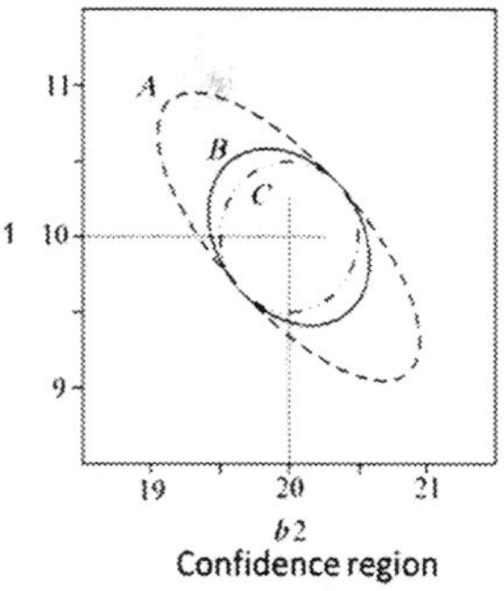


FIGURE 14.2 Example of a D-optimal design.

this case minimizes the covariance of parameter estimates for a specified model. Other optimality criteria include:

- A-optimal: Minimizing the average variance of the model parameter estimates.
- V-optimal: Minimizing the average prediction variance of the model over specific points.

## System characterization and data utilization

In modern powertrain engineering activities, extensive use is made of model-based methods. The motivation for this depends on the application but is primarily to reduce the time and cost. The means by which this is achieved is through the reduction of the dependence on data and expensive data collection experiments, through interpolation and extrapolation.

Models that are deployed in an interpolative manner assist the engineer in understanding what the most likely system behavior in between measurement points. It should be remembered that all data collected and stored from UUT experiments is discrete in nature, that is, it represents individual and very definite points within the continuum of the engines analogue operating space. This fact is a consequence primarily of the data acquisition equipment which even when sampling a continuous signal will capture and store the information collected digitally, that is, using discrete values of finite precision. Even if it were possible to store the measurement data in an analogue form, it would be foolish to attempt to since this would take an infinite amount of time on a UUT even with one single analogue actuator. The position of the actuator would have to be varied along the real number line in infinitesimally small increments as its outputs were measured.

Models used to extrapolate behavior are predictive in that they describe behavior of the system that has not yet been observed. These models often make use of an engineer's knowledge of the physics of the system under study embedded in mathematical equations. These equations will always contain parameters that require measurement, for example, volume, mass, specific gravity of fuel, etc. and coefficients that are used to modify the equations in such a way that they correspond more directly with observed reality. Creating these models requires selection of the appropriate mathematical equations, data collection to determine parameter values and coefficients, and finally further data collection to validate the final model's accuracy against observed behavior.

Since models form such an integral part of modern ICE development activities, a short exposition of the main types of models deployed is given in the following text.

## Dynamic modeling approaches

A model is a representation of reality using mathematical constructs. Physics-based models use a combination of physical first principles to approximate reality. Empirical models identify the parameters of mathematical equations using experimental test results. Consequently, models can either be physics based, experimentally based or a combination of the two. This allows the classification of models into **white-box** models, **gray-box** models, and **black-box** models.

White-box models are purely physics-based models where all parameters are known. Gray-box models are a combination of physical and experimental models. Black-box models are purely experimental based. The three model types show a significant difference in complexity, simulation speed, cost, and model fidelity. White-box models have, due to their physics-based differential equations, a high complexity and a high fidelity. However, the differential equations often require advanced, iterative solvers which result in a very slow simulation speed compared to the other modeling approaches. Black-box models, on the other hand, are very cost intensive, since they require a comparatively large amount of test data. However, the low complexity allows a very high simulation speed. Gray-box model offer a compromise between the two extremes.

## Example: engine system modeling

For the application of modeling an ICE, it is common practice to divide the overall system model into an air-path and a cylinder model. The purpose of air-path modeling is to model the pressures, temperatures, and the air mass flows through the intake and exhaust system. In-cylinder modeling encapsulates the in-cylinder processes, such as torque production and exhaust gas temperature estimation. However, when a real system is described by physical equations, it is necessary to make some assumptions or simplifications of the system, which means that it is often the case that the air-path and the cylinder can be modeled with a different fidelity.

### *White-box models*

The term "White-Box" refers to the fact that the entire model structure is visible which means that every equation used to describe a specific element of the system is based on physical principles. The overall process of the internal combustion engine is very complex and in order to describe this process with physical equations, it is first of all necessary to break it down into partial problems which can be physically described and formulated mathematically. Each problem is then solved by applying first principals from, for example, thermodynamics, fluid dynamics, chemical reactions, mechanics, and kinematics. The overall process model is finally formed by connecting the individual solutions for each of the partial problems with the other. The entire system can be represented as a combination of five elements, which are cylinder, plenum, pipe, restriction, and turbocharger. These elements can be modeled with different modeling techniques, depending on the desired model fidelity, complexity, simulation speed, and cost for parameterization.

### *Black-box models*

The term "Black-Box" indicates that the physical system behavior is not directly visible from the mathematical equation which is used to represent the system process. The models do not include any physical laws and a model of the process is obtained purely from measurements. This model type is also known as system identification and the result of the identification process is an experimental model. Black-box models can be divided into stationary and dynamic models.

#### Stationary models

The simplest stationary experimental models are nonparametric models such as grid-based look-up tables. However, the data points increase exponentially with the number of systems inputs. Consequently, map-based models are only practical for systems with one or two inputs. A more advanced solution is parametric models such as polynomials and splines, neural networks, and fuzzy models. The model parameters are identified from experimental data using parameter estimation techniques such as linear and nonlinear least squares and maximum likelihood.

#### Dynamic models

Experimental models for nonlinear dynamic systems can be separated into models with a special structure such as Hammerstein models, Wiener models, Volterra series and models with a general structure such as local linear models and multilayer perceptrons. The model parameters of the dynamic models are identified from experimental data with the parameter identification methods listed previously.

A summary of the development process of black-box models can be stated with three consecutive steps:

### Design of experiments and data collection

The first step is the definition of the model inputs/outputs, and the operating range of the ICE that the model will have to be identified and validated. Once the inputs and outputs as well as the operating range is known, DoE methods can be applied to define the experimental UUT operating test points. The purpose of DoE is to minimize the required number test points and maximize the information content of each test point. In other words, DoE allows an efficient estimation of the experimental model. As previously mentioned, three popular DoE methods are used in ICE testing, which are classical/full factorial, space filling, and optimal.

### Data modeling

Once the experimental data are available, a suitable mathematical model is fitted to the data to explain the relation between the inputs and the outputs.

### Validation and verification

The final step of the model development process is the validation of the model. Several statistical measures such as root-mean-square-error (RMSE) are used to determine the prediction accuracy of the model. It is important for the validation to use a set of experimental test points, which were not included in the data, used to fit the model. If the accuracy is not sufficient, the modeling process is repeated until the model achieves the desired results. Eventually additional experimental test data must be collected.

### *Gray-box models*

Gray-box models are the combination of white-box and black-box models. They are often known as semi-physical models since they combine physics-based models with experimental models. The combination of both modeling techniques allows a compromise between complexity, fidelity, cost, and simulation speed. Highly complex subsystems, which require complicated and time intensive solvers, can be replaced with experimental models. On the other hand, making use of simple physics-based equations to describe the general dynamic system behavior means keeping the cost for parameterization at an acceptable limit. The replacement of complex subsystems with experimental models also allows to increase the model accuracy of UUT responses which are very difficult to model solely with physical equations (such as exhaust gas emissions).

## Mean value engine modeling

The most popular way to model the dynamic behavior of an ICE is the mean value engine model (MVEM) approach. All ICE variables are described as the mean over one engine cycle instead of on a crank angle basis; for this reason, an MVEM is also called a cycle average model. The model is, therefore, only able to capture phenomena with a duration longer than one engine cycle (720 degree crank angle for a 4 stroke engine). Traditionally, MVEMs can be classified as gray-box models since usually the entire cylinder is represented with experimental models. Although some examples do exist that allow a physical representation of the cycle average cylinder variables in which case the model could also be seen as a white-box type. However, pipes and plenums along the air path are lumped together into two to six main volumes, which are modeled using filling and emptying dynamics. Each volume is treated as a storage for mass and energy, which are defined by the levels of pressure and temperature inside the control volume. The volumes are considered to be zero-dimensional (0D) which means that it is assumed that all properties are homogenous over the entire volume. Restrictions are modeled using quasi-steady 1D isentropic flow and the turbocharger is represented with a look-up table or a regression model.

The experimental cylinder model is what really separates the MVEM from the 0D and 1D physical-based, crank angle resolved white-box models. The air mass flow into the cylinder is modeled using a combination of simple physics corrected by an experimental model.

The main application field of MVEMs is the air-path control of modern engines during transient operation. MVEMs are often used as an observer for cylinder air charge estimation during transient engine operation. Another application is to use an MVEM as open-loop prediction of intake manifold pressure and temperature and air mass flow during fast throttle transients. This allows to avoid the response delay of pressure and temperature and mass air flow sensors during very fast transients. In addition, the predictor can be used to predict the air mass flow ahead of time for engines with electronically controlled throttle. This offers the advantage that the amount of fuel that has to be injected can be calculated in advance.

The validity of mean value models is a discussion point. There is an open question regarding the application of MVEMs for transient engine operation or not. This question arises as consequence of the fact that all of the empirical models contained in an MVEM are based on steady-state measurements. Three dynamic phenomena are not able to be modeled by a simple MVEM approach:

- *Inertial effects*—Pumping fluctuations continuously accelerate and decelerate the air mass inside the induction system. Therefore, during the acceleration, the gas inside the intake system has to acquire kinetic energy which is then later released during the deceleration of the gas. This energy balance could affect the cylinder air charge at the end of the induction stroke depending on valve timing and UUT operation point.

- *Wave effects*—Disturbances in the system initiated by the boundaries (e.g., intake valve closing) travel back and forwards through the induction system at the speed of sound and are reflected at open and closed pipes. This phenomenon can be used to improve the engine breathing performance and is called the wave effect. The volumetric efficiency maps of the mean value models account for that phenomenon but during transients, the waves do not have time to establish themselves, which in theory cause an error in the volumetric efficiency map.
- *Friction effects*—Mean value models only account for flow resistance that separate the main volumes of the air-path. Flow resistance in the pipes and especially in the junctions between plenum and runners are neglected in MVEMs. During transient operation, these flow resistances could have an additional impact on the filling and emptying of the manifold state equations and, therefore, lead to errors in the predicted cylinder air mass flow.

## Part 2: Machine-learning fundamentals

### Machine-learning concepts

The primary difference between ML and conventional modeling lies at the level of process, and not the ultimate goal of the modeling task. This will be explored further later in the text but for now, it is important to know that there are no DoEs in a standard ML solution and that the model's structure and parameterization are totally derived from the data. It follows that the models they produce are "Data-Driven." In order to understand the different categories of data-driven modeling approaches, it is important to clearly understand the breakdown between ML concepts. There are many ways of categorizing this, but the simplest distinction can be made between supervised and unsupervised learning:

- *Supervised learning*—Encapsulates all techniques that are trained using complete data, meaning that the data has strictly defined inputs and outputs. These models are essentially curve-fitting exercises (albeit complex and sophisticated) that allow inferences and *predictions* to be made for a new observation based on patterns in the training set.
- *Unsupervised learning*—In this the model is agnostic to outputs and searches for patterns among variables in an existing data set. This is used to establish subpopulations within the data set, a process often referred to as *clustering*.

Beyond this distinction, there are further options to subcategorize. For example, many supervised techniques have significant overlap with black-box modeling techniques that already exist in engineering system identification approaches and constitute forms of regression analysis. However, they are also deployed in qualitative predictor models called classifiers. The distinction is outlined in further detail later, with guidance on types of ML problems to be faced, along with which techniques are most suitable.

## Overview of inference models—regression and classification

A great many consumer-facing ML applications are examples of supervised inference models. They are used in one of two ways: as a regression model or a classification model. The core functionality of both model types is identical, with the only difference being the way the output is expressed:

- *Regression models*—Infer the most probable value of a *continuous variable*.
- *Classification models*—Deliver an output as a probability of a *categorical variable*. This is declared as a "true" or "false" statement.

There are different metrics for model performance of each type. In regression models, accuracy is defined by how close the predicted value of the model is to a true value for the same inputs. Metrics include RMSE, mean-absolute-error, and coefficient of determination ($R^2$), which are well established in engineering practice. In a classification model the output is expressed as a probability of the desired class being expressed. A value of $>0.5$ is interpreted as "class = true" and a value of $<0.5$ is interpreted as "class = false." Accuracy in this case is defined simply as whether the interpreted Boolean statement ("true" or "false") corresponds to the real class information for the inputs in test data. Regression modeling is a familiar concept to most engineers, and so is a better place to start in understanding the core concepts of ML without being confused by niche use cases.

## Machine learning as regression: solving "hard" problems

Regression analysis is an important tool in system identification, used to characterize physical processes for applications that include control and simulation/predictive modeling.

The difference between a ML approach and classic regression is that the latter uses data to define the parameters of a known model structure (e.g., slope and intercept), but the former also implicitly defines the structure of the model in a data-driven way. Some approaches such as neural networks also include the benefit of being universal approximators [10]. This proven quality means that the algorithm can approximate any possible function, regardless of complexity. This is not contingent upon sophisticated tuning of the algorithm, is demonstrated by the most rudimentary of neural networks, and is uniquely helpful for modeling problems for which there is little or no prior knowledge of the system behavior that can be used to predefine what kind of fit is required.

For many engineering problems having "no prior knowledge" is rare; at the least a brief interrogation of the system will likely yield some information (such as the presence/nature of nonlinearities and time-variant behavior). In most engineering processes legacy results form the basis of experience brought into new programs. This requirement for a first-principles approach forms much of the critique aimed at ML approaches. However, this assumes that such a well-structured experiment is possible. For some problems there are additional complexities, namely, high dimensionality (a large number of input variables).

High-dimensional problems are "hard" to model and investigate experimentally. This is because it is expensive (with respect to time and resource) to independently vary each input and monitor the output response. These systems are also complex and not easily visualized. Consider a problem with 12 independent inputs and 1 output and assume that there is enough reliable data to characterize the system. Given some level of complexity, this data will still be difficult to interpret solely because visual space is limited to three dimensions. There are approaches to visualizing high-dimensional trends in data, but it is commonly understood that as dimensions increase, visual interpretation becomes extremely difficult.

Any exploration of high-dimensionality leads to a discussion of popular *classifiers*.

## Modern classification problems with deep learning

Facial recognition is perhaps the most ubiquitous example of classification ML. The output is straightforward: the probability that a human face exists in the image (plus some sophistication around the location of the face in the image, delivered by the functionality of convolutional neural networks). However, the input layer of the model has dimensionality dictated by the number of pixels in the image. For a $400 \times 400$ image: 160,000 pixel. Each pixel in an image has three color values (RGB - Red, Green, Blue) that can take a value between 0 and 255. Therefore the total number of inputs is $160{,}000 \times 3 = 480{,}000$.

A number of 480,000 inputs and 1 output is perhaps the most extreme example of a multiple input–single output model (MISO). For problems of this magnitude, most ML practitioners will defer to deep learning approaches. These methods comprise variations upon the artificial neural network formula, wherein the model structure connecting the inputs and outputs is represented graphically by many hidden layers (Fig. 14.3). Increasing the number of layers in a network structure improves the granularity of feature that it may search for. This is called "representation" and is often colloquialized as follows:

- A neural network is trained on images containing images of people and must identify the presence of glasses on a face.
- The first layer of the network identifies the coarsest representation, such as a human figure against a background.
- The next layer identifies the face of the identified human figure.
- It is then the third layer that identifies facial features and smaller objects, such as glasses.

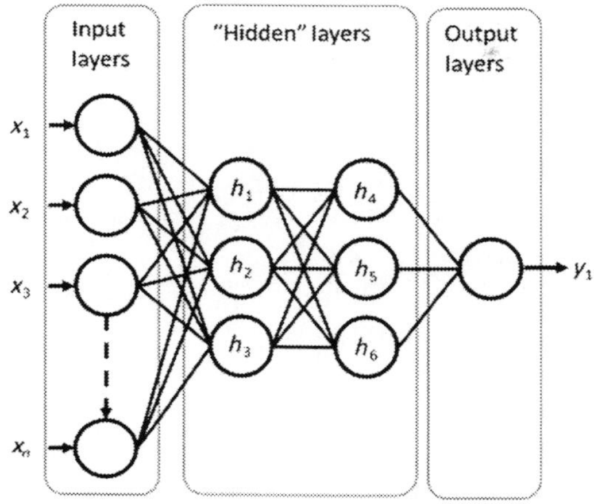


**FIGURE 14.3** Graphical representation of a multilayer neural network. The example takes multiple inputs and provides a single output, mediated by two hidden layers comprising three nodes each. Each node bears an activation function with a single weighting parameter.

## Overview of unsupervised learning—large-scale data mining

The inference approaches discussed thus far are strictly predictive in their application. Success is defined only by

*For any set of inputs, can I accurately predict the output?*

There is almost zero emphasis placed upon helping the user learn from the training data. Unsupervised approaches seek to achieve this in a number of different ways. The common feature among them is that they establish general patterns governing their training data and either provide descriptions of those patterns or detect when a new observation deviates from the observed patterns. As such, there is no strictly defined set of input-to-output relationships.

Unsupervised methods provide more scope for engineering process improvement than supervised modeling approaches, as they offer the chance to extract information from data, such that it can use it to encode knowledge. These three terms have very specific definitions, which must be accounted for:

- *Data* is an accumulation of quantitative or qualitative observations upon which a statistical framework can be imposed.
- *Information* is the formal expression of data when it is viewed through a contextual lens. Information allows us to make increasingly authoritative claims about a system based on the data we have observed by reducing uncertainty.
- *Knowledge* is the encoding of information from multiple sources into reproducible rule-sets and universal truths describing systems, their processes and behaviors. Knowledge takes observations from information

and makes them accessible and learnable by others, who have not seen the original data from which it was derived (universal truths).

Deriving knowledge directly from raw data is still a challenge in the development of complex systems such as powertrains. There is an entire subfield of data science dedicated to exploring this topic called "causal inference" [11]. The initial step of deriving useable information from data has several useful techniques associated with it.

## Informing classifiers with clustering

The example shown in Fig. 14.4 shows clustering with respect to two variables, and so it can be shown in a two-dimensional plot. The value of clustering extends into higher dimensions, where the process of identifying subpopulations is less intuitive. There are many different clustering algorithms, among which are *K*-means, DBSCAN (Density-based spatial clustering of applications with noise), *C*-medoids, Gaussian mixture model, and fuzzy clustering. At their core these techniques share similar approaches. The methodology described below is an overview of the most basic among them, *K*-means clustering.

1. Establish some initial position of cluster "center" locations in the data space. This may be done at random or relative to high-density areas of the data.
2. Calculate distance metrics for a point in the data to the proposed cluster centers. The data point is assigned to the cluster center to which it is closest.
3. Concurrently with step 2, the cluster centers are updated at each step by computing the mean position of all data points that belong to them.
4. Iterate steps 2 and 3 until all data points belong to clusters. At this point the cluster centers and size may be finalized and investigated.

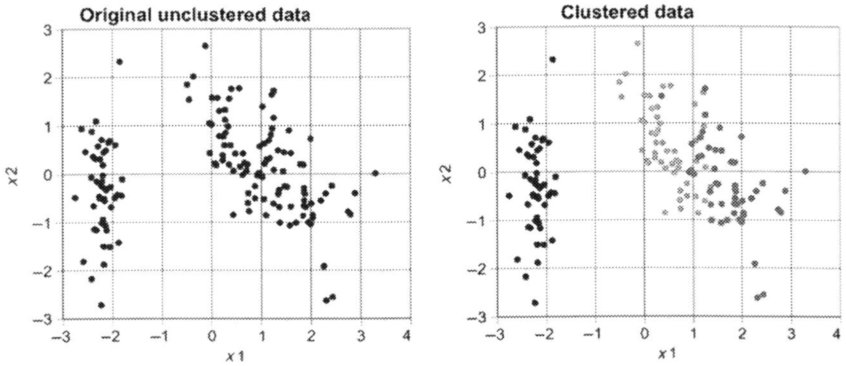


**FIGURE 14.4** Original data without clustering on left, cluster formations shown on right. *Courtesy: Kistler Group.*

This is a single example of a clustering algorithm. Factors that differ between techniques include:

- $K$-Means requires the number of clusters to be defined in advance, whereas other algorithms such as DBSCAN do not.
- The ability to approximate arbitrarily shaped and nonlinearly separable structures.
- Different distance metrics for determining cluster locations and membership.

These differences mean that the clusters formed by each algorithm may be drastically different. Therefore it is always advisable to test multiple approaches.

Users should be aware that there is no *ground truth* in clustering. This arises from fact that there is no "test data" which can be used to validate the performance of the model due to the lack of a strict output variable. This in turn makes it difficult to measure whether the clusters determined by an algorithm are "meaningful." As such, the techniques are best used when there is a strong intuition that subpopulations exist. Aside from the uncertain assignment of class labels to clusters, the best use cases of clustering are those where deviation is monitored in complex data.

Clustering is deployed in financial sectors for fraud detection in transaction behaviors, or for metadata analysis of documents to determine fakes. These are anomaly detection exercises. Once clusters of "normal" behavior are established within large data sets, new observations that fall outside the boundaries of the clusters may be flagged as anomalous. Distance metrics in each variable may then be used to identify how far from "normal" the observation is, and along what axes. Clustering approaches are also similarly used in social media analytics to monitor the migration of individual actors from one behavioral cluster to another. This transition may then be used to infer types of content to be promoted on user homepages. This highlights the utility of clustering (and unsupervised approaches generally) in applications where new data samples are being continuously generated and the behavior of the system is constantly evolving (i.e. non-stationary).

The final distinguishing feature of cluster learning relative to supervised approaches is that it may be used in an *online* and *adaptive* way, dropping and adding new observations to the model in real time. By comparison, supervised learning models are typically trained from scratch when the model no longer represents new and emerging test observations. The "online" application of clustering inference has a unique engineering application, and that is its use in *adaptive modeling*. These techniques are used to update the model structure *recursively*. Often this is achieved via a series of local models based on simplified rules expressed in fuzzy logic (i.e., if air-fuel ratio is high and cylinder temperature is high, then $NO_x$ will be high—a formulation known as a linguistic variable). Partitioning of the input data space for such models may be achieved by splitting it into regions populated by clusters of data [exemplified by the dynamic evolving neural-fuzzy inference system (DENFIS) [12]]. This ensures that the local models constructed have a level of confidence that cannot be guaranteed by using uniform partitioning approaches. The flexibility of DENFIS and similar adaptive learning systems is unique among system identification techniques and is particularly potent for online applications.

## Dimensionality reduction

There are occasions in data science wherein the data is so large and complex that even beginning to model or learn from it can be daunting. Ensuring that the data is well formatted and preprocessed can be a worthwhile investment of time and help to alleviate some of the difficulty. However, for gaining a deeper understanding of which variables to include in an analysis, there are some useful *unsupervised* learning analytical tools that can be deployed.

Principal component analysis (PCA) has its roots in classical statistics and is the most commonly used tool for dimensionality reduction in large data sets. Dimensionality reduction is the process of compressing the points in a data cloud from their initial high-dimensional space (with dimensions equal to the number of variables collected in the data) down to a smaller number of dimensions while still preserving the variability of the data and patterns captured within it. Dimensionality reduction has two primary applications:

1. *Visualization*—PCA can compress all data points from high dimensions into a two or three-dimensional spaces, the axes of which are referred to as principal components. The contribution of each feature (variable) in the data set may then be plotted in this space as a vector. The magnitude of each vector determines the contribution of the feature to the overall variance. The direction of the vector is dictated by its correlation with the principal components, but more importantly with other features in the data set. In a two-dimensional representation (shown in Fig. 14.5), vectors for features that are positively correlated are colinear and large in magnitude, whereas vectors for features that are negatively correlated are orthogonal and large in magnitude. This is a straightforward way of identifying relevant features in the data, that is, those which would characterize a model of the system. This may also contribute to a feature selection process.
2. *Reduced-order modeling*—It is possible to build data-driven models based on principal components, rather than the original features of the data set. This can result in lower fidelity models but has utility for simulations of highly complex systems. In such scenarios the sensitivity of parameter changes for a full-order model can be problematic, due to the wide array of operating conditions that need to be characterized. In this case, a surrogate model based on PCA can be used at a significantly reduced computational overhead and time investment.

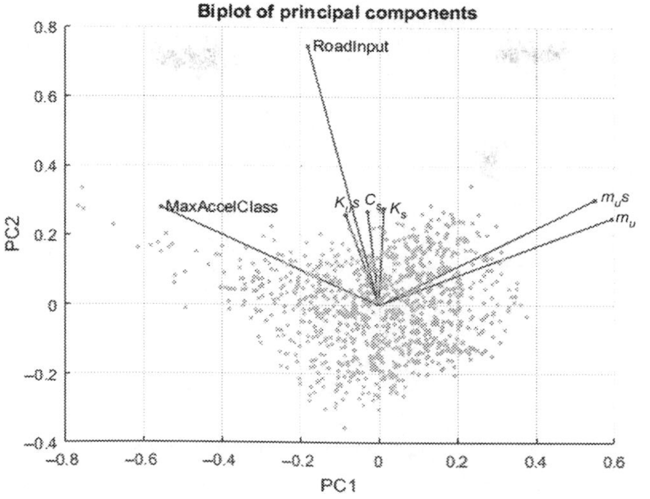


**FIGURE 14.5** Example of principal component visualization for a quarter-car suspension model. Road input displacement and sprung mass acceleration account for much of the variance, with the sprung/unsprung masses also prevalent.

## How to identify a machine-learning problem?

For those engineers who want to explore new tools and workflows that may help make their job easier, expectation management can help determine the value of the time investment. This is vital for anyone who is new to the application of data science, owing to its position as a current "on trend" topic and the common perception that it is a solution for many industrial problems.

If this sounds pessimistic, it is also worth noting that many of the field's core techniques are oriented toward problems and data types that are not represented well in powertrain testing. Issues such as the effective data mining of events in time series and the robustness of models produced by black boxes are non-trivial and would require modified, if not completely new sets of techniques. A new practitioner of ML will find value in characterizing the data problems they face, more so than investing heavily in learning the nuances and advantages of specific algorithms.

Most problem-solving is data driven; observations are made about reality and then rules are established that describe that behavior. Corrections are made to a policy or system that rectify/avoid disadvantageous behaviors, or optimize a reward/minimize a loss. This is true in engineering, but also in fields as diverse as business, finance, marketing, politics, etc. However, not all data-driven problems are ML problems, in the sense that the techniques laid out in this chapter are not always optimal solutions.

For example, ML finds the most success in data-abundant scenarios wherein a feature space is naturally explored. Search engine optimization is served well for this reason, as the data of millions of users is generated at minimum cost daily. It is this "naturalistic" data upon which models can be constructed. Conversely, using a black-box ML approach to solve a problem for which new data must be generated from the ground up may prove to be inefficient, particularly for complex systems. Complexity arises in higher order models because the number of operating conditions increases dramatically with dimensionality. Characterizing a complex system in a purely black-box manner with no model simplification, prior assumptions, or imposed boundaries requires overwhelmingly expensive and time-consuming data collection tasks. This is one of the drivers behind the engineering tendency toward simplified models (in addition to ease of tuning/adjustment and computational run-time). However, if a large reserve of data can be identified which does characterize a behavior (such as repeat runs of a legislative drive cycle/endurance test or real driving data), this may be used in previously unforeseen ways using ML approaches but with some given creativity.

A major advantage of ML is large-scale data handling at low computational cost. Many powertrain testing tasks involve running the same repeat maneuvers that last seconds over a duration of hours or days. Sometimes these tests will be for validation purposes, and, therefore, the primary objective is to identify anomalous or outlying behaviors. This is easy enough for a human operator in theory but hours of time series visual analysis is not a compelling or interesting task—and will thus be prone to the effect of human error at some point. For the pattern-learning capabilities of unsupervised ML, such tasks may be automated. Clustering approaches may be used here, but some diligence is required in making them work for ICE test data. This is primarily because ICE data exists in the *time domain*.

## Part 3: Machine learning: data preprocessing

### Understanding data structure

Statistical learning from powertrain test data presents its own unique set of challenges. Much of the data collected takes the form of high frequency ($\sim > 1$ Hz) time series signals. This is a necessity for monitoring dynamic processes and the effectiveness of control strategies, particularly during rapid transient events. However, many ML frameworks are built on data that is not expressed in time series, but instead in *relational databases*.

It is quite common knowledge and many engineers will already understand the standard shape of a database: a collection of records in tabular format (sometimes called "relations") with columns that denote features of the data, and rows that are individual observations of the data set. Beyond this, relations may incorporate reference features such as order numbers or identification numbers that may be searched across the database. These features amount to data that is heavily structured. Data not only may be accessed *en masse* for training models but also may be searched, filtered, or reorganized via queries typically in SQL (Structured Query Language) without fundamentally altering the original data itself.

The core shape of relational databases will be familiar to those who have worked with time series data, as they are also expressed in a tabular format. There is one key structural difference however: the reference feature of entries within a time series is timestamps, whose order must be immutable. This means that the value of time series is tied to the order of its time stamps, such that it is lost if the data is in any way filtered or reorganized.

The primary function of a time series is to observe the *dynamics* of a system, such that the observation at $T=t$ only yields maximum information if it is considered relative to an observation at $T=t-k$. The ordered nature is also essential for observing *periodic behaviors* or for *anomaly detection*. However, statistical learning methods are agnostic to this property and are usually trained on static observations with no temporal dependency, so the dynamics must be incorporated within the training data if they are deemed to be important.

Table 14.2 demonstrates the problem with "raw" time series data. Here two features are collected: the gas temperature through an unspecified component and ICE torque at intervals of time "*a*." Broadly it seems that the temperature is dependent upon torque, but on closer inspection, the rate of change of the torque is also a factor (Table 14.3). A model of this relationship is, therefore,

**TABLE 14.2** Time series of temperature and ICE torque.

| Time (s) | Temperature (°C) | Torque (Nm) |
|---|---|---|
| $t$ | 250 | 230 |
| $t+a$ | 471 | 300 |
| $t+2a$ | 234 | 180 |
| $t+3a$ | 216 | 150 |

**TABLE 14.3** Data from Table 14.2 but with added torque dynamics.

| Time (s) | Temperature (°C) | Torque (Nm) | Rate of change of torque (Nm/s) |
|---|---|---|---|
| $T$ | 250 | 230 | – |
| $t+a$ | 471 | 300 | $70/a$ |
| $t+2a$ | 234 | 180 | $-120/a$ |
| $t+3a$ | 216 | 150 | $-30/a$ |

dynamic, and to fully characterize it, the training data must reflect this. To incorporate the dynamics the training data at each time step will normally include the data from the same variable over the previous k time steps, which characterize the rate-of-change information. Input variables over previous time steps are often referred to as "lagged variables."

Such models are often referred to as *auto-regressive*. When considered outside of a strict modeling perspective, the addition of dynamic variables means that the time series can now be filtered or rearranged while still preserving valuable information. This has value for less sophisticated statistical techniques (such as determining what percentage of a drive cycle is spent at certain operating points, or how dynamic the maneuvers are on average).

## Training/test data split for best assessment of performance

A well explored problem in data science is making stable models that generalize well. This means that they can make accurate predictions beyond the test data used to validate the model. There are two modeling characteristics that account for poor generalizing: *bias* and *variance*. The reader should note that these terms are analogues for the concepts of *accuracy* and *precision* explored in conventional system modeling, but with subtly different implications.

- *Bias/inaccuracy*: The tendency of a model to fit too closely to a subset of patterns or behaviors that do not describe the whole system well. A biased model will uniformly produce inaccurate predictions.
- *Variance/imprecision*: A high-variance model encodes patterns that are just noise in the training data, making them extremely sensitive to inputs and imprecise in their predictions.

It is important to note that bias and variance are both by-products of training data and validation rigor, and that they are not inherent in model structures. This is one of the primary risks in data-driven approaches to problem solving: data is not created divorced from context. It is always collected by humans using processes and tools designed by humans.

When training a data-driven model, there is always some degree of a bias-variance trade-off, as biased models always underfit with insufficient training data and high-variance estimations are characteristic of overfitting due to excess training data. The trade-off is, therefore, affected by how the training-test data split is defined. This is analogous to the two-step verification process of traditional modeling.

Training data is used in the learning routine of an algorithm to define the model. The inputs of the test data set are fed to the trained algorithm to produce an estimate, which is compared against the real output to determine error and overall model accuracy. It is part of data science practice to ensure that the test and training data are derived from the same sample. This is because producing a bespoke sample for use as test data is more likely to have methodological bias based on how/when/where it is collected and, therefore, belong to a different distribution entirely.

Validating a model using one test set and one training set results in an accurate result for a small number of test points but provides no metric as to how well the model generalizes to other unforeseen data. This is because there is no way of telling if the model is over- or underfit, as we only have one set of predictions to work with.

*k*-Fold cross validation adds an extra step to the training and validation process: splitting the data set into *k* even sized "folds" and then training and validating the model *k* times, each using a different partition as the test data. It allows all of the data to be used for both validation and training purposes. This is visually explained in Fig. 14.6.

*Note:* Cross-validation, also called rotation estimation, is any of various similar model validation techniques used for assessing how the results of a statistical analysis will generalize to an independent data set. In the context of cross-validation the test set may also be referred to as the validation set.

*k*-Fold cross validation, therefore, presents *k* sets of validation results that may be compared to each other (or averaged). This adds subtlety to our assessment of model performance. Inconsistent accuracy across multiple folds shows that the model is having difficulty approximating the function, perhaps due to overcomplication of the model (overfitting) and, therefore, high variance, whereas accuracy that is consistent but low across all folds says that the model is oversimplified (perhaps due to insufficient training data) and is, therefore, biased. A model that is accurate across all folds consistently represents the best generalizing behavior.

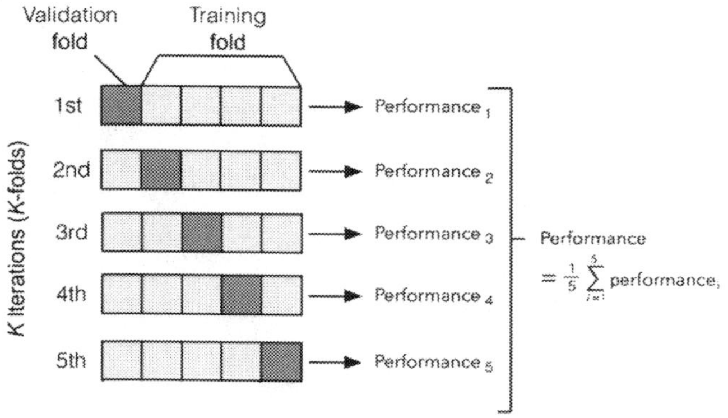


**FIGURE 14.6** All permutations of a fivefold cross validation.

When performance is deemed satisfactory in cross validation, the final model may be trained on the whole data set and tested on new samples.

It is important to note that the bias-variance trade-off is also affected by the number of folds used in cross validation. As a rule, a smaller number of folds induces bias (this is intuitive, fewer folds means a smaller training set) and many folds will produce highly varied predictions. General best practice suggests that $5 < k < 10$, but this may be optimized case-by-case.

In conclusion, cross validation is a superb tool for accessing the performance of a data-driven model with a large feature space, makes the best use of available data for validation purposes (especially when the data sample used is "small") and facilitates useful insights as to the generalization of the model.

## Feature selection and problem definition

For high-dimensional, low sample problems, feature selection is required to avoid the curse of dimensionality:

> *As the number of dimensions increases, the volume of the feature space increases dramatically. Data in such a space is sparse. This makes it hard to build a model due to the difficulty of obtaining statistical significance from sparse data.*

Generally, it is accepted that the curse haunts data *wherein number of observations is less than the number of dimensions*, though choice of algorithm can help to mitigate it. Feature selection also lowers the threat of the curse, by extracting features of low value and increasing the sample size-to-dimension ratio.

The idea of feature selection is intuitive: select only those needed to produce a working model of the system. However, as has previously been identified in this text, many data-driven problems are characterized by low prior knowledge of the system. Under these conditions, understanding of the data may be sufficiently low that even the relevant features are unknown. Data collection may also be so broad, sprawling and untargeted that redundant features are also collected. The definition of these terms is as follows:

- *Relevant features*—These are highly correlated to the output variable(s). Irrelevant features may be invariant within a data set or have no noticeable correlation to other features.
- *Redundant features*—These may be well-correlated with the desired output but are also broadly correlated with other features in the set. Redundancies can, therefore, be highly dependent on other features (and, therefore, low in value for modeling) or deemed to be proxies. By example: two sensor channels that record identical values, with the only distinction being that one is a record of the raw data and one bears filtered values. The use of both to characterize behavior is unwarranted.

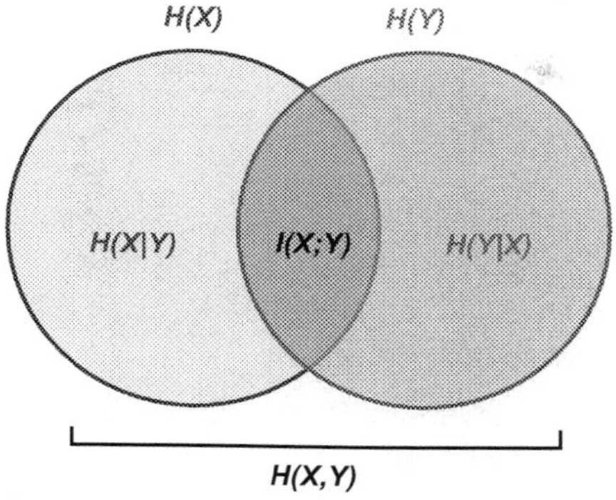


**FIGURE 14.7** Venn representation of two features *X* and *Y* as a subtractive breakdown of entropies *H*(*X*) and *H*(*Y*) producing mutual information *I*(*X*:*Y*).

A typical feature selection policy is minimum redundancy, maximum relevance. This may be achieved via a mutual information (MI) criterion. This is a statistical assessment of the relationship between two variables by comparison of their probability distributions. It is an approach that is more flexible and general than many correlation approaches and observes the "amount of information" that is contained in each about the other as new observations appear. The result of the calculation is the overlap of Shannon entropies (rate of information acquisition from a data set/source), visually represented in Fig. 14.7. *Shannon entropy H* is given by the formula:

$$H = -\sum_{i} p_i \log_b p_i$$

where $p_i$ is the probability of character number $i$ appearing in the stream of characters of the message. This is a quantification of an abstract concept, the details of which are beyond the scope of this text but is useful for identifying both relevance and redundancy.

Relevance is ranked by how much MI is shared between the target/output variable and each other measured feature in the data set. If a 20 feature subset is required, the process is as simple as selecting the 20 features that have the highest MI with the output variable.

While relevance is determined by maximization of MI between the output and inputs, redundant features are those that have the highest mean MI with every other feature in the set (excluding the output).

The process for using MI as a criterion for feature selection can vary slightly: relevant features may first be selected and then redundancy evaluated based upon the subset, or a combined criterion may be used to access all features for relevance and redundancy concurrently. Regardless of the process used, an MI approach to feature selection is one that is computationally effective and divorced from context, such that application knowledge is less essential.

The MI criterion comes from a purely statistical approach based only on the data used to train a model. It is possible, however, to use heuristic approaches to feature selection that incorporate the performance of a given model type as a factor.

A simple and intuitive example is sequential feature selection. The process is as follows:

- Build a single-input, single-output model once for every feature in the set of features n under consideration.
- Keep the feature $X1$ that performs best according to a desired accuracy metric.
- Build $n - 1$ two-input, single-output models using $X1$ and each of the remaining features. Once again, keep the feature $X2$ that performs best alongside $X1$.
- Repeat the process of acquiring features up to $Xk - 1$, where the $k$th run is the one in which no model accuracy improvement is seen.

This approach is valuable when the modeling technique is already decided upon and optimizes specifically for this model. The MI criterion contrasts by being agnostic to the modeling method. It should be noted that sequential selection occurs by training and testing models proportional to the number of features. Therefore a heuristic approach becomes computationally intractable given a very large number of features and a large data set. However, it is the more optimal approach in identifying the number of features required, in addition to which features may be used.

## Part 4: Data science, software tools, workflow, and maintenance

### Data-utilization workflow overview

In the Introduction of this chapter the term "Data Utilization" was introduced. This is a concept novel to this text and is used to describe any of the tools (traditional or ML adjacent) that extract value or information from data, whether in a direct modeling capacity or a data mining one. While data utilization is always the end goal, the peripheral processes that enable it have also been discussed. In Fig. 14.8 a broad summary of such a workflow is shown.

Where the required subtasks are as follows:

- *Selection*: Data acquisition, determining pre-existing knowledge.
- *Preprocessing*: Cleaning and focusing of data, finding related events and data points.
- *Transformation*: Putting the cleaned data in a format suitable for analysis (reducing dimensions, matching units).
- *Data mining*: Usage of statistical algorithms for generating additional knowledge.
- *Evaluation*: Evaluation by the engineer, preparation for communication to stake holders.

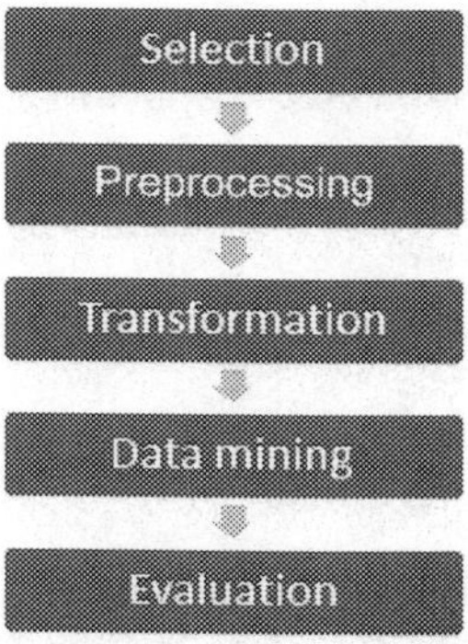


**FIGURE 14.8** Workflow steps in the application of data science to engineering data sets.

Data needs to be collected prescreened, preprocessed but most importantly, the goal or output of the exercise should be clearly defined where possible—applying data science algorithms to large data sets costs valuable computational time and resources. Therefore a carefully thought out experiment is a far more efficient approach compared to just applying the chosen algorithms to see what happens!

In categorizing data-utilization tools, we have variously discussed approaches to conventional modeling and the breakdown of supervised and unsupervised ML. Here we present further subcategorization for comparison in a list that is by no means exhaustive:

1. *Outlier detection* (anomaly detection)—Engineering data can contain outliers, and these are often considered as noise, or erroneous in nature. However, they can be significant in assessing overall data quality, their removal should be undertaken carefully.
2. *Cluster analysis*—Techniques that are used to classify objects or data having common attributes into groups known as clusters. Note that there is normally no prior information about the group or cluster membership for any of the objects. Although this is not always true with engineering data where some basis of relationships can be known.
3. *Classification/prediction*—The former is the problem of identifying to which of a set of categories a new observation belongs, on the basis of training data. For the latter the algorithm derives the model or a predictor from this training data. The model predicts a categorical label beyond the space of the training data.
4. *Association analysis*—Used to find interesting relations between variables in data sets. It identifies the frequent individual items in the database and extending them to larger and larger item sets as long as those item sets appear sufficiently often in the data mass. The results can be used to determine association rules showing general trends in the data.

5. *Regression analysis* (regression)—A very common statistical approach that is foundational to modeling. Regression analysis is a set of statistical methods used for the quantification of relationships between a dependent variable and one or more independent variables.
6. *Data summarizing*—Consolidation of the analysis methods used to create a report, of documentation that resolves the data science into real world, context applicable information that can be used by the engineer to solve, or at least understand, the problem as defined in the first place during the experimental definition.

The last step is probably the most significant as it is the consolidation of the advanced analysis into a real world meaning. This could be display objects—used in reports or documents. Or data that is exported to other tools or workflows for further processing. Data science approaches can provide new insight to data sets that are otherwise impossible to make sense of. The summarizing and reporting may not always be presented to engineers though. Therefore presentation of the data has to be appropriate in length, detail, style and content to the intended audience and should always be prefixed with an executive summary.

## Software life cycle management and maintenance

Where tools are widely distributed within an organization, management of the roll-out of updates including centralization of methods and workflows is essential. In many large original equipment manufacturers, this will be a task dedicated to a team, person, or job function. It is often the case that tools that are widely distributed can become tightly integrated into workflow and operational processes, as is often the case with computer-aided engineering or design software tool suites. This then creates a risk to productivity and efficiency should a tool fail—and the topic of software maintenance and support is an essential discussion to be had within a facility so that appropriate budgeting can be put into place. It is often the case the IT departments are in control of the management of software tools, even though they are not the users and have no idea of the application or importance of such software in a facility. As such, the maintenance and update costs are generally supported from a different budget than the general facility capital investment. Good communication between the relevant departments to ensure that tools are maintained and robust as well as evolving with the general facility is essential. The typical costs for software maintenance of a perpetual licence can be broken down into three main areas:

- *Maintenance*—Hotfixes and bug fixes, generally within a main release (v1.X). Ensuring that the software remains functional and reliable may include small extensions and limited additional functionality.
- *Subscription*—Normally main version updates (i.e., V1.2 to V1.3) that are often deployed in response to operating system updates of the host system in

the facility (i.e., Windows 8 to Windows 10), to ensure reliability and compatibility of the software for the test environment computers. It is normally the case that major features additions and functionality updates are included. A subscriber also often has access to the software manufacturer development requirements listing and development process, in order to give their input for specific needs in the software, as well as being able to benefit from the community of users by proactively receiving updates derived from other users' issues and experience. There is a general move toward subscription only based licenses (i.e. non-perpetual), since it provides the seller with a predictable income stream over time, the benefit to the purchaser is debatable, other than lower cost entry to gain access to the software in the first place.
- *Support*—Connection to the providers support network that could range from a simple phone or email contact—to a multilevel support network with ticketed issue tracking and guaranteed response times. This support could include not only reporting of bugs and user issues but also application-based support, to help solve a domain specific task. This requires considerable application specific knowledge and capability within the support network at the software manufacturer, which can be a challenge for them to supply and maintain. Note that support is generally included in subscription only (non perpetual) licences.

The costs of such support models are often very variable, along with the actual scope. In the authors experience the software support contract is a highly negotiable deliverable that requires careful discussion with the software supplier. As a general rule, the cost model is based as a percentage of the list price of the software, including all options and extensions, paid annually. A typical figure being 20%, which includes all the three elements mentioned previously. It is often the case that the software users want to fragment this deliverable in order to optimize costs, for example if a local support person is within the company, then the support element may not be required.

It is important for the facility engineer to understand and appreciate the value of maintaining the facility software, which is as important as maintaining the test facility hardware. Especially where software tools are mission critical—for example, test bed control software, ECU calibration tools. It is also important to understand the negotiation levers that are available to optimize these annual costs such that they can be justified internally and signed off easily:

- Each element of software support (maintenance/subscription/support) has a value of approximately one-third (of the annual maintenance factor). Only what is required should be purchased and this should be possible from the provider—who must be able to deliver value to the user from support contract.
- The percentage value on which the support is based is negotiable and varies with suppliers, also whether this factor applies to the purchase cost

(fixed) or the list price (variable, increasing yearly) of the software, as originally purchased.
- A multiyear contract is often attractive for the software supplier and should allow cost optimization discussions and possibilities between them and the user.
- Typical response times should be stated by the supplier and reporting of issues and response performance should be possible with a "ticketed" system—this must be included in the contract such that performance levels can be monitored (whether they are included in the contract or not) with the potential to optimize cost based on actual performance.
- Additional value that can be offered will help to justify the software maintenance contract, for example, if there is little use of the support element, then the supplier should include some other "value" to justify the contract—that is, training or support days, user workshops, etc.
- If software tools in the facility are coming from the same supplier, then "bundling" for the purpose of optimizing facility costs should be a priority discussion for the facility engineer and also the software supplier.

It is important for the facility that a software maintenance contract provides much more than just a "software insurance policy." Most suppliers of software include in their terms and conditions that they accept no third-party responsibility should their software fail or give wrong answers. It is, therefore, important that real value is achieved by putting a maintenance contract in place that protects the facility from mission critical failures and ensures that the software is used effectively in the working environment to increase efficiency.

## Appendix

### The future of data in powertrain engineering design and development

Data science is a topic that is prone to inexhaustible jargon, truisms, and codes of practice, as is the case in many technical fields. The content of this chapter will not make subject matter experts, or completely demystify the language of the average industry data scientist but instead will highlight when and why a data science approach is warranted. With the fundamentals discussed the next task is to look to the future, leveraging the advances of recent history to augment engineering practices with insightful data approaches.

The data revolution of the 2000s and 2010s has been facilitated by greatly expanded connectivity and high performance computing innovations, that are able to generate, transmit, and store data globally into distributed repositories: this is colloquially referred to as Cloud Computing and has led to the rise of a commodity known as "Big Data."

## The Vs of big data

These are an ever-expanding set of criteria for assessing the utility of large-scale data. Five of the core principles are:

- *Variety*: The breadth of the data in terms of the different structured and unstructured types collected, in addition to the factors/features observed.
- *Value*: The value of the data is determined by how they are used, and with what tools they are analyzed. Collecting vast volumes of data on a very narrow set of conditions and failing to deploy valid analytics means reduced insight and overall benefit.
- *Velocity*: The rate at which data are generated, stored, and analyzed. High velocity not only means more abundant and, therefore, voluminous data but also means a higher rate of information gain given adequate tools. Velocity must be accounted for in any information architecture; however, as in some use cases, real time access and synchronicity (time-matching of data from different channels) is essential.
- *Veracity*: A measure of certainty about the data's origin, and how trustworthy they are. Insights drawn from untrustworthy data lead to uncertain or even erroneous conclusions.
- *Volume*: The depth of data accumulated, wherein more is always deemed to be better given that value can still be extracted.

Compared to other technology-focused industries, the automotive industry is slow to adopt a data-driven approach to product development predicated on analysis of big data. This is perhaps due to a perceived lack of necessity; most development resources are directed to satisfy global standards for emissions and other performance metrics before product release. Despite this, the industry is trending toward more diverse sources of data and methodologies for storing them.

Automotive manufacturers of the present are beginning to incorporate telecommunications features such as in-car entertainment that can receive software updates and Wi-Fi hotspots via 4G and 5G. This investment is driven by the consumer experience but opens doors to engineering advances, such as powertrain control unit software and calibration download via cellular communications.

If such a framework is available for delivering information to a road-going vehicle remotely, it is only a small leap of imagination, and a relatively large hardware investment, to reverse the path of information transfer: "vehicle to cloud." There already exists a proof of concept: the aggregation of mobile data from road users to characterize daily traffic conditions in applications such as Google Maps. The collection of more data based on how, when, where, and why customers use their vehicles after purchase has obvious connotations for making business decisions. At the technical level there is scope for the collection of performance-oriented data from the powertrain to facilitate statistical analysis of fault codes, reduce uncertainty around models of physical processes and assess the ramifications of certain control decisions.

Observations from real-world conditions may translate into a change in the way ICE testing activities are performed in a dynamometer setting. "Problem maneuvers" or driving behaviors in real world conditions may be identified that are not reflected in typical test cycle definitions. These may be used to inform test profiles for validation and verification, in addition to satisfying emissions requirements such as Real-Driving Emissions. These opportunities are tied to the intrinsic properties of large data as captured in the "Five Vs". There is a richness in naturalistic data that results from its abundance and variety which is simply not observed in predefined experimentation.

In addition to "fleet accumulated data," there is also a drive for testing organizations to centralize data storage and analytics generated by multiple sources. While this presents some security challenges (particularly in segregating and firewalling data from different customers), the benefits lie in ease of access to large data and the computational power that can be leveraged. Modern analytics software is chiefly directed at visualizing and comparing individual test runs against each other, while providing simple metadata about each channel recorded (mean value, maxima, minima, etc.). Cloud storage and computation facilities present the opportunity for more sophisticated analytics across multiple test files, new visualization opportunities, and more robust project tracking.

## References

[1] J. Gama, Knowledge Discovery From Data Streams, FL: CRC Press, 2010.

[2] I. Goodfellow, Y. Bengio, A. Courville, Deep Learning, MIT Press, 2016.

[3] M. Hayes, Statistical Digital Signal Processing and Modeling, John Wiley & Sons, 1996.

[4] K. Murphy, Machine Learning: A Probabilistic Perspective, MIT Press, 2012.

[5] L. Ljung, System Identification, Theory for the User, Prentice-Hall, 1999.

[6] C. Vallaude, C. Berthou, D. Pingal, and S. Ficheux, Particle Number Round Robin Test, UTAC, 2009.

[7] J. Andersson, B. Giechaskiel, R. Muñoz-Bueno, R. Sandbach, P. Dilara, Particle Measurement Programme (PMP) Light-duty Inter-laboratory Correlation Exercise (ILCE_LD) Final Report, Institute for Environment and Sustainability, 2007.

[8] U. Kirchner, R. Vogt, M. Maricq, F.M. Co, Investigation of EURO-5/6 level particle number emissions of European diesel light duty vehicles, in: SAE Technical Paper 2010-01-0789, 2010.

[9] Sougawa, Y., Koseki, K., Kawaguchi, H., Nishimura, J. et al., Improvement of Repeatability in Tailpipe Emission Measurement with Direct Injection Spark Ignition (DISI) Vehicles, SAE Technical Paper 2002-01-2710, 2002.

[10] K. Hornik, M. Stinchcombe, H. White, Multilayer feedforward networks are universal approximators, Neural Networks 2 (1989) 359–366.

[11] J. Pearl, D. Mackenzie, The Book of Why: The New Science of Cause and Effect, Penguin, 2018.

[12] N. Kasabov, DENFIS: dynamic evolving neural-fuzzy inference system and its application to time-series prediction, IEEE Trans. Fuzzy Syst. 10 (2002).

## Further reading

Beattie, T., R.P. Osborne, W. Graupner, Engine test data quality requirements for model based calibration: a testing and development efficiency opportunity, in: SAE Technical Paper 2013-01-0351, 2013.

R.A. Fisher, The Design of Experiments, Oliver & Boyd., 1935.

D. Salsburg, The Lady Tasting Tea: How Statistics Revolutionized Science in the Twentieth Century, W.H. Freeman, New York, 2001.

Chapter 15

# Measurement of liquid fuel, oil, and combustion air consumption

## Chapter Outline



## Part 1: Fluid fuel and oil consumption measurement

Note: The section of Chapter 7 entitled "In-cell fluid fuel systems" should be studied before this section on fuel consumption measurement as it is important to understand the importance of the role of the liquid fuel supply system.

### Introduction

The general, vehicle driving public notice and may record cumulative fuel and oil consumption over a distance driven or period of time and, until the arrival of instruments capable of measuring "instant" air and liquid flows, test facilities followed the same practice.

Standard test methods can still record cumulative consumption over a period of time or they measure the number of revolutions over which a known mass of liquid fuel is consumed. These test methods and the

**Engine Testing. DOI: https://doi.org/10.1016/B978-0-12-821226-4.00015-2**

instruments used are still viable, but additionally and increasingly test engineers are required to record the actual, transient, consumption during Real Drive Emissions test and other rapidly changing operational conditions.

Modern automotive internal combustion engine (ICE) designs incorporate a number of fuel spillback, fuel return, and vapor recovery strategies that make the design of test cell consumption systems quite complex. Liquid fuel conditioning and consumption measurement in the test cell is no longer a matter of choosing the appropriately sized volumetric measuring instrument, but rather of designing a complex pipe circuit incorporating instruments, heat exchangers, vapor traps, and pressure regulators as shown diagrammatically in Chapter 7, Energy Storage, Fig. 7.5.

The great problem of liquid fuel handling and consumption measurement is vapor and air, whether entrained in the liquid flow or as a bubble creating an air lock somewhere in the system; the more complex the tubing systems and the higher the liquid temperatures, the worse are the problems. All systems must be designed and installed to minimize vapor formation and allow release of any air or vapor that does get into them.

For all liquid fuel consumption measurement, it is critical to control fuel temperature within the metered system so far as is possible; therefore most modern cells have well-integrated temperature control and measurement systems installed as close to the engine as is practical. The condition of the fuel returned from the engine can cause significant problems as it may return at pulsing pressure, considerably warmer than the control temperature and containing vapor bubbles. This means that the volume and density of the fluid can be variable within the measured system, and that this variability has to be reduced so far as is possible by the overall metering system and the level of measurement uncertainty taken into account.

## Measurement instruments for liquid fuel consumption [1]

Whatever type of instrument is used in the measurement of liquid fuels they must be capable of handling the range of liquids used without suffering chemical attack to any of its internals and seals. It is also vital that the meters and the local pipe systems in which they are installed should be capable of complete draining to avoid cross contamination when changing fuels; the alternative is to use cells dedicated to only one fuel type.

There are two generic types of fuel gauge intended for *cumulative measurement* of liquid fuels on the market:

- *volumetric gauges*, which usually measure the number of engine revolutions, at a constant power output, taken to consume a known volume of fuel either from containment of known volume or as measurement of flow through a measuring device and

- *gravimetric gauges*, which usually measure the number of engine revolutions, at a constant power output, taken to consume a known mass of fuel.

Fig. 15.1 shows a volumetric gauge in which an optical system gives a precise time signal at the start and end of the emptying of one of a choice of calibrated volumes. This signal actuates a counter, giving a precise value for the number of engine revolutions made during the consumption of the measured volume of fuel.

These devices are still made in both manual and automated forms and prove to be robust and reliable; however, repeatability of results depends upon the elimination of vapor or air bubbles in the fuel supply.

Fig. 15.2 shows a gravimetric gauge designed to meter a mass rather than a volume of fuel, consisting essentially of a vessel mounted on a weighing cell from which fuel is drawn by the engine. The signals are processed in the same way as for the volumetric gauge, a typical base specification being:

- measuring ranges: 0–150, 0–200, 0–360 kg/h or 10–200, 50–500 g;
- fuels: petrol and diesel fuels and biofuels; and
- computer interface: serial RS232C, 1–10 V analog plus digital *I/O* or RS 485.

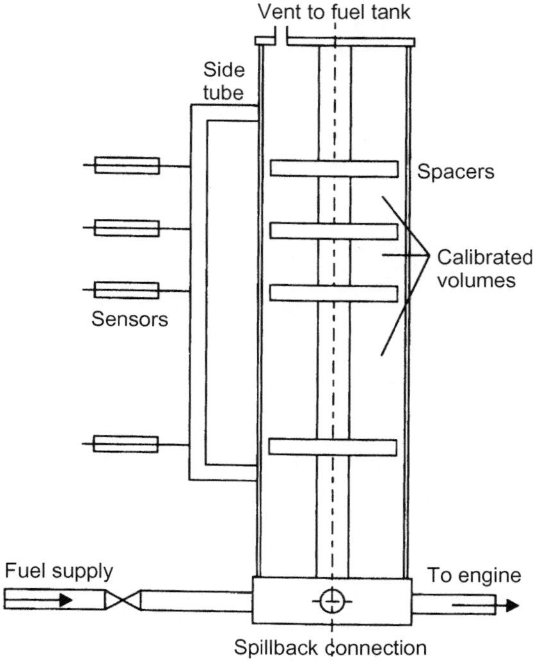


**FIGURE 15.1** A diagrammatic representation of a volumetric fuel consumption instrument capable of integration with automated control system.

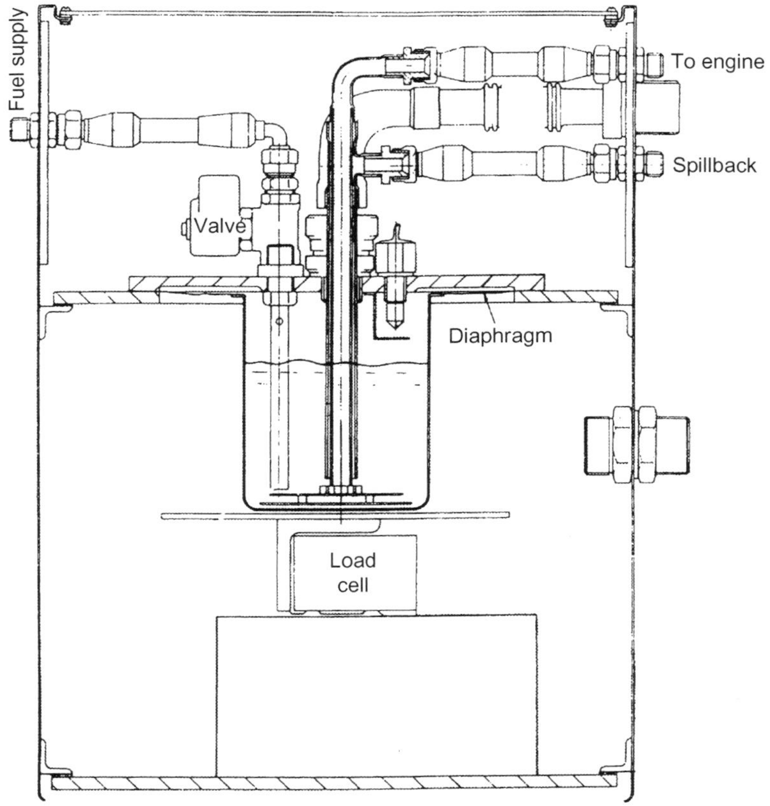


**FIGURE 15.2** Section through a typical gravimetric fuel gauge or "fuel balance."

These gravimetric gauges, also known as fuel balances, have the advantage over the volumetric type that the metered mass, between minimum and maximum, may be chosen at will and a common measuring period may be chosen, independent of fuel consumption rate. The specific fuel consumption is derived directly from only three measured quantities: the mass of fuel consumed, the number of engine revolutions during the consumption of this mass, and the mean torque. The disadvantage of a fuel balance is that it cannot be used continuously in the same way as a flowmeter, but the inherently greater accuracy of cumulative fuel consumption measurement is due to the averaging of three direct measurements, while rate measurements involve four or, in the case of volumetric meters, five measured quantities.

A further type of gravimetric fuel gauge exists, in which a cylindrical float is suspended from a force transducer in a cylindrical vessel. The change in flotation force is then directly proportional to the change of mass of fuel in the vessel.

All gauges of this type have to deal with the problem of fuel spillback from fuel injection systems, and Fig. 15.3 shows a circuit incorporating a gravimetric fuel gauge that deals with this matter. When a fuel consumption measurement is to be made, a solenoid valve diverts the spillback flow, normally returned to the header tank during an engine test, into the bottom of the fuel gauge. It is not satisfactory to return the spillback to a fuel filter downstream of the fuel gauge, since the air and vapor always present in the spilled fuel lead to variations in the fuel volume between fuel gauge and engine, and thus to incorrect values of fuel consumption.

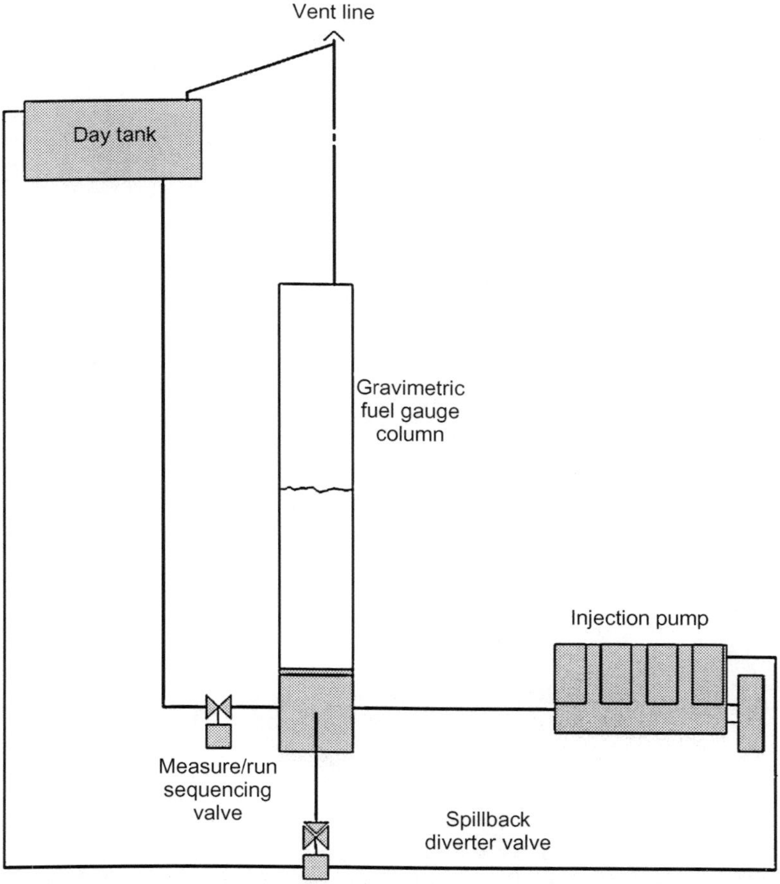


**FIGURE 15.3** The circuit with associated control valves required when using a gravimetric fuel gauge with an engine having fuel spillback.

## Mass flow or consumption rate meters

There are many different designs of rate meter on the market and the choice of the most suitable unit for a given application is not easy. The following factors need to be taken into account:

- absolute level of accuracy
- turndown ratio (see note later)
- sensitivity to temperature and fuel viscosity
- pressure difference required to operate
- wear resistance and tolerance of dirt and bubbles
- analog or impulse-counting readout
- suitability for stationary/in-vehicle use.

*Note*: The *Turndown ratio* is the width of the operational range of a device, defined as the ratio of the maximum capacity to minimum capacity. Sometimes referred to as "rangeability," it is important part of the specification when choosing any measuring instrument for a specific range of test applications, from a dynamometer to a flow meter, because it varies greatly between different devices and different models within a range of devices.

### *Positive-displacement flowmeters*

Positive-displacement meters provide high accuracy, ± 0.1% of actual flow rate is claimed, and good repeatability. Their accuracy is not affected by pulsating flow and many designs do not require a power supply nor lengths of straight pipe upstream and downstream in their installation. Their main disadvantage is the appreciable pressure drop required to drive the metering unit, which may approach 1 bar. The process fluid must be clean because they operate with small clearances; particles greater than 100 μm must be filtered out upstream of the unit.

Several designs of positive-displacement (volumetric) fuel gauges make use of a four-piston metering unit with the cylinders arranged radially around a single-throw crankshaft. Crankshaft rotation is transmitted magnetically to a pulse output flow transmitter. Cumulative flow quantity and instantaneous flow rate are indicated, and these meters are suitable for in-vehicle use. A turndown ratio as high as 100:1 is claimed in some designs.

### *Coriolis effect flowmeters*

The market-leading mass flowmeters now tend to make use of the Coriolis effect, in which the fuel is passed through a pair of vibrating U-tubes (Fig. 15.4).

When the design consists of two parallel tube loops, flow is divided into two streams by a splitter near the meter's inlet and is recombined at the exit. The tubes are vibrated by an electromagnetic driver that forces the tubes

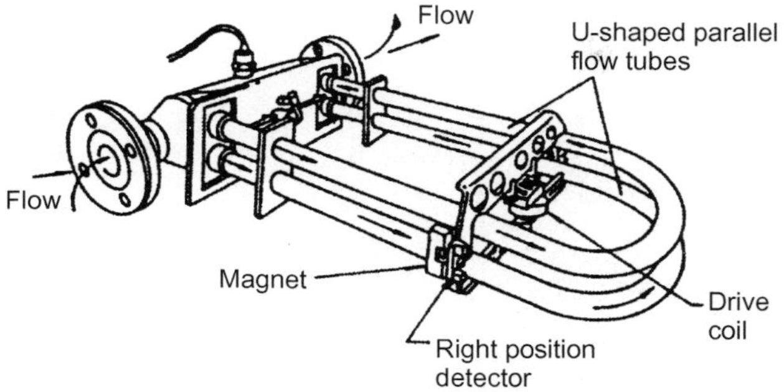


**FIGURE 15.4** Diagram of the internal circuit of one type of "Coriolis effect" flowmeter.

toward and away from one another at a fixed frequency. Two sensors, positioned on opposite sides of the loop, detect the position, velocity, or acceleration of the tubes, creating a sinusoidal voltage output. When there is no flow through the tubes, the vibration caused by the driver results in identical displacements at the two sensing points. When flow is present, the Coriolis forces act to produce a secondary twisting vibration, resulting in small phase differences in the relative motions that are detected at the sensing points. The deflection of the tubes caused by the Coriolis force only exists when both axial fluid flow and tube vibration are present, and the meters are designed so that identical phased vibration detected by the detectors equals zero flow and the difference in phase, induced by flow, is calibrated to read as the volume of liquid flowing through the unit.

Coriolis effect devices, when built into an electronically controlled, fuel mass flowmetering package, have a very high turndown factor and can have (depending on seal materials, etc.) multifuel density capacity, allowing a single unit to cover a wide range of engines, typically, in the automotive range, from a small three-cylinder gasoline engine up to 600 kW diesel units. The electronic controls also allow flow to be updated at rates measured in kilohertz. These meters should be installed so that they will remain full of liquid and so air cannot get trapped inside the tubes, but they are insensitive to fluid swirl. Therefore there is no requirement for straight runs of piping upstream or downstream of the meter to condition the flow.

Calibration of meters using the Coriolis effect is not straightforward and requires a special rig, or the unit being returned to the manufacturer's service base.

## Measurement of fuel spillback

If it is ever required to measure spillback *as a flow separate from engine consumption*, it may be necessary to use two sensing units, one in the flow line and the other in the spill line; this increases cost and complexity.

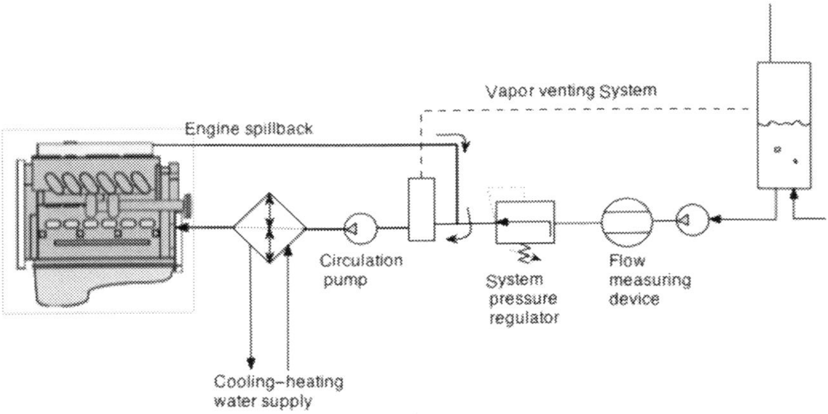


**FIGURE 15.5** A simplified schematic diagram of an engine fuel system in which the spillback is recirculated and flow into the loop recorded as engine consumption.

In most cases engine tests are required only to record engine fuel consumption and therefore the fuel meter is on the supply side of a recirculating loop circuit containing a volume of fuel from which the engine draws and to which spillback is returned, as shown in Fig. 15.5. Change in the volume of fuel in the recirculating loop, from whatever cause, will be recorded as consumption, which is why stable, continuously circulating fuel has to be ensured by a pump of a greater flow rate than consumption. The recirculating fuel in the loop has to be temperature conditioned and fitted with some means of detecting excessive aeration so that a venting sequence can be initiated.

## Fuel consumption measurements: gaseous fuels

Metering of gases is a more difficult matter than liquid metering, partially because the pressure difference available to operate a meter is limited. The density of a gas is sensitive to both pressure and temperature, both of which must be known when, as is usually the case, the flow is measured by volume. The traditional domestic gas meter contains four chambers separated by bellows and controlled by slide valves. Successive increments in volume metered are quite large, so that instantaneous or short-term measurements are not possible.

Other meter types, capable of indicating smaller increments of flow, make use of rotors having sliding vanes or meshing rotors. In the case of natural gas supplied from the mains flow, measurement is a fairly straightforward. Rotary meters are used that, with each turn, move a specific quantity of gas through the meter. The operating principle is similar to that of a Roots blower; the rotational shaft may produce electrical pulses for recorder or may drive an odometer-like counter.

Measurement of liquefied petroleum gas consumption is less straightforward, since this is stored as a liquid under pressure and is vaporized, reduced in pressure, and heated before reaching the engine cylinders. The gas meter must be installed in the line between the "converter" and the engine, and it is essential to measure the gas temperature at this point to achieve accurate results.

## Effect of fuel condition on engine performance

Note that fuel storage conditions and reticulation is covered in Chapter 7, Energy Storage, and that the effect of the condition of the combustion air is discussed later in this chapter.

In the case of a spark-ignition engine the maximum power output is more or less directly proportional to the mass of oxygen contained in the combustion air, since the air/fuel ratio at full throttle is closely controlled.

In the case of the diesel engine the position is different and less clear-cut. The maximum power output is generally determined by the maximum mass rate of fuel oil flow delivered by the fuel injection pump operating at the maximum fuel stop position. However, the setting of the fuel stop determines the maximum volumetric fuel flow rate, and this is one reason why fuel temperature should always be closely controlled in engine testing.

Liquid hydrocarbon (HC) fuel has a high coefficient of cubical expansion, lying within the range 0.001–0.002/°C (compared with water, for which the figure is 0.00021/°C). This implies that the mass of fuel delivered by a pump is likely to diminish by between 0.1% and 0.2% for each degree rise in temperature, not a negligible effect.

A further fuel property that is affected by temperature is its viscosity, significant because in general the higher the viscosity of the fuel, the greater the volume that will be delivered by the pump at a given rack setting.

*Note*: The viscosity of oils and fuels are commonly classed by their *kinematic viscosity* which is the measure of a fluid's resistance to flow under the weight of gravity. The unit of measure of kinematic viscosity is *Centistokes* (cSt).

The *dynamic viscosity* of a fluid is the measure of its resistance to flow when an external force is applied. The unit of measure for dynamic viscosity is *Centipoise* (cP).

This effect is likely to be specific to a particular pump design; as an example, one engine manufacturer regards a "standard" fuel as having a viscosity of 3 cSt at 40°C and applies a correction of 2.5% for each centistoke departure from this base viscosity, the delivered volume increasing with increasing viscosity.

As a further factor, the specific gravity (density) of the fuel will also affect the mass delivered by a constant-volume pump, though here the effect is obscured by the fact that, in general, the calorific value of an HC fuel falls with increasing density, thus tending to cancel the change out.

## Measurement of lubricating oil consumption

*Note*: Oil consumption and blowby will be found to be abnormally high if a newly built engine has been run in so that the cylinder bores have become "glazed." Once running in the test bed, a new or rebuilt engine should quickly be subjected to a running-in routine which puts on variable loads over a wide speed range. Most OEMs and engine builders will have their own routines to prevent such engine damage but problems with some other part of the test system has in the past led to engines running for long periods at idle, early in their test life, which has proved disastrous to their subsequent performance.

Oil consumption measurement is of increasing importance since it is considered to be an important component in the production of engine exhaust emissions and in particulates, but it is one of the most difficult measurements associated with engine testing within the test cell.

Oil consumption rate is very slow relative to the quantity in circulation. It is also quite variable between engine designs and highly dependent on the engine's operating regime.

Oil consumption of 2010 versions of US-produced heavy-duty diesel engines was believed to be about 1:1000 of fuel consumption while in modern passenger car engines is usually less than 0.05 % of actual fuel consumption meaning that the running of a test to simulate vehicle mileage of 2000 mi would generally be considered as being too expensive for wide use.

As, in the last decade, development of the ICE has reduced the internal friction and increased the use of turbo charging, there has been a commensurate small increase in oil consumption in some designs.

All this means that, provided the piston ring/cylinder bore is well bedded-in, test durations would have to be of long duration before meaningful results, from measurements based on changes in oil volume, are obtained.

There are several oil consumption meter systems on the international market based on a procedure that is an automated "drain and weigh" method. By connections into the engine's sump drain hole or dipstick hole, the sump is arranged to drain to a separate receiver, the contents of which are monitored, either by weighing or by depth measurement. An alternative method is to draw off oil down to a datum sump level and weigh it before and after a test period.

Difficulties with all such systems include the following:

- The quantity of oil adhering to the internal surfaces of the engine is very sensitive to temperature, as is the volume in transit to the receiver. This can lead to large apparent variations in consumption rate.
- Volumetric changes due to temperature soak of the engine and fluids.
- Apparent consumption is influenced by fuel dilution and by any loss of oil vapor.
- The rate of oil consumption tends to be very sensitive to conditions of load, speed, and temperature. There is also a tendency to medium-term variations in apparent rate, due to such factors as "ponding up" of oil in return drains and accumulation of air or vapor in the circuit.

Near real-time continuous measurement of oil consumption is now possible by the detection of products of its combustion in the exhaust gas stream. One method is based on lubricant labeling using radiotracer compounds, where oil consumption measurement is achieved by trapping and monitoring tracer residues in the exhaust line. The use of radio-traceable isotopes is a technique that is increasingly used in tribology research but is restricted in its use to specialist laboratories.

An alternative near real-time oil consumption measurement is achieved through the monitoring of the $SO_2$ concentration in the exhaust stream. The Da Vinci Lubricant Oil Consumption (DALOC) system [2] uses an ultraviolet fluorescence detector and pretreats the exhaust sample in order to remove NO, which would create unwanted fluorescence, and convert any unburnt sulfur species into $SO_2$. Clearly the fuel used must be of a known, constant, and very low sulfur content, so that variations in readings are truly indicative of lubricating oil consumption. As with exhaust emission benches, the DALOC system can be repeatedly calibrated using an $SO_2$ calibration gas.

## Measurement of crankcase blowby

In all reciprocating ICEs there is a flow of gas into and out of the clearances between piston top land, ring grooves, and cylinder bore. In an automotive gasoline engine these can amount to 3% of the combustion chamber volume. Since in a spark-ignition engine the gas consists of unburned mixture that emerges during the expansion stroke too late to be burned, this can be a major source of HC emissions and represents a loss of power. Some of this gas will leak past the rings and piston skirt as blowby into the crankcase. It is then vented back into the induction manifold and to this extent reduces the HC emissions and fuel loss but has an adverse effect on the lubricant. A second possible source of blowby gas is from the oil drain a turbocharger system feeds into the crankcase void; to isolate this source and to measure the volumes produced requires a separate metering device.

**TABLE 15.1** Types of blowby meters.

| Type | Typical accuracy | Dirt sensitivity | Lowest flow (L/min) |
|---|---|---|---|
| Positive impeller | 2% FS | Medium | Approx. 6 |
| Hot wire | 2% | High | Approx. 28 |
| Commercial gas meter | 1% FS | High | Claimed 0.5 |
| Vortex | 1% | High | Approx. 7 |
| Flow through an orifice | 1% | Low | Claimed 0.2 |

Note that a "flow through an orifice" device usually has an inferior turndown ratio to most other types. *FS*, Full scale.

Blowby flow is highly variable over the full range of an engine's performance and life; therefore accurate blowby meters will need to be able to deal with a wide range of flows and with pulsation. Instruments based on the orifice measurement principle, coupled with linearizing signal conditioning (the flow rate is proportional to the square root of the differential pressure), are good at measuring the blowby gas in both directions of flow that can occur when there is heavy pulsation between pressure and partial vacuum in the crankcase.

Within the gas flow there may be carbon and other "dirt" particles; the sensitivity of the measuring instrument to this dirt will depend on the application. Sensitivity is shown in Table 15.1.

Crankcase blowby is a significant indicator of engine condition and should preferably be monitored during any extended test sequence. An increase in blowby can be a symptom of various problems such as incipient ring sticking, bore polishing, deficient cylinder bore lubrication, or developing turbocharger problems.

## Part 2: Measurement of air consumption and gas flows

The accurate measurement of the air consumption of an ICE is a matter of some complexity but is of great importance and since the theory also has relevance to gas flow measurement in emissions testing the subject needs covering in some detail.

### Properties of air

Air is a mixture of gases with the following approximate composition given in Table 15.2.

**TABLE 15.2** The major constituents of dry air, by volume.

| Gas | | Volume | |
|---|---|---|---|
| Name | Formula | ppm by volume | % |
| Nitrogen | $N_2$ | 780,840 | 78.08 |
| Oxygen | $O_2$ | 209,460 | 20.95 |
| Argon | Ar | 9340 | 0.93 |
| Carbon dioxide (as of January 1, 2020) | $CO_2$ | 413.45 | 0.04 |
| Neon | Ne | 18.18 | 0.0018 |
| Helium | He | 5.24 | 0.00052 |
| Methane | $CH_4$ | 1.87 | 0.00019 |
| Krypton | Kr | 1.14 | 0.00014 |
| Water vapor not included in above dry air contents | | | |
| Water vapor | $H_2O$ | 0–30,000 | 0–3 |

The amount of water vapor present depends on local temperature and prevailing atmospheric conditions. It can have an important influence on engine performance, notably on exhaust emissions, but for all but the most precise work its influence on air flow measurement may be neglected.

The relation between the pressure, specific volume, temperature, and density of air is described by the gas Eqs. (15.1) and (15.2):

$$pV = nRT \tag{15.1}$$

In SI units, $p$ is measured in pascal, $V$ in cubic meter, $n$ in mole, and $T$ in Kelvin, and $R$ has a value of $8.314\ \text{J/(K mol)} \approx 2\ \text{cal/(K mol)}$, or $0.08206\ \text{L} \cdot \text{atm/(mol K)}$.

$$\rho = \frac{m}{V} \tag{15.2}$$

In SI units, density $\rho$ is measured in in $\text{kg/m}^3$, mass $m$ in kg, and volume $V$ in $\text{m}^3$.

A typical value for air density in temperate conditions at sea level would be $\rho = 1.2\ \text{kg/m}^3$.

## Air consumption, condition, and engine performance

The ICE is essentially an "air engine" in that air is the working fluid; the function of the liquid fuel is merely to supply heat. There is seldom any particular technical difficulty in the introduction of sufficient fuel into the working cylinder, but the attainable power output is strictly limited by the charge of air that can be aspirated.

It follows that the achievement of the highest possible volumetric efficiency is an important goal in the development of high-performance engines, and the design of inlet and exhaust systems, valves and cylinder passages represent a major part of the development program for engines of this kind.

The standard methods of taking into account the effects of charge air condition as laid down in European and American Standards are complex, difficult to apply and are mostly used to correct the power output of engines undergoing acceptance or type tests. A simplified treatment, adequate for most routine test purposes, is given next.

The condition of the air entering the engine is a function of the following parameters:

1. pressure
2. temperature
3. moisture content
4. impurities

For the first three factors, standard conditions, according to European and American practice, are:

Atmospheric pressure 1 bar (750 mmHg)
Temperature 25°C (298K)
Relative humidity 30%

## *Variations in engine performance due to atmospheric pressure*

Since the mass of air consumed tends to vary directly with the density, which is itself proportional to the absolute pressure, other conditions remaining unchanged. Since the standard atmosphere = 1 bar, we may write:

$$\rho_t = \rho_n P \tag{15.3}$$

where $\rho_t$ is the density under test conditions (kg/m$^3$), $\rho_n$ is the density under standard conditions (kg/m$^3$), and $P$ is the atmospheric pressure under test conditions (bar).

It follows from this that a change of 1% or 7.5 mmHg corresponds to a change in the mass of air entering the engine of 1%. For most days of the year the (sea level) atmospheric pressure will lie within the limits 750 mmHg ±3%, that is between 775 mmHg (103 kPa) and 730 mmHg (97 kPa), with a corresponding percentage variation in charge air mass of 6%.

It is a common practice to design the test cell ventilation system to maintain a small negative pressure in the cell, to prevent fumes entering the control room, but the level of depression should not exceed about 50 mm water gauge (50 Pa), equivalent to a change in barometer reading of less than

0.37 mmHg. Clearly, in these circumstances, the effect on combustion air flow, if the engine is drawing air from within the cell, is negligible.

Barometric pressure falls by about 86 mmHg (11.5 kPa) for an increase in altitude of 1000 m (the rate decreasing with altitude). This indicates that the mass of combustion air falls by about 1% for each 90 m (300 ft) increase in altitude, a very significant effect, but this is only entirely true if the temperature is constant because the ambient air temperature will decrease by 6.5°C for every 1000 m gained in altitude; this is called the standard (temperature) lapse rate.

Since the highest drivable roads in the world do not exceed around 5800 m (19,000 ft), automotive test facilities do not need to exceed these extreme conditions in expensive altitude chambers, whereas for aeronautical facilities the sky is the limit and component test chambers rated at 0–100,000 ft (30,000 m) are available. The standards covering such work include Mil-STD-810 and RTCA-DO-160, which regulate aviation components and equipment subjected to rapid decompression.

Variations in charge air pressure have an important knock-on effect: The pressure in the cylinder at the start of compression will in general vary with the air supply pressure and the pressure at the end of compression will change in the same proportion which can have a significant effect on the combustion process.

## *Variations in engine performance due to air temperature*

Variations in the temperature of the combustion air supply have an effect of the same order of magnitude as variations in barometric pressure, within the range to be expected in test cell operation.

Air density varies inversely with its absolute temperature. The temperature at the start of compression determines that at the end. In the case of a naturally aspirated diesel engine with a compression ratio of 16:1 and an air supply at 25°C, the charge temperature at the start of compression would typically be about 50°C. At the end of compression the temperature could be in the region of 530°C and increasing to about 560°C for an air supply temperature 10°C higher. The level of this temperature can have a significant effect on such factors as the $NO_x$ content of the exhaust, which is very sensitive to peak combustion temperature. The same effect applies, with generally higher temperatures, to turbocharged engines, all of which has to be taken into account during any ICE calibration exercise.

## *Variations in engine performance due to relative humidity*

Compared with the effects of pressure and temperature the influence of the relative humidity of the air supply on the air charge is relatively small, except at high air temperatures. The moisture content of the combustion air does, however, exert a number of influences on performance. Some of the thermodynamic properties of moist air have been discussed under the heading of psychometry, but certain other aspects of the subject must now be considered.

The important point to note is that unit volume of moist air contains less oxygen in a form available for combustion than the same volume of dry air under the same conditions of temperature and pressure. Moist air is a mixture of air and steam: while the latter contains oxygen, it is in chemical combination with hydrogen and is thus not available for combustion.

The European and American Standards specify a relative humidity of 30% at 25°C. This corresponds to a vapor pressure of 0.01 bar, thus implying a dry air pressure of 0.99 bar and a corresponding reduction in oxygen content of 1% when compared with dry air at the same pressure. Fig. 15.6 shows the variation in dry air volume expressed as a percentage adjustment to the standard condition with temperature for relative humidity ranging from 0% (dry) to 100% (saturated).

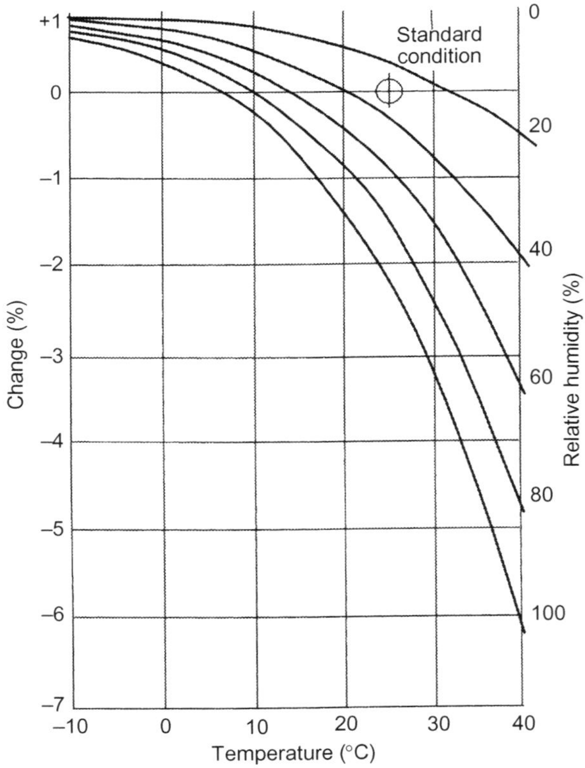


**FIGURE 15.6** Variation of air charge mass above and below standard condition with relative humidity at different temperatures.

The effect of humidity becomes much more pronounced at higher temperatures; thus at a temperature of 40°C the charge mass of saturated air is 7.4% less than for dry air. The effect of temperature on moisture content (humidity) should be noted particularly, since the usual method of indicating specified moisture content for a given test method, by specifying a value of relative humidity, can be misleading. Fig. 15.6 indicates that at a temperature of 0°C the difference between 0% (dry) and 100% (saturated) relative humidity represents a change of only 0.6% in charge mass; at −10°C it is less than 0.3%. It follows that it is barely worthwhile adjusting the humidity of combustion air at temperatures below (perhaps) 10°C.

The moisture content of the combustion air has a significant effect on the formation of $NO_x$ in the exhaust of diesel engines. The SAE procedure for measurement of diesel exhaust emissions gives a rather complicated expression for correcting for moisture content. This indicates that, should the $NO_x$ measurement be made with completely dry air, a correction of the order of +15% should be made to give the corresponding value for a test with moisture content of 60% relative humidity.

## Corrections to the charge mass of air

The various adjustments or "corrections" to the charge mass are cumulative in their effect.

The following example illustrates the magnitudes involved.

Consider:

1. a hot, humid summer day with the chance of thunder and (Column A)
2. a cold, dry winter day of settled weather (Column B).

The conditions are summarized in Table 15.3.

**TABLE 15.3 The adjustment factors by which the observed power in examples A and B should be multiplied to indicate the power to be expected under "standard."**

| Condition | A | B |
|---|---|---|
| Pressure | 0.987 bar, 740 mmHg | 1.027 bar, 770 mmHg |
| Temperature (°C) | 35 | 10 |
| Relative humidity (%) | 80 | 40 |
| Pressure | (1/987) = 1.0135 | (1/1.027) = 0.9740 |
| Temperature | (308/298) = 1.0336 | (283/298) = 0.9497 |
| Relative humidity from Fig. 15.6 | +3.48% = 1.0348 | 0% |
| Total adjustment | 1.0840, +8.4% | 0.9233, −7.7% |

We see that under hot, humid, stormy conditions the power of a powertrain, without compensating control mapping, may be reduced by as much as 15% compared with the power under cold anticyclonic conditions. The power adjustment calculated previously is based on the assumption that charge air mass directly determines the power output, but this would only be the case if the air/fuel ratio were rigidly controlled (as in spark-ignition engines with precise stoichiometric control). In most cases a reduction, for example, in charge air mass due to an increase in altitude will result in a reduction in the air/fuel ratio, but not necessarily in a reduction in power. The correction factors laid down in the various standards take into account the differing responses of the various types of engine to changes in charge oxygen content.

It should be pointed out that the effects of *water injection*, (sometimes used by the engine tuning fraternity), are quite different from the effects of humidity already present in the air. Humid air is a mixture of air and steam; the latent heat required to produce the steam has already been supplied. When, as is usually the case, water spray is injected into the air leaving the turbocharger at a comparatively high temperature, the cooling effect associated with the evaporation of the water achieves an increase in charge density that outweighs the decrease in oxygen associated with the resulting steam.

### *Variations in engine performance due to impurities*

Pollution of the combustion air can result in a reduction in the oxygen available for combustion and can alter the chemistry of exhaust emissions. In the test cell the likely source of such pollution may be the ingestion of exhaust fumes from other engines or from neighboring industrial processes such as paint plant. Care should also be taken in the siting of air intakes and exhaust discharges to ensure that, even under unfavorable conditions, there cannot be unintended exhaust recirculation. Exhaust gas has much the same density as air and a free oxygen content that can be low or even zero, so that, approximately, 1% of exhaust gas in the combustion air will reduce the available oxygen in almost the same proportion.

## Measuring engine air consumption

### *The airbox method*

The simpler methods of measuring air consumption involve drawing the air through some form of measuring orifice and measuring the pressure drop across the orifice. It is a good practice to limit this drop to not more than about 125 mm $H_2O$ (1200 Pa). For pressures less than this, air may be treated as an incompressible fluid, with much simplification of the air flow calculation.

The velocity $U$ developed by a gas expanding freely under the influence of a pressure difference $\Delta p$, if this difference is limited as mentioned above, is given by:

$$\frac{\rho U^2}{2} = \Delta p, \quad U = \sqrt{\frac{2\Delta p}{\rho}} \tag{15.4}$$

Typically, airflow is measured by means of a sharp-edged orifice mounted in the side of an air-box, coupled to the engine inlet and *of sufficient capacity to damp out the inevitable pulsations in the flow* into the engine, which are at their most severe in the case of a single-cylinder four-stroke engine (Fig. 15.7). In the case of turbocharged engines the inlet airflow is comparatively smooth and a well-shaped nozzle without an air-box will give satisfactory results.

The airflow through a sharp-edged orifice takes the form sketched in Fig. 15.8. The coefficient of discharge of the orifice is the ratio of the transverse area of the flow at plane $a$ (the vena contracta) to the plan area of the orifice. Tabulated values of $C_d$ are available in BS 1042, but for many purposes a value of $C_d = 0.60$ may be assumed.

*Note: Notation used in air consumption calculations are listed at end of the chapter.*

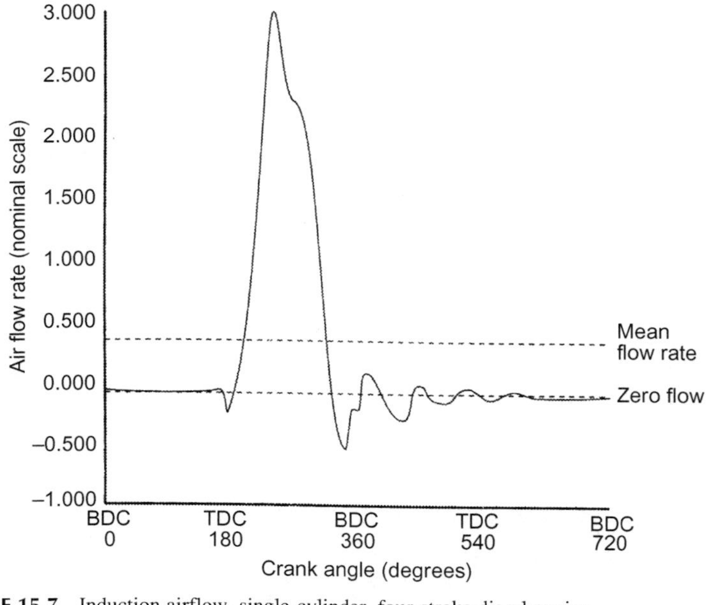


**FIGURE 15.7** Induction airflow, single-cylinder, four-stroke diesel engine.

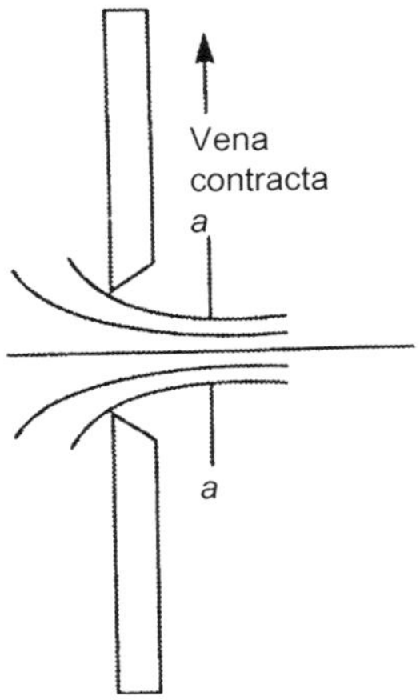


**FIGURE 15.8** Flow through a sharp-edged orifice forming the end wall of an air-box of sufficient capacity to damp out the flow pulsations. The vena contracta ($a-a$) is the point in a fluid stream where the diameter of the air stream is the least, and fluid velocity is at its maximum.

We may derive the volumetric flow rate of air through a sharp-edged orifice as follows:

*Flow rate = coefficient of discharge × cross-sectional area of orifice × velocity of flow.*

$$Q = C_d \frac{\pi d^2}{4} \sqrt{\frac{2\Delta p}{\rho}} \tag{15.5}$$

Noting from the air density equation that:

$$\rho = \frac{p_a \times 10^5}{RT_a}$$

And assuming that $\Delta p$ is equivalent to $h$ mm $H_2O$, we can write:

$$Q = C_d \frac{\pi d^2}{4} \sqrt{\frac{2 \times 9.81h \times 287T}{p_a \times 10^5}} \tag{15.5a}$$

$$Q = 0.1864 C_d d^2 \sqrt{\frac{hT_a}{p_a}} \text{ m}^3/s \tag{15.5b}$$

To calculate the mass rate of flow, note that:

$$\bar{m} = \rho Q = \frac{p_a Q}{RT_a}$$

Giving, from Eqs. (15.5a) and (15.5b):

$$\dot{m} = C_d \frac{\pi d^2}{4} \sqrt{\frac{2 \times 9.81 h \times 10^5}{287 T_a}} \quad (15.6a)$$

$$\dot{m} = 64.94 C_d d^2 \sqrt{\frac{h p_a}{T_a}} \quad (15.6b)$$

Eqs. (15.5b) and (15.6b) give the fundamental relationship for measuring airflow by an orifice, nozzle, or venturi.

## Sample air-box flow calculations and selection of orifice size

If $C_d = 0.6$, $D = 0.050$ m, $H = 100$ mm $H_2O$, $T_a = 293$K (20°C), $p_a = 1.00$ bar, then $Q = 0.04786$ m$^3$/s (1.69 ft$^3$/s) and $\dot{m} = 0.05691$ kg/s (0.1255 lb/s).

To assist in the selection of orifice sizes, Table 15.4 gives approximate flow rates for orifices under the following standard conditions:

$H = 100$ mm $H_2O$, $T_a = 293$K (20°C), $p_a = 1.00$ bar.

A disadvantage of flow measurement devices of this type is that the pressure difference across the device varies with the square of the flow rate. It follows that a turndown in flow rate of 10:1 corresponds to a reduction in pressure difference of 100:1, implying insufficient precision at low flow rates. It is a good practice, when a wide range of flow rates is to be measured, to select a range of orifice sizes, each covering a turndown of not more than 2.5:1 in flow rate.

### *Example of approximate calculation of engine inlet airflow*

The air consumption of an engine may be calculated from:

$$V = \eta_v \frac{V_s}{K} \frac{n}{60} \text{ m}^3/s \quad (15.7)$$

where $K = 1$ for a two-stroke and $K = 2$ for a four-stroke engine.

**TABLE 15.4** Approximate airflow rates for orifice sizes 10–150 mm.

| Orifice diameter (mm) | $Q$ (m$^3$/s) | $\dot{m}$ (kg/s) |
|---|---|---|
| 10 | 0.002 | 0.002 |
| 20 | 0.008 | 0.009 |
| 50 | 0.048 | 0.057 |
| 100 | 0.19 | 0.23 |
| 150 | 0.43 | 0.51 |

For initial sizing of the measuring orifice $\eta_v$, the ratio of the volume of air aspirated per stroke to the volume of the cylinder may be assumed to be about 0.8 for a naturally aspirated engine and up to about 2.0 for turbocharged engines.

If we consider a single-cylinder, naturally aspirated, four-stroke engine of swept volume 0.8 L, running at a maximum of 3000 rev/min:

$$V = 0.8 \times \frac{0.0008}{2} \times \frac{3000}{60} = 0.016\ \mathrm{m}^3/s$$

A suitable measuring orifice size, from Table 15.4 would be 30 mm.

### *Connection of air-box and orifice plate to engine inlet*

It is essential that the configuration of the connection between the air-box and the engine inlet should model as closely as possible the configuration of the air intake arrangements in service. This is because pressure pulsations in the inlet can have a powerful influence on engine performance, in terms of both volumetric efficiency and pumping losses.

The resonant frequency of a pipe of length $L$, open at one end and closed at the other, equals $a/4L$, where $a$ is the speed of sound, roughly 330 m/s. Thus an inlet connection 1 m long would have a resonant frequency of about 80 Hz. This corresponds to the frequency of intake valve opening in a four-cylinder four-stroke engine running at 2400 rev/min. Clearly such an intake connection could disturb engine performance at this speed; therefore the intake connection should be as short as possible.

### *Proprietary air flowmeters*

*Note*: As with the selection of all instrumentation, it is necessary to carefully read specifications and ensure that accuracy is stated in a common format. The accuracy of different types and sizes of flowmeters from different manufacturers will be specified in one, or all of three ways:

- as a percentage of actual reading,
- as a percentage of calibrated span, and
- as percentage of full-scale units.

In addition, the accuracy should be stated at minimum, mean, and maximum flow rates.

### *Viscous flowmeters*

The viscous flow air meter was for many years the most widely used alternative to the air-box and orifice method of measuring airflow. In this device the measuring orifice is replaced by an element consisting of a large number of small passages, generally of triangular form. The flow through these passages is substantially laminar, with the consequence that the pressure difference across the element is approximately directly proportional to the velocity of flow, rather than to its square, as is the case with a measuring orifice.

This has two advantages. First, average flow is proportional to average pressure difference, implying that a measurement of average pressure permits a direct calculation of flow rate, without the necessity for smoothing arrangements. Second, the acceptable turndown ratio is much greater.

The flowmeter must be calibrated against a standard device, such as a measuring orifice.

## *Lucas–Dawe air mass flowmeter*

This device depends for its operation on the corona discharge from an electrode coincident with the axis of the duct through which the air is flowing. Airflow deflects the passage of the ion current to two annular electrodes and gives rise to an imbalance in the current flow that is proportional to air flow rate.

An advantage of the Lucas–Dawe flowmeter is its rapid response to changes in flow rate, of the order of 1 ms, making it well suited to transient flow measurements, but it is sensitive to air temperature and humidity and requires calibration against a standard.

## *Hot wire or hot film anemometer devices*

The principle of operation of these popular devices is based on the cooling effect caused by gas flow on a heated wire or film surface. The heat loss is directly proportional to the air mass flow rate, providing the flow through the device is laminar. The advantages of these designs are their reliability and good tolerance to contaminated airflows, they also have a very short time response, which makes them ideal to measure rapidly fluctuating phenomena such as turbulent flows. Their disadvantages include the possibility of condensation of moisture on the temperature detector and coating or material buildup on the sensor. Calibration at site is virtually impossible so they are supplied with a maker's certification. Probably the best known device of this type in European test cells is the Sensyflow range made by ABB.

## *Engine cylinder-head flow benches*

An air *flow bench* is a device used for testing the internal aerodynamic qualities of an *engine* component primarily for testing the intake and exhaust ports of cylinder heads of ICE.

They work on the principle of a measured volume of air being drawn through the unit under test (UUT) and the pressure drop created being recorded by a calibrated manometer.

Flow benches are much favored by motorsport engine tuners and can be built by any competent workshop; however, the variability in quality and design of these units makes any objective correlation between different units very difficult. Commercially available units range significantly in complexity and price, but industrial standard flow benches such as those made by companies such as SuperFlow are the best known and are designed for use on a wide number of engine components in addition to cylinder heads. The top-of-the-range designs can be used the test the flow through not only engine gas ports, but other components including catalytic converters, and are capable of integrating signals from multiple local velocity probes and from devices measuring air swirl and tumble, thus giving a detailed analysis of airflow through the UUT.

The steady state of pressure and known temperature conditions used in flow bench measurement bear no relation to the dynamics of air, or exhaust gas flow in a running engine; however, changes in engine breathing through reworking of the intake ducting, valve layout, and head design can be judged and recorded by the use of such devices.

## *Notation used in engine air consumption calculations*

| | |
|---|---|
| Atmospheric pressure | $p_a$ (bar) |
| Under test conditions | $p$ (bar) |
| Atmospheric temperature | $T_a$ (K) |
| Test temperature | $T_t$ (°C) |
| Density of air | $\rho$ (kg/m$^3$) |
| Under standard conditions | $\rho_n$ (kg/m$^3$) |
| Under test conditions | $\rho_t$ (kg/m$^3$) |
| Pressure difference across orifice | $\Delta p$ (Pa), $h$ (mm $H_2O$) |
| Velocity of air at contraction | $U$ (m/s) |
| Coefficient of discharge of orifice | $C_d$ |
| Diameter of orifice | $d$ (m) |
| Volumetric rate of flow of air | $Q$ (m$^3$/s) |
| Mass rate of flow of air | |
| Constant, 1 for two-stroke and 2 for four-stroke engines | $K$ |
| Engine speed | $n$ (rev/min) |
| Number of cylinders | $Nc$ |
| Swept volume, total | $V_s$ (m$^3$) |
| Air consumption rate | $V$ (m$^3$/s) |
| Volumetric efficiency of engine | $\eta_v$ |
| Volume of air-box | $V_b$ (m$^3$) |
| Gas constant | $R$ (J/kg K) |

## References

[1] BS 7405:1991, 'Guide to selection and application of flowmeters for the measurement of fluid flow in closed conduits'.

[2] Da Vinci Emissions Services, Ltd. 12665 Silicon Drive, San Antonio, Texas 78249. http://www.davinci-limited.com/index.php/products/dafio-da-vinci-fuel-in-oil

## Further reading

Stone, C., Air Flow Measurement in Internal Combustion Engines, SAE Technical Paper 890242, 1989.

Yilmaz, E. et al., The Contribution of Different Oil Consumption Sources to Total Oil Consumption in a Spark Ignition Engine, Society of Automotive Engineers, Inc, 2002.

Chapter 16

# The combustion process and combustion analysis

## Chapter outline



Engine Testing. DOI: https://doi.org/10.1016/B978-0-12-821226-4.00016-4

## Introduction

Combustion pressure analysis is a task that is normally undertaken during the early development phase of an internal combustion engine (ICE). With the drive toward electrification, it may seem that combustion-related measurements could be a diminishing topic—eventually to be extinguished altogether. There is much discussion around what the leading technology will be for future propulsion systems but an underlying agreement in the community of experts is that powertrain concepts, that include an ICE in some form, in combination with an electric propulsion mechanism, is currently the best way to produce the lowest harmful emissions when considering the whole average vehicle life cycle. In addition, a powertrain which includes an ICE is the easiest way to deal with range anxiety—and the easiest technology to bring to market quickly on existing and new vehicle platforms.

In order to address this—there is still much ongoing research into fundamentals of combustion and combustion processes. The continuously increasing processing power of engine controllers (ECUs) mean that complex combustion systems, requiring real-time control, become an increasingly valid option—particularly with respect to the cost of after treatment systems required for vehicles to achieve increasingly lower emission limits demanded by legislation. Dealing with emissions at root cause level is a topic of high interest to engine manufacturers—low temperature combustion, in conjunction with advanced ignition mechanisms, can provide excellent technology pathways for the ICE to stay as a relevant technology.

However, to develop such systems, this level of research work requires an "eye" into the combustion chamber, even more so than was previously required, in order to be able to understand energy conversion and efficiency at a fundamental, combustion process level. This, therefore, defines that combustion pressure measurement and measurement chain technology is still a valid and essential tool in the powertrain development process—and is, therefore, still a relevant topic for those involved in specifying or working with test facilities that include an ICE.

The importance of in-cylinder measurements to the engine development process should be explained in some detail. Converting chemical energy from the combustion of hydrocarbon-based fuel into heat, which in a confined space (combustion chamber) develops pressure, this pressure acts upon the engine piston surface, creating a mechanical force that is used to drive a crankshaft via a crank slider mechanism, thus converting linear force into torque. The efficiency of this process is of great importance and interest, and test facilities can easily measure the "system" input via mass flow measurement of fuel and air. System output can be established via the dynamometer that can accurately derive output power from rotational speed and torque. Heat rejection to the facility environment, cooling systems, and exhaust can also be measured with high confidence. However, unless we can establish the losses inside the engine, within the internal processes (thermal, gas exchange), then it is difficult to understand where to focus to improve the overall engine efficiency.

This is the reason why combustion pressure measurement is important—it provides an insight into the in-cylinder processes that allows an engine engineer to understand where losses occur in that energy conversion process—and where to focus development efforts. The high-pressure combustion data provides information of total energy delivered by the fluid in the cylinder, also the rate at which the energy is being released. This is essential information that the engineer needs to ensure the cylinders are well balanced and contributing equally (ensuring flow systems are fully optimized) and that energy is released at the correct time for maximum efficiency in the heat to work conversion process (optimized fuel path and ignition timing). In addition, the in-cylinder high-pressure information can be used for cycle-based thermal efficiency calculations—and can also be used to optimize simulation data such that "virtual sensor" values from the simulation code can be established accurately due to the quality of physical models that have been optimized with actual measurement data (typical values of interest are those which cannot be measured—such as in-cylinder residual gas content and trapped masses).

The high-pressure measurement provides useful information about the gas-exchange phase. In between the working parts of the engine cycle waste gas must be expelled and fresh charge induced. In that part of the cycle, the engine behaves as a gas pump, which absorbs work. So, understanding this phase, and optimizing it to reduce flow losses is critical to engine efficiency—this process can be fully characterized from cylinder pressure data—along with the potential to understand frictional losses as well.

As complementary measurements, where accurate in-cylinder analysis will be undertaken, a "three pressure measurement" is also often carried out. In particular, on thermodynamic single cylinder research engines (TSCREs) where detailed combustion studies are undertaken, and new combustion concepts are developed with an isolated cylinder. This technique allows the Engineer to fully understand the combustion-related effects, avoiding the influences created by the gas path or other adjacent cylinders and air path components. The three-pressure technique involves pressure measurements in the air path, in order to be able to establish mass flows and gas dynamic effects, in combination with in-cylinder flows. This allows a full understanding of complex air path components such as turbochargers, after treatment systems and their overall effect on the transient response of the engine.

Fuel path measurements are also a common requirement in conjunction with cylinder pressure. Modern "rail" based fuel injection systems for SI and CI applications operate at extremely high pressures, fluid dynamic effects and resonances can have a significant effect on injected fuel quantity and control systems have compensating functions to offset these effects that are operating point specific. Hence the need for accurate measurement of, absolute and dynamic fuel pressure at high fidelity.

There are significant new challenges that current and next-generation ICE engine technology brings. The ability to capture and store data, which allows characterization of normal and abnormal combustion modes, is a prerequisite. However, some of these abnormal events are stochastic in nature and of short duration. This then requires a measurement system that is capable of storing very large data volumes in order to capture very specific events and to allow a reaction on those events before a critical situation. This can be handled with modern day data acquisition systems but how to make sense of the huge mass of data generated is a major challenge.

ICE combustion systems are becoming more sophisticated, with more components to accommodate in and around the combustion chamber. This creates a challenge for sensor technology. Less space to accommodate sensors and sensor adaptor solutions, along with increasing influencing factors that affect data quality around the installation and operating environment, mean that sensor selection and understanding the boundary conditions of the sensor and its installation is now even more critical than was previously considered.

In summary, we can state "combustion, air and fuel path related pressure measurements are still entirely relevant knowledge for those seeking to understand in greater depth powertrain test environments and that the challenges for the measurement system are increasing along with the evolution of the modern powertrain."

The user demand has always been for more and faster calculating power to execute the algorithms and return results and statistics as quickly as possible. Each year physically smaller and faster data processing and analysis equipment is brought to the market, but the need for training in its cost-effective use does not diminish.

For the combustion pressure analysis systems, important features are:

- Real-time or online calculation and display of all required results for screen display or transfer to the test bed or host system through software or hybrid interfaces.
- Features in the user interface for easy parameterization of result calculations, or for definition of new, customer-specific result calculation methods without the requirement for detailed programming knowledge.
- Flexibility in the software interface for integration of the combustion pressure analyzer to the test bed, allowing remote control of measurement tasks, fast result data transfer for control mode operation, and tagging of data files such that test bed and combustion data can be correlated and aligned in postprocessing.
- Intelligent data model that allows efficient packaging and alignment of all data types relevant in a combustion measurement (e.g., crank angle data, cyclic data, time-based data, and engine parameters).

Most importantly, engine indicating and combustion pressure analyzers are always very sensitive to correct parameterization and the fundamentals to the test regime, in relation to high-pressure in-cylinder signals are:

- correct determination of TDC (particularly for IMEP);
- correct definition of the polytropic exponent (particularly for heat release); and
- correct and appropriate method of zero-level correction.

Cylinder pressure measurements play the key role in any engine indicating and combustion analysis work, but since modern engine research requires more than just cylinder pressure to be measured against crank angle, pressure transducers are also available for the following roles in engine indicating work:

- intake and exhaust manifold pressure for R&D work, optimized to handle the temperature and pressure ranges of each work environment;
- fuel line pressure for R&D work for pressure pulse measurement up to 3000 bar; and
- cylinder pressure for the in-service monitoring of non-automotive engines, optimized for durability in the run-time application, for thousands of hours of use.

## "Exact" determination of true top dead center position

The test engineer in possession of modern equipment continues to be faced with what appears to be a rather prosaic problem: the determination of the engine's top dead center point. This is a more serious difficulty than may be at first apparent. A typical combustion pressure analyser records cylinder pressure in terms of crank angle, with measurements at each 10th of 1 degree of rotation. However, to compute indicated work it is necessary to transform the cylinder pressure–crank angle data to a basis of cylinder pressure–piston stroke. This demands a very accurate determination of crank angle at the top dead center position of the piston. If the device records top dead center 1° of crankshaft rotation ahead of the true position the computed IMEP will be up to 5% greater than its true value. If the device records TDC 1° late computed IMEP will be up to 5% less than the true value.

Top dead center can be accurately determined, for a particular running condition, by using a specifically designed capacitance sensor, which is mounted in a spark plug hole of one cylinder while cranking, or running the engine on the remaining cylinders. The TDC position thus determined is then electronically recorded by using the angle of rotation between sensor determined TDC and a reference pulse, of a rotational encoder attached to the crankshaft. Another technique makes use of the cylinder pressure peak from a cylinder pressure trace, but without specific TDC transducers and the required software to match the shaft angle, precise determination of geometrical TDC is not easy. The usual and time-honored method is to rotate the engine to positions equally spaced on either side of TDC, using a dial gauge to set the piston height, and to bisect the distance between these points.

The rotation position of TDC determined this way ignores dynamic effects and even then, has to be fixed on the flywheel in such a way as to be recognized by a pickup.

This traditional method leads to several sources of error:

- Difficulty of carrying out this operation with sufficient accuracy.
- Difficulty of ensuring that the signal from the pickup, when running at speed, coincides with the geometrical coincidence of pin and pickup.
- Torsional deflections of the crankshaft, which are always appreciable and are likely to result in discrepancies in the position of TDC when the engine is running, particularly at cylinders remote from the flywheel.

In high-performance engines the TDC position does not remain entirely constant during the test run due to physical changes in shape of the components; in these cases a TDC position specific to a particular steady state may have to be calculated.

## Combustion process basics

At a fundamental level the performance of an ICE is largely determined by the events taking place in the combustion chamber and engine cylinder. These, in turn, are influenced by the following factors:

- configuration of the combustion chamber and cylinder head;
- flow pattern (swirl and turbulence) of charge entering the cylinder, in turn, determined by design of induction tract and inlet valve size, shape, location, lift, and timing;
- ignition and injection timing, spark plug position and characteristics, injector and injection pump design, and location of injector;
- compression ratio;
- air/fuel ratio;
- fuel properties;
- mixture preparation;
- exhaust gas recirculation;
- degree of cooling of chamber walls, piston, and bore; and
- the presence or absence of content or particulates from previous cycles.

While it is beyond the scope of this publication to discuss in detail the physics and chemistry of combustion itself, it is useful for a person involved with powertrain test facilities to possess a basic appreciation of the combustion systems used in past, present, and future ICEs. The reason is that the development targets and boundary conditions, with respect to combustion related measurements for each combustion system, create an impact on the equipment used, particularly sensors and installations. So, for those readers and students not familiar with the details of the combustion process of the spark-ignition (SI) or compression-ignition (CI) engines, they are summarized below, along with other, more recent combustion system developments that could be encountered more often during the lifetime of this publication.

## Combustion in the conventional spark-ignition engine

A mixture of gas and vapor is formed, either in the induction tract, in which the fuel is introduced either by a carburetor or injector, or in the cylinder in the case of gasoline direct injection (GDI).

Combustion is initiated at one or more locations by an electric arc at an in-cylinder location, it is correctly phased according to the engine operating point—triggered by the engine control unit (ECU).

After some delay the control of which still presents problems, a flame is propagated through the combustible mixture at a rate determined, by factors such as fuel chemistry and by the motion of the in-cylinder charge.

Heat is released progressively with a consequent increase in temperature and pressure. During this process the bulk properties of the fluid change as it is transformed from a mixture of air and fuel to a volume of combustion products. Undesirable effects, such as preignition, excessive rates of pressure rise, late burning and detonation or "knock," may be present.

Heat is transferred by radiation and convection to the surroundings. Mechanical work is performed by expansion of the products of combustion. The development of the combustion chamber, inlet and exhaust passages, and fuel supply system of a new engine involve a vast amount of experimental work, some on flow rigs that model the geometry of the engine, most of it on the test bed.

## Combustion in the conventional compression-ignition engine

Air is drawn into the cylinder, generally without throttling, but frequently with pressure charging. Compression ratios range from about 14:1 to 22:1, depending on the degree of supercharge, resulting in compression pressures in the range 40–60 bar and temperatures from 700°C to 900°C. Fuel is injected at pressures that have increased in recent years, with current, state-of-the-art common rail (CR) FIE (fuel injection equipment) using fuel injection pressures over 2000 bar. Charge air temperature is well above the auto-ignition point of the fuel. Fuel droplets vaporize, forming a combustible mixture that ignites after a delay that is a function of charge air pressure and temperature, droplet size, and fuel ignition quality (cetane number).

This simplified description is of a single injection event; however, most automotive CI engines now deploy multiple injection events per combustion cycle. Fuel subsequently injected (after the initial premixed combustion phase) is ignited immediately, and the progress of combustion and pressure rise is to some extent controlled by the rate of injection. Air motion in the combustion chamber is organized to bring unburned air continually into the path of the fuel jet. Heat is transferred by radiation and convection to the surroundings. Mechanical work is performed by the expansion of the products of combustion.

The choice of combustion "profile" involves a number of compromises. A high maximum pressure has a favorable effect on fuel consumption but increases $NO_x$ emissions, while a reduction in maximum pressure, caused by retarding the combustion process, results in increased particulate emissions (commonly known as the $NO_x$ - Soot trade off). Unlike the SI engine, the diesel must run with substantial excess air to limit the production of smoke and soot. A precombustion chamber engine can run with a lambda ratio of about 1.2 at maximum power, while a DI engine requires a minimum excess air factor of about 50%, roughly the same as the maximum at which a SI engine will run. This characteristic has led to the widespread use of pressure charging, which achieves an acceptable specific output by increasing the mass of the air charge.

Note that large industrial or marine engines use DI; in the case of vehicle engines the indirect injection or prechamber engine is now redundant, abandoned in favor of DI due to better fuel consumption and cold starting performance of the DI engine. Compact, very high pressure, piezoelectric injectors, and electronic control of the injection process have made these developments possible.

## Advanced combustion modes

A detailed description of all combustion systems is beyond the scope of this publication; however, an overview of the most important attributes for the current scope of advanced systems for spark and compression ignition systems is given below.

### *Gasoline direct injection (GDI)*

This widely adopted technology and a current standard for spark ignition engines, often used in conjunction with a turbocharger and this combination, formed the base technology for "downsized" engine that were on-trend prior to RDE legislation. Fuel is injected directly into the combustion chamber, but this can be done during the intake stroke, or during compression depending upon the engine load demand. At part load the engine is de-throttled and a heterogenous mixture is used to increase efficiency and reduce fuel consumption. At higher load a homogenous mixture formation is used and the engine operated closer to a standard gasoline combustion system.

### *Miller/Atkinson cycle*

Miller cycle engines use late inlet valve closing in combination with turbo charging to extend the expansion phase compared to the compression phase. The Atkinson cycle is similar but without turbocharging.

These technologies are often seen used in combination with hybrid vehicle powertrains because their trade-offs can be offset with the use of electric propulsion—such that the ICE can operate at the most optimum speed and load point (known as load point shifting). The technology can be applied to both spark-ignition and compression ignition engines.

### *Homogenous charge compression ignition/premixed charge compression ignition*

Homogenous charge compression ignition and premixed charge compression ignition is used to decrease $NO_x$ and soot engine-out emissions and achieved by a reduced combustion temperature and a well prepared mixture formation. However, the system requires cycle-by-cycle control and the mixture state defines the start of combustion—which is challenging for the engine control system.

## Part 1: Combustion pressure measurement and analysis

Many special techniques have been developed for the study of the combustion process, one of the oldest being direct observation of the process of flame propagation, using high-speed photography through a quartz window. More recent developments include the use of flame ionization detectors to monitor the passage of the flame and the proportion of the fuel burned; laser Doppler anemometry is also used in advanced research.

But the standard tool for the study of the combustion process in a normal engine-development process is the cylinder pressure indicating system, the different attributes of which are described later in this chapter. Along with cylinder pressure, a variety of other quantities must be measured, using appropriate sensors either on a time base or synchronized with crankshaft position. They may include:

- fuel line pressure and needle lift
- cylinder pressure
- ionization signals
- crank angle
- time
- ignition system events
- inlet and exhaust pressures
- various (quasistatic) temperatures

Each signal calls for individual treatment and appropriate recording methods. Data acquisition rates have increased in recent years and systems are available at the time of writing with a sampling rate of up to 2 MHz on all available channels.

Combustion analysis aims at an understanding of all features of the process, in particular of the profile of energy release in the cylinder during the working stroke. The test engineer concerned with engine development should be familiar with both the principles involved and the considerable problems of interpretation of results that arise.

One of the aims of combustion analysis is to produce a curve relating mass fraction burned (in the case of a SI engine) or cumulative heat release (in the case of a diesel engine) to time or crank angle. Derived quantities of interest include rate of burning or heat release per degree crank angle and the analysis of heat flows based on the laws of thermodynamics.

## Combustion measurement base calculations and theory

### *Introduction*

From the acquired data (nominally in-cylinder pressure and volume/crank angle) a significant number of derived calculations are typically undertaken to establish the performance of the engine combustion system and the combustion process. This section should be considered as a high level overview. There are many excellent published works which cover combustion analysis and thermodynamics in more detail, the most important and commonly referenced are stated in the reference section of this chapter.

### *Calculation of cylinder volume*

Prior to the measuring procedure, several engine specific variables need to be known and used in order to calculate cylinder volume, these are:

- cylinder bore ($B$)
- crank radius ($a$)
- connecting rod length ($l$)
- compression ratio ($r_c$)

To avoid unnecessary repetition of calculations, cylinder volume ($V$) is calculated using Eqs. (16.1)–(16.4) for each crank angle division as required according to the measurement table and resolution.

$$V = V_c + \pi\left(\frac{B}{2}\right)^2 (l + a - s) \tag{16.1}$$

where $s$ is the distance between the crank axis and the piston pin axis and calculated by:

$$s = a\cos\theta + \sqrt{(l^2 - a^2\sin^2\theta)} \tag{16.2}$$

Swept volume ($V_s$) and clearance volume ($V_c$) are calculated thus:

$$V_s = 2a\pi \frac{B^2}{4} \tag{16.3}$$

$$V_c = \frac{V_s}{r_c - 1} \tag{16.4}$$

A shaft encoder mechanism will supply both crank angle and trigger marks to the measurement system. The crank degree marks can be used to initiate analog–digital samples and hence set the acquisition frequency which is then engine speed dependent.

The accuracy of the volume calculation is limited by the accuracy to which the clearance volume can be measured or determined. In addition, the accuracy of the crankshaft displacement measurement (via the angle encoder) also has a significant effect on the accuracy of instantaneous cylinder volume determination. Note that clearance volume is not required for MEP-related calculations, as it is only the change in volume relative to pressure that is considered here.

### *Pressure correction zero-level, or pegging*

The process of pressure zero level correction is also known as pegging. It is required for dynamic pressure only measuring piezo-electric sensor input channels (normally in-cylinder pressure) and consists of a method of calculating and correcting for the measured signal voltage applied to the measurement system input channel, at a crank angle where absolute in-cylinder pressure is known. There are several pegging methods available, the simplest and most common is to set in-cylinder pressure at Inlet Bottom Dead Centre (BDC) equal to inlet manifold pressure. This method is known to work well for normally aspirated engines, though is unsuitable for engines with highly tuned intake systems or those with forced induction. Another approach is to measure a correction pressure value, in the manifold, or in-cylinder at BDC via access in the cylinder wall, with a second, piezoresistive (absolute) sensor. For situations where an absolute reference pressure is not available or possible, a pegging calculation based upon the polytropic equation $PV^n = C$ can be used. The change in pressure experienced during the compression stroke due to the volume changing from $V_1$ to $V_2$ can be written as shown in the following equation:

$$\Delta p = p_1 \left[ \left( \frac{V_1}{V_2} \right)^n - 1 \right] \tag{16.5}$$

where $p_1$ and $V_1$ are the cylinder pressure and volume at crank angle$_1$ and likewise for $p_2$, $V_2$ at crank angle$_2$. Rearranging this equation with $p_1$ as the subject allows the derivation of a pressure value at position$_1$ where pegging can then be implemented on the data from that cycle.

## *Cyclic-based work*

Cylinder pressure data can be used to calculate the work transfer from the gas to the piston. This is generally expressed as the indicated mean effective pressure (IMEP) and is a measure of the work output for the swept volume of the engine. The result is a fundamental parameter for determining engine efficiency as it is independent of speed, number of cylinders, and displacement of the engine. The gross IMEP calculation is basically the enclosed area of the high-pressure part of the $P-V$ diagram and can be calculated as the integral of the pressure volume curve divided by the swept volume [Eqs. (16.6) and (16.7)]. IMEP can also be presented in net form. The gross vales only accounts for the work done on, or by, the piston during the compression and expansion strokes, the net value also includes the work done by the piston in the exhaust and inlet strokes [Eq. (16.8)].

*The IMEP for a single cylinder can be computed thus:*

$$\text{IMEP(Gross)} = \frac{1}{V_s} \oint p dV \tag{16.6}$$

*This can be explained by the following equation:*

$$\text{IMEP(Gross)} = \frac{\text{(area of high pressure loop)}}{V_s} \tag{16.7}$$

This effectively gives energy released or gross MEP (gross work over the compression and expansion cycle, GMEP).

*Net IMEP for a single cylinder is derived by subtraction in the following equation:*

$$\text{IMEP(Net)} = \frac{\text{(area of high pressure loop)} - \text{(area of low pressure loop)}}{V_s} \tag{16.8}$$

Integration of the low-pressure (or gas exchange) part of the cycle gives the work lost during this part of the process (pumping losses, also known as PMEP). Subtraction of this value from the gross IMEP gives the net IMEP (NMEP) or actual work per cycle [Eq. (16.9)]. Therefore it can be stated that:

$$\text{NMEP} = \text{GMEP} - \text{PMEP} \tag{16.9}$$

where GMEP is the gross indicating mean effective pressure (area of power loop), PMEP is the pumping indicating mean effective pressure (area of pumping loop), and NMEP is the net indicating mean effective pressure.

The most important factor to consider when measuring IMEP is the TDC position (see the "'Exact' Determination of True Top Dead Center Position" section in this chapter).

Note that, in order to optimize the calculation, it is not necessary to acquire data at high resolution. Measurement resolution of a maximum of 1 degree crank angle is sufficient for MEP calculations; higher resolution than this does not improve accuracy and is a waste of system resources. Also important for accurate IMEP calculations are the transducer properties, mounting location, and stability of the transducer sensitivity during the engine cycle.

*If several cycles are analyzed, the coefficient of variation (CoV) of IMEP can be determined as shown in the following equation:*

$$\text{IMEP}\cdot\text{CoV} = \frac{\sqrt{\sum_1^n \left(\text{imep}-\overline{\text{imep}}\right)^2/(n-1)}}{\overline{\text{imep}}} \tag{16.10}$$

## *Polytropic coefficient*

In a purely polytropic expansion or compression process, the relationship shown in the following equation will be obeyed throughout:

$$PV^n = C \tag{16.11}$$

where $P$ is the pressure, $V$ is specific volume, $n$ is the polytropic index, and $C$ is a constant.

The polytropic process equation can be used to compute a crank angle based curve of $C$ (where $n$ can be estimated). This curve shows a strong response during the compression/expansion process where combustion takes place and can be used to identify start and end points (Fig. 16.1).

The polytropic coefficient itself can be determined between two data points on the pressure curve (Fig. 16.2). It is of particular interest during compression and expansion processes and can be derived as a single result

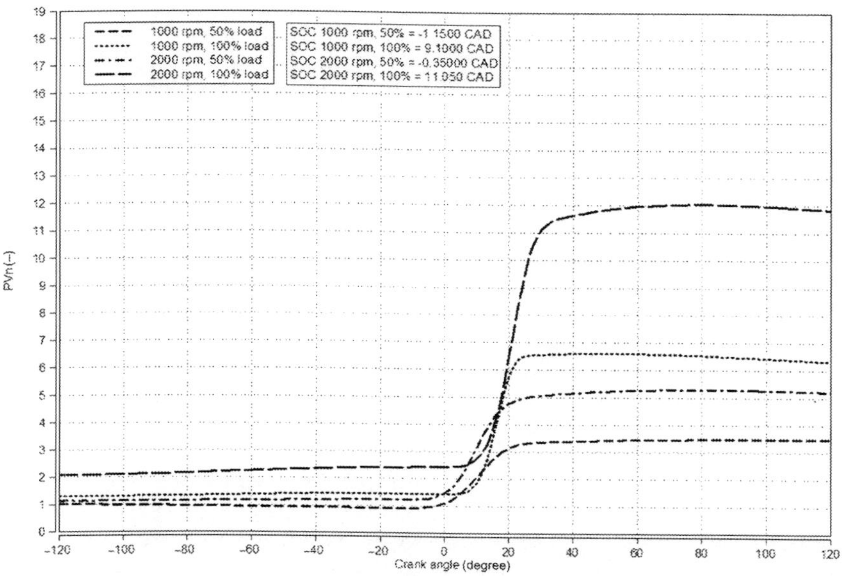


**FIGURE 16.1** Typical curve of "C" plotted against crank angle for 4, fired, engine operating points, this curve can be used to observe and characterize the start of combustion.

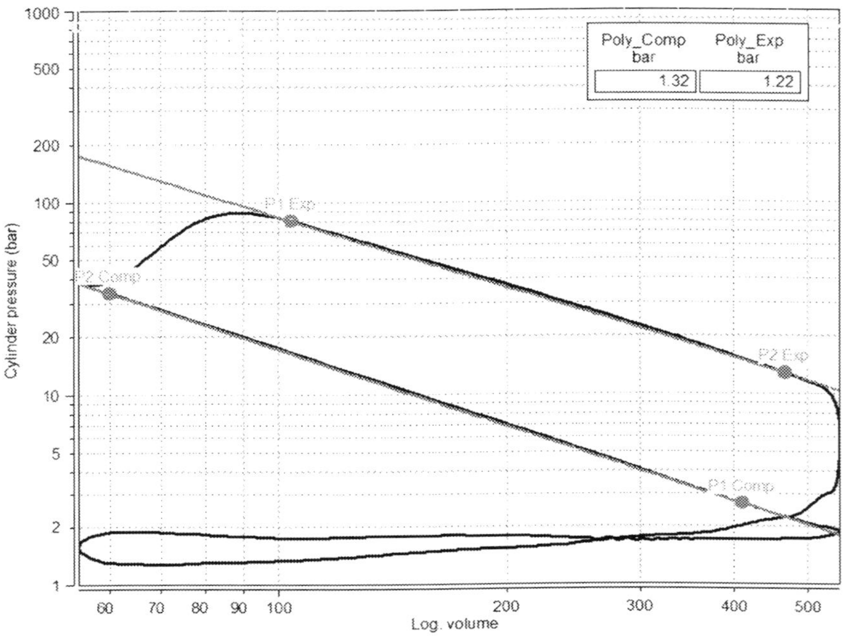


**FIGURE 16.2** Polytropic value derived from a fired pressure curve, displayed as a logarithmic pressure–volume diagram. Values shown are derived from the curve between the marker points for compression and expansion processes, respectively.

between two points, or as a continuous curve versus crank angle. The calculation is shown in the following equation:

$$X = \frac{\ln P_1 - \ln P_2}{\ln V_1 - \ln V_2} \qquad (16.12)$$

where $X$ is the polytropic coefficient, $P$ is the cylinder pressure at crank angle location 1 or 2, $V$ is the cylinder volume at 1 and 2, respectively

Most practical thermodynamic processes can be considered polytropic with values ranging between the theoretical limits of 1 and 1.4. The value can also be considered as defining the gradient of the compression and expansion lines on a logarithmic pressure/volume diagram.

### *Instantaneous energy release*

The cylinder pressure data, in conjunction with cylinder volume, can be used to extrapolate the instantaneous energy release from the combustion event with respect to crank angle. There are many well-documented theories and methods for this, but all are fundamentally similar and stem from the original work carried out by Rassweiler and Withrow [1]. The calculations can range from the quite simple with many assumptions, to very sophisticated models with many variables to be defined and boundary conditions to be set.

The fundamental principle to the calculation of energy release relies on the definition of the compression and expansion process via a polytropic exponent (ratio of specific heats under constant pressure and volume conditions for a given working fluid). If this can be stated with accuracy, a motored pressure curve can be extrapolated for the fired cycle. The fired curve can then be compared with the motored curve, and the difference in pressure between them, at each angular position, and with respect to cylinder volume at that position, allows derivation of energy release during the engine cycle progression.

A typical simplified algorithm to calculate energy release is shown in the following equation:

$$\Delta Q \frac{K}{\kappa - 1}(\kappa \cdot p \cdot (\Delta V) + V \cdot (\Delta p)) \tag{16.13}$$

where $K = 100$, $\kappa = 1.32$ for gasoline engines, $\kappa = 1.35$ for gasoline direct injection engines, and $\kappa = 1.32$ for diesel engines.

From such a calculation the engine engineer has the ability to gain an energy release or mass burn fraction (MBF) curve in real time. This knowledge, of the rate at which fuel is oxidized, can assist with understanding the entire energy conversion process in the cylinder. Most algorithms deploy a simplistic zero or one-zone heat-release approach—and this is generally deemed to be sufficient for a qualitative analysis and is also computationally efficient.

The calculation shown above (Fig. 16.3) is relatively simple—it does not account for wall or blow-by losses in the cylinder. Also, it assumes that the

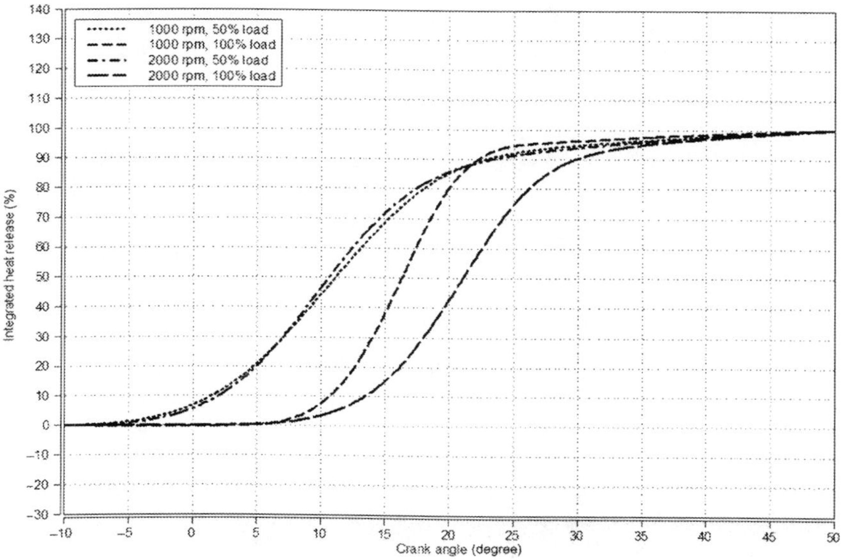


**FIGURE 16.3** Integrated heat release values plotted against crank angle for four engine operating points. The 50% conversion value is widely used as a metric optimizing combustion phasing during ignition timing calibration for gasoline engines.

polytropic index is constant throughout the process. These compromises can affect accuracy in absolute terms, but in relative measurement applications, such simplified calculations are well proven and established in daily use. Under conditions where engine development times are short, algorithms must be simplified such that they can be executed quickly. Most important for the engine developer are the results that can be extracted from the curves, which give important information about the progress and quality of combustion.

## *Gamma*

Known as the specific heat ratio of the working fluid (or ratio of specific heats for constant pressure and constant volume conditions), it is an important parameter with respect to fast heat release calculations that has a significant effect on the peak and shape of the heat release curve (Fig. 16.4). A too low value of gamma produces a heat release value which is too high and a heat release rate which is negative after the completion of combustion. The opposite effect being the case where gamma is too high, although the effects are nonlinear. Gamma can be used as a tuning parameter for the heat release calculation. For example, the use of a temperature dependent value for gamma allows optimization of the heat release calculation for different engine and combustion system operating conditions.

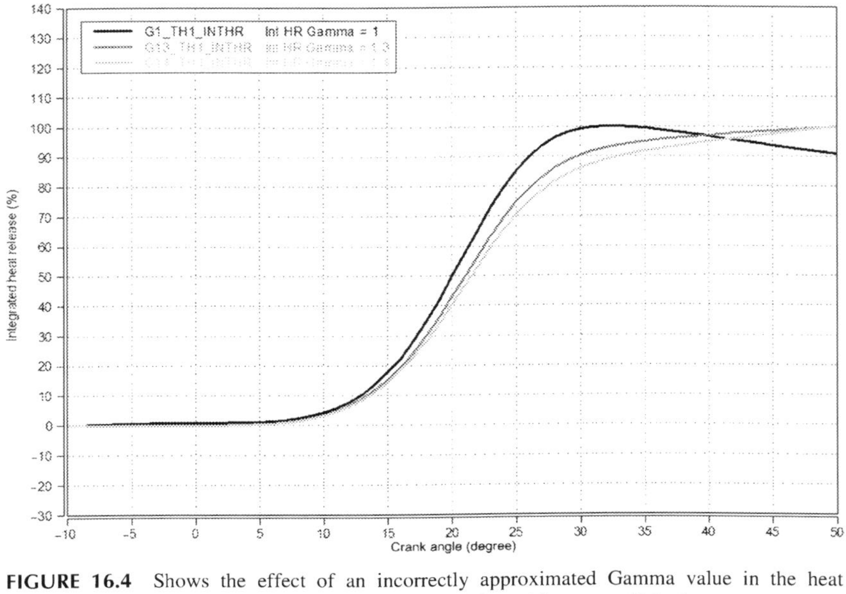


**FIGURE 16.4** Shows the effect of an incorrectly approximated Gamma value in the heat release calculation. End of combustion values derived from this curve will be inaccurate.

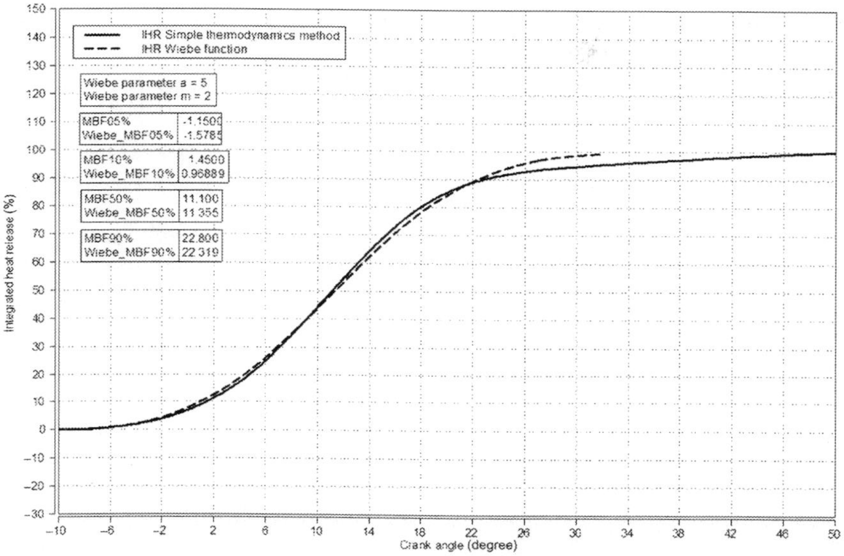


**FIGURE 16.5** The Wiebe function (also known as Vibe) is a simple calculation used by engineers to simply model and predict the energy release curve based on some simple tuning parameters. There are no physics involved in this simple model.

### *Wiebe function*

A simple but widely adopted algorithm to model the typical energy release curve in the combustion process. It creates a characteristic curve (Fig. 16.5) based on some simple parameterization that matches the S shape of the heat release integral. It is simplistic approach, providing a robust output that can be used for relative comparisons. The Wiebe function describes the mass burned fraction curve and is computed as shown in the following equation:

$$x(\theta) = 1 - \exp\left\{ -a\left(\frac{\theta - \theta_{\text{soc}}}{\Delta\theta_{\text{dur}}}\right)^{m+1} \right\} \tag{16.14}$$

where $x(\theta)$ is the mass burned fraction at angle $\theta$, $\theta_{\text{soc}}$ is the angle at start of combustion, $\Delta\theta_{\text{dur}}$ is the combustion duration angle, $a$ is the constant, typical value 5, $m$ is the constant, typical value 2.

## Part 2: Sensors and measurands

The sensors, signal conditioning, and measurement system are quite specific to the combustion pressure measurement application, with many necessary features that should be understood by the engineer working in the test environment. In addition, the technology used has some attributes which are quite different to other measuring systems normally encountered in and around the test environment. Understanding the fundamentals will allow the engineer to effectively source and implement a system—and be able to deploy it efficiently, including handling of the "consumable" parts of the system.

## Sensors

### *High pressure in-cylinder sensors*

Almost exclusively, for in-cylinder engine pressure measurements, piezoelectric sensing technology is deployed. There are many other sensing technologies available—but the piezoelectric sensor has some specific properties that mean it is by far the best choice for the application.

There are some fundamental properties of this type of sensor technology that need to be understood with respect to the powertrain testing environment. First, it should be clear that this measuring technology is capable of a "dynamic" pressure measurement only. *The measuring element only generates electrical charge (i.e., an output) when the stress in the crystal structure is changing (i.e., a force is applied).* Once in a "static" state, the crystal generates no charge. The consequence of this being that a piezoelectric pressure sensor measures change in pressure and not absolute pressure. For in-cylinder pressure measurement applications this is appropriate, as the measurement system is able to convert the raw signal from relative to absolute pressure via signal processing and software routines—but it should be remembered that the raw pressure value measured (prior to processing) is dynamic.

The piezoelectric measuring element has a physical property of high stiffness—this means it has a high-natural frequency, a wide dynamic range and is extremely durable. Severe overloading of the sensor will almost always destroy the mechanical parts of the sensor first and not the actual measuring element. Thus, sensor overload capability is high and often, recalibration of the sensor is all that is required to bring it back to useable condition after an overpressure event (assuming no mechanical damage). Due to the hardness of the measuring element, there is minimal displacement of the diaphragm when applying force to the element—the sensor is, therefore durable, as it has no significant moving parts.

Most importantly the piezoelectric sensor has a very wide dynamic range and is able to measure very large pressure levels but is also sensitive enough to measure very small pressure changes. This is ideal for combustion applications where large pressures in the high-pressure part of the cycle along with very small pressure fluctuations during the gas exchange part of the cycle need to be established.

The basic construction of all piezoelectric sensors centers around a piezoelectric measuring element within a cylindrical housing that forms the body of the sensor itself (Fig. 16.7). Different manufacturers have different ideas about the details of the construction, but in simple terms, one end of the measuring element is connected to the charge transfer interface, which transports the

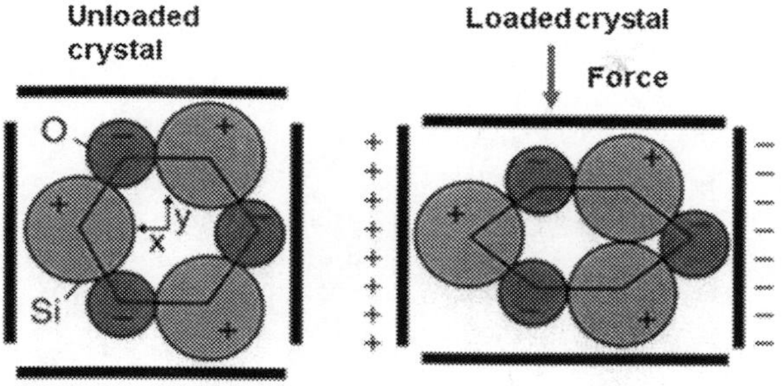


**FIGURE 16.6** The direct piezoelectrical effect used in combustion pressure sensors. *Courtesy: Kistler Group.*

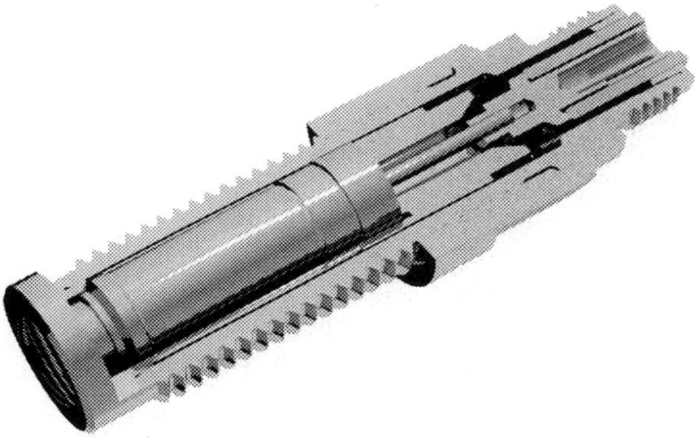

**FIGURE 16.7** Typical construction details of piezoelectric sensor for combustion pressure measurements. *Courtesy: Kistler Group.*

charge signal to the external connector. The other end of the measuring element is connected to the rear of the diaphragm membrane via the force transfer mechanism. The operation involves the application of a basic property of piezoelectrical material; that is, when a force is applied to the crystal structure, the subsequent stress in the material generates a proportional electrical charge on certain surfaces (direct piezoelectric effect)—and this is the signal that is converted to a voltage and measured by the data acquisition device (Fig. 16.6).

Another consideration compared to other measuring devices is the type and magnitude of the sensor output signal. The actual signal is a tiny electrical charge, which is converted and amplified with a signal conditioner (charge amplifier—discussed later). Effective handling and transfer of this raw signal relies on exceptionally high-impedance cabling and connectors, along with very high insulation resistance for the signal cables. In test cell environments, the level of cleanliness required for handling this equipment is not normally expected and can be difficult to achieve.

## Selecting a sensor and installation concept

This can be quite a complex decision which is a trade-off between the application requirements, the available space for installation, and the amount of effort expended for the installation. The first two parameters depend upon the chosen sensor, the latter has a direct relationship to measurement quality (more invested in installation = better quality data).

### *Location of the sensor*

This is a decision made of compromises; the sensor has a task to measure a globally representative pressure change from inside the cylinder, which can be used to represent to total activity in the combustion chamber, the problem being that in-cylinder flows are highly dynamic and sometimes these dynamics are of interest, and sometimes not. Modern passenger car sized ICEs generally have compact bore sizes, along with complex valve and port designs with not much space to provide access points to the in-cylinder pressure. When creating a sensor bore, some best practice can be considered. Central locations are favorable for global pressure values, but for phenomena such knock in SI engines, which occurs at the periphery of the cylinder, they can be less favorable. However, readers should note that knock is a complex topic with very little consensus of opinion regarding the best way to measure and characterize it. This means that only the following general advice can be given about sensor locations:

- avoid areas of high flow—around valves openings;
- avoid areas of high temperature (higher than cylinder average)—around exhaust valve ports; and
- avoid areas of "squish," where considerable gas dynamics can occur.

### *Intrusive sensor installation*

The sensor installation has a great impact on the data quality as much as the sensors proprieties; a good sensor can be neutralized by a poor installation. The choice for an engineered bore can be via direct or adaptor-based installation, but in both cases, the availability of a precision machine shop, with the services of a skilled machining technician, are essential. For a direct installation the sensor manufacturer will provide machining dimensions and tolerances for the machined bore (Fig. 16.8); these must be adhered to while taking into account space and material properties. The sensor is installed directly into a bore in the cylinder head material (Fig. 16.9). The advantage of this approach is that the sensor body is in direct contact with the cylinder head, so that heat dissipation to the surroundings is optimal which helps with sensor temperature related sensitivity shift. The disadvantage is that stress in the cylinder head material (induced by pressure in the cylinder) can be transferred to the sensor body, inducing a strain related signal that is super imposed onto the

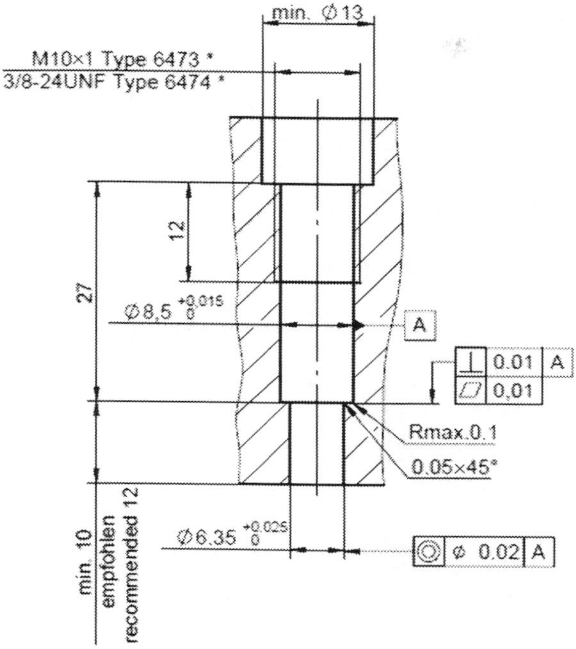


**FIGURE 16.8** Typical bore machining requirements for a sensor installation. *Courtesy: Kistler Group.*

**FIGURE 16.9** Installation examples from left to right—flush direct mounting, measuring spark plug, direct mounting front sealing. *Courtesy: Kistler Group.*

pressure signal from the sensor output. This latter effect is a considerable modern-day problem where engine components are light weight, manufactured with minimal safety factor tolerances. Cylinder head flexing may not be a problem for the engine manufacturer as long as material stress limits are not exceeded and durability is maintained but for the sensor installation and user, this is an issue to be aware of.

An alternative to direct installation is to use an adaptor—this is a machined sleeve in the form of a tube that is installed into the cylinder head that provides a sensor bore and seating. An adaptor is often used where there is no access to the combustion chamber directly through the material of the cylinder head. This means that access must cross fluid bearing channels and passages in the cylinder head. Obviously, the adaptor has to be bigger than the sensor, so its physical size must be accommodated and in addition, sealing the fluid channels after installing the adaptor has to be achieved. It should also be noted that the adaptor must not significantly impede any coolant flow otherwise artificial "hot spots" will be created in the combustion chamber that could affect the combustion process. It should also be noted that an adaptor sleeve reduces the heat flow from the sensor through to the surrounding material such that using an adaptor normally causes an increase in operating temperature for a given sensor, in a given engine and operating condition. This is not necessarily a problem but should be observed. If in any doubt, sensor manufacturers can provide temperature dummies (a sensor body equipped with two fast response thermocouples, located to give measuring element and diaphragm temperatures during operation) in order to validate a new sensor installation. An advantage of the adaptor-based approach is the quality of the sensor sealing interface. As the adaptor is normally produced by the sensor manufacturer, all requirements with respect to the quality of the sealing faces at microscopic tolerance levels can be fulfilled. This is essential for any sensor installation but is difficult when creating a direct installation; the sealing surface compliance to the specifications is down to the machinist skill and the material quality. In many cases, this is created suboptimally and the effect of this has a considerable effect on sensor life and data quality.

The *theoretically* ideal location for the sensor diaphragm is to be exactly flush with the surrounding cylinder head surface. The reason being is that there will then be no dead volume in front of the diaphragm which can provide a "passage" or "pipe" that can generate acoustic oscillations under certain conditions. Ideally, the sensor should be mounted in a perpendicular location; but this is not always possible, an angled installation creates a recessed mounting position. The sensor design itself also has an impact on this factor since not all sensors have sealing arrangements that allow a flush diaphragm position and the installation location will dictate how optimally the diaphragm can be positioned. Note that the critical frequency for any passage volume can be easily modeled and approximated and it is not always the case that the resonance condition will create a

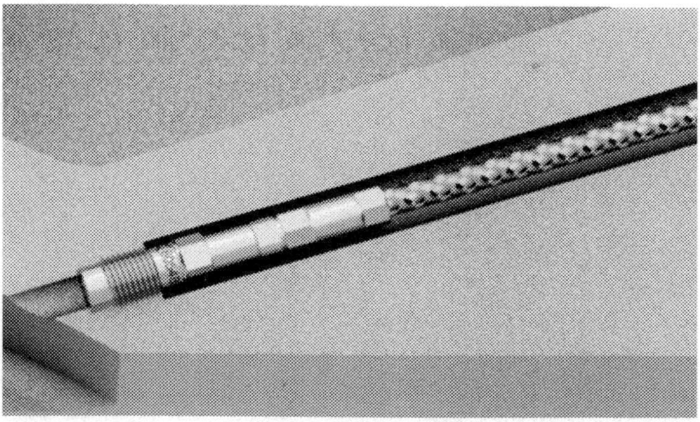

**FIGURE 16.10** Sensor installation with recessed diaphragm mounting situation. *Courtesy: Kistler Group.*

problem as it may be away from the operating points of interest. Alternatively it could be possible that the high frequency components on the signal are not of interest anyway, so that signal processing and filtering can be used to remove all of these components, including the pipe oscillation.

The recession of the diaphragm can have a positive effect because can it reduce the temperature of the diaphragm by reducing the exposure to the flame—and thus reduce the effect of cyclic heating. A small recession in the range of 1–2 mm is considered a suitable option and can reduce error sources due to intracycle heating of the diaphragm (Fig. 16.10).

## *Nonintrusive sensors and adaptors*

An alternative method for accessing the in-cylinder pressure signature is available, with reduced installation effort, using instrumented engine components that already access the combustion chamber. For Gasoline engines, this is a spark plug (Fig. 16.11), for Diesel engines, this would mean a "glow-plug" or preheating device (Fig. 16.12). This involves the installation of a piezoelectric sensor into a modified version of the original equipment part, the whole unit being manufactured by the sensor manufacturer. The user then replaces the original part with the instrumented version and can then gain a pressure signal for measurement without any significant engine modifications, but this raises the question:

"Why would you bother with an engineered installation 'intrusive' method if you could use this 'nonintrusive' method that seem to be considerably less effort?"

The reason is data quality; although these "nonintrusive" methods seem attractive, there are compromises with respect to the possible location of the sensor, the type of sensor, and the performance of the instrumented component compared to the original part. All these factors have an effect on the

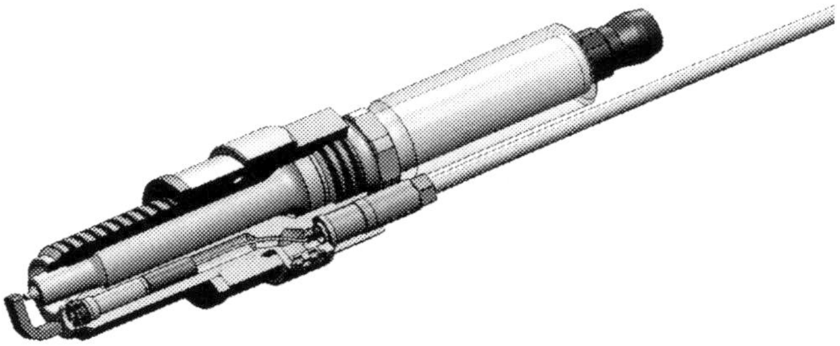

**FIGURE 16.11** MSP with integrated sensor. *MSP*, Measuring spark plug. *Courtesy: Kistler Group.*

**FIGURE 16.12** Cutaway showing glow plug adaptor and installed sensor. *Courtesy: Kistler Group.*

data and signal quality and there is no doubt that the best installation is one made and engineered specifically for the purpose. However, nonintrusive methods do have a place in the development process where certain aspects of data quality can be compromised in exchange for the reduced installation effort and efficiency.

For gasoline engines applications, pressure measuring spark plugs include an integrated measuring element, the design of which varies with the manufacturer. As there are so many different variations in spark plugs for modern engines, it is often the case that an exact match between the original equipment and the measuring spark plug cannot be achieved with respect to spark position and/or heat range. In this instance it is then up to the test engineer do decide which type to specify, with the support of the sensor manufacturer. For gasoline engine with direct injection, spark position is essential as the heterogeneous in-cylinder conditions, at certain engine operating points, requires that the electrical arc must be in exactly the right place to ensure reliable combustion under all conditions. It can be the case that the original spark plug is often indexed for this purpose but, due to the installation of the sensor in the measuring spark plug, the same arc position cannot always be achieved. This does not always cause an issue, but it is important to note this, along with any compromises made with respect to heat range and operating temperature of the spark plug.

Another issue to be noted is installation of a spark plug with additional measurement cabling. Often it is necessary to install high-voltage extension leads in order to accommodate the external cabling for the sensor—many engines have as standard coil-on-plug designs that reduce high-voltage cables (as they are mounted directly onto the spark plug). Therefore introducing high-voltage extension leads between the spark plug and coil can introduce leakage paths and cause misfiring during testing. In many cases though, careful handling can counter all these issues—and it is possible for measuring spark plug to produce good data for general engine calibration applications later in the development cycle, where detailed thermodynamic analysis is not required.

For nonintrusive measurement in diesel engine applications the preheating device is replaced with an instrumented version which has the heating element replaced with a pressure sensing probe (although devices containing heater and pressure sensor combined have been produced, they are not specified for research level measurements). The lack of heating function is normally an accepted compromise as it only becomes relevant for cold-starting tests, which are mostly done when the prototype engine is equipped with accurately machined sensor bores, earlier in the development process. A glow-plug profile sensor adaptor is produced which has exactly the same dimensions as the original glow plug. This forms a mounting for a standard design sensor probe that is installed in the adaptor. The tip of the adaptor has "holes" that allow the cyclic combustion pressure to be applied to the sensor diaphragm. There are several factors to note:

- The adaptor is manufactured to the specific dimension of the installation bore (details that must be provided to the sensor supplier for its design and manufacturing)—and not the original equipment manufacturers published dimensions for the glow plug. These latter dimensions relate to a mass produced part where production tolerances are not sufficiently accurate. This is important as the body of the adaptor must be a close fit into the bore, and the sealing angle of the adaptor has to be within tolerance as these are the interfaces between the adaptor body and its surroundings that form the heat flux path that manages the sensor operating temperature. This also means that in theory, the adaptor is manufactured for a specific bore in a specific engine. In practice though, adaptors are generally passed around and not dedicated to use in a specific bore or engine. The impact of this is that sensor life can be reduced when compared to an optimized installation for a specific bore.
- The installation of the sensor in the adaptor can be optimized during the adaptor design process—the sensor position can be recessed more to reduce operating temperature, but this increases the potential for pipe oscillation effects. The sensor can be moved closer to the tip, increasing signal quality but also operating temperature. In addition, tip-less sensors are available—these reduce the sensor operating temperature and have no real downside apart from a small potential effect of slight cylinder volume change. The most important message is to choose the best trade-off from the available designs according to the application and development focus.

## Sensor types and designs

In combination with the installation, the sensor itself would be selected according to the measurement application. There are several attributes relating to the sensor construction that should be considered when selecting a sensor, namely, cooling and sealing arrangements. These properties have a significant effect on the design of the sensor itself but also bring some specific benefits in the general trade-off between data quality and installation effort; the test engineer must consider these attributes as well.

### *Sensor cooling*

Sensors fall broadly into cooled or uncooled types. Cooling means that the sensor construction is such, that cooling passages are provided in and around the measuring element, this allows a cooling media to flow in and out of the sensor body itself in order to carry away heat and maintain the measuring element at a constant temperature, irrespective of the operating condition of the engine. The benefit is quite clear with respect to sensor temperature management and stability—and for many years, cooled sensors were considered the benchmark for gaining the most accurate measurement data. However, over the last 10 years, considerable progress in crystal material technology and production has improved the performance of uncooled sensors such that their performance is now sufficient for applications where the most detailed thermodynamic analysis is required from the measurement data. So cooled sensors (Fig. 16.13) are a matter of choice or preference, rather than necessity.

When selecting a sensor, it is normally the case that a cooled sensor is physically larger—and thus cannot always be accommodated in some applications,

**FIGURE 16.13** Cooled sensor construction showing internal passages for cooling media. *Courtesy: Kistler Group.*

where space constraints exist. In addition, a cooling system must be provided to ensure a steady flow of the correct cooling media at the appropriate flow rate and pressure. A simple pump and fluid receptacle are definitely suboptimal; therefore, a suitable cooling system must provide a flow of cooling fluid at constant temperature (nominally, 50°C). This flow must be free of any pressure pulsations that could be created by the pump as flow pulsations can be superimposed onto the raw sensor signal. This is due to the fact that the cooling fluid flows directly around the sensor measuring element and cross-talk can and does occur. The cooling system should also have monitoring capability such that flow disruption or failure can be signaled to a master or host system, to warn the user before the sensor is destroyed by overheating.

Cooled sensors also need careful handling. They can be difficult to handle during installation and removal due to cooling fluid pipework, such that damage during use can easily occur. The cooling tube extensions on the sensor are quite delicate and need correct installation and sealing (Fig. 16.14). The cooling pipes must be integrated securely and sealed with a suitable compound such that there is minimal possibility of leakage. Fluid leaking into the sensor bore or installation can damage electrical connections and leads. Damage to the engine can occur if cooling fluid leaks into the engine and is not evacuated before fired operation.

Another factor to consider is thermal shock—it is often thought that cooled sensors are better with respect to intracycle drift but this is not always the case. This is due to the greater temperature difference that occurs across the sensor diaphragm due to the cooling media on one side. This greater temperature difference (compared to an uncooled sensor) can cause increased deformation of the diaphragm due to the cyclic heating effect and thus greater intracycle sensitivity change. It is, therefore, very important that the test engineer understands fully the sensor data sheet—and the stated parameters given by the sensor manufacturer with respect to thermal shock performance.

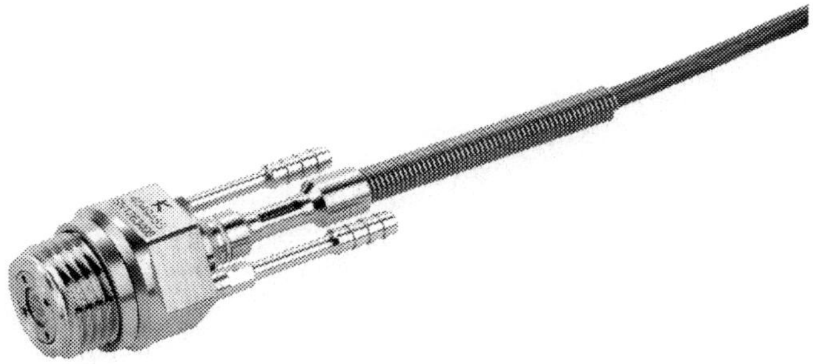

**FIGURE 16.14** A cooled sensor for low-pressure measurements showing cable and cooling connections, often cooling tubes have to be removed in order to disconnect the cable. *Courtesy: Kistler Group.*

Uncooled sensors are simpler construction and normally rely on stable crystal technology along with the general design of the sensor in order to reduce temperature related drifts. Intracycle drift (also known as thermal shock) is due to membrane deformation due to cyclic heating. However, general drift of sensitivity due to bulk temperature change is related to the change of sensor preload due to different materials thermal expansion coefficients (and not so linked with the crystal material). Therefore a well-designed sensor can have good thermal sensitivity response (i.e., low) irrespective of the crystal material. Again, this is where it is important to properly understand manufacturer data sheet information. Uncooled sensors can be constructed in smaller dimensions and thus more easily accommodated in modern engine designs. The main challenge being to create a measuring element that has sufficient sensitivity to provide a good quality signal at smaller signal amplitudes such that detailed analysis of signals with a low peak to peak deviation can be undertaken.

### Sensor mounting

The sensor can be mounted directly or indirectly but is secured either by a thread, or alternatively, plug-in sensor designs are available. In this latter case there is no thread and the sensor is effectively "clamped" into the installation via a threaded device as part of the installation mechanical arrangement. The benefit of this design is that the sensor body is effectively decoupled from the mounting bore completely, with a consequence that strain is not transferred to the sensor mechanically. The disadvantage is that the space requirement for a given sensor dimension is greater (more installation space needed) and that the level of precision required to create the bore is much greater as the concentric positioning of the sensor in the bore is essential.

### Reinforced versions

Sensor suppliers can offer versions of their sensors which are designed to be capable of use in extreme working conditions, for example, knocking or preignition, that can occur in spark ignition engines running on gasoline or gaseous fuels. Knocking can be extremely damaging and phenomena such as "super knock" can destroy a sensor in a few engine cycles due to extreme pressure rises and high instantaneous pressures. In order to combat this—the sensor can be equipped with a reinforced diaphragm (normally thicker or different material, or both) which provides greater resistance to extreme pressures, but there is a trade-off. The diaphragm is stiffer and has greater thermal inertia—thus thermo-shock recovery is less favorable and sensitivity is reduced. This has an effect on IMEP and PMEP accuracy and, thus, reinforced sensors should only be considered for measurement activities later in the development cycle, where highly accurate thermodynamic analysis is not required.

### Flame guards

The function of a flame guard (Fig. 16.15) is to reduce the thermal load on the sensor diaphragm. The flame guard keeps the extreme flame temperature

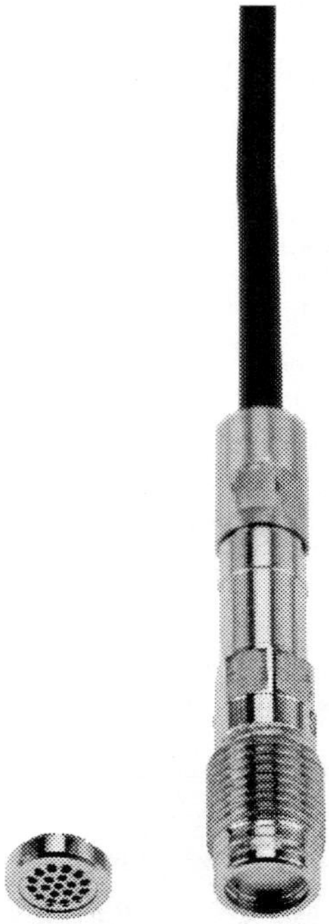

**FIGURE 16.15** Clip-in flame guard (left) for front sealing sensor. *Courtesy: Kistler Group.*

away from the diaphragm; the temperature of the sensor diaphragm is subsequently reduced. Under heavy knock conditions the flame guard damps the high-pressure amplitudes and minimizes pressure gradients. However, with long-term operation at low engine load, (with low cylinder head temperature and high soot emission), the flame guard can become contaminated—a damping effect can then occur which causes a phase shift of the measured pressure curve. Thus any benefits gained in data quality can be offset by increased operational maintenance.

## Sensor selection and specifications

It is important to be able to understand and interpret that data given by the sensor manufacturer on the data sheet (Tables 16.1 and 16.2). Some of the specified parameters are quite straight forward to process as they are similar to those which are typically stated for any sensor—for example, range, overload, and sensitivity. These would be the basic parameters for selection based upon the measuring application requirements. Natural frequency and acceleration sensitivity are important for dynamic measurements—in general though, piezo-electric engine sensors have a capability here that is beyond the application requirements. Some sensor designs possess technology

**TABLE 16.1 Information from a sensor data sheet—5 mm front sealed sensor.**

| | |
|---|---|
| Measuring range | 0–250 bar |
| Overload | 300 bar |
| Lifetime | ≥ 108 load cycles |
| Sensitivity | 19 pC/bar nominal |
| Linearity | ≤ ±0.3% FSO |
| Natural frequency | ~160 kHz |
| Acceleration sensitivity | ≤0.0005 bar/g axial |
| Linearity | ≤ ±0.3% FSO |
| Natural frequency | ~160 kHz |
| Acceleration sensitivity | ≤0.0005 bar/g axial |
| Shock resistance | ≥2000 g |
| Insulation resistance | ≥ $10^{13}$ Ω at 20°C |
| Capacitance | 7.5 pF |
| Operating temperature range | −40°C to 400°C |
| Thermal sensitivity change | ≤1% 20°C–400°C |
| | ≤ ±0.25% 250°C ± 100°C |
| Load change drift | 1.5 mbar/ms max gradient |
| Cyclic temperature drift[a] | ≤ ±0.5 bar |
| Thermo shock error[b] | $\Delta p \leq \pm 0.3$ bar |
| | ΔIMEP ≤ ±1.5% |
| | $\Delta P_{max} \leq \pm 1\%$ |
| Thread diameter | M5 × 0.5 front sealed |
| Cable connection | M4 × 0.35 negative |
| Weight | 2.2 g without cable |
| Mounting torque | 1.5 Nm |

[a] *at 7 bar IMEP/1300 rpm, diesel engine.*
[b] *at 9 bar IMEP/1500 rpm, gasoline engine.*
Source: Reproduced from PRESSURE SENSORS FOR COMBUSTION ANALYSIS, Product Catalog of AVL GmbH, publicly available at www.avl.com.

specifically aimed to reduce acceleration sensitivity—which is important for some extreme applications such as high-performance motor sport engines, or measurement on components of relatively low mass. However, it should be noted that active damping technology within the sensor can deoptimize the

**TABLE 16.2 Information from a data sheet—5 mm front sealed sensor.**

| | | |
|---|---|---|
| Measuring range bar 0–300 | bar | 0–300 |
| Calibrated ranges (23°C, 200°C, 350°C) | bar | 0–100, 0–150 |
| | | 0–200, 0–300 |
| Overload bar 350 | bar | 350 |
| Sensitivity (at 23°C) | pC/bar | −17 |
| Natural frequency (measuring element) | kHz | ≈185 |
| Linearity (at 23°C) | %FSO | ±0.3 |
| Tightening torque, greased | Nm | 1.5 |
| Shock resistance (half sinus 0.2 ms) | g | ≥2000 |
| Acceleration sensitivity | | |
| Axial | mbar/g | 0.8 |
| Radial | mbar/g | 0.2 |
| Sensitivity shift | | |
| 23°C–350°C | % | ±1.0 |
| 200°C ± 50°C | % | ±0.4 |
| Operating temperature range | °C | −20 to 350 |
| Temperature, min./max. | °C | −40 to 400 |
| Thermal shock error (at 1500 L/min, IMEP = 9 bar) | | |
| $\Delta p$ (short-term drift) | bar | ±0.25 |
| ΔIMEP | % | ±1.5 |
| $\Delta P_{max}$ | % | ±1.0 |
| Insulation resistance (at 23°C) | Ω | $\geq 10^{13}$ |
| Capacitance sensor | pF | 8 |
| Connector, sapphire | | Insulator M3 × 0.35 |
| Protection rating, with cable Type 7 (IEC 60529) | IP | 65 |
| Weight sensor | g | 1.5 |

Source: Reproduced from Kistler data sheet, publicly available at www.kistler.com.

sensor in other respects, for example, less space inside the sensor body, thus, smaller crystal and lower sensitivity.

Temperature range and sensitivity are important factors—in general, inter-cycle temperature effects are due to the sensor construction and not the crystal technology. The sensor can change sensitivity over its operating range and normally two performance values are stated—percentage error over the full working range and percentage error in a range around a nominated value. The former value is just a general statement of the sensor sensitivity to temperature. The latter value is a nominated average working temperature for the sensor (according to the type and design), with a typical tolerance band that represents typical working conditions from idle to full load. This value is far more representative of the sensor performance during real-life conditions—as such, this is a more useful value for the test engineers selection criteria.

The most important parameters stated relate to the thermodynamic performance of the sensor. This is characterized by a dynamic test in an engine (done by the sensor manufacturer), comparing the measured data with data from a master, reference sensor (mounted in the same cylinder). This reference sensor is a very carefully selected and controlled unit (held by the sensor manufacturer)—specifically designed for precision measurements with little or no sensitivity to external factors. This reference sensor is carefully monitored and is not generally based on any production sensor design but is the unit against which production quality is measured and maintained by the sensor manufacturer.

The production sensors are tested and compared to the reference by the manufacturer in a TSCRE, at a single operating point. This operating point is normally low speed (i.e., long cycle time) which is a worse case for cyclic heating of the sensor membrane. The reference and test sensors are held in a specially designed adaptor that isolates them from large temperature fluctuations and strain effects—but ensures that the sensors both receive the same global pressure value from the combustion chamber. The test environment is controlled carefully and is maintained continuously in a stable operating condition. The test measures a statistically representative number of cycles for analysis. The data is then processed in order to produce the data sheet values—but also other parameters (not always published) which completely define the sensor performance under dynamic conditions. There are several values that are published that define the thermodynamic performance of the sensor; these are derived from the measured curves (Table 16.3).

**TABLE 16.3 Typical published sensor performance metrics for Thermoshock.**

| Thermal shock error at 1500 rpm/9 bar IMEP | | |
|---|---|---|
| **Metric** | **Unit** | **Value** |
| $\Delta p$ (short term drift) | bar | $< \pm 0.2$ |
| ΔIMEP | % | $< \pm 1$ |
| $\Delta P_{max}$ | % | $< \pm 1$ |

Source: Reproduced from Kistler data sheet, publicly available at www.kistler.com.

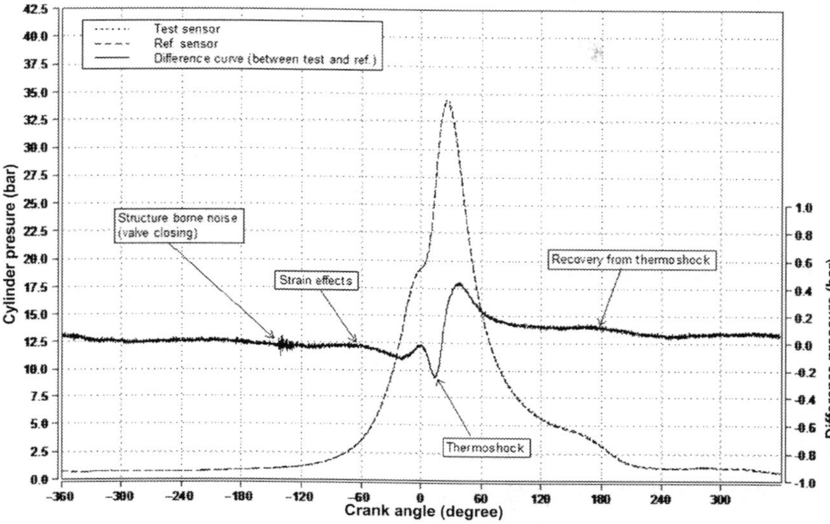


**FIGURE 16.16** Measured and reference pressure are averaged and then subtracted, providing the difference curve.

Delta IMEP and $P_{max}$ is a comparison of the calculated results between test and reference sensor—and thus the error on these results due to thermoshock alone. $P_{max}$ (maximum pressure) error has a sensitivity in the same range as the pressure error due to thermoshock. IMEP can have greater error sensitivity—depending upon the angle at which the thermoshock deviation occurs and how long it lasts for in the high pressure part of the cycle (Fig. 16.16). If the sensor suffers from extreme thermoshock—and does not recover within the expansion part of the cycle—a significant effect in the accuracy of pumping work (PMEP) will also be observed as the measured pressure deviations in this part of the cycle can be in the same range as the thermoshock error. Hence, thermoshock performance and characterization are important for the sensor manufacturer, as well as the sensor user. The values shown on the data sheet as "delta pressure—short term drift" is the maximum deviation between the measured and reference pressure at any point in the averaged cycle.

While linearity, thermal sensitivity shift at nominal operating temperature range and thermodynamic specifications ($\Delta$IMEP, $P_{max}$, and $P_{shortterm}$) are key attributes; there are other attributes not stated on the datasheet that are important, for example, the sensors resistance to mechanical strain. This is important as modern engines have lightweight component design—as such, mechanical deformation occurs and is acceptable within material limits. In addition, due to advanced design and sophisticated control systems—the engine operates very close to its design limits and components such as cylinder heads can flex during high load. For the production vehicle, this is acceptable and managed—but for the sensor, mounted in the cylinder head, this can be an issue whereby deformation of the head causes strain of the sensor body, which can affect the measuring element preload and provide a measured strain component that can be superimposed onto the measured pressure curve. This is clearly an issue for the sensor performance and sensor manufacturers have specific design concepts to provide antistrain performance, but a metric for this is not generally stated, so it would be the user's responsibility during the selection process to gain this information from the sensor manufacturer (strain performance is characterized during the sensor development process). This is particularly important for direct mounted sensor installations.

## Summary of best practice in using and maintaining the sensor and installation

The following list is not exhaustive but covers the main points with respect to sensors and installations in service:

- The installation bore should be created and machined with the sensor manufacturers recommended tools only. Often step drills are required to form the correct bore profile, along with specific tapping and reaming tools.
- Machining dimensions and tolerances supplied by the sensor manufacturer must be observed, also the correct use of any specified sealing compounds or adhesives for adaptor installations.
- There will be specific tools for mounting and dis-mounting the sensor, along with specific mounting torques for installation—the latter is imperative for the sensor to give an accurate reading as it must be subjected to the correct preload condition in operation.
- When the sensor is removed from the engine, it must be stored in its case, along with the cable, all exposed connectors must be sealed with the supplied caps to prevent contamination and charge leakage.
- Consider the use of sensor dummies so that the sensor can be removed from the engine when measurements are not required.
- The sensor must be handled with care—if the sensor is dropped—as a minimum it will need recalibration. Worse case, the crystal is brittle and can fracture rendering the sensor useless.
- On removal the sensor can be visually inspected and cleaned; combustion deposits may be cleaned with a soft brush and isopropanol. The front part of the sensor (membrane) cannot be cleaned using mechanical means such as brushing, sand blasting, and grinding as this will irreparably damage the diaphragm and therefore the sensor. Ultrasonic cleaning is also not recommended.
- The insulation resistance of the sensor with piezoelectric cable should be checked with a specific insulation tester for piezoelectric components.

The requirement at room temperature is 10E13 Ω. If the value is too low, all piezoelectric connectors should be disconnected, and the insulators cleaned with electronic cleaning spray.

- Sensor recalibration is recommended every 200 hours of operation or after each disassembly of the sensor from the cylinder head—the installation bore and sealing surface should be checked—clean the bore thoroughly with cleaning spray. Using a borescope with integrated illumination, inspect the measurement bore paying particular attention to the front sealing face. There must be no visible surface roughness or chatter marks. The sealing surface must have generally an even appearance. If the sealing face is found to be suboptimal, it is possible to recondition the sealing surface with the correct tools and knowledge.

### *Accessories*

There are a number of accessories and equipment required in a working environment where piezoelectric sensors are used. These are available from the sensor manufacturer and should be supplied and kept is a suitable case—ready for use by the test engineer or technician in order to carry out repair and maintenance.

- Cleaning and dusting aerosol sprays—must be suitable for high-insulation resistance electrical connections
- All required installation keys/sockets according to the sensor/adaptors used
- All required machining tools for sensor/adaptor installations according to type
- Sensor dummies—in order to remove the sensor when measurement is not required
- Insulation tester—suitable for piezoelectric connections and cables
- Bore scope—for sealing face and measurement bore inspections
- Maintenance tools—reamers for sealing face repairs, taps for thread repairs

## Low pressure gas path sensors

For measurement applications where, detailed engine thermodynamic analysis will be undertaken—a three pressure analysis is required, and this requires pressure sensors mounted in the inlet and exhaust system of the engine. This allows a much greater understanding of the flow regimes in and out of the combustion chamber, along with the in-cylinder pressure. With some knowledge of engine parameters, mass flows can be established, and this data can then be used to optimize simulation models—in order to establish virtual values, for example, trapped masses and residuals (Fig. 16.17).

The requirement in this instance is to be able to measure absolute pressure, as opposed to the dynamic pressure value given by a piezoelectric measuring chain. Therefore different sensor technology is deployed—most commonly the piezoresistive sensor. This sensor deploys a strain based, bridge circuit similar to a normal strain gauge but the measuring element is miniaturized and

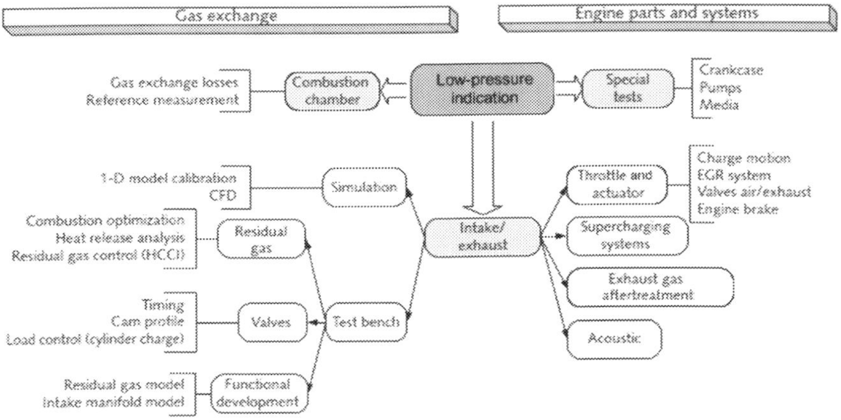


**FIGURE 16.17** Applications requiring low pressure measurement technology in the air path. *Courtesy: Kistler Group.*

mounted on a semiconductor base. The measuring element contains four resistors for a Wheatstone bridge. A change in electrical resistance occurs when mechanical stress is applied. In contrast to piezoelectric effect the piezoresistive effect, is a passive measurement principle and this enables the capability to measure and establish the static component of a signal.

The main advantage of this technology is that the semiconductor-based circuit is much more sensitive—having a much higher gauge factor than a conventional strain-gauge arrangement. This disadvantage is that this semiconductor measuring element is very temperature sensitive. Thus the raw signal must be corrected and linearized. This can be done using analogue or digital electronic technology—the latter giving greater accuracy and lower drift over time. Note that analogue-based sensor and amplifier combinations are calibrated together as a chain—and as such cannot be separated between calibrations.

There are two main classifications for piezoresistive sensors—exposed measuring element (DCE—direct chip exposure) and oil-filled types. In sensors with DCE the element is covered with a coating to protect it from the measurement media and increase durability. This allows a very compact design of sensor with high dynamic response. Due to the direct exposure of the measuring element to the media—lifetime tends to be engine operating and test condition dependent.

The alternative types use media-separated technology. The pressure acts via an oil-filling onto the measuring element—this is separated from the media by a steel diaphragm. This allows the sensor to be used in harsh applications and ensures media compatibility with practically all gases and liquids used in the powertrain testing environment. However, dynamic response is lower than the DCE type as the oil-filling provides a degree of damping.

Normally installations are made in the inlet and exhaust flow path and for both, an important consideration for the test engineer is where to locate the sensor, and if multiple sensor installations are required? (i.e., 1 per instrumented cylinder). As with in-cylinder pressures, a site must be chosen where a representative global pressure value can be achieved, without significant gas dynamic effects that could corrupt the raw data. The main issue is finding a suitable location bearing in mind the above, along with the position of flow system components (Fig. 16.18).

For the exhaust path an additional requirement may cooling. Cooled sensors are available, also cooled adaptors (Fig. 16.19) that can also provide a

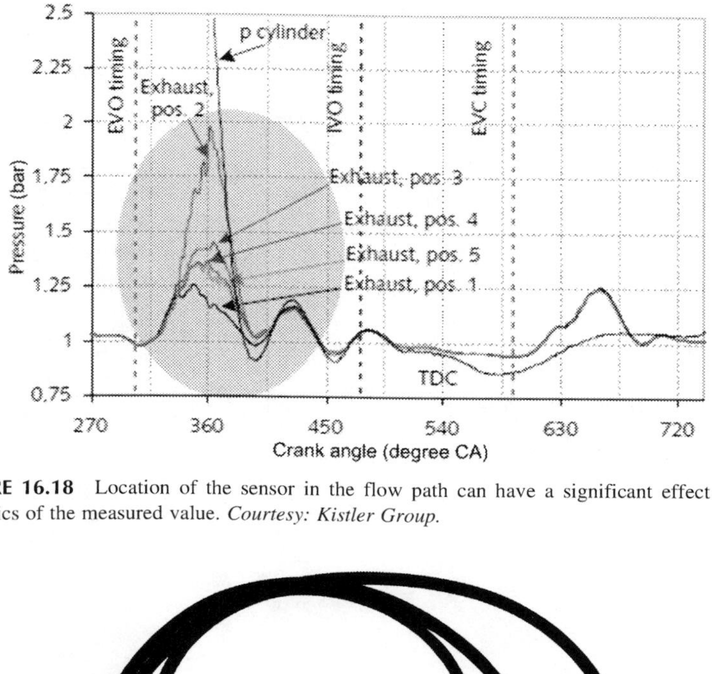


**FIGURE 16.18** Location of the sensor in the flow path can have a significant effect on the dynamics of the measured value. *Courtesy: Kistler Group.*

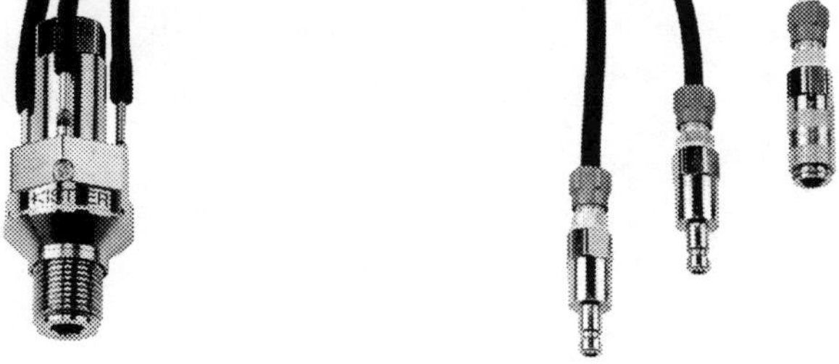

**FIGURE 16.19** Cooled switching adaptor for optimizing low pressure measurement procedures. *Courtesy: Kistler Group.*

switching function. The switching function allows control of exposure of the sensor to the measurand—which is normally exhaust gas—so this provides the ideal situation that the sensor diaphragm may only exposed when measuring. Another benefit of the switching adaptor is that it allows accurate zero-point correct during a measurement routine (zeroing prior to each measurement). This allows for highly accurate, absolute pressure measurements which are essential for detailed energy balance calculations.

## Mounting of sensors in the gas path

Many of the discussion points around piezoelectric sensor installation and handling is exactly the same for piezoresistive sensors. Sensors can be directly mounted or installed in an adaptor. The benefits of the latter being:

- Precise sensor bore inside the adaptor.
- Machining of the bore for the adaptor is simpler.
- The adaptor has the required strength to ensure that the sealing part can resist wear. This allows the sensor to be mounted and dismounted repeatedly without restrictions.

### *Location*

In order to limit the influence of the pressure waves and transient flow in the flow path between the combustion chamber and the low pressure measuring point, the sensor should be installed as close as possible to the combustion chamber (Fig. 16.20). At position where access to a mean, representative

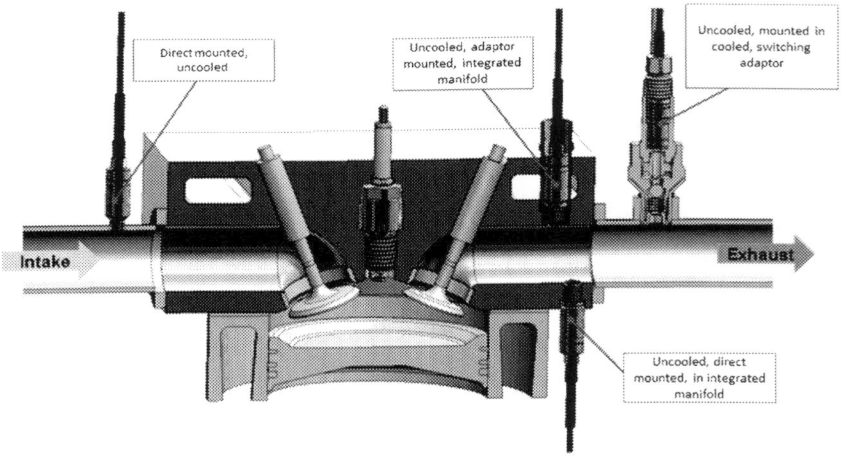


**FIGURE 16.20** Example gas path mounting situations, engine with integrated exhaust manifold. *Courtesy: Kistler Group.*

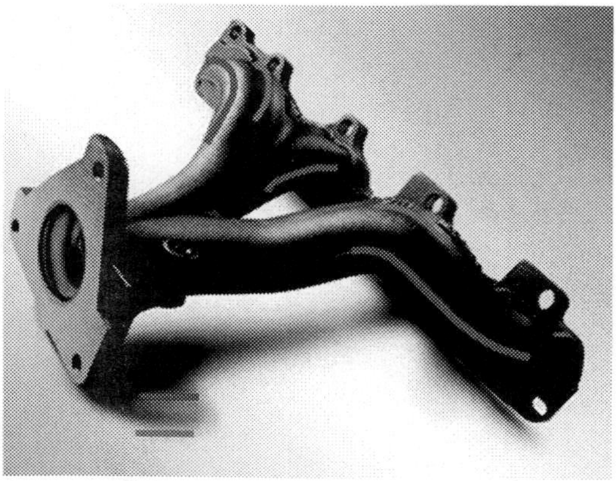

**FIGURE 16.21** Light gray = suitable location for representative flow, dark gray = not recommended. *Courtesy: Kistler Group.*

pressure trace of the flow is expected—not in the vicinity of separation zones, supercritical flows, or extreme pressure distribution (e.g., close to valve head/stem, transition cylinder-exhaust). An ideal arrangement would be a flush installation at a perfect right angle to the flow.

In the event of a suboptimal location, for example, the sensor is located at a pipe radius; then a pressure differential can occur between the inner and outer walls of the manifold due to the change of the direction of flow. It is for that reason that the sensor should be mounted between the outer and inner radius, in order to be able to acquire a representative mean pressure curve (Fig. 16.21). Ideal mounting situations can be difficult to find and compromises must often be made. For example, engines where an exhaust manifold is integrated into the head (in order to reduce exhaust gas temperatures). This provides a challenge to be able to install sensor through the manifold water jacket in order to locate the sensor close to the exhaust port.

Any pressure measurement can be challenging; therefore when selecting the location for a gas path sensor, consideration must be given not only to the physical size of the adaptation but perhaps more significantly to the geometry. It should be noted that studies on the influence of the absolute pressure on the calculated residual gas fraction show that a precision of better than ± 10 mbar is necessary. Using a switching adaptor allows precise zero-point correction of the piezoresistive pressure sensor measurement chain—such that a reference precision of ± 1 mbar is achievable.

## Fuel rail pressure and shot measurement

Common rail fuel (CR) systems have enabled significant reduction of exhaust emissions and, of particular importance to the passenger diesel car market, lowering engine noise and the "diesel rattle." CR fuel injection systems along with electronic management have allowed the flexible division of the injected fuel quantity into multiple but separate events during any one combustion cycle in order to approximate a rate-based injection approach. As such the pre-, main, and post injections, variable in timing and duration, produce engine performance that can better match the requirements of the driver while remaining within permitted emission limits over the full range of vehicle use. A passenger car or truck diesel engine that met Euro 3 emissions levels would typically require two or three fuel injections per combustion cycle, but the same engines having to meet Euro 4 and Euro 5 limits might require four or more injection events. Recent control strategies might require two early or pilot injections, to reduce combustion noise, followed by the main injection, which is then followed by one or two post injections to help control of particulate emissions. A further late injection may follow late in the cycle to enable the regeneration of exhaust gas after treatment devices.

Each injection pulse in this complex, multiinjection strategy creates a pressure wave within the CR pipe, waves that can combine to create transient pressures, above (additive) or lower (subtractive) than that regulated by the fuel injection system. This means that, for example, a main injection event with a duration of 500 ms that would be calibrated to inject 12 $mm^3$ of fuel could inject 1 or 2 $mm^3$ (more or less), depending on the coincidence of pressure waves within the rail at the time of injection. In other words, when more than one injection per combustion cycle takes place, the size of the second injection is influenced by the first injection or possibly by injection events of other cylinders. Some of the injection volumes are as small as 1 $mm^3$, so the pressure wave effect must be understood during the combustion analysis and engine calibration process.

General layout of high-pressure fuel injection systems for both gasoline and diesel applications is similar and a convergence of technology can be observed over recent years (Fig. 16.22). The main difference being in the control system with respect to injector operating strategy but more importantly—pressure ranges.

For gasoline direct injection applications, 200–300 bar is the normal expected range; for diesel applications this is increased by an order of magnitude (2–3000 bar). In both cases an understanding of the fluid dynamics at this high pressure is essential in order to be able to calibrate the pressure wave compensation function in the engine controller which ensures the target fuel quantity is always delivered. For pressure measurements, piezoresistive technology is commonly utilized, but the sensor internal design is appropriate for the pressure ranges encountered.

Installation of these sensors and their adaptors needs to be undertaken carefully due to the high pressures involved. The adaptor (Fig. 16.23) must be mounted on a straight section of the fuel line. A separate section of pipe allows the adapter to be easily installed and when not required, the adapter

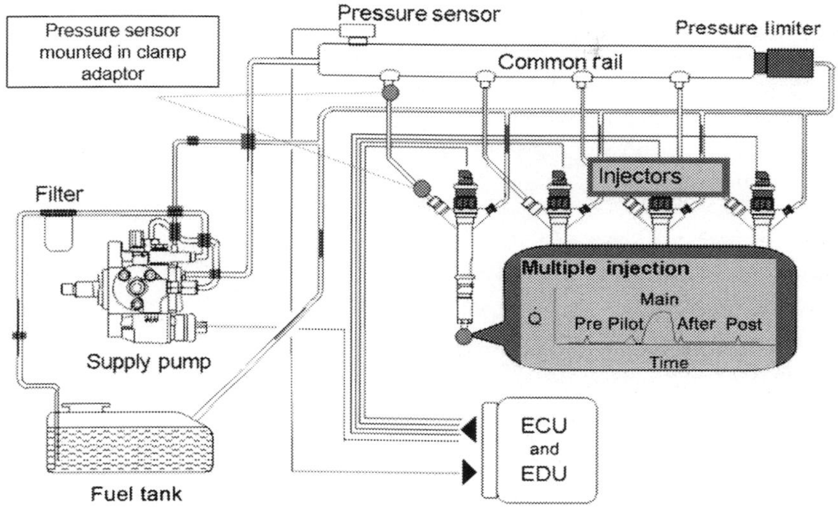


**FIGURE 16.22** General layout of FIE system for vehicle applications showing location of pressure sensors.

**FIGURE 16.23** An injection pipe bridge adaptor to allow mounting of the sensor at the required location on the injector feed pipe. *Courtesy: Kistler Group.*

section can then be replaced with an unmodified section. The adapter can only be mounted once—once removed the high-pressure sealing cannot be guaranteed a second time. Once the adaptor is mounted the pipe needs to be drilled at high speed to minimize the particle size of the metallic residue—avoiding damage to the opposite wall of the fuel pipe. Once mounted it is, it is advisable to flush the pipe to remove all of the metallic residue.

For fuel volume measurement—standard fuel consumption devices measure the volume or mass of fuel consumed over many injection cycles. There are instruments available that are able to measure the volume of individual injection events down to volumes of 0.3 $mm^3$. Such devices are normally used in conjunction with single-cylinder research engines or spray chambers.

## Angle encoder (volume) sensor

The pressure-related measurements in-cylinder—and in the air and fuel path—are normally indexed to the crank position and cylinder volume, because the goal behind this measurement technique is to gain a detailed understanding of the conversion processes occurring in the engine in real time. For this, thermodynamic calculations and observations are of great interest, particularly during the engine test and optimization procedures. In order to execute these calculations the cylinder volume, as well as related pressures in the combustion chamber and gas flow must be known. At the time of writing, there are no industrialized solutions, usable in the engine measurement application, that could measure the cylinder volume in real time. The established method is to use an encoder system, to gain engine position and rotation angle. Then use this information, along with some basic engine parameters and a crank-slider equation to establish the instantaneous in-cylinder volume and change in volume. The crank slider equation is discussed earlier in this chapter in the "Calculation of cylinder volume" section.

The encoder signal is, therefore, fundamental to a combustion pressure and analysis measurement—it could be considered as the engine volume sensor. It does, however, have some shortfalls that must be noted for the application:

- Almost exclusively, incremental encoders are used which means some mechanism to calibrate the encoder position to the correct engine position must be available and there is always a chance of slippage or error.
- Machine encoders are widely available, however encoders that are robust enough to be located on an engine and able to survive the temperature and vibration that is typical during measurements and operation are less common, and more expensive.
- Using an encoder assumes that the derived volume is the same for all cylinders and no dynamic effects that affect cylinder volume or the encoder per cylinder can be accounted for—for example, flexing of the cylinder head and torsional effects of the crankshaft system.

But given the above constraints, angle encoder derivation of cylinder volume is a standardized approach. The most commonly adopted method is to use incremental displacement information, along with reference information, so that absolute position can be obtained. This can be implemented via:

- use of an externally mounted encoder system and
- use of an engine crank position sensor signal (OEM fitted equipment).

The externally mounted optical encoder is a popular choice for engine research and development tasks (Fig. 16.24) as typically, a prototype engine is being tested that has many modifications or bespoke adaptations—such that fitting an encoder system is not too disruptive. The most common encoder design is attached to the front pulley via a bespoke flange and is anchored to the engine mechanically. There are a number of boundaries relating to the mechanical installation that must be observed—for example, the anchor point, encoder maximum run out and vibration specifications. These are provided by the manufacturer and must be observed, if not, the life of the encoder will be considerably shortened.

The electrical interface is also important, a typical encoder produces a number of pulses per revolution (known as crank degree marks—CDM), plus a one per revolution, reference pulse (known as a trigger pulse—TRIG) (Fig. 16.25). These signals can be ground referenced or floating—so that the interface between the measurement device and encoder system must be set correctly along with the correct definition of which edge to trigger on for both Trigger and CDM. Getting this configuration correct in the working

**FIGURE 16.24** Classic design encoder for engine nondrive end mounting. *Courtesy: Kistler Group.*

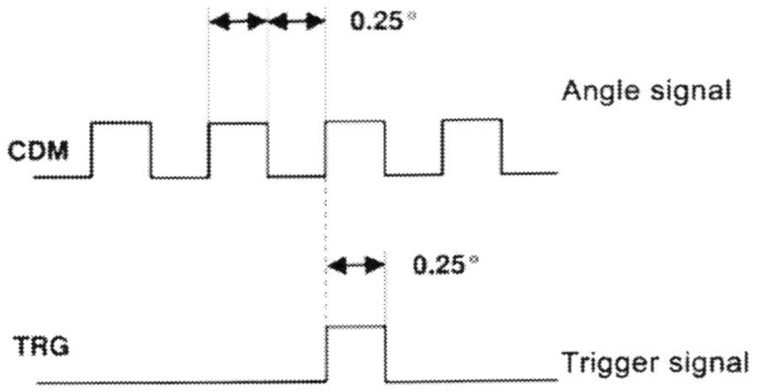


FIGURE 16.25 Typical encoder trigger and crank degree mark profile.

environment, especially where equipment from different manufacturers is used, is a challenge not to be underestimated. To add further complication it is often that case that a pulse multiplier exists or is required. Often 0.1 degree resolution is required for measurement (for appropriate signal resolution even at low speeds) but the encoder disk may only have 1.0 or 0.5 degree marks. In this case, pulse multiplication devices (external or internal) are often needed and these can cause further "interfacing" problems between the measurement device and encoder system. An important point to note is that pulse multipliers often struggle where torsional resonance at test operating points occur—the reason being that resonance in the crank system causes high instantaneous speed variation that can saturate the encoder sampling circuit. In this case it may be necessary to avoid certain steady-state speed/load points in order to gain reliable measurements.

Encoder systems with externally mounted disks are available; these are used where no front-end engine drive exists, or is not accessible. For example, motorcycle engines, high-performance race car engines. The disk itself is normally a bespoke laser cut design. The number of slots depends upon size. The devices are often lower inertia when compared to the front-end mounted encoder and as such can operate under more demanding conditions of speed, temperature, and vibration. However, the disk and sensor head are exposed and as such, more maintenance is required in use to ensure reliable operation—for example, removal and cleaning of the disk and sensor head.

The alternative is the possibility to use an existing crank position sensor (Fig. 16.26). These devices are fitted by the engine manufacturer in order to provide a crank position signal for the ECU. The signal generated can vary considerably but an often-used configuration is a number of edges per revolution, often 60, with missing teeth providing the reference information, often 2 teeth are missing. This gives a nomenclature of 60 − 2. Depending upon the sensor, the signal can be inductive (analog) or digital (square wave) and this can be converted in a signal processer to generate crank degree marks at the required resolution, along with a reference pulse via

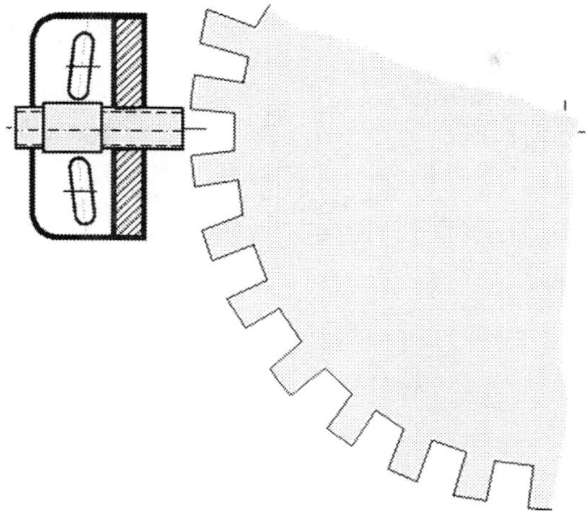

**FIGURE 16.26** Typical engine control unit pulse wheel and sensor location, the missing tooth for reference is not shown in the diagram.

extrapolation or interpolation, depending upon the methodology used in the signal processor.

The advantage of this approach is that no additional equipment is needed for crank position sensing for the measurement system—reduced cost and installation time, and potentially more reliable. However, the trade-off is that these that these original equipment sensors are not as accurate as a purpose designed angle encoder as they are mass produced devices with greater manufacturing tolerances than precision measurement equipment. Also, the signal processing creates virtual crank marks in between actual, physical marks. Depending upon the ratio of physical to virtual marks, inaccuracies can occur during transient operation—as the risk for the system to incorrectly interpolate marks during highly transient speed changes is high, due to latencies in the signal processor. However, this is a popular technique for in-vehicle measurements as the preparation time prior to measurement is reduced considerably and the reduced accuracy is accepted as in-vehicle development work is often later in the development cycle where detailed studies requiring more accuracy have already been completed and the task is final calibration and vehicle integration work.

## Cabling and signal interfaces

Correct type and layout of electrical cabling is an important topic in test environments and there are some specific topics in relation to combustion measurements and piezoelectric measurement chains that must be considered by the test or facility engineer. As a general rule, the sensor is connected to a signal conditioning device, which then provides an analog voltage to the measurement system. The cable between the piezoelectric sensor and the charge amplifier (i.e., the signal conditioner for piezoelectric signals) has some quite specific attributes—although visually it looks like a standard analog cable. The piezoelectric charge signal is small when compared to a normal voltage or current signal encountered in a test environment. The signal is susceptible to "leakage" that occurs over contaminated surfaces therefore the cable and interfaces must be maintained scrupulously clean. In addition, connectors (sensor and signal conditioner) must be covered with protective caps when not in use. Failure to observe this laboratory level of cleanliness will cause signal drift and saturation of the amplifier on a regular basis which is extremely frustrating for an extended measurement campaign, for example, engine mapping. The design of the high-impedance cabling is such as to prevent charge leakage, but artificial charge generation is also an issue due to cable movement (tribo-electrical effect). This means that cables must be secured and not allowed to flex or move during measurements. Not so easy to achieve in a test cell where the engine moves due to torque reaction and cables can move due to air flows around the engine (e.g., spot coolers). Generally high impedance cables have a unique internal construction to prevent the above mentioned problems (Fig. 16.27) the innermost wire of high-impedance cables is insulated with PTFE which reduces drift effects to the absolute minimum. In addition, a special graphite sheathing minimizes the tribo-electric noise, as such they are normally quite expensive—and should not be confused with standard BNC cables (this often happens). Note that the length of the cable is not an issue—there is no current or voltage of any significance in the signal—however, a longer cable is more likely to be susceptible to electromagnetic interference.

Cabling between the signal conditioning and measurement device is in most cases, for analogue voltage signals between −10 and +10 V. Normal precautions and best practice apply to theses cables to ensure good quality, noise free signal transfer.

*Ground loops* must also be considered. These are discussed in Chapter 4, Electrical Design Requirements of Test Facilities, and can be a significant issue for piezoelectric measurement chains as problems can be extremely

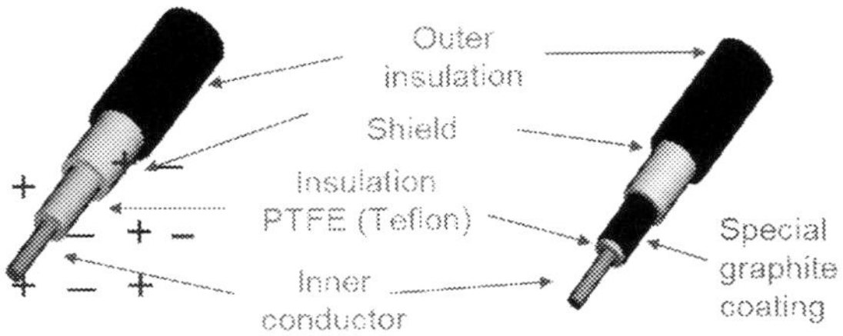


**FIGURE 16.27** Design attributes of high impedance cable for piezo-electric charge signals. The right hand side shows high-impedance version with additional conductive layer to prevent tribo-electrical charge build up due to cable movement. *Courtesy: Kistler Group.*

difficult to detect. It is, therefore, prudent to take appropriate precautions to prevent them and hence avoid their effects. The simplest way is to ensure all connected components are at a common ground through interconnection with an appropriate cable; this is more difficult to implement if equipment is not permanently installed at the test cell.

Ground isolated transducers and charge amplifiers can be used that will eliminate the problem, but it is important to note that the system is matched correctly (i.e. all ground isolated, or all grounded). If used in a mixed configuration the charge circuit may not be complete and hence will not function!

Modern amplifiers are physically small, which allows mounting very close to the actual sensors, and this means short cables between sensors and amplifier and, therefore, the best protection from electromagnetic interference. The direct connection of piezo sensors to the amplifier (avoiding extension cables) reduces the possibility of signal drift and the potential of problems with contamination on intermediate connectors.

## Part 3: Signal conditioning and measurement systems

The raw signal from a sensor often needs conditioning and amplification before measurement can take place and data acquired. The most common situation is to acquire signals in analog voltage form, such that they can be digitized and stored in a signal processor, via an analog-to-digital converter in the measuring device. The signal conditioning device is, therefore, matched to the sensor type and converts the raw signal to a suitable, analog voltage level.

## Charge amplifiers

Piezoelectric combustion pressure sensors are always used with appropriate, external signal conditioning in combustion measurement applications. The signal conditioning required for a charge signal is a charge amplifier which consists of a high-gain inverting voltage amplifier with high insulation resistance electronics to prevent charge leakage.

The purpose of the charge amplifier is to convert the high-impedance charge input into a usable output voltage.

Note that an output is produced only when a change in state is experienced (i.e., the connected sensor is producing a charge), thus the piezoelectric transducer and charge amplifier cannot perform a true static measurement (although quasistatic is possible with an appropriate time constant). This is not an issue in a normal combustion measurement, as it is a highly dynamic measurement chain to adequately capture all aspects of the combustion pressure change phenomenon. Special considerations are required for calibration of the measurement chain and in data analysis. While the modern charge amplifier measurement chain is a robust and well-proven measurement, and the technique is almost universally adopted for combustion pressure measurement, there are important issues and measurement characteristics to consider:

- *Drift*. This is defined as an undesirable change in output signal over time that is not a function of the measured variable. In the piezoelectric measurement system there is always a cyclic temperature drift that is a function of the actual transducer design. Electrical drift in a charge amplifier can be caused by low insulation resistance at the input.
- *Time constant*. The time constant of a charge amplifier is determined by the product of the range capacitor and the time constant resistor. It is an important factor for assessing the capability of a piezoelectric measurement system for the measurement of very slow engine phenomena (cranking speed) without any significant errors due to the discharge of the capacitor. Many charge amplifiers have selectable time constants that are altered by changing the time constant resistor.
- *Filtering*. Electrical filtering is generally provided in the charge amplifier to eliminate certain frequencies from the raw measured data. Typically, band-pass filters are used to remove unwanted frequency components, according to the measurement application.

Drift and time constant simultaneously affect a charge amplifier's output. One or the other will be dominant. There are a number of methods to counteract electrical drift, and most modern charge amplifier technology contains an electronic drift compensation circuit. If it is possible to maintain extremely high insulation values at the sensor, amplifier input, cabling and associated connections, leakage currents can be minimized, and high-quality measurements can be made. The equipment must be kept clean and free of dirt and grease to laboratory standards. A normal engine test cell environment does not provide this level of cleanliness and therefore this method of preventing drift is not always a practical proposition. Operation of the amplifier in short-time constant mode means that the drift due to input offset voltage can be limited to a certain value and thus drifting into saturation can be prevented.

The use of a high-pass filter though will allow optimization of the input range on the measurement system to ensure the best possible analog-to-digital conversion of the signal. Low-pass filters can be implemented to remove unwanted high-frequency, interference signal content from the measurement signal and prevent aliasing—an example being structure-borne noise signals from the engine that can be transmitted to the transducer.

It is important to remember when using filters that a certain phase shift will always occur and that this can cause errors that must be considered. For example, any phase shift has a negative effect on the accuracy of the IMEP determination when a low-pass filter is used. The higher the engine speed, the higher the lowest permitted filter frequency. As a rule, in order to avoid unacceptable phase shift, the main frequency of the cylinder pressure signal should not be more than 1% of the filter frequency.

## Piezoresistive amplifiers

Piezoresistive amplifiers are basically strain gauge amplifier circuitry. A stabilized current supply is maintained across the measurement bridge within the sensor—and a corresponding voltage signal is returned to the amplifier as a response to a sensor measurement. This signal is filtered and amplified, then output as a scaled analog voltage for the data acquisition system. As previously mentioned, the piezoresistive element is extremely temperature sensitive—therefore the amplifier must linearize this raw output signal effectively—this is done internally via electronic circuitry which is adjusted during calibration. Therefore the sensor and amplifier, once calibrated, should be considered a complete "chain"—and, as such, cannot be separated without corrupting the calibration of the chain. This is important for piezoresistive measuring chains with analog temperature compensation.

A more recent development is digital compensation—in this case the temperature calibration is held in the connector of the sensor using a Transducer Electronic Data Sheet (TEDS) and read by the amplifier when the sensor is connected. This allows a separation between sensor and amplifier that is more efficient to deploy in a working environment. Note that piezo-resistive sensors that are effectively transmitters can be used. These include the signal conditioning circuit within the sensor body itself and thus provide a linearized, voltage output as a function of pressure. Note though that they require a power supply for the sensor circuitry, and tend to be less accurate, with a limited temperature range of operation.

## Other amplifier types

There are several other amplifier and signal types that are often encountered alongside a typical combustion and gas path measurement set up. This list is not exhaustive but the most common signals are listed first:

- *Ignition/injection signals*: Using a current clamp to measure the drive current to single or collective ignition/injection components allows the possibility to understand complex pulse timings and durations. Modern engines have component driver circuits where many combinations of pulse duration, timing, and events are configurable and these need to be measured and analyzed as they have a direct influence on the combustion process. For these "timing" signals—data acquisition rates needs to be higher than the standard analog channels, often these signals are acquired as a multiplexed signal, which is then demultiplexed by the measurement system. Thus the overall channel count can be optimized (all timing signals can be acquired with 1 channel). A current clamp which uses a hall effect current sensor is often used for signal detection, this is normally converted to an analog signal for measurement within the clamp itself.
- *Lift/displacement sensors*: Historically, needle lift in fuel injectors and valve lift in engines were of interest for high speed crank angle

measurements. This is much less common today as electrical signals for injector and valve actuation are normally available. However, amplifiers that can utilize hall effect or LVDT (linear variable differential transformer) technologies, providing a voltage output in response to motion may still be encountered in some markets or regions.

- *Bridge circuit*: Less common but it is feasible that some engine components may be instrumented with strain gauges and that this information may be required on a crank angle basis—valve train components are the most obvious.
- *General voltage*: Various transmitters may be used that produce a voltage output that may need filtering or amplifying—as this is often not possible on the measurement system raw inputs—hence a voltage amplifier may be used.
- *Temperature*: Temperature measurement sensors normally require specific conditioning according to type (e.g., cold junction for a thermocouple). Although temperatures are normally slow changing values, not needed to be acquired at the same speed as a combustion measurement however, slow channels are sometimes recorded synchronously with calculated results to provide a time-based channel for additional operational parameters.
- *Acceleration/vibration*: Currently less common but perfectly valid to require vibration information at the same time as combustion data. To be able to characterize transfer paths from combustion event to surface response. Piezoelectric accelerometers use charge amplifiers and are suitable for high-temperature measurements. More commonly IEPE (Integrated Electronics Piezo-Electric) sensors are used and these require a specific circuit and input channel in order to drive the inbuilt signal amplifier—this is an emerging application that is expected to develop, extending the area of application for combustion/high-speed measurement systems.
- *High-voltage/current*: Electrification of the powertrain needs measurement of electrical power phenomena, and energy flows around the vehicle system. At the time of writing there are several suppliers of combined combustion/e-power measurement systems. These require access to current and voltage signatures, DC and AC, at various points in the powertrain. This is also an emerging topic and it is likely that interfaces to electrical systems will developed beyond the simple probes and clamps currently available.

## Sensor operational data acquisition and management (Transducer Electronic Data Sheet)

Sensor identification is a useful additional capability for the signal conditioning system. The international standard for this smart sensor technology is IEEE 1451.4 also known as TEDS. The technical implementation involves the use of a nonvolatile memory chip, installed in the sensor or connector. When the sensor is connected to a TEDS enabled amplifier, sensor specific parameters can be accessed to automatically configure the channel also to ensure correct connection and channel identification for the device. This technology is not limited to any particular sensor type many standard types of sensor are covered by the protocol which is seen in many application areas of industry. The template embedded in the sensor allows various levels of information to be stored.

At the most basic level sensor type, serial number and sensitivity are stored. In addition, calibration information can also be retained within the TEDS memory. An overview of the sensor data and application for TEDS, as used in combustion pressure measurement applications, is given below:

- *Basic sensor information*: Sensor serial number, type, and sensitivity. Allowing semiautomatic channel parameterization and system security as the correct calibrated value will automatically be used. In addition, the software can detect the sensor serial number and store this in the set-up parameters so that change of sensor can be detected before a measurement is started (e.g., wrong sensor connected to a channel).
- *Calibration information*: This extension allows the storage of the first-production calibration, along with the most recent calibration. This is extremely useful information as some general rules of thumb can be applied in order to gauge the sensors suitability prior to deploying it in a measurement campaign. Comparison with production and last calibration values, deviation tolerances of 3%–5% can be used to show the sensors suitability for use, or need for recalibration.
- *Runtime/operating duration*: A more specific feature that goes beyond the standard IEE TEDS templates. It is possible that an intelligent charge amplifier can record operating cycles and time during measurements, then write this back to the TEDS chip on completion of a measurement sequence. This then provides the possibility to understand the sensors working life duration, also time stamps for each calibration event in time and working cycles.
- *Sensor duty cycle and working life profile*: Another combustion measurement specific metric, beyond the standard template level of information. An intelligent amplifier can calculate some basic metrics that relate to the sensor operating environment—namely, maximum pressure or maximum pressure rise per cycle. These metrics can be classified in ranges and counted as event frequencies. This then provides a duty cycle profile and an understanding of the sensors loading, and severity of events in its working life. This helps with the management of the available sensors and their suitability for deployment in specific measuring tasks.

Analysis of the available TEDS data can provide a working overview or dashboard on the users sensor pool current status, readiness of sensors for deployment, upcoming calibration needs, as well as specific sensor information to understand the sensor life cycle and suitability for tasks.

Operational data from sensors is an essential tool for effective and efficient deployment of sensors within the working environment. Increased efficiency and reduction in test downtime due to sensor failures in service are easily achievable with this level of data. For large sensor data pools, data mining and machine learning can be deployed to recognize usage or failure patterns in order to have quick access to relevant metrics for improving sensor working environments. For example, areas of high sensor failure might identify a need for process or system improvements within the test environment (i.e., user needs training on sensor handling).

## Part 4: Overview of combustion pressure measurement and analysis systems

### Hardware

The measurement device used for combustion pressure analysis has evolved over the years from simple mechanical indicators, through analog oscilloscope technology capable of capturing images, through to high-speed digital sampling devices that can calculate, store, and transfer many important combustion-related results. Fundamentally, the device consists of a measurement system with various input and output interfaces, along with a signal processor and interface to a standard personal computer.

The control and analysis software to operate the system resides on the PC and provides the main user interface. The main measurement device then connects to the various sensors and provides interfaces to host systems.

The main system inputs and outputs are:

- Analog inputs: These are the most important, taking the signal from the amplifier as a voltage and, subsequently, digitizing and storing the data—which is in the form of measured curves. The input range is typically $-10$ to $+10$ V, the dynamic range can be from 12 to 18 bit, with sampling in the range of 1–2 MHz at the time of writing.
- Encoder: Various types supported that are typically encountered (TTL, LVDS) with trigger and CDM inputs, or Crank position sensor inputs. Tooth pattern and multiplication is often available in the hardware.
- Digital inputs/outputs: Often digital signals are acquired for timing, or timer-counter applications, normally running at sampling speeds greater than the analog inputs—to ensure capturing timing events/sedges at high resolution and maximum precision.
- Digital interfaces: Once the raw data is captured, calculation of results which characterize the pressure curve is done, and often statistics of those results are also calculated. This data is often combined with other acquisition system data models (test bed automation, calibration tool) so that standard digital interfaces between systems are supported (e.g., XCP, CAN, and DCOM).

### Software

The measurement system is provided with a software package that runs on PC hardware. This software allows the user to configure the system prior to starting a measurement, creating a reusable parameter file that can be task oriented and reused. The software is used during measurement runtime to view the data during acquisition and supports various display objects that the user can fully configure according to the application and their preferences.

All combustion measurement systems support a standard set of well-established calculations that engineers use to characterize and understand combustion from the cylinder pressure trace. But most systems also support a degree of freedom for the user such that they can implement their own calculation routines and algorithms via a formula compiler interface. The software allows measurements to be event triggered, or triggered by the user manually or automatically via a host system (test bed controller).

Data storage concepts are also configurable (pre- and posttrigger buffer, measurement duration). As measurements are undertaken the data is transferred and stored on the PC as a file (note that there is no standard file format for combustion data which tends to have a complex model that must include crank angle, cyclic, and time-based data). On completion of measurements the software supports data viewing for initial analysis using standard display objects. It may also be possible to use this software for a deeper analysis where more detailed display capability is required along with a library of advanced signal analysis modules. Also, the possibility to adjust raw data and "recalculate" results postmeasurement.

## Data, signal processing, and analysis flow

The system measures and stores the raw curves that are generated by the engine sensors—but there are specific considerations that have to be handled in the path from sensor to data (Fig. 16.28). Considering the signal flow:

1. Pressure in the cylinder or gas path is converted from a physical phenomenon to a signal at the sensor, this is transferred via cabling to the amplifier.

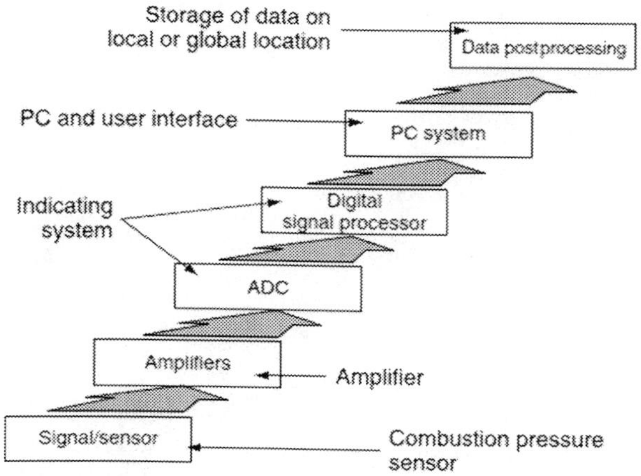


**FIGURE 16.28** Work and signal processing flow for digital, combustion pressure measurements.

2. The amplifier converts and conditions the signal, applying filtering and compensating any drift—the output is an analog voltage which is transferred via standard cabling technology to the measurement system.
3. The system has to compensate the raw signal in two ways:
   a. Zero-level correction: In particular, for piezoelectric signals that are dynamic pressure values, these must be converted to absolute pressure values for the display and storage. There are routines in the measurement system for this which basically performs a Y offset on the data, on a cyclic basis, in real time. This is known as zero-level correction or pegging.
   b. TDC phasing: The measurement system uses the angle encoder signal to convert to data domain into a series of sequential, engine cycle measurements with an abscissa of crank angle. Therefore the incoming data has to be correctly phased with respect to cylinder firing order and actual crank position (phasing). A calibration routine in the software allows correct global determination of crank position relative to encoder reference position during the set-up of the measurement and this calibration is stored in the parameter file that is subsequently created.
4. The encoder signal defines the crank position, this processed in real time and used in combination with the engine geometry parameters to calculate the cylinder volume table in order to allow the system to be able to calculate the main thermodynamic result values (using correctly phased pressure and volume information).
5. The raw data is now corrected and digitized—standard results can be calculated by the signal processor; from this, statistics of those results can also be derived, and these must all be written to the system memory or transferred to the PC memory or hard disk—this is a philosophy that is specific to the system manufacturer.
6. The calculated results and statistics are available in real time at the system interface boundaries, ready for transferring externally to other data acquisition of control systems.

The above signal flow is the same for all typical inputs seen at combustion pressure measurement systems; there are some variations for the specific signals and these are managed in the parameterization and definition of the signal type. For example, absolute pressure sensors do not need zero-level correction, ignition signals need higher throughput sampling and possibly demultiplexing, but they do not need acquisition over the full cycle measurement window.

## Calculations for combustion analysis test results

Calculations of critical results from the large store of raw combustion data is an essential part of the analysis and forms the basis of intelligent data reduction during run time. The actual calculated results, which are derived from the raw measured data and formula based curves, can be classified logically into direct and indirect results.

Direct results are derived directly from the raw cylinder pressure curve. Most of them can also be calculated before the end of the complete engine cycle. Generally, any signal error will create a result in a calculation error of similar magnitude. Typical direct result calculations are:

- maximum pressure and position
- maximum pressure rise and position
- knock detection
- combustion noise analysis
- injection/ignition timing
- cyclic variation of above values

These results will be acquired from data with acquisition resolution appropriate to the task. Resolution of pressure and pressure rise will typically be at 1 degree crank angle; but certain values will need higher resolutions due to higher frequency components of interest (combustion knock or noise) or the need for accurate determination of angular position (injection timing).

Indirect results are more complex and are generally thermodynamics based. They are derived by calculation from the raw pressure curve and associated data. They are reliant on additional information, for example, cylinder volume and polytropic exponent. These results are much more sensitive to correct set-up of the equipment and the error in a signal acquired is multiplied considerably when passed into the calculation. Hence, the result error can be greater by an order of magnitude. Example indirect results are:

- heat release calculation (dQ, integral)
- IMEP
- pumping losses, PMEP
- combustion temperature
- burn rate calculation
- friction losses
- detailed energy conversion
- gas exchange analysis
- residual gas calculation
- cyclic variation of above values

Using digitally based data acquisition systems, these results, although derived through complex calculations, do not place a heavy demand on the measurement hardware, nor do they require high-resolution sampling with respect to crank angle. Typically 1 degree for thermodynamic results is sufficient, with analog-to-digital dynamic range of 14–16 bit.

As a facility Engineer or operator, a detailed understanding of the results may not be required. However, some appreciation of what the meaning of a result is (in context)—and what impact it has on the development of the engine or combustion system is useful. Table 16.4 gives and overview of the most significant results.

**TABLE 16.4 Calculated results and meaning.**

| Type | Result | Relevance |
|---|---|---|
| Thermodynamic | IMEP | Indicated torque and cyclic energy |
| | Gross, net | Thermal efficiency |
| | PMEP | Friction and mechanical losses |
| | | Gas exchange losses |
| | CoV IMEP | Combustion stability indicator |
| | 0%–5% MBF | Early flame growth (SI), start of combustion |
| | 50% MBF | Combustion phasing, calibration strategy target |
| | 10%–90% MBF | Turbulent flame growth (SI), combustion duration |
| Direct | $P_{max}$/$AP_{max}$ | Component loading |
| | $R_{max}$/$AR_{max}$ | Noise metric |
| | Knock peak and analysis | Gasoline abnormal combustion |
| | Combustion noise | Diesel abnormal combustion |
| | Ignition delay | Diesel fuel property |
| | AEAP, AIAP | Metrics to characterize sensor performance with respect the thermoshock |

*CoV*, Coefficient of variation; *IMEP*, indicated mean effective pressure; *MBF*, mass burn fraction; *SI*, spark-ignition.

## Part 5: Integration of combustion analysis equipment within the test cell

In many test cells in the world, combustion analysis equipment may be a temporary visitor, brought in when required and operated in parallel to the test bed automation system. Temporary installations may give problems and inconsistencies due to signal interference and poor event alignment in separate data acquisition systems. It is now common to find dedicated engine calibration cells with combustion analysis and other high-speed data recording equipment as a permanent part of an optimized installation.

There may be fewer than 10 specialist suppliers of complete combustion pressure measurement hardware and software suites worldwide and probably only half of these are also specialist engine in-cylinder transducer manufacturers. Any search of the internet will result in discovery of many universities and private test houses that have assembled their own analysis systems from commercially available components and even more who have used standard software tools such as the MATLAB suite to develop their own diagnostics packages. Currently there is no ASAM standard for the storage of raw crank angle data, so it tends to be stored within structured binary files designed by the makers of the analysis equipment.

As with many other aspects of engine testing, the correlation of results produced on different systems, some of which are not well integrated, ranges from difficult to self-deluding. Whatever the level of sophistication, the physical equipment involved in engine calibration and combustion pressure analysis work can be considered as a high-speed data acquisition, storage, manipulation, and display system. It will consist of three subsystems:

- Engine-rigged transducers, some of which may require special engine preparation such as machining of heads and mounting of an optical speed encoder.
- Interconnection loom, which, in the case of mobile systems, will have to pass through the firewall between cell and control room.
- Data acquisition and analysis equipment mounted in the control room, either fully integrated with the test control system or at least sharing a common time and crank position flag.

## Occasional use of combustion analysis in test cells

The operator of an engine test required to carry out occasional map modification work or performance testing will use combustion analysis equipment that is often, although not always, mounted on a wheeled trolley. Test sequences designed to support engine indicating work are often run under manual control to locate the problem areas of engine performance that require analysis.

The running of the required cable looms and large multiway plugs, from cell to control desk, can give fire safety problems when having to be passed through the cell to control room firewall and resealed; consideration should be given to the permanent installation of such looms in the case of multiples of cells that each have occasional use of the same combustion pressure measurement equipment.

Typically, engine indicating plant will have a "pseudo-oscilloscope" display, quite separate from the cell control system display; therefore desk space for two operators and an absolute minimum of two flat display screens should be catered for in the control room layout.

The application would be standard combustion high pressure analysis along with subsystem measurements in the fuel and air path. If the system is occasionally used, this would mean a use case for diagnostics, or where specific information is required in order to gain a deeper understanding of some aspect of combustion. Often this is the case in Motorsport applications where intermittently, combustion measurements are made to validate simulation or gain a deeper understanding in order to optimize the engine and increase the performance.

A typical system for this application would perhaps be from 4 to 8 channels, standard combustion analysis data acquisition procedures at engine steady state, with a deeper analysis completed offline.

## High-end engine calibration cells

Let us consider the operational problem faced by the cell operators of an OEM test cell carrying out the optimization of an engine control map. There are three distinct subsystems producing and displaying data while the engine is running:

- The test cell automation system that is controlling the engine performance and monitoring and protecting, by way of alarm logic, both engine and test facility operation.
- The engine ECU that is controlling the fueling and ignition using engine-derived parameters to calculate essential variables.
- The combustion analysis system that is using special (nonvehicular) transducers to directly measure cylinder pressure against crankshaft rotation and is able to derive and display such values as:
    - mean effective pressures, IMEP, PMEP, etc.
    - location of peak pressure and pressure rise rates
    - mass fraction burn parameters
    - injector needle movement and injected fuel mass
    - heat release, etc.

As is the case with exhaust emissions equipment, the set-up and operation of combustion pressure measurement equipment and analysis of the data requires training and experience. The test equipment industry has produced integrated suites of instrumentation and software that, to a degree, can reduce the skill levels required to design the test sequences and to efficiently locate the boundary operating conditions of the test engine, but appropriate training remains essential.

Cells that carry out engine calibration work such as that described above have to be designed from the beginning as an integrated system if they are to work efficiently. All the data having different and various native abscissa such as time base, crank angle base, and engine cycle base, communicated via different software interfaces (COM, ASAP, CAN bus, etc.), have to be correctly aligned, processed, displayed, and stored, which is not an easy task. In fully automated beds it is common to have upwards of six large flat screen monitors on the operator's desk.

A system would typically have 16 channels and be used across the prototype development phase in order to optimize the engine performance and emissions, as well as subsystem characterization and component design validation. The system would have connectivity with the test bed host in order to transfer essential combustion metrics in real time, such that online design-of-experiment and mapping procedures have all the required combustion related data for the experiment execution during runtime. For example, model-based mapping and calibration procedures where combustion phasing information and knock protection metrics are particularly important during the measurement routine along with emissions, fuel, and air consumption. Remote control of the combustion measurement system is also initiated in order to save raw data at specific test points. Important is the ability to connect the raw data to the operating or measurement point in the data structure. Also, for transient measurements, the time base for the test bed recorder must be linked to the combustion data that is converted also to time domain data (but still retaining crank angle or cyclic reference). This requires a coherent structure for the test bed data model (which is normally a database) that includes logical linking to combustion data (which is normally file based). This connectivity between data types is an often underestimated task.

## In-vehicle measurement

Traditionally, the combustion pressure measurement was a task early in the engine development cycle, gradually migrating to the main prototype development phases and becoming a more commonplace requirement in a test cell. Over recent years, as measurement technology has reduced in size and cost, in combination with the increased complexity of control unit calibration (often finalized in the later stages of development with the unit under test fitted in a vehicle and tested on road or track). The demand for combustion measurement system that is useable in-vehicle has established as a clear market requirement with approximately 20% of the market being for in-vehicle use. The specific attributes required for such a system, apart from the ability to carry out the same measurement tasks as a test cell-based system are:

- Compact size and form factor that allows the unit to be easily placed anywhere in the vehicle and secured safely, quickly (e.g., on a seat, secured by a seat belt). Amplifier and signal conditioning should be fully integrated with no intermediate connection interfaces.
- Ability to operate on vehicle-based DC power supply with wide range input that does not require a separate supply from an instrument battery or DC/DC converter—the system must be able to continue to measure at cold start, cranking voltage levels (10 V or less).
- Must have wide operating temperature range—the system will be used in-vehicles in climatic chambers or test tracks at temperature extremes. Temperature of $-20$ to $+40^{\circ}C$ is a minimum range.

- Must be able to use operate with engine system crank position sensor (as well as standard encoders).
- Connectivity to the application system for control and data transfer—the system must connect and be compatible to the ECU calibration system for transferring data, controlling measurements and analyzing data.

Typical systems available are between 4 and 8 channels, including additional digital channels for timing measurements (ignition and injection). Bespoke interfaces to the most common calibration tools have been developed, also using a universal interface—for example, XCP—is an appropriate option. There are some aspects with respect to the typical user (i.e., an engine calibration engineer) that should be considered for this requirement:

- The calibration engineers only wants to use one PC for everything so they will not have to carry two computers on test trips. Essential is that the combustion measurement software does not have a significant overhead on the computer it runs on, as this is normally the same computer which the calibration tool also runs (calibration tools have a high PC load).
- The calibration engineer will only use one tool for data analysis (that which is supplied with the calibration tool). Therefore data models must be fully integrated, and the combustion data must be available in a format that can be read directly into the calibration data analysis tool. This means converting the combustion data to a time domain and using a standard format (ASAM MDF—Measure Data File).

## Reference

[1] G.M. Rassweiler, L. Withrow, Motion pictures of engine flames correlated with pressure cards, SAE J. Trans. 42 (1938) 185–204.

## Further reading

This suggested further reading represents papers and publications considered as the most useful with respect to combustion measurement best practice and techniques, and as such, are the most often cited in literature. However, it is not an exhaustive list and further papers can be located with good content to support the user of combustion pressure measurement equipment.

C.A. Amann, Classical combustion diagnostics for engine research, SAE Trans. 94 (1985).

C. Amann, Cylinder pressure measurement and its use in engine research, SAE Trans. (1985). Available from: https://doi.org/10.4271/852067.

W.L. Brown, Methods for evaluating requirements and errors in cylinder pressure measurement, SAE Int. (1967). Available from: https://doi.org/10.4271/670008.

M.F.J. Brunt, G.G. Lucas, The effect of crank angle resolution on cylinder pressure analysis, SAE Int. (1991). Available from: https://doi.org/10.4271/910041.

M.F.J. Brunt, C.R. Pond, J. Biundo, Gasoline engine knock analysis using cylinder pressure data, SAE Int. (1998). Available from: https://doi.org/10.4271/980896.

M.F.J. Brunt, C.Q. Huang, H.S. Rai, A.C. Cole, An improved approach to saving cylinder pressure data from steady-state dynamometer measurements, SAE Int. (2000). Available from: https://doi.org/10.4271/2000-01-1211.

M. Brunt, A. Emtage, Evaluation of IMEP routines and analysis errors, SAE Int. (1996). Available from: https://doi.org/10.4271/960609.

M.F.J. Brunt, K. Platts, Calculation of heat release in direct injection diesel engines, SAE Int. 01 (1999) 161–175.

M.F.J. Brunt, C.R. Pond, Evaluation of techniques for absolute cylinder pressure correction, SAE Int. (1997). Available from: https://doi.org/10.4271/970036.

M.F.J. Brunt, A.L. Emtage, Evaluation of burn rate routines and analysis errors, SAE Int. (1997). Available from: https://doi.org/10.4271/970037.

M.F.J. Brunt, A.L. Emtage, H. Rai, The calculation of heat release energy from engine cylinder pressure data, SAE Int. 26 (1998).

M.D. Checkel, J.D. Dale, Computerized knock detection from engine pressure records, SAE Int. (1986). Available from: https://doi.org/10.1556/AAlim.2015.0002.

R.S. Davis, G.J. Patterson, Cylinder pressure data quality checks and procedures to maximize data accuracy, SAE Int. (2006). Available from: https://doi.org/10.4271/2006-01-1346.

R.H. Kuratle, B. Maerki, Influencing parameters and error sources during indication on internal combustion engines, Trans. SAE (1992). Available from: https://doi.org/10.4271/920233.

D.R. Lancaster, R.B. Krieger, J.H. Lienesch, Measurement and analysis of engine pressure data, SAE Trans. 84 (1975).

G.J. Patterson, R.S. Davis, Geometric and topological considerations to maximize remotely mounted cylinder pressure transducer data quality, SAE Int. J. Eng. 2 (2009). 2009-01–0644.

H.S. Rai, et al., Quantification and reduction of IMEP errors resulting from pressure transducer thermal shock in an S. l. engine, in: SAE Technical Paper 1999-01-13, 1999.

A.L. Randolph, Cylinder-pressure-based combustion analysis in race engines, in: SAE Technical Paper, 1994, Available from: https://doi.org/10.4271/942487.

A.L. Randolph, Methods of processing cylinder-pressure transducer signals to maximize data accuracy, SAE Int. (1990). Available from: https://doi.org/10.4271/900170.

A.L. Randolph, Cylinder-pressure-transducer mounting techniques to maximize data accuracy, SAE Int. (1990). Available from: https://doi.org/10.4271/900171.

D.R. Rogers, Engine Combustion: Pressure Measurement and Analysis, SAE International, Warrendale, PA, 2010.

H. Zhao, Laser Diagnostics and Optical Measurement Techniques in Internal Combustion Engine, SAE International, Warrendale, PA, 2012.

# Chapter 17

# Engine exhaust emissions

## Chapter Outline



Engine Testing. DOI: https://doi.org/10.1016/B978-0-12-821226-4.00017-6

## Introduction and the future of Emission testing

Burning hydrocarbon fuels produces heat and the by-products of combustion, most of which are toxic to mammalian life and injurious to the environment; the more hydrocarbons a process burns the more harmful by-products are produced. We are now scheduled to be entering the last decades of the development of liquid hydrocarbon fueled land transport. Meanwhile, it is the task of automotive engineers to reduce, so far as is possible, the role of the internal combustion engine (ICE) as a primary source of pollutants. The recommended future developments of emission legislation, post Euro 6, are listed below (Tables 17.1–17.3) while the main body of the Chapter deals with the current (2020) situation and describes the basic chemistry, the tools, and the facilities required in ICE emission measurement.

The detailed roles of the test facilities, the instrumentation, and the engineers involved in the measurement of exhaust emissions can very crudely be divided into two: those who carry of legislatively required testing and those who are doing research into the physics and chemistry of vehicle emissions for the purposes of reducing their toxicity and consequentially aid vehicle development.

Those involved solely in vehicle type approval and homologation will only test for pollutants listed in the legislation and use facilities designed and equipped to reliably and repeatedly carry out such tests.

**TABLE 17.1 Part 1 of post-Euro 6 emission legislation recommendation by the International Council on Clean Transportation.**

| | What to regulate |
|---|---|
| Limits | • Introduce fuel- and technology-neutral emission limits<br>• Tighten the emission limits to harmonize with other markets<br>• Introduce application-neutral emission limits |
| Ultrafine particles | • Lower the size cutoff for particle counting from 23 to at least 10 nm<br>• Develop a methodology to measure volatile and semivolatile particles<br>• Include emissions that occur during filter regeneration<br>• Make particulate number (PN) standards fuel- and technology-neutral<br>• Investigate the feasibility of PN tailpipe measurements |
| Unregulated pollutants | • Set limits for ammonia emissions<br>• Set limits for $CH_4$ and $N_2O$ emissions and account for them in the $CO_2$ standards<br>• Set limits for aldehyde emissions<br>• Regulate all VOCs and just HC<br>• Set emission limits for brake wear particles<br>• Consider limits for $NO_2$ emissions |

*VOCs*, Volatile organic compounds.

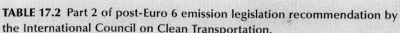

**TABLE 17.2 Part 2 of post-Euro 6 emission legislation recommendation by the International Council on Clean Transportation.**

| | How to regulate it |
|---|---|
| Evaporative emissions | • Tighten the evaporative emissions limit<br>• Introduce an onboard refueling emissions standard<br>• Increase the temperature during hot-soak, prior to the two-day diurnal test<br>• Introduce requirements for leak monitoring in OBD provisions |
| Low-temperature test | • Low-temperature emission limits should be techonology-neutral<br>• Set low-temperature limits for a wider set of pollutants<br>• Tighten the current low-temperature limits<br>• Develop a new low-temperature test procedure<br>• Monitor the greenhouse gas emissions over the low-temperature test |
| On-road CO | • Introduce not-to-exceed limits for CO during RDE testing<br>• Reduce the laboratory limit for CO<br>• Introduce limitations for fuel enrichment as an auxiliary emissions strategy |
| RDE test | • Extend the upper boundary condition for RDE driving dynamics<br>• Eliminate the lower boundary condition for RDE driving dynamics<br>• Revise the vehicle speed requirements during RDE tests<br>• Extend the cumulative elevation gain boundary condition<br>• Extend the temperature range for RDE testing and revise the correction factors<br>• Adjust trip requirements to allow shorter urban sections and cold-start driving<br>• Remove boundary conditions that reveal that an RDE test is taking place<br>• Eliminate the RDE evaluation factor for adjusting emissions downward |

*OBD,* Onboard diagnostics; *RDE,* Real-Driving Emission.

There is a fairly mature market in the supply of complete sets of laboratory-based emissions equipment, which, packaged with their control and data handling systems, produce results in a form and to a standard compliant with the chosen legislation and class of vehicle. This is true of both exhaust and evaporative emissions and can be used, within the procedures, for the testing of hybrid electric vehicle and battery electric vehicles; test facility owners therefore tend to buy complete packages. Building a laboratory by adopting a "pick and mix" policy of instrumentation acquisition demands a very high level of integration and specialized skills.

**TABLE 17.3** Part 3 of post-Euro 6 emission legislation recommendation by the International Council on Clean Transportation.

| | How to guarantee it |
|---|---|
| Durability | • Extend the definition of useful life for durability demonstration<br>• Establish the whole-vehicle test as the only durability demonstration option<br>• Extend the age/mileage requirements for in-service conformity to the full useful life<br>• Set a minimum emission warranty program<br>• Set an emission defect tracking and reporting program<br>• Develop in-service conformity testing for $CO_2$, fuel consumption, and electric range<br>• Develop a battery durability test |
| OBD and OBM | • Align OBD requirements with those of California and China<br>• Introduce OBM of pollutant emissions<br>• Set OBD threshold limits for PN and reduce the threshold limits for other pollutants<br>• Strengthen the antitampering provisions |
| Market surveillance | • Develop a methodology for fleet screening to identify noncompliant vehicle models<br>• Develop a remote sensing standard and establish a database of measurements<br>• Clarify the criteria for failure of market surveillance tests<br>• Issue defeat device guidance<br>• Extend the scope of market surveillance beyond pollutant emissions |

*OBD*, Onboard diagnostics; *OBM*, onboard monitoring.

Mobile emissions equipment [portable emissions measuring system (PEMS)], now required by the Real-Driving Emission (RDE) tests, is a less mature technology than its laboratory-based equivalent, that fact plus the inherent variability of the tests mean that correlation of results with different instrumentation packages is proving more difficult. Whilst the legislative RDE tests and instrumentation are seeking to confirm that emissions *do not exceed* (DNE) cumulative limits over a prescribed route, the fast response units now available ($T_{10\%-90\%} < 10$ ms) are revealing the details and exact location of transient events within the route.

The equipment used by research engineers looking at the total emissions from vehicles (not just the ICE) needs to be capable of dealing with a wider range of vehicle systems and have to examine a far greater range of both gaseous and particulate pollutants, including brake and tire dust.

Before diving into the complexities of ICE exhaust emission testing and the international legislation seeking to control those atmospheric pollutants, it is necessary to review the processes that produce them.

# Part 1: The chemistry and physics of ICE exhaust emissions

## The vital role of fuel quality in exhaust emission control

A fundamental prerequisite for meeting the limits set by emission legislation in 2020, and to maintain the efficient operation of the required exhaust after-treatment devices, is having fuels of consistent quality with as near as possible zero sulfur content.

ICE emission testing work is often concerned with "chasing" very small improvements in fuel consumption and reductions in transient emissions; these differences can easily be swamped by variations in the composition and calorific value of the fuels, in different seasons and from different crude oil feedstocks. Similarly, comparisons of results from identical test cycles will be invalid if they have used different batches of the "same" retail fuels, or fuels that have been allowed to deteriorate at differential rates.

The EU's fuel quality improvement initiatives, up to the current EN590 limits for diesel, have set the time line for the improvement of fuel quality, which has resulted in region-wide supply of both gasoline and diesel fuel (highway and off-road) with near-zero sulfur content.

In America the US-EPA Tier 3 Motor Vehicle Emission and Fuel Standards consider the vehicle and its fuel as an integrated system and starting in 2017, set new vehicle emissions standards tied to lower "sulfur" content of gasoline.

Countries in the world that remain on lower emission standards do so largely because of lack of access to ultralow sulfur and lead-free fuels and/or refineries capable of producing them.

EN590 is a very demanding standard for diesel fuels, which are produced with both seasonal blends, and with up to 7% FAME (fatty acid methyl ester) content. A critical requirement is the water content, set at 200 mg/kg of fuel, which equates to just 0.02% water contamination. Biodiesels are naturally hygroscopic therefore adherence to the standard will require regular fuel testing, possibly fuel treatment and regular tank cleaning.

### *Critical tests require critical analysis and records of the fuels used*

Test facilities carrying out emissions research and development work need access to a number of storage tanks to avoid unintended "cocktails" and need to have strict disciplines covering fuel delivery, storage, and recording.

The marine industry, the ICE exhaust emissions of which are rather belatedly being subjected to stricter legislation from the International Maritime Organization, the United States and EU have to use, from January 1, 2020, cleaner fuels. The limit for sulfur in heavy fuel oil, used on board ships operating outside designated emission control areas (ECAs), has been reduced to 0.50% m/m (mass by mass) from 3.50% m/m. Bunkering facilities have now to provide very low–sulfur fuel oil to meet the specification although heavy residual fuel oil will reportedly still be available for ships fitted with exhaust "scrubbers". The sulfur limit for fuel oil used by ships operating in ECAs is now 0.10% m/m and will require the use, and onboard separation of, marine gasoil.

*Hydrotreated vegetable oil (HVO)* became more widely commercially available in 2020 and has been advertised as a "drop-in" substitute for fossil-diesel.

HVO is being referred to as a "renewable diesel fuel" to differentiate it from "biodiesel," the latter term being reserved for the fatty acid methyl ester (FAME) fuels. It is reported that this fuel, which is a straight-chain paraffinic hydrocarbon with the general formula $C_nH2_{n+2}$ is free from any sulfur or aromatics content, gives a substantial reduction in PN, HC, and CO emissions but has little effect on $NO_x$. Since the cetane number is higher ($\approx$ 80–99) and the lubricity is lower than conventional diesel, it would indicate that, in order to fully optimize engine performance using HVO, a recalibration of some existing engine designs would be required. Two clear benefits of HVO is that it has a cold filter plugging point or waxing point, which at around $-30°C$, is much lower than other diesel fuels, making its storage much less challenging.

## Basic chemistry of internal combustion engine emissions

### Complete and incomplete combustion of fuel

The two chemical processes that are shown in Fig. 17.1 are the complete and incomplete combustion of the hydrocarbon fuel in air (treated here as an oxygen/nitrogen mixture). The production of carbon dioxide in this process is of concern since it is classed as a "greenhouse" gas but can only be reduced by an increase in the overall efficiency of the engine and vehicle, thereby producing lower fuel consumption per work unit.

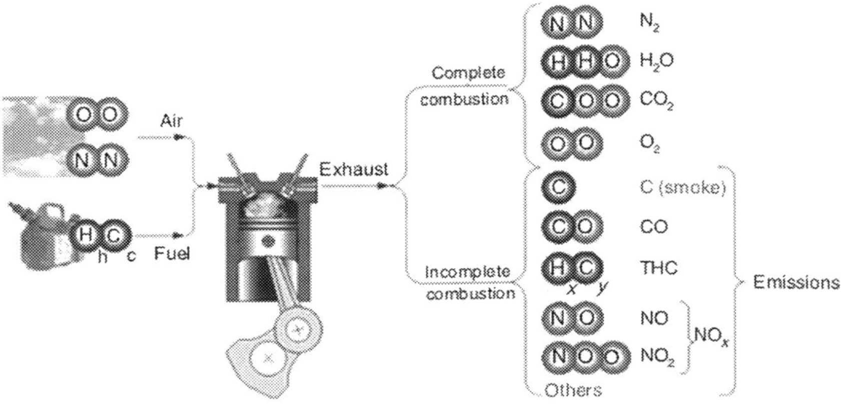


**FIGURE 17.1** The *gaseous* components of the complete and incomplete ICE combustion processes.

The emission gases and particulates covered by legislation are produced by incomplete combustion, including:

- carbon monoxide (CO), a highly toxic, odorless gas;
- carbon (C) experienced in the form of smoke and soot particles;
- hydrocarbons (HCs) or total hydrocarbons (THCs) formed by unburnt fractions of the liquid fuel and oil; and
- nitric oxide (NO) and nitrogen dioxide ($NO_2$), together considered as $NO_x$.

## Smoke and particulates (PM)

Particulates, when they appear to the human observer, are called "smoke." Smoke colors are indicative of the dominant source of particulate and are used as a crude diagnostic aid:

- black = "soot" or more accurately carbon, which typically makes up some 95% of diesel smoke either in the majority elemental, or organic form;
- blue = hydrocarbons, typically due to lubricating oil burning due to an engine fault;
- white = water vapor, typically from condensation in a cold engine or coolant leaking into the combustion chambers. White smoke is not detected by conventional tailpipe smoke meters; and
- brown = $NO_2$, maybe visibly detected in exhaust of heavy fuel engines.

## Particulate classifications

Unlike gaseous emissions, particulates are not a well-defined chemical species. Note that an *aerosol* includes both the particles and the gas (usually air) in which they are suspended.

The particulates produced within an ICE will consist of both volatile and non-volatile particles and it is only the latter species that are required to be measured in mass and number by emission legislation (see "Catalytic stripper" later).

Since ICE emission particles are highly variable in morphology and density, their "sizing" gives a problem that has been solved by the use of two slightly differing ways of expressing their "equivalent diameter."

### *Mobility diameter*

The mobility equivalent diameter is the diameter of a spherical particle with the same mobility (drag) as the particle in question.

### *Aerodynamic diameter*

The aerodynamic diameter of an irregular particle is defined as the diameter of the spherical particle with a density of 1000 kg/m$^3$ (1 g/cm$^3$) and the same settling velocity as the irregular particle.

While both equivalents are used in automotive engineering, the practical preference of many is mobility diameter, while general classification and pharmaceutical companies typically use aerodynamic diameter and apply it to particulate pollutants or to inhaled drugs to predict where in the respiratory tract such particles deposit.

This edition, in common with many authors, has divided particles into the following general categories based on size:

- *coarse* or $PM_{10}$: particles of an aerodynamic diameter $\leq 10\ \mu m$,
- *fine* or $PM_{2.5}$: particles with an aerodynamic diameter $\leq 2.5\ \mu m$,
- *ultrafine*: particles of aerodynamic diameters $< 0.1\ \mu m$ or 100 nm, and
- *nanoparticles*: particles having equivalent diameters of <50 nm.

The physics and chemistry covering the content and development of particles between being formed in the combustion chamber and circulated in the atmosphere is complex. It is a process that has to be emulated within the emission analysis systems that are designed to measure particulates. The main phases of particle formation are shown in Fig. 17.2, it is the products of these processes that form the gases and particles that may be absorbed by life forms "on the street."

Changes in the physical properties of solid particles from combustion chamber to the street depend on the processes to which they are subjected. Nucleation occurs in the combustion chamber, where ignition and combustion parameters have significant impacts on this process. Agglomeration of

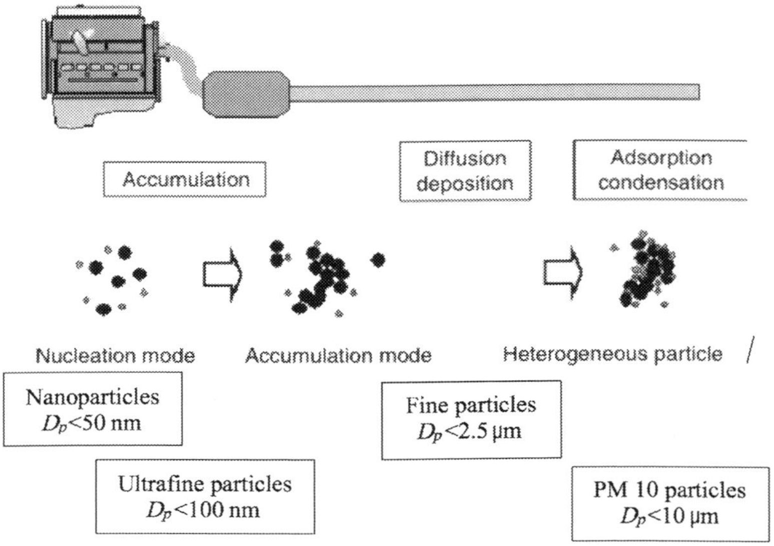


**FIGURE 17.2** The major phases of particulate formation from combustion chamber to the tailpipe.

hydrocarbons with other compounds takes place as a result of cooling and diluting of the exhaust gas in the exhaust system. The increase in the size of the particles depends mainly on their degree of accumulation and agglomeration, which is affected by the number and concentration of the particles and the coagulation time.

Individual diesel exhaust particles may be highly complex, heterogeneous, coagulations of compounds, as shown in Fig. 17.3, the individual aerodynamic diameters of which range from 10 to 80 nm, but particles with dimensions of 15–30 nm can be most common, while some particles emitted may be around 1 μm in size, which challenges the tools used to measure their presence (see "Particulate detection and measurement" later in this chapter).

Interest in particulate emissions from gasoline engines is relatively recent and arises primarily from the widespread shift to gasoline direct injection

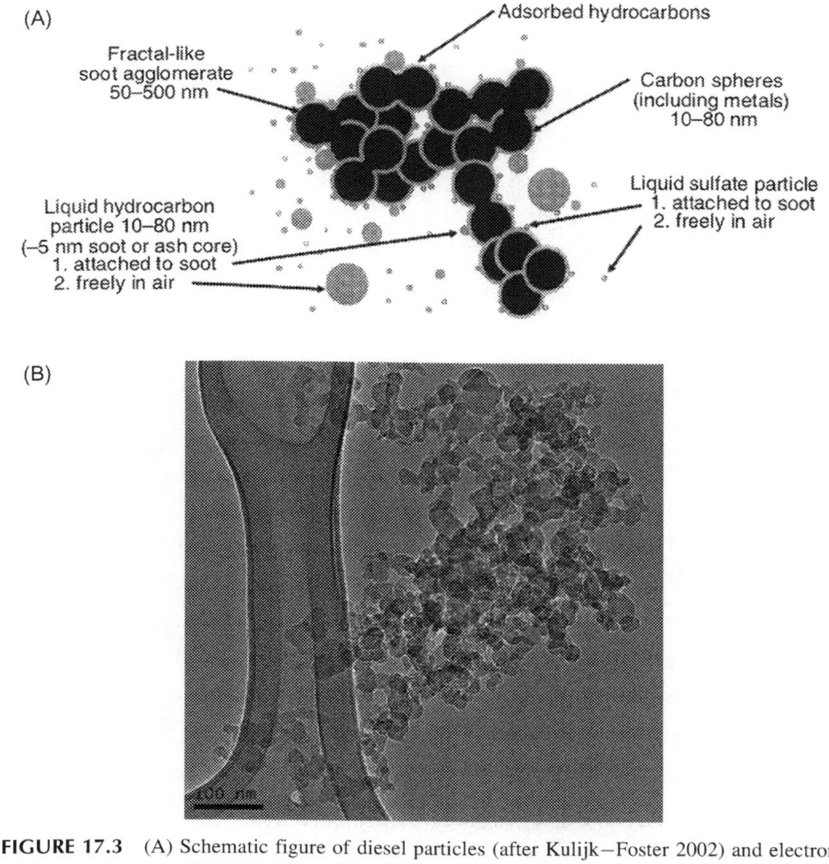


**FIGURE 17.3** (A) Schematic figure of diesel particles (after Kulijk–Foster 2002) and electron microscope photograph of the same (B). *Photograph supplied by Cambustion after work by Khalid Al Qurashi, EMS Energy Inst, Penn State University.*

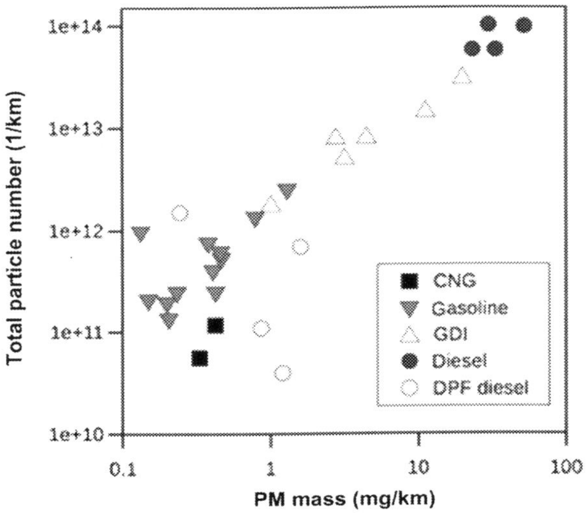


**FIGURE 17.4** A graphical representation of the mass and number of exhaust emission particulate typically produced by different engine and fuel types. *Adapted and redrawn from original by Majewski & Jääskeläinen.*

(GDI), which enables improved fuel consumption and $CO_2$ emission. Gasoline engine particulate emissions have been less studied to date than those for diesel engines, but many of the same harmful effects are assumed to be associated with them.

Diesel engines have had the highest particulates emission rates for many years but with the introduction of diesel particulate filters (DPFs), such emissions were reduced to levels comparable to port fuel—injected (PFI) gasoline engines. However, to reduce fuel consumption and $CO_2$ emission in light-duty vehicles (LDVs) has resulted in the widespread shift away from PFI to GDI technology for gasoline engines. Now (2020) particulate emissions from gasoline vehicles have increased significantly compared to diesels with DPFs, while particulate emissions from PFI gasoline engines are now low. Fig. 17.4 shows the relative positions in 2019.

Non-exhaust particulate emissions, originating from other components such as brakes and tires, as well as volatile compounds emitted from the materials in new vehicles are not quantified by any legislatively described process but will increasing be the subject of research and testing.

## Gaseous emissions from spark-ignition, gasoline-fueled engines

Perhaps the most characteristic feature of exhaust emissions, particularly in the case of the spark-ignition engine, is that almost every step that can be devised to reduce the amount of any one pollutant has the effect of increasing some other pollutant.

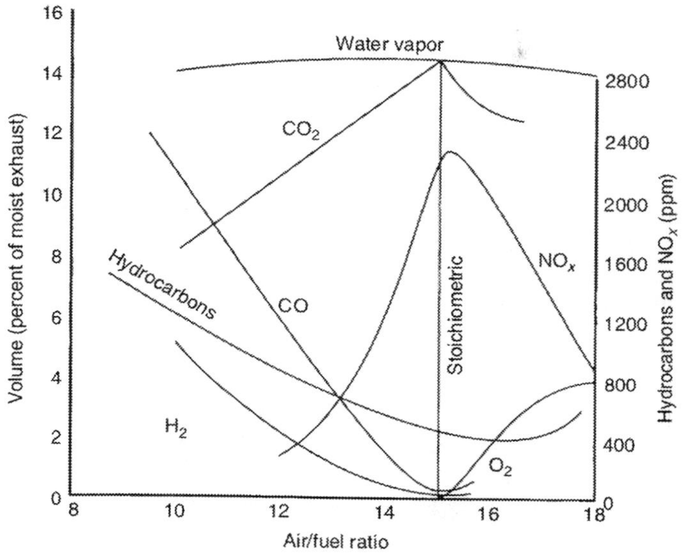


**FIGURE 17.5** Graphical representation of the relationship between exhaust emissions and air/fuel ratio for gasoline engines. *Mixture strength.* A loose term usually identified with air/fuel ratio and described as "weak" (excess air) or "rich" (excess fuel). *Air/fuel ratio.* Mass of air in charge to mass of fuel. *Stoichiometric air/fuel ratio.* The ratio at which there is exactly enough oxygen present for complete combustion of the fuel. Most gasolines lie within the range 14:1–15:1 and a ratio of 14.51:1 may be used as a rule of thumb. *Excess air factor or "lambda" ($\lambda$) ratio.* The ratio of actual to stoichiometric air/fuel ratio. The range is from about 0.6 (rich) to 1.5 (weak). *The equivalence ratio* $\phi$ is the reciprocal of the lambda ratio and is preferred by some authors.

Fig. 17.5 shows the effect of changing air/fuel ratio on the emission of the main pollutants: CO, $NO_x$, and unburned hydrocarbons. It will be apparent that one strategy is to confine combustion to a narrow window, around the stoichiometric ratio, and a major thrust of vehicle engine development since the mid-1990s has concentrated on the so-called stoichiometric engine, used in conjunction with the three-way exhaust catalyzer, which converts these pollutants to $CO_2$ and nitrogen.

A spark-ignition engine develops maximum power with a rich mixture, $\lambda \sim 0.9$, and best economy with a weak mixture, $\lambda \sim 1.1$, and before exhaust pollution became a matter of concern it was the aim of the engine designer to run as close as possible to the latter condition except when maximum power was demanded. In both conditions the emissions of CO, $NO_x$, and unburned hydrocarbons are high.

Note that a zirconia-type $O_2$ sensor can measure both the excess and deficiency of oxygen by comparing air with the combustion reaction in the stoichiometric air–fuel mixture.

The catalyzer requires precise control of the mixture strength to within about $\pm 5\%$ of stoichiometric and this requirement has been met by the introduction of fuel injection systems with elaborate arrangements to control fuel/air mixture (lambda closed-loop control). In addition, detailed development of the inlet passages in the immediate neighborhood of the inlet valve and the adoption of four-valve heads has been aimed at improving the homogeneity of the mixture entering the cylinder and developing small-scale turbulence: this improves the regularity of combustion, itself an important factor in reducing emissions. The recent and wide introduction of GDI engines has produced improvements in fuel economy and CO production but significantly increased the production of nanoparticles—just one more example of conflicting influences coming into play.

Spark plug design and location, duration, and energy of spark, as well as the use of multiple spark plugs, all affect combustion and hence emissions, while spark timing has a powerful influence on fuel consumption as well as emissions. Delaying ignition after that corresponding to best efficiency reduces the production of $NO_x$ and unburned hydrocarbons, due to the continuation of combustion into the exhaust period, but increases fuel consumption.

### *Exhaust gas recirculation (EGR)*

EGR, involving the deflection of a proportion of the exhaust into the inlet manifold, reduces the $NO_x$ level, mainly as a result of reduction in the maximum combustion temperature. The level of $NO_x$ production is very sensitive to this temperature, which has discouraged development of the so-called adiabatic engine, and also acts against the pursuit of higher efficiency by increasing compression ratios.

EGR control calibration requires that the volume and content of the recirculated portion of the exhaust flow is measured as a distinct channel for the gas analyzer system (see Fig. 17.12).

An alternative line of development is concerned with the lean-burn engine, which operates at $\lambda$ values of 1.4 or more, where $NO_x$ values and fuel consumption are acceptable but HC emissions are then high and the power output per unit swept volume is reduced; also light load and idling running conditions tend to be irregular. The lean-burn engine depends for its performance on the development of a stratified charge, usually with in-cylinder fuel injection.

Development of exhaust aftertreatment systems (covered later in this chapter) is a continuing process and particularly concerned with reducing the time taken for the catalyzer to reach operating temperature, vitally important because emissions are particularly severe during cold starts.

## Gaseous emissions from diesel engines

The diesel engine presents rather different problems from the spark-ignition engine because it always operates with considerable excess air, so that CO emissions are not a significant problem, and the close control of air/fuel ratio, so significant in the control of gasoline engine emissions, is not so important. On the other hand, particulates are much more of a problem necessitating postprocessing, and $NO_x$ production is substantial.

The historical indirect injection (prechamber) engine performed well in terms of $NO_x$ emissions; however, the fuel consumption penalty associated with indirect injection has resulted in a general move to direct injection (DI). $NO_x$ emissions are very sensitive to maximum cylinder temperature and to the excess air factor. This has prompted the use of EGR to become universally required for engines meeting the equivalent of Euro 4 emission standards and above.

The development of the "common rail" fuel systems is associated with a sharp increase in injection pressures, now commonly between 1500 and 2000 bar, and up to 2500 bar (250 MPa, 36,000 psi). The latest common rail diesels now feature piezoelectric injectors for increased precision allowing the "shaping" of the fuel injection or the use of a multipulse injection strategy. For test facilities such pressures represent a risk unless great care is taken in the avoidance of any leaks caused by the modification or instrumentation of such systems as they can result in a highly flammable atomized fuel spray escaping into the cell.

## Part 2: Exhaust emission legislation

### *Legislation introduction*

Since so much of the operational strategy of automotive test facilities is determined by current and planned environmental protection legislation, it is necessary for all test engineers to be aware of the legislative framework under which their home and export markets work.

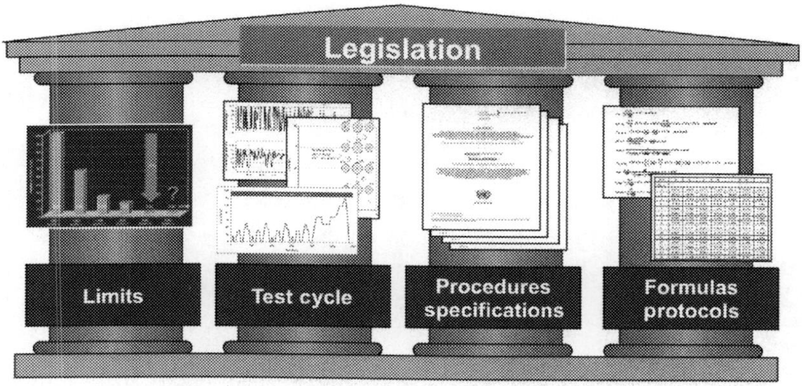


**FIGURE 17.6** The four pillars of emission legislation.

Emissions are regulated in world regions by different legislative packages:

- Europe and the countries following EU legislation based now on the Worldwide Harmonized Light Vehicle Test Procedure (WLTP) and RDE test. South Korea has adopted the WLTP for light diesel vehicle and Japan is moving toward adoption of it.
- The United States and some Central and South American countries. The Environmental Protection Agency (EPA) authority regulates engine emissions and the air quality in general, which is based on the *Clean Air Act*. Fuel economy standards in the United States are developed by the National Highway Traffic Safety Administration. The State of California has the right to adopt its own emission regulations, where the engine and vehicle emission regulations are defined by the California Air Resources Board (CARB).
- China that combines some elements of both the US and EU test. From July 2020 China 6a standards have been adopted. These standards incorporate elements of Euro 6 and US Tier 2 regulations for tailpipe and evaporative emissions. China 6b that will apply from July 2023 will include mandatory RDEs testing modeled after the EU RDE regulation with a few enhancements and modifications.

Like any specialist technology, the papers and press articles produced by exhaust emission experts are made somewhat impenetrable to the non-specialist by the very wide use of acronyms; a feature made worse by the legalese form of English used by the authors of Environmental Legislation, together with even more acronyms. A brief explanation of the most important acronyms and terms used is included at the end of this chapter.

*Note*: The European standards should always be designated by Arabic numerals for light-duty vehicles as in "Euro 6," and in Roman numerals for heavy-duty vehicles as in "Euro VI."

Although emission instrumentation has undergone significant development and levels of pollutants produced by vehicles and engines are much reduced, the basic form of much legislative testing had remained very similar to the original EPA methodology of the 1970s until the introduction of real drive emissions (RDE) legislation was introduced by the EU in 2016.

In spite of the introduction of RDE, all engine emission legislation still consists of the four component parts shown diagrammatically in Fig. 17.6:

- *Test limits*, which define the maximum allowed emission of the regulated components in the engine exhaust. For LDVs the limit is expressed in mass per driving distance (g/km), while for heavy-duty and off-road (including marine) engines, the limits are expressed in mass per unit of work (g/kWh). Limits for particulates are expressed both in mass (mg/km) and number (number/km, often written as #/km).

- *Test cycles*, which describe the operation of the tested vehicle or engine. For LDVs the current (2020) cycle is contained within the WLTP; however, from January 1, 2020 real-life, "free-form" driving has been introduced (see RDE later). For heavy-duty and off-road engines where only the engine is tested on a dynamometer, the test cycle defines a speed and torque profile over the test time.
- *Test procedures*, which define in detail how the test is executed, which exhaust sample collection and measurement methods are used. They state which test systems have to be used and define the test conditions. Probably the best known example of a legislative procedure is the use of sample bags in which an accumulation of diluted sample of exhaust gases resulting from a drive cycle is stored for analysis of content and concentration.
- *Formulas*, which are used in calculating the results, and the protocols covering the manner in which they are reported.

The details within the first of the pillars are defined by the democratic or bureaucratic processes in the originating world regions. These are processes that are by definition uncoordinated and that have created situations where implementation dates of similar legislation across the world are often in some doubt.

There is now legislation covering every type of machine powered by an ICE, from grass mowers and chainsaws to the largest off-road tractor units and cruise ships.

In the decade leading up to 2020 it became clear that the published fuel consumption figures for new cars did not relate to the experience of most vehicle purchasers, therefore trust in the way in which the figures were generated was widely degraded. Finally trust in the whole system of vehicular emission measurement and control was irreparably lost with the exposure of the VW "Dieselgate" scandal that, in September 2015 revealed the major manufacturer, was not just "gaming the system," but had installed a well-engineered and tested system to defeat emission controls.

In the years since the US EPA served a Notice of Violation (NOV) on Volkswagen Group, the legislative framework within which automotive test engineers work has undergone major changes and consequentially their workload, in the area of emission measurement, has increased by an order of magnitude.

In reaction to Dieselgate and the long-acknowledged inadequacies of emission legislation and its implementation, the European parliament issued regulation EC 715/2007, which, in addition to covering test procedural matters, specifically deals with the subject of devices designed to defeat emission controls in three separate articles:

- In Article 4 the manufacturers are obligated to take all technical measures to ensure that tailpipe and evaporative emissions are limited in line with the regulations throughout the normal life of the vehicle.

- In Article 5 the use of any type of emission control defeat device or system is prohibited, with very specific short-term exceptions listed.
- In Article 13 each member state is charged with the duty of laying down penalties, for manufacturers under their jurisdiction, who infringe the provisions of Article 5. It charges member states to implement penalties that must be "effective, proportionate, and dissuasive."

The use of defeat devices is also forbidden in US legislation and the language used is virtually identical to that of the EU. However, up to date, these regulations have been implemented very differently either side of the Atlantic as a result of the very different institutional settings that have reflected poorly on the EU.

The generally perceived inadequacies of the emission test procedures led to the UN ECE GRPE (Working Party on Pollution and Energy) group developing the Worldwide Harmonized Light Vehicles Test Cycles (WLTC), which are chassis dynamometer tests for the determination of emissions and fuel consumption from LDVs and replaced the much more benign NEDC test cycle.

Note: *While the acronyms WLTP and WLTC are sometimes used interchangeably, the WLTP procedures define a number of* procedures *that are in addition to the chassis dynamometer* WLTC *test cycles that are needed to type approve a vehicle.*

*The* WLTP *procedure is a global, harmonized standard for determining the levels of pollutants, $CO_2$ emissions, and fuel consumption of ICE-powered, hybrid and fully electric vehicles. Besides EU and EEA countries WLTP is the standard fuel economy and emission test also for India, South Korea, and Japan. It also ties in with Regulation (EC) 2009/443 to verify that a manufacturer's new sales-weighted fleet does not emit more $CO_2$ on average than the target set by the European Union, which is currently set at 95 g of $CO_2$ per kilometer for 2021.*

As further assurance that the legislative testing of emissions and fuel consumption reflected, as near as possible the real-life experience of the car driving citizenry, the RDEs test has been introduced to complement the WLTC test. This introduces mobile test equipment into the legislatively required tool kit in the form of a portable emissions measurement system (PEMS) and takes the test engineer out of the cell and onto the open road.

## Worldwide Harmonized Light Vehicle Test Procedure and its implications on tax legislation

It is the intention of many governments around the world to use taxation as a means of influencing vehicle purchasing decisions and the range of vehicles manufactured as part of a drive to reduce air pollution. To quote the UK government: "Understanding the impact of WLTP on $CO_2$ emissions and zero emission mileage of Ultra Low Emission Vehicles (ULEV)s is particularly important, due to the incentives offered through the tax system for these cars. *The government is committed to ensuring these incentives continue*."

This comes at a time of potentially crippling investment demands in the switch to Electric Vehicle (EV) production, development, and testing in the supply chain. The EU is reportedly making some concessions to limit the damage to the automotive industry in 2020 caused by tax penalties imposed on manufacturers for missing fleet emission targets but it remains clear that the test engineer is going to be at the center of a paradigm shift in the profile of light vehicle market.

In addition to the legislated EU penalties to be imposed on manufacturers, in the event of them failing to meet the sales-weighed $CO_2$ limits, set in the WLTP legislation, there are tax implications for individuals within Europe, a trend that will surely spread worldwide. In Europe the taxation levied on citizens and drivers of company supplied vehicles is a matter for individual states and are often based on their vehicle's $CO_2$ emissions, as a rough proxy for engine and vehicle price. There will be an increase in the published $CO_2$ emissions for every vehicle as a direct result of the changes in test cycles from the older NEDC to the current WLTC, the estimated factors are shown in Table 17.4.

**TABLE 17.4** Relationship between Worldwide Harmonized Light Vehicle Test Procedure (WLTP) and New European Driving Cycle (NEDC): $CO_2$ emissions for cars.

| Light vehicles | Engine size (L) | Ratio WLTP/NEDC |
|---|---|---|
| Petrol/gasoline | All | 1.22 |
| | <1.4 | 1.24 |
| | 1.4–2.0 | 1.15 |
| | >2.0 | 1.07 |
| Diesel | All | 1.2 |
| | <1.4 | 1.26 |
| | 1.4–2.0 | 1.21 |
| | >2.0 | 1.14 |

*WLTP*, Worldwide Harmonized Light Vehicle Test Procedure.
Original data source: European Commission Joint Research Centre (2017).

Moreover WLTP $CO_2$ emission limits are "model specific" so the addition of air-conditioning, or a different wheel sizes, may change an individual car's reported $CO_2$ emissions. This will greatly increase the number of unique $CO_2$ taxable values even within a single "family" of light vehicle models, which, together with the apparent increase in emissions resulting from the new procedures, will mean that taxation, manufacturer's model designations, and marketing strategies will undergo significant changes during the early 2020s.

## Exhaust emission limits

The history of vehicle exhaust emission test procedures and the limits set by legislative bodies working in the United States and Europe and around the world can be studied through sources recommended at the end of this chapter [1,2].

International and National limits applied to all classes of road vehicles, no-road vehicles, ICE-powered "estate management" tools and shipping, can be found on relevant websites and are too extensive for inclusion in this volume. However, it must be appreciated that the generations of increasingly stringent, legislatively applied emission limits start in 1966 with emission standards that were promulgated by the State of California in 1966 while the Euro 1 emission limits for passenger cars (91/441/EEC) were enacted in 1992.

European emission legislation, from 2020 based on the WLTP and RDE tests, is being implemented by all UNECE members (EU-28, Norway, Iceland, Switzerland/Liechtenstein, Turkey, and Israel) according to a common time line. Other countries that have signed the WLTP agreement, such as China, Japan, South Korea, Russia, India, and the United States, are currently (2020) monitoring developments and implementation and have yet to publish any implementation program.

The position in the United States has always been complicated by the existence of two legislative systems overseeing exhaust emissions: the federal system through the Environmental Protection Agency (EPA) and the California Air Resources Board (CARB), which are part of a nine-state coalition that released in 2018 a zero emission vehicle action plan. A complicated legislative position is made worse by the swings in opinion within the changing federal administrations. US test facility engineers must assume that although the rate of travel may vary, the direction toward every lower ICE emissions and the use of World Harmonized Test Procedures will remain and that industry and public opinion will lead the way.

Table 17.5 lists the pollutant limits allowed for passenger cars in EU legislation between 2005 and the current year 2020, which will be most relevant to most readers and from which the incremental tightening of limits over the period can be observed.

**TABLE 17.5** European emission standards 2005 to 2021 for passenger cars (category M) in grams per kilometer (g/km).

| Tier | Date of type approval | Date of first registration | CO | THC | Non-Methane Hydrocarbons (NMHC) | $NO_x$ | $HC + NO_x$ | PM | PN (no./km) |
|---|---|---|---|---|---|---|---|---|---|
| **Diesel** | | | | | | | | | |
| Euro 4 | January 2005 | January 2006 | 0.50 | – | – | 0.25 | 0.30 | 0.025 | – |
| Euro 5a | September 2009 | January 2011 | 0.50 | – | – | 0.180 | 0.230 | 0.005 | – |
| Euro 5b | September 2011 | January 2013 | 0.50 | – | – | 0.180 | 0.230 | 0.0045 | $6 \times 10^{11}$ |
| Euro 6b | September 2014 | September 2015 | 0.50 | – | – | 0.080 | 0.170 | 0.0045 | $6 \times 10^{11}$ |
| Euro 6c | – | September 2018 | 0.50 | – | – | 0.080 | 0.170 | 0.0045 | $6 \times 10^{11}$ |
| Euro 6d-TEMP | September 2017 | September 2019 | 0.50 | – | – | 0.080 | 0.170 | 0.0045 | $6 \times 10^{11}$ |
| Euro 6d | January 2020 | January 2021 | 0.50 | – | – | 0.080 | 0.170 | 0.0045 | $6 \times 10^{11}$ |
| **Petrol (gasoline)** | | | | | | | | | |
| Euro 4 | January 2005 | January 2006 | 1.0 | 0.10 | – | 0.08 | – | – | – |
| Euro 5a | September 2009 | January 2011 | 1.0 | 0.10 | 0.068 | 0.060 | – | 0.005[a] | – |
| Euro 5b | September 2011 | January 2013 | 1.0 | 0.10 | 0.068 | 0.060 | – | 0.0045[a] | – |
| Euro 6b | September 2014 | September 2015 | 1.0 | 0.10 | 0.068 | 0.060 | – | 0.0045[a] | $6 \times 10^{11}$ |
| Euro 6c | – | September 2018 | 1.0 | 0.10 | 0.068 | 0.060 | – | 0.0045[a] | $6 \times 10^{11}$ |
| Euro 6d-TEMP | September 2017 | September 2019 | 1.0 | 0.10 | 0.068 | 0.060 | – | 0.0045[a] | $6 \times 10^{11}$ |
| Euro 6d | January 2020 | January 2021 | 1.0 | 0.10 | 0.068 | 0.060 | – | 0.0045[a] | $6 \times 10^{11}$ |

*THC*, Total hydrocarbon.
[a] *Applies only to vehicles with gasoline direct injection engines.*

The inclusion of the identical limits for particulate matter from 2013 for diesels and 2015 for GDI engines is of importance to the test industry as it has required the development of new detecting measuring technologies and the miniaturization of existing ones in the form of portable emission measuring instrumentation. It has also accelerated the development and integration of gasoline particulate filter (GPF).

## Real-Life Driving Emission (RDE) and conformity factors

The RDE does not replace the WLTP chassis dynamometer–based test but complements it. RDE serves to confirm WLTP results in real life, thereby ensuring that cars do not exceed the required pollutant limits on the road.

Under RDE a vehicle, in its full model-version trim, is driven on public roads and exposed to a wide range of different conditions while the exhaust pollutants it produces are being measured and recorded by a Portable Emissions Measurement System (PEMS). It is important for test engineers to develop an RDE test route that complies with the regulatory requirements and has a reasonable expectation of having similar traffic conditions so as to lower the risk of expensive failed trips. This can also allow a model of all and part of the route to be rerun on a chassis dynamometer in order to resolve problems of peak emissions met on the route.

Because of the inherent variability of an RDE test drive and the fact that PEMS units are not as accurate as a full and stabilized laboratory system, the concept of a conformity factor was introduced. The conformity factor is defined as a "not-to-exceed" (NTE) limit so is the number by which the current $NO_x$ emission limit would be allowed to be exceeded in the RDE. The EU parliament has requested that conformity factors should be as low as possible and kept under review. A conformity of 1 means that the Euro 6 $NO_x$ limit of 8 mg/km for a diesel car must be met over a full 2-hour RDE test.

One of the RDE trip requirements is for measurement of cumulative positive altitude gain that is limited to 1200 m/100 km; this is intended to prevent bias of results due to a trip with either excessive energy demand caused by driving uphill, or excessive energy dissipation caused by long periods of driving downhill.

All ICE-powered cars, including all hybrid types, are subject to this procedure, albeit the hybrid procedures require greater distance to be driven (see next). In addition to the requirements listed in Table 17.6, the driving conditions must also include year-round temperatures and additional vehicle model payloads.

**TABLE 17.6** Specification of the Real-Driving Emission trip requirements (Regulation 2016/427).

| | Unit | Urban | Rural | Motorway | Notes |
|---|---|---|---|---|---|
| Speed ($V$) | km/h | $V \leq 60$ | $60 < V$ | $90 \leq V \leq 145$ | $V > 100$ for at least 5 min in motorway |
| Distance | % of the total distance | 29–44 | $33 \pm 10$ | $33 \pm 10$ | |
| Minimum distance | km | 16 | 16 | 16 | |
| Average speed ($V_{AVG}$) | km/h | $15 \leq V_{AVG} \leq 40$ | – | – | |
| Number of stops | # | Several > 10 | – | – | |
| Maximum speed | km/h | 60 | 90 | 145 | |
| Total test time | min | Between 90 and 120 | | | |
| Elevation difference | m | 100 | | | Between start and end points |

## Worldwide Harmonized Light Vehicle Test Procedure used on hybrid vehicles

Hybrid vehicles have to complete the complete the WLTP test several times. They start up with a full battery and the full cycle is repeated until the battery is empty. The calibration strategy used in each model will mean that the ICE operates for a longer time each cycle during which the emissions are measured. This is followed by a test run with an empty battery in which the drive energy originates solely from the ICE and regenerative braking. This process is used to determine not only fuel consumption and $CO_2$ emissions but also the electrical range and total range. The $CO_2$ value to be determined is then calculated as the ratio of the electrical range to the total range. At the same time, a so-called utility factor (UF) is introduced, which represents the proportion of the total vehicle distance traveled electrically. In the case of pure electric vehicles, a UF of 100% applies and in the case of traditional ICE vehicles the UF is 0%.

## Principles of measurement and analysis of gaseous and particulate emissions

### Note concerning instrument response times

All the mentioned instruments mentioned in the following text are required to operate under one of two quite different regimes, depending on whether the demand is for an accurate measurement of a sample collected over a fairly long time interval as in the case of the various statutory test procedures, or for high-speed measurement in real time, as is required in development work.

These two sets of requirements are conflicting: analyzers for steady-state work must be accurate, sensitive, and stable and thus tend to have slow response times and to be well damped. Analyzers for transient and RDE work do not require such a high standard of accuracy but must respond very quickly, preferably within a few milliseconds and in varied ambient conditions. A prime objective of instrument development in this field is to solve the problem of a lag between the emission measurement and the exact road position at which it was taken.

Fast response gas and particulate analyzers are revealing in ever greater detail the transient phenomena in the creation and development of exhaust components. When used in miniaturized forms within PEMS units, fast response instruments are illustrating not only the transient powertrain performance but also how the driver and road characteristics, such as 'speed-humps' or 'sleeping policemen' contribute to increased air pollution.

#### *Removal of volatile components from, and measurement of, particulate mass (PM) and number (PN)*

As mentioned earlier in the chapter, ICE particulates will consist of both volatile and nonvolatile particles of which only the non-volatile are required to be counted and measured in European automotive emission regulations, which address solid particles above 23 nm.

Therefore before legislative testing the volatiles have to be stripped out of the gas flow; this is achieved using one of two main types of volatile particle remover (VPR).

An instrument type referred to as a thermodenuder is based on chemical adsorption of evaporated volatile and semivolatile material. It has a dilution stage using particle-free air then a heater that brings volatile content into a gaseous phase, followed by the absorber that may consist of active charcoal material. A problem that has been met is the renucleation of semivolatile particles downstream of these "thermodilution" VPRs.

An alternative strategy is used in the device commonly known as a catalytic stripper [3] that is essentially a heated oxidation catalyst (CAT) with some sulfur storage capacity. The main advantage of the calibrated span over other VPRs is that volatile species are either oxidized or chemically adsorbed on the CAT surface; hence, the potential for subsequent renucleation is minimized.

## Particulate mass (PM)

PM, measured by deposits on filter paper and visible smoke, has been part of type approval and aftermarket emission legislation for some years. Euro 5 legislation also laid down permitted maximum limits for diesel engine particulate number and Euro 6 includes the same limits for DI gasoline engines, as shown in Table 17.5.

Particulate samplers measure the actual mass of particulates trapped by a filter paper during the passage of a specified volume of diluted exhaust gas. The relevant standard is typically contained in BS ISO 16183:2002. There are a number of such instruments on the market, all of which require climatically controlled laboratory test filter weighing facilities.

## Particle number (PN)

Due to developments in laser and microphone technology, particulate counting, based on either a laser-based condensation particle counter (CPC) or the photoacoustic principle, is now practical; very sensitive, reliable, and cost-effective instruments are now being produced.

The CPC is an optical particle counter; wherein the particles have first to be enlarged by a condensing vapor to form easily detected droplets; typically, such CPC units have a detection cutoff diameter of 23 nm.

The measurement principle behind the photoacoustic instrument is based on the periodic irradiation of an absorbing sample (soot particle), in the near infrared (IR) region of the spectral range. The resulting periodic heating and cooling of the absorber induces a periodic expansion and contraction of the carrier gas; this produces pressure waves to develop around the particles that are detected as sound waves. The reflection of the pressure waves, from the ends of a resonant measuring cell, the length of which is tuned to the irradiation period, causes a standing wave to form. The more particles that are inside the measurement tube, the greater the amplitude will be of the standing wave; this is picked up by a supersensitive microphone and is directly proportional to the soot concentration. The measurement of soot concentrations at 5 $\mu g/m^3$ at data rate of 10 Hz is specified by manufacturers; however, the numerical results produced by instruments of differing types from different manufacturers cannot readily be correlated one to the other [4].

Nanoparticle size, number, and mass can be measured using a transient engine particle analyzer [3] that works by positively charging each particle entering its tubular classification section in a high-efficiency particulate air–filtered airflow. Particles are detected at different distances down the tube depending upon their charge and aerodynamic drag.

*Opacity meters* measure the opacity of the undiluted exhaust by the degree of obscuration of a light beam. These devices are able to detect particulate levels in gas flow at lower levels than the human eye can detect; they are widely used in $Q_A$ testing and in service on ships. The value output is normally the percentage of light blocked by the test flow with zero being clean purge air and 100% being very thick black smoke. A secondary output giving an absorption factor "*k*" $m^{-1}$ may be given, which allows some degree of comparison between devices since it removes the effect of different distances between light source and sensor.

*Smoke meters* perform the measurement of the particulate content of an undiluted sample of exhaust gas by drawing it through a filter paper of specified properties; the filtered soot causes blackening on the filter paper, which is detected by a photoelectric measuring head and evaluated in the microprocessor to calculate the result in the form of "filter smoke number" specific to the instrument maker or $mg/m^3$.

## Principles of measurement and analysis of gaseous emissions

The description of instrumentation that follows lists only the main types of instrument used for measuring the various gaseous components of exhaust emissions. Developments continue in this field, notably in the evolution of instruments having the shortest possible response time. Legislation prescription of approved instrumentation and test methods always lags behind their development, therefore producing, as mentioned in the introduction of this chapter, a dichotomy between test equipment used in research and that used for type-approval testing.

### *Nondispersive infrared analyzer*

Nondispersive IR (NDIR)–based analyzer is widely used to measure CO, $CO_2$, HC, and NO.

These are called "nondispersive" because all the polychromatic light from the source passes through the gas sample before going through a filter in front of the sensor.

The $CO_2$ molecule has a very marked and unique absorbance band of IR light that shows a dominant peak at the 4.26-μm wavelength that the instrument sensor is tuned to detect and measure. By selecting filters sensitive to other wavelengths of IR, it is possible to detect other compounds such as CO and other hydrocarbons at around 3.4 μm (see "Fourier transform infrared analyzer" section). Note that the measurement of $CO_2$ using an NDIR analyzer is cross-sensitive to the presence of water vapor in the sample gas.

### *Fourier transform infrared analyzer*

This operates on the same principle as the NDIR but performs a Fourier analysis of the complete IR absorption spectrum of the gas sample. This permits the measurement of the content of a large number of different components. The method is particularly useful for dealing with emissions from engines burning alcohol-based fuels, since methanol and formaldehyde may be detected.

### *Chemiluminescence detector*

Chemiluminescence is the phenomenon by which some chemical reactions produce light. The reaction of interest to exhaust emissions is:

$$NO + O_3 \rightarrow NO_2 + O_2 \rightarrow NO_2 + O_2 + \text{photon}$$

The nitrogen compounds in exhaust gas are a mixture of NO and $NO_2$, described as $NO_x$.

In the detector the $NO_2$ is first catalytically converted to NO and the sample is reacted with ozone, which is generated by an electrical discharge through oxygen at low pressure in a heated vacuum chamber. The light is measured by a photomultiplier and indicates the $NO_x$ concentration in the sample.

A great deal of development work continues to be carried out to improve chemical reaction times, which are highly temperature dependent, which are currently quoted for a 10%–90% response as around 2–10 ms.

### *Flame ionization detector*

The FID has a very wide dynamic range and high sensitivity to all substances that contain carbon. The operation of this instrument depends on the production of free electrons and positive ions that takes place during the combustion of hydrocarbons. If the combustion is arranged to take place in an electric field, the current flow between anode and cathode is closely proportional to the number of carbon atoms taking part in the reaction. In the detector the sample is mixed with hydrogen and helium and burned in a chamber that is heated to prevent condensation of the water vapor formed. A typical "sample to measurement" response time is 1–2 seconds but fast FID, cutter FID, and gas chromatograph (GC)-FID have been developed, which are miniaturized FID instruments capable of a response time measured in milliseconds. FID devices developed for RDE work have a response time of $\geq 1$ ms and may thus be used for in-cylinder and exhaust port measurements. Cutter FID and GC-FID analyzers are used to measure methane ($CH_4$). In the fast FID nonmethane hydrocarbons are catalyzed into $CH_4$ or $CO_2$ and measured.

Very high–speed FID devices using miniaturized detectors are now being used in ICE development work with performances that are able to identify millisecond transients against crank angle [5].

### *Paramagnetic detection analyzer*

PMD analyzers are used to measure oxygen in the testing of gasoline engines. They work due to the fact that oxygen has a strong paramagnetic susceptibility. Inside the measuring cell, the oxygen molecules are drawn into a strong

**TABLE 17.7 The differing magnetic susceptibilities of the major exhaust gas components relevant to PMD analysis.**

| Component | Magnetic susceptibility ($O_2 = 100$) | Component | Magnetic susceptibility ($O_2 = 100$) |
|---|---|---|---|
| $O_2$ | + 100.0 | $CH_4$ | − 0.71 |
| $N_2$ | − 0.40 | $C_2H_2$ | − 0.45 |
| NO | + 45.2 | $C_2H_4$ | − 0.51 |
| $NO_2$ | + 2.89 | $C_2H_6$ | − 0.84 |
| $N_2O$ | − 0.44 | $C_3H_8$ | − 0.85 |
| $NH_3$ | − 0.98 | $C_4H_{10}$ | − 0.44 |
| $H_2O$ | − 1.65 | $C_5H_{12}$ | − 0.81 |
| CO | − 0.32 | $C_6H_{14}$ | − 0.75 |
| $CO_2$ | − 0.44 | $C_6H_{14}$ | − 0.80 |
| $CH_3COOH$ | − 0.49 | $C_7H_{16}$ | − 0.79 |

inhomogeneous magnetic field where they tend to collect in the area of strongest flux and physically displace a balanced detector, the deflection of which is proportional to oxygen concentration. Since other gases such as those being measured like $NO_x$ and $CO_2$ show some paramagnetic characteristics, shown in Table 17.7, the analyzer has to be capable of compensating for this interference.

### *Mass spectrometers*

These devices, not yet specified in any emission legislation, are developing rapidly and can distinguish most of the components of automotive engine exhaust gases. "Selected ion flow tube mass spectrometry" (SIFT-MS) is a technique [4] that has been used to analyze the diesel exhaust gas emissions on-line and in real time and in this form SIFT-MS analyzes the alkanes, alkenes, and aromatic hydrocarbons within the exhaust gases.

Currently mass spectrometers remain an R&D tool but may represent the future technology of general emission measurement.

## Portable emission measurement systems (PEMS)

To avoid optimization or "gaming the system" of pollutant control devices for a test-specific cycle, a randomization of the test conditions has been introduced in the form of the RDE test procedure, which is based on the PEMS equipment and driving on public roads. PEMS-based emission limits will be applied for $NO_x$ and for PN (CO only for monitoring). HC emissions are not included in the RDE test procedure. A cold-start phase is included in the test. The RDE test procedure is currently published as regulation EU 2018/1832. It contains a revised data processing methodology, provisions for in-service conformity (ISC) control based on RDE, updates for hybrid vehicles, a revised evaporative emission control systems (EVAP) procedure (see next), and provisions for the onboard fuel and electrical energy consumption metering.

There is no mandatory specification for a PEMS unit and while there is a minimum requirement (see next), there are several units currently available that greatly exceed these basic criteria. Three main types of PEMS testing task that exist are:

- official EU-RDE tests that have to be conducted during type approval of new vehicle models and include mandatory measurements of listed parameters (EU-RDE-TA);
- official EU-RDE tests conducted on vehicles that had done required mileages, for ISC testing (EU-RDE-ISC); and
- unofficial PEMS tests that do not necessarily conform to EU-RDE LDV regulations and are conducted for purposes of vehicle development, problem sequence reruns, or 'defeat device' screening.

The list of test parameters measured are given in Table 17.8 in which "M" = mandatory and "O" = optional.

Note that loss of, or incorrect acquisition of, exact location data via GNSS can invalidate a test drive so it is recommended to duplicate variables captured by GPS using other sensors. Regulation 2016/427 point 4.7 requires that vehicle speed shall be determined by a specific sensor, GNSS or electronic control unit (ECU).

### *Legislatively required physical attributes of a portable emissions measuring system unit*

PEMS units used in EU-RDE testing should have the following minimum characteristics:

- To be easy to install, being small and lightweight enough to be handled by one person.
- To have a working battery life of between 2.5 and 3 hours.
- Testing accuracy must be within 10% (or better) of a 1065 PEMS, which was the original regulatory model specification.
- Once deployed at a field site, the PEMS unit should have the ability to be testing vehicles within 30 minutes (assuming that the required onboard power pack has been charged).
- To measure and record the concentration of $NO_x$ (or NO and $NO_2$), CO, CO2 gases, and particulate number (PN) in the vehicle exhaust.
- To measure exhaust mass flow rate.
- To measure and record weather data, vehicle position via GNSS and optionally engine data from the ECU.

**TABLE 17.8** Test parameters for all vehicle (internal combustion engine and hybrid) portable emissions measuring system (PEMS) units used during EU-Real-Driving Emission (RDE) and research testing.

| Parameter | Unit | EU-RDE-TA | EU-RDE-ISC | Other PEMS tests | Test device |
|---|---|---|---|---|---|
| $CO_2$ concentration | ppm | M | M | O | Analyzer |
| CO concentration | ppm | M | M | O | Analyzer |
| $NO_x$ concentration | ppm | M | M | O | Analyzer |
| PN concentration | No./m$^3$ | M | M | O | Analyzer |
| Exhaust mass flow rate | kg/s | M | M | O | Exhaust mass flow, EFM |
| Ambient humidity | % | M | M | O | Sensor |
| Ambient temperature | K or °C | M | M | O | Sensor |
| Ambient pressure | kPa | M | M | O | Sensor |
| Vehicle speed | km/h | M | M | O | Sensor, GNSS, or ECU |
| Vehicle latitude | deg: min:s | M | M | O | Global Navigation Satellite System, GNSS |
| Vehicle longitude | deg: min:s | M | M | O | GNSS |
| Vehicle altitude | m | M | M | O | GNSS or sensor |
| Engine fuel flow | g/s | O | O | O | Sensor or ECU |
| Engine intake airflow | g/s | O | O | O | Sensor or ECU |
| Intake air temperature | K or °C | O | O | O | Sensor or ECU |
| THC concentration | ppm $C_1$ | – | – | O | Analyzer |
| $CH_4$ concentration | ppm $C_1$ | – | – | O | Analyzer |
| NMHC concentration | ppm $C_1$ | – | – | O | Calculated |
| Exhaust gas temperature | K or °C | M | M | O | Sensor |

(*Continued*)

**TABLE 17.8** (Continued)

| Parameter | Unit | EU-RDE-TA | EU-RDE-ISC | Other PEMS tests | Test device |
|---|---|---|---|---|---|
| Engine coolant temperature | K or °C | O | O | O | Sensor or ECU |
| Engine speed | rpm | O | O | O | Sensor or ECU |
| Engine torque | Nm | O | O | O | Sensor or ECU |
| Fault status | – | O | O | O | ECU/OBD |
| Regeneration status | – | O | O | O | ECU |
| SOC of battery | % | O | O | O | ECU |
| Hybrid battery current | A | O | O | O | ECU or power analyzer |
| Hybrid battery voltage | V | O | O | O | ECU or power analyzer |

*ECU*, Electronic control unit; *EFM*, Exhaust Mass Flow; *GNSS*, Global Navigation Satellite System; *ISC*, in-service conformity; *OBD*, onboard diagnostic; *SOC*, state of charge; *THC*, total hydrocarbon.

The analyzer specifications can be found under EU-RDE regulations: 2016/427 (Appendix 2), 2016/646, 2017/1151 and 2017/1154.

*Note*: There are a number of important safety considerations in using PEMS equipment that, in addition to safe handling, cover the protection of the driver and the prevention of exhaust gasses leaking into the vehicle cabin. A comprehensive guide to installing PEMS equipment will be found in the equipment supplier's guides and in Ref. [6].

## Exhaust "aftertreatment" catalysts and filters, operation and testing

Fig. 17.5 graphically shows how automotive engineers are "boxed in" by chemical restraints when attempting to reduce pollutants created by the combustion of hydrocarbon fuels in any type of ICE. Given the transient variability of the engines operating range, the variability of pollutant production is inevitable and the only strategy available is to "clean up" the exhaust gas flow before it exits the tailpipe. Post-combustion exhaust treatment consists of devices that either convert gaseous components into less toxic compounds or filter out and trap particulates before periodically burning them off. Since, unlike the near stoichiometric operation of a gasoline engine, a diesel fuel engine operates with excess air, the makeup of their exhaust gases differs as does their post-combustion treatment. The following paragraphs review the various exhaust after treatment modules.

## Gasoline three-way catalytic converters

Spark-ignition engines in modern powertrains are fitted with three-way catalytic converters that carry out three simultaneous chemical reactions:

- the reduction of $NO_x$ to nitrogen and oxygen,
- the oxidation of carbon monoxide to carbon dioxide, and
- the oxidation of unburnt hydrocarbons to carbon dioxide and water.

These three reactions generally occur most efficiently when the catalytic converter receives exhaust from an engine running between $\lambda$ 14.6 and $\lambda$ 14.8 for gasoline and require very tight closed-loop control in order to maintain optimum efficiency over an acceptably long service life. In some designs variable air injection takes place between the first ($NO_x$ reduction) and second (HC and CO oxidation) stages of the converter.

The reactions depend on the composition of the exhaust gas and critically on the temperature of the internal surfaces of the converter. The "light-off" temperature at which the reactions start is around 300°C and the highest conversion rate requires temperatures in the 400°C−700°C range, but prolonged exposure to temperatures over 800°C will lead to thermal degradation. The problem of overheating has led to designs in which the three-way catalytic converters are built as two separate units: the first reduction unit, which is optimized for the highest temperature operation and fast light-off, is installed close to the engine and the second, physically larger, oxidation unit is installed further downstream under the vehicle floor in full air-cooling flow and designed to have as low a light-off temperature as possible.

The period between a cold-start and light-off temperature being reached is when the majority of polluting emissions are recorded; many passive and active strategies are being developed to reduce both the light-off temperature and the engine warm-up time.

Catalytic converters all have the same three major components installed within the containment "can," which are:

- A core or substrate. This is commonly made up of a ceramic honeycomb, but folded stainless-steel foil is also used.
- A washcoat. This is a coating on the walls of the core designed to create the maximum possible surface area on which the chemical reactions may take place. It is usually formed by an alumina and silica mixture.

- The catalytic material deposited on the washcoat will be one or more precious metals. Platinum and rhodium are the reducing CATs, used in the first stage of three-way converters, while platinum and palladium act as oxidizing CATs used in the subsequent stages.

A three-way CAT may be deactivated by meltdown of the substrate, thermal aging, or by chemical poisoning. Testing of the thermal and chemical stability of catalytic units may be carried out either in an engine test cell as part of a vehicle exhaust system, or in a special thermal test bench, or in a laboratory furnace at high temperature in a varied controlled gas atmosphere.

The use of an engine test cell for the CAT aging and testing process can be a long and therefore expensive procedure, but it is the closest to real life in service. It is commonly carried out by Tier 1 suppliers, in custom-built test cells using client's engines.

A common form of damage suffered by a CAT, particularly the first $NO_x$ reduction stage, results from the exposure to a strongly oxidizing gas flow at high temperatures; conditions that can be encountered during vehicle braking, with the fuel cut, after running at high power; these are conditions that can be repeatedly simulated in an engine test cell.

The high-temperature gas flow test bench may be the best way to simulate meltdown. Testing of CAT poisoning can use an engine cell or a flow bench. There are a large number of compounds known to cause problems, but the sources are sometimes difficult to locate and include agents such as the gasoline additive methylcyclopentadienyl manganese tricarbonyl (MMT); other common sources are lubrication oil additives, silicone, and lead compounds. Silicon contamination that may be present, albeit rarely nowadays in fuel, but certainly in engine coolants and test facility tubing will poison not only a CAT but can damage the lambda sensor; specialist CAT aging test cells and some emission research facilities ban the use of silicon tubing.

An important feature of the catalytic converter is the ability to store oxygen during the inevitable and small excursions of the lambda ratio into a "lean" phase; this allows continuous full operation during a short "rich" phase. The correct and continuous operation of the lambda sensor is crucial to the whole closed-loop emissions control system and, like catalytic converters, they are vulnerable to thermal degradation and poisoning. The one sensor, fitted before the first CAT, is being replaced in many systems by two-sensor systems (pre- and post-CAT) and, in Super Ultra-Low Emissions Vehicles (SULEV), by three or more sensors that use a complex cascade control.

## Diesel oxidizing catalysts

The diesel oxidizing CAT (DOC) is employed to remove CO and hydrocarbons using the oxygen in the exhaust (5%–15% of the gases) and oxidize NO to $NO_2$ (Fig. 17.8).

Example of hydrocarbon oxidation:

$$C_{12}H_{26}(g) + 18\frac{1}{2}O_2(g) \rightarrow 12CO_2(g) + 13H_2O(g)$$

To improve the hydrocarbon performance at low operating temperatures, zeolites (a class of tectosilicates) can be added to the DOC substrate:

Oxidation of NO:

$$2NO(g) + O_2(g) \rightleftharpoons 2NO_2$$

Like other CATs the DOC consists of a metal canister containing a honeycomb-like substrate through which the exhaust gases flow. It is installed after the exhaust manifold and turbocharger.

It uses oxygen in the (lean) diesel engine exhaust gas stream to convert carbon monoxide to carbon dioxide and hydrocarbons to water and carbon dioxide and reaches a peak efficiency of 90% at around 500°C (Fig. 17.7). However, diesel engines present a particular challenge in the area of DOC temperatures because current turbodiesel engines rarely generate temperatures over 200°C at the exhaust outlet in inner-city driving, a problem made worse by the operation of the automatic engine stop feature. As modern catalytic converters generally have a minimum light-off temperature of around 200°C, the exhaust gas temperature often drops below this level. Electrically heated catalytic converters, which can be activated at any time via the ECU, have been developed to solve this problem and can significantly reduce emissions during the cold-start phase and during intermittent operation.

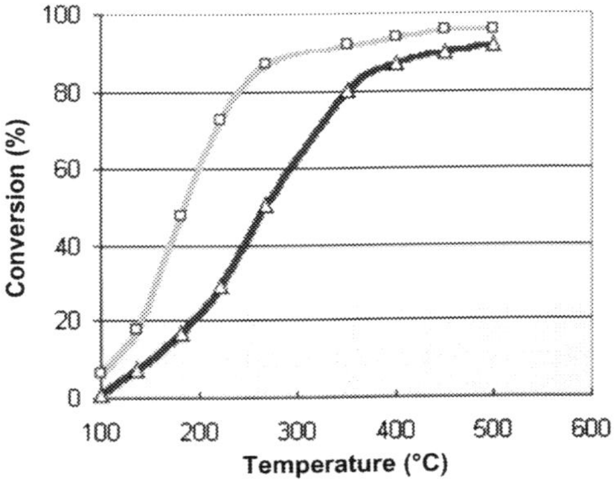


**FIGURE 17.7** Graph of operating temperature range of the catalytic conversion of carbon monoxide (top and LH line) and hydrocarbons (bottom RH line) in a DOC. *DOC*, Diesel oxidizing catalysts.

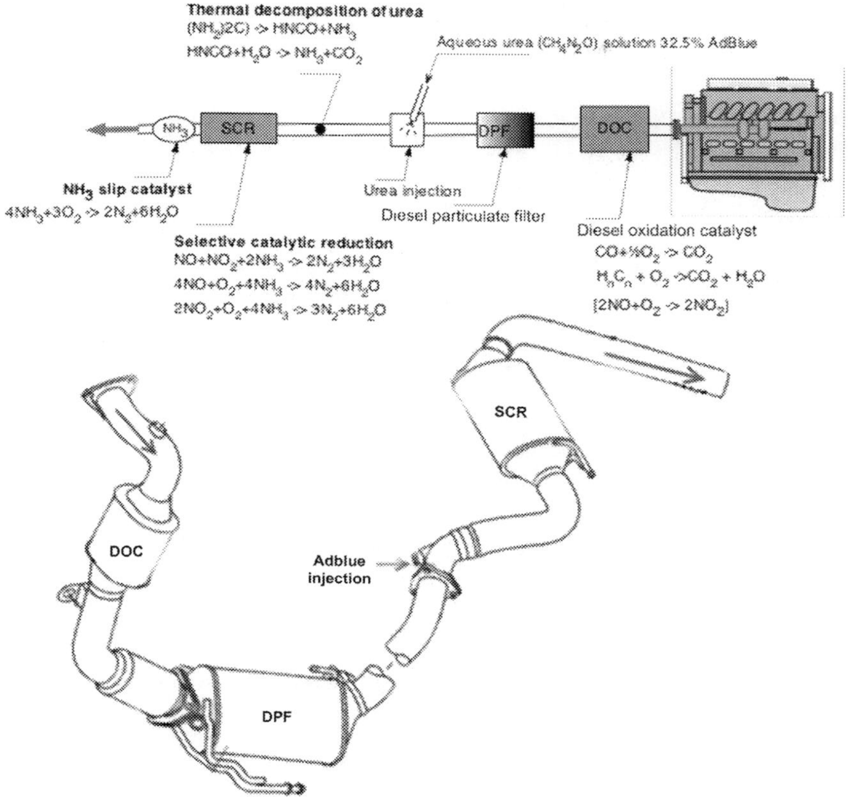


**FIGURE 17.8** (Top) Diagram showing different modules and reactions, in their normal relative positions, that may be found in the CI (diesel) engine exhaust systems of vehicles of both HD and LV classifications. (Bottom) A typical diesel passenger car exhaust layout, presilencer (muffler) unit.

Some of the in-cell and vehicle testing in the last few years has been related to "retrofit" DOC kits and test results found that removal of around 80% of the soluble organic fraction produces reductions in total PM emissions of some 30%–50%. The backpressure imposed by a DOC may be around 2.5 kPa (10″ of water) on a large truck engine but has the virtue of being constant rather than the increasing backpressure of a DPF as it loads up.

## Diesel particulate filter

The DPF has a honeycomb structure that is usually made from silicon carbide. Testing of these devices has revealed how the efficiency of any DPF may be as low as <40% when completely new and clean and how it rapidly increases up to full efficiency as the soot load builds causing a decrease in the filter's pore size. Note that filtration efficiency of a particulate filter is normally a stable 99.99% at 1 g/L soot load but in reality, most modern designs have stable 99.99% FE at 0.5 g/L. Note that measuring soot load as g/L takes into account the wide variety of filter volumes currently in use.

However, by the nature of their role, there is a progressive increase of exhaust backpressure as the particles are captured within its matrix to the point where regeneration has to take place.

A DPF will develop a buildup of two types of particulate: *soot*, which can be "burnt off" during the regeneration process, and *ash* that only makes up a very small fraction of the incoming soot but can be formed during regeneration, which remains in the DPF indefinitely and can build up to high levels. Ash properties can vary greatly between different vehicles because of differences in the engine operating conditions, lubricating oil, regeneration strategy, and the type of DPF. These differences in ash properties, discovered either during service life or by testing, play an important role in how the ash affects the performance of the DPF and also how easily, if at all, it can be removed.

The in-vehicle methods of regeneration are in three different modes:

- *Passive* or spontaneous regeneration requires the unit's internals to be at a high enough temperature during regular use to "burn off" the deposits, usually a minimum of 350°C and aided by the presence of $O_2$ and some $NO_x$ in the gas flow.
- *Active* or dynamic regeneration is triggered when the exhaust backpressure across the DPF is sensed to have reached a first critical point, whereupon the ECU initiates a fuel injection routine designed to increase the exhaust temperature in the DPF to over 600°C in order to oxidize the particulate deposits.
- *Service* regeneration is carried out by a service agent and requires an ECU and onboard diagnostic (OBD) scanner unit, which will initiate a special engine running mode that injects pulses of fuel into the exhaust so as to raise the DPF temperature.

Failure of DPF by "clogging" leading to replacement is a common cause of complaint by some passenger car and delivery van drivers. Failures are usually caused by ignoring the display warning lights after a series of stop–start, short journeys that prevent the active regeneration cycle from starting or completing.

Simulating partial blocking levels and the clogging failure mode within an engine test cell has proved problematic and time-consuming, since the rate of particulate deposit is quite slow from an engine running legislative cycles and any acceleration of the process requires a modified control map

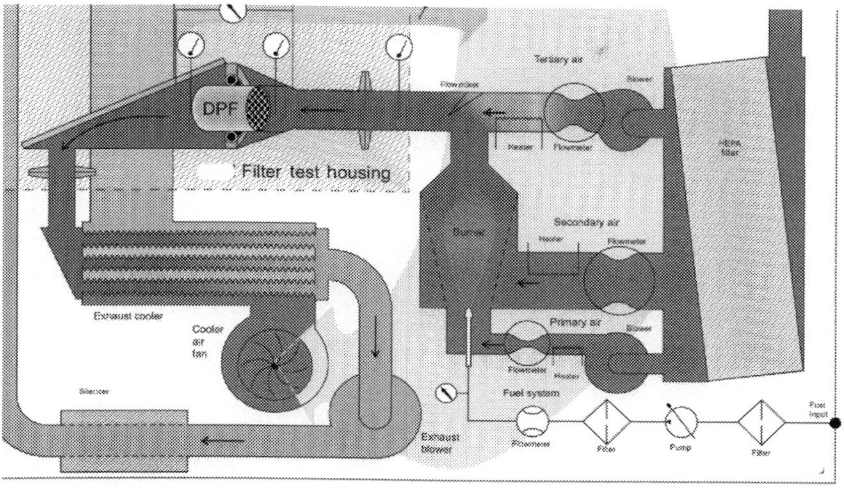


**FIGURE 17.9** Diagrammatic representation of a diesel particulate filter test system. *Reproduced with permission of Cambustion Ltd.*

in the ECU. This has led to the development of DPF and GPF test benches such as those produced by Cambustion [7] (see Fig. 17.9) a sample of typical test results are shown diagrammatically in Fig. 17.10.

## Gasoline particulate filter

The GPF may be either a separate unit or canned together with the TWC. It has a honeycomb structure that is usually made from a synthetic ceramic "cordierite." While the porosity of the GPF is higher than that of a DPF, because its substrate is lighter, and this allows the gas to move more easily, it also means the GPF is more fragile than a DPF.

As in DPF there are three particulate trapping mechanisms: interception, impaction, and diffusion. The trapping mechanisms depend on the particle and pore sizes and, also like the DPF, filtration efficiency increases as deposits build up.

While the smallest particles are trapped by diffusion, the larger particles are trapped by interception and impaction.

Unlike the DPF the regeneration of the GPF rarely has to be actively generated. At stoichiometric conditions under high load, the GPF core temperature may be around 650°C but the soot will only burn off with the leaner mixture that will occur during deacceleration, in the presence of $O_2$ and some $NO_x$.

Testing so far has indicated that the GPF is a durable and efficient technology that controls PN well below the regulatory limit (Fig. 17.10).

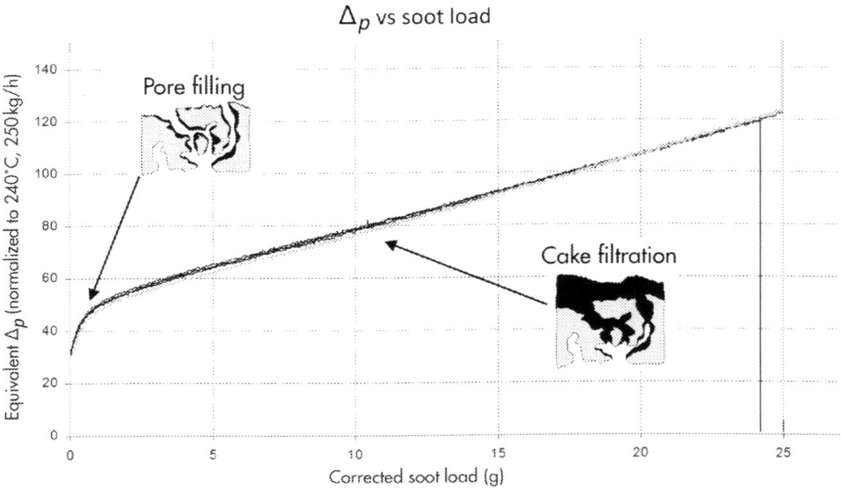


**FIGURE 17.10** A graph showing increasing backpressure plotted against soot load. The increasing soot load (caking) decreases the effective pore size and increases the particulate capture efficiency from <40% at 0 g/L load to 99.9% at around 1 g/L load. *Image supplied by Cambustion UK.*

## Selective catalytic reduction

Selective catalytic reduction (SCR) CATs are similar in basic construction to the three-way type. Currently the most common and effective SCR is the ammonia-SCR system, which reacts ammonia ($NH_3$) with the $NO_x$ to form nitrogen ($N_2$) and water. There are three reaction pathways, which are shown under the "SCR" graphic in Fig. 17.8.

The alternative hydrocarbon-SCR (lean $NO_x$ reduction) systems use hydrocarbons as the reductant. The hydrocarbon may be that occurring in the exhaust gas (native) or it may be added to the exhaust gas. This has the advantage that no additional reductant fluid source needs to be carried, but these systems do not currently have the performance of ammonia-SCR systems.

Ammonia-SCR systems use AUS32, which is a nontoxic 32.5% solution of urea ($CH_4N_2O$) in demineralized water sold in Europe under the trade name of "AdBlue."

AdBlue is a trademark held by the German Association of the Automobile Industry (VDA) and is produced to ISO 22241 specifications, which is important as impurities or contamination will damage the SCR. In North America the fluid is currently called diesel exhaust fluid. In all cases the solution is injected in metered doses upstream of the SCR CAT.

A few words of caution to test houses using urea injection:

- AdBlue, like any AUS32 fluid, is very susceptible to contamination from foreign matter and absorption of ions from containment material; it is also slightly corrosive for some metals, so it must be treated strictly according to instructions and kept in approved containers. Contamination can lead to SCR deactivation.

- There is a developing, and so far, apparently legal, trade in diluting and repackaging AdBlue then selling it as "automotive urea solution." Test results are completely invalidated if a solution of unknown concentration is used, so the lesson is to buy only from authorized dealers supplying fluid in original AdBlue containers.

AdBlue is held in a tank/pump unit and dosed upstream of the SCR system at a rate equivalent to 3%–5% of diesel consumption. Vehicle SCR systems, both HD and LV, that are designed to meet Euro 6 and (US EPA 2010 USA) LEV-Bin 5 limits are currently in use throughout Europe, Japan, Australasia, and the United States.

All European truck manufacturers currently offer SCR-equipped models. Retrofitted kits are available, but their fitting requires an amount of optimization and testing to be carried out.

Research into the dosing of the SCR system and calibration of the ECU requires accurate measurement of the liquid volume being injected. Devices such as the AVL "Shot To Shot" PLU 131 injector flow measuring device are available for this purpose.

Note that AdBlue can freeze at temperatures below −11°C, which means that any external storage tanks, tubing, pumps, etc., need to be trace heated.

## Ammonia slip catalyst

Ammonia ($NH_3$) is a toxic compound and a precursor in the formation of atmospheric secondary aerosols. The particulate matter that is formed, namely, ammonium nitrate and ammonium sulfate, is also associated with other adverse health effects.

Ammonia escaping into the environment from vehicles and known as "ammonia slip" occurs when some of the ammonia passing through the SCR is unreacted. This occurs when ammonia is overinjected into gas stream, gas temperatures are too low for ammonia to react, or CAT has degraded. This is now recognized by OBD systems as a fault and legislative limits have been applied. To deal with the problem ammonia slip CAT (ASC) units have been developed that are installed, close coupled, to the downstream side of the SCR. The ASC has to be very selective to $N_2$ meaning that most of the $NH_3$ slip is converted to $N_2$ rather than $NO_x$.

## Marine exhaust gas cleaning systems or scrubbers

Exhaust scrubbing systems have had to be developed and employed to treat exhaust from marine engines, auxiliary engines, and boilers, both onshore and onboard vessels. They also have to be installed on land installations and test facilities of the large diesels using heavy fuels.

Due to the continued, if declining, use of residual (bunker) oil and sulfur content still present in MARPOL 2020–specified fuels, marine scrubbers of several different types are very widely used. On the basis of their operation, marine scrubbers can be classified as wet, dry, or hybrid types. Whatever the type used they are all capable of operational misuse whereby the pollutants are removed from the atmosphere and dumped into the ocean.

Wet scrubbers may use seawater, fresh water with added calcium/sodium sorbents or pellets of hydrated lime as the scrubbing medium. The wash water produced is discharged into the open sea after being treated in a separator to remove any sludge from it, the sludge is put into a holding tank for later discharge to onshore treatment systems. Mist eliminators are used in scrubbing towers to remove any acid mist that forms in the chamber by separating droplets that are present in the inlet gas from the outlet gas stream.

Dry scrubbers consist of packed beds of gas-pollutant removal reagents, such as limestone to increase the contact time between the scrubbing material and gas. The packed beds slow down the vertical flow of water inside the scrubbers increase the exhaust gas cooling and the acidic water neutralization process. Dry scrubber systems consume less power than wet systems as they do not require circulation pumps; however, they weigh much more the reactions involved, which are:

$$SO_2 + Ca(OH)_2 \rightarrow CaSO_3 + H_2O(\text{calcium sulfite})$$

$$CaSO_3 + \frac{1}{2}O_2 \rightarrow CaSO_4(\text{calcium sulfate})$$

$$SO_3 + Ca(OH)_2 \rightarrow CaSO_4 + H_2O(\text{gypsum})$$

## Testing of post-combustion automotive exhaust systems and devices

The majority of ICE test cells used in the testing of CATs and exhaust systems are specifically designed for the purpose or may now be a full powertrain test cell.

All such cells need to have a large and easily adaptable floor space able to contain whole-vehicle exhaust systems and deal with high heat input into the cell air.

The various modules forming the exhaust system have to be installed in the appropriate positions relative to each other and the engine.

The high-temperature catalytic units need to use the actual vehicle connection to the engine and while some modification of the pipes joining other components can take place, great care has to be exercised with the pipe section joining of the AdBlue injection unit and the SCR due to the critical mixing with the gas flow taking place.

The position of the exhaust components downstream of the highest temperature units are, in the vehicle, carefully arranged to expose them to an optimum cooling airflow when the vehicle is moving. In the test cell it may be necessary either to shield exhaust sections or to use spot cooling in order to obtain optimum performance and avoid overheating. It is important that in rigging the exhaust components the forces applied by their mounting and containment does not exceed that for which they are designed; where practical, the in-vehicle fixing points must be used with the loads imposed as in the vehicle.

In test work designed specifically to provide exhaust at the flows and temperatures required to simulate the key life regime of the catalytic converters and filters, the ICE tends to be considered as a gas generator and the test sequences are designed accordingly.

*Catalytic aging* cells invariably run continuously and are therefore designed for unmanned or limited supervision running, very rarely now ICE based, they take the form of a packaged gasoline-fueled, burner-based test system similar in concept to the particulate filter test systems discussed earlier in this chapter. Stringent fire safety precautions are required.

## Measurement and containment of evaporative emissions

Evaporative emissions considered here are the volatile organic compounds (VOCs) emitted by a vehicle during all its operating and resting conditions that are not derived from the combustion process. They are a complex phenomenon that depends on multiple factors, including typical regional ambient conditions that differ around the world making globally harmonized test procedures somewhat unrealistic. A group of VOCs, collectively known as BTEX, comprising benzene, toluene, ethylbenzene, and xylene, is not specifically or individually covered by vehicle emission legislation but is frequently monitored within the urban environment. The BTEX suites are found in gasoline; but their largest sources are industrial chemical processes and they are ingested by ICEs. These pollutants react with nitrogen oxides in the presence of sunlight to form ground-level ozone and other secondary pollutants. Ozone is a significant ingredient of smog, it will inflame and damage human airways; it will also attack the double bonds in the rubbers used in low-level vehicle components such as CV joint gaiters.

Evaporative emissions occur in several forms and conditions as listed:

1. Hot soak losses. Hot soak emissions are usually attributed to the evaporation of the gasoline from the fuel system immediately following the engine being shut off after a trip in which it reached full running temperatures.
2. Diurnal losses. Evaporative emissions coming from the fuel system as a consequence of diurnal temperatures fluctuation while the vehicle is parked. This is a highly variable seasonal and regional feature.

3. Permeation. Hydrocarbons escaping the vehicle's fuel system by permeation through the plastic and rubber components. Some research shows that the presence of ethanol in fuel may increase permeation losses.
4. Running losses. Emissions from the fuel system and ICE while the vehicle is being driven.
5. Refueling emissions. Vapors present in the tank displaced by the liquid fuel entering into the tank through the filler neck, the measurement of which presents problems to the test engineer.
6. Leaks. OBD-II systems have routines to check the integrity of the vapor part of the fuel system and will produce an "engine fault" condition if a leak is determined.

All modern vehicles are fitted with Evaporative Emission Control System (EVAP), the main component of which is a canister filled with granules of activated charcoal, which are able to store, up to a saturation limit, hydrocarbon vapors. The canister is connected to the engine air inlet via a tube and "purge valve" that only becomes active once the lambda control is operating. As a result of the suction from the engine inlet the vapors are drawn out of the canister (it is purged) and form part of the fuel flow. The canister is open to atmosphere via a normally open vent valve allowing air replacement to the purged vapors, it is closed as part of an ECU implemented system pressure test (Fig. 17.11).

The role of ethanol in increasing evaporative emissions has been studied extensively, particularly in the United States, and has led to US EPA and

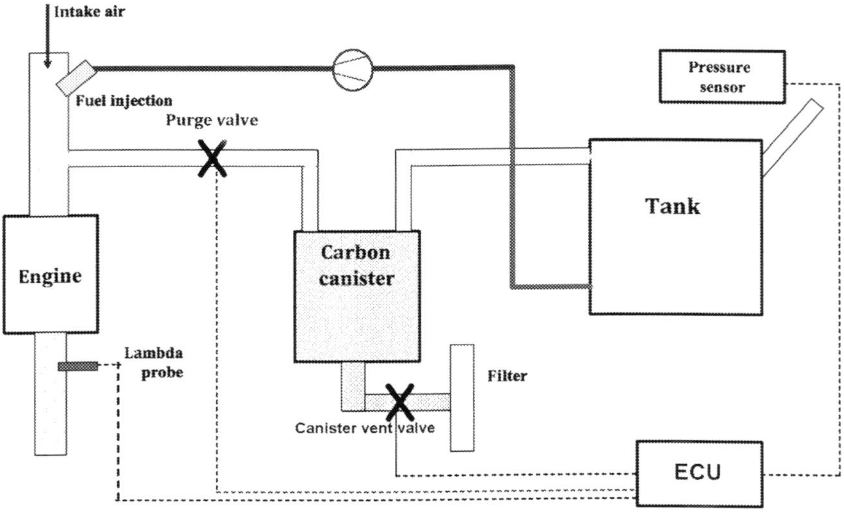


**FIGURE 17.11** Simplified diagram of a typical vehicle EVAP system. *EVAP*, Evaporative emission control systems.

CARB legislation covering the subject being the strictest in the world. The effects of ethanol as a 5% additive is that it raises the DVPE (dry vapor pressure equivalent) of the combined fuel leading to faster saturation of the vehicle's EVAP and potential leaks.

As with some exhaust emissions, the most critical driving pattern for real-world evaporative emissions is when the vehicle is driven for very short distances at low speeds. This may lead to a situation in which the carbon canister, being purged only for short periods and with very low purging flow rates, can become saturated and ineffective. The vast majority of in-service failures of the system (OBD-II error code "PO446") are caused by saturation caused by cold, city driving, or by overfilling, and leakage through breakage.

### *Testing evaporative emission with a sealed housing for evaporative determination*

The sealed housing for evaporative determination (SHED) typically consists of a "box" fitted with a vehicle access door capable of being closed with a gas-proof seal, lined with stainless steel, and constructed from materials that are chemically inert at the temperatures used in testing. This box or container, together with its external control cabinet, is invariably installed within a larger building and connected to the electrical and ventilation services of that building. A typical light vehicle SHED unit will have internal measurements of 6.5 m long by 3 m wide by 2.5 m high and will be equipped with systems to ensure the safe operation while handling potentially explosive gases.

Legislation covering evaporative emissions is subject to regional change as limits are tightened and altered to reflect real-life situations, such as a 48-hour diurnal test cycle that represents a vehicle being parked for a non-working weekend.

Present legislation requires the test vehicle to be subjected to a diurnal temperature range of between 20°C and 35°C (98/69/EC) or 18.3°C and 40.6°C (CARB, 2001), but modern units should be able to extend that range.

Special, smaller than standard, SHED units for testing of motorcycles and fuel system assemblies are available. There are also special units that allow refueling to be carried out within the sealed environment without a human being present.

The withdrawal of polluted air for analysis and its replacement with treated air in the controlled environment of the SHED requires special control and some explosion-proof air handling equipment.

## Personnel, systems, and devices for analysis of post-combustion exhaust gasses

The great majority of ICE performance test work is concerned with the taking of measurements that are in principle quite simple even though great skill may be needed to ensure the measurements give the right answer. Where emissions measurements are concerned, we are forced to move into a totally different and very sophisticated field of instrumentation engineering. The apparatus makes use of subtle and difficult techniques borrowed from the field of physics and the chemical laboratory requiring staff trained specifically in the science involved and their application. Emission specialists are a well-defined subtribe in the automotive test engineering family.

Emission measurement technology is having to continually to adapt to demanding new legislative requirements such as EPA 40 CFR Part 1065/1066, the heavy-duty Euro VI and light-duty Euro 6d. R&D work requires the quantitative analysis of nonlegislative gases such as $NH_3$, $N_2O$, and $NO_2$. Driven by RDE testing there are also demands for instrument miniaturization, faster analysis, and reduced total cost of ownership.

The integration of the various exhaust component analyzers with the ICE control and data acquisition system is vitally important, which tends to lead to most test facilities doing homologation work to use fully packaged emission "benches" coming from one major supportive supplier. However, the wider integration is equally important since there are significant subsystems that need to be housed in engine or vehicle emission test facilities, each of which requires some special health and safety considerations due to the handling of hot and toxic gases within a building space. These will be considered later in this chapter.

The full flow of exhaust gases has to be safely ducted and samples for analysis have to be removed in the correct amounts at critical points in any vehicular aftertreatment system.

It is important that water, which is a product of the combustion process, is not allowed to condense within the sampling system because some of the compounds produced by the partial combustion of fuels are soluble in water, so it is possible to lose a significant amount of these compounds in the condensation on the equipment surfaces.

To prevent condensation in critical parts of the exhaust sampling system, they will be electrically heated; for example, the sampling system for THCs will consist of heated probe, sampling line filter, and sampling pump, all of which are required to prevent the temperature of the equipment being lower than the dew point of the exhaust gas.

*Note that it is very desirable, for all but the simplest garage-type emission apparatus, for every test organization have one or more technicians who are specially trained to take technical and logistical responsibility (ownership) for the maintenance and calibration of the whole emissions measurement chain, including the bulk gas storage.*

## Laboratory class air for sample dilution

In addition to the synthetic air used as a bottle calibration gas (see next), many emission test laboratories will require a reticulated supply of "oil-free" air. Manufacturers of emission analysis equipment will quote the quality of air required for sample dilution and systems purging as having to meet ISO 8573-1: 2010 Class 1.4.1. The classes in this legislation are presented in the order of contaminants: *P*articles: *W*ater: *O*il, and in decreasing order of purity from 0 to *X*, where:

*P* is the purity class for particles (Classes 0–7, *X*).

*W* is the purity class for water content (Classes 0–9, *X*).

*O* is the purity class for oil content (Classes 0–4, *X*).

It is important for those involved with gas analysis to note that ISO 8573-1 does not include purity classes for gases or for microbiological contaminants. As with all calibration gas distribution systems it is vital that "pure" air reticulation systems are made up of tubing with low particle shedding and water permeation properties.

## *Calibration, span gases, and storing gas distribution system*

In order to function correctly exhaust gas analyzers have to be calibrated regularly, using gases of known composition. The principle of calibration is similar to that of any transducer in that the zero point is set, in this case by purging with pure nitrogen (sometimes referred to as "zero air"), then the 100% value is set with a gas of that composition, known as a "span gas." Good practice dictates that a gas of intermediate composition is then used as a check of system linearity. The whole routine may be highly automated by a calibration routine built into the analyzer control unit. After calibration the gas supply system is purged of the gases, particularly of $NO_x$, to prevent "cross talk" and degrading in the lines.

Many R&D analyzers are capable of part spanning in order to change their measuring sensitivity and can be calibrated by using known span gases, precisely diluted with zero air using a device known as a "gas divider." Alternatively, they may be calibrated directly using a bottled span gas of the correct composition. If using the gas divider method, it is important that the same gas ($N_2$ or artificial air) is used in the dilution as that which was used as the zero gas.

In addition to calibration or span gases, the facility will require operational gases such as hydrogen, plus an oxygen source as fuel for any flame ionization device (FID) and, in some cases, test gases for such routines as a critical flow check [see constant-volume sampler (CVS)].

Table 17.9 shows a possible list of gases, including a set of span gases, required in a modern emissions test facility. All these gases have to be stored at high pressure, usually between 150 and 200 bar, and distributed, at a regulated pressure of about 4 bar, to the analysis devices within the facility. The pipework system used in the distribution has to be chemically inert and medically clean to prevent compromising the composition of the calibration gases. For a basic single-fuel emission cell, the minimum number of 40- or

**TABLE 17.9 A list of typical gases that may be stored and used in the operation of and engine emission calibration cell.**

| Gases | Used as | Concentration | For analyzer type |
|---|---|---|---|
| Synthetic air (2) | Burner air | 21% $O_2$, 79% $N_2$ | FID |
| $H_2$/He mixture | Burner gas | 40% $H_2$, 60% He | FID |
| $C_3H_8$ in synth. Air | Test gas | 99.95% | External bottle with CFO system |
| $O_2$ | Operating gas | 99.98% purity | CLD |
| CO in $N_2$ | Span gas | 50 ppm | $CO_{low}$ |
| $C_3H_8$ in synth. Air | Span gas | 100 ppm | FID |
| $C_3H_8$ in synth. Air | Span gas | 10,000 ppm | FID |
| NO in $N_2$ | Span gas | 100 ppm | CLD |
| NO in $N_2$ | Span gas | 5000 ppm | CLD |
| CO in $N_2$ | Span gas | 500 ppm | $CO_{low}$ |
| CO in $N_2$ | Span gas | 10% | $CO_{high}$ |
| $CO_2$ in $N_2$ | Span gas | 5% | $CO_2$ |
| $CO_2$ in $N_2$ | Span gas | 20% | $CO_2$ |
| $O_2$ in $N_2$ | Span gas | 20% | $CO_2$ |
| Synthetic air (1) | Zero gas | Oil free | FID |
| $N_2$ | Zero gas | CO free | NO, $CO_{low}$, $CO_{high}$, $CO_2$, $O_2$ |

*CFO*, Critical flow orifice; CLD, chemiluminescence detector.

50-L gas storage bottles containing individual gases is probably 12, and a typical diesel/gasoline certification cell may have upward of 20 gas lines feeding its analyzer system.

*Note*: A 40-L gas bottle operating at 150 bar has a capacity of about 6000 L of gas at the analyzer inlet pressure and a 50-L bottle, which is the normal size for synthetic air and nitrogen, 7500 L.

It is not uncommon in mixed fuel facilities with two cells to have 20 calibration bottles and 4 operating gases connected at any one time.

Each gas bottle has to be delivered, connected, disconnected, and removed in its cycle of use on-site; therefore the positioning of the gas store has some important logistical considerations.

Clearly the distance between the gas bottles, the pressure regulation station, and analyzers directly affects the cost of the gas distribution system. The practical requirement of routing upward of 12 quarter-in. or 6 mm OD stainless-steel pipes on a support frame system within a new facility is an important design and cost consideration. The length of gas lines may be sufficient to create an unacceptable pressure drop, which may require transmission at an intermediate pressure and regulation local to the analyzers.

The storage of the gas bottles should be in a well-ventilated secure store where the bottles are not subjected to direct solar heating and thus remain under 50°C. For operational reasons in very cold climates the bottle store needs to be kept at temperatures above freezing that allows bottle handling and operation without regulator malfunction.

The store commonly takes the form of one or more roofed "lean-to" structures on the outside wall of the facility, in temperate climates having steel mesh walls and always positioned for easy access of delivery trucks. Storage of gas bottles above ground within buildings creates operational and safety problems that are best avoided.

Pressure regulators may be either wall mounted with a short length of high-pressure (up to 200 bar) flexible hose connecting the cylinder or, more rarely, directly cylinder mounted. The regulators handling $NO_x$ gases need to be constructed of stainless steel; other gases can be handled with brass and steel regulators.

The gas lines are usually made of stainless steel specifically manufactured for the purpose; sometimes, and always in the case of those transporting $NO_x$, the lines are electropolished internally to minimize degradation and reaction between gas and pipe. Pipes have to be joined by orbital welding or fittings made expressly for the purpose. The pipes should terminate near to the emission bench position in a manifold of isolating valves and self-sealing connectors for the short flexible connector lines to the analyzer made of inert Teflon or PTFE, ideally not more than 2-m long. Local gas storage regulations and planning permission requirements may apply to the construction of a bulk gas bottle store.

> *Note*: Orbital welding uses a gas/tungsten arc welding process that prevents the oxidation and minimizes surface distortion of the pipe's interior.

## *Gas analyzers and "emission benches"*

Although often used interchangeably with "emission bench," the term "analyzer" should refer to an individual piece of instrumentation dedicated to a

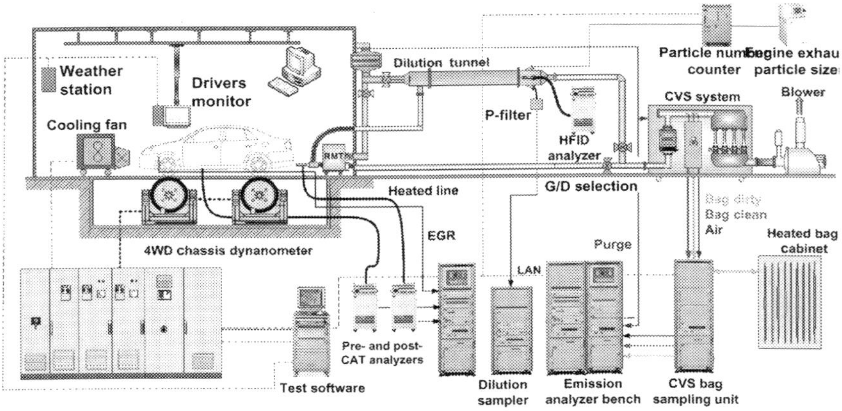


**FIGURE 17.12** The instrument layout and interconnection in an emissions laboratory for testing vehicles to Euro 6 and SULEV legislation in addition to R&D work. Chamber climatic controlled temperature range from −35°C to +60°C. *Diagram reproduced with permission of BOSMAL Sp.*

single analytical task installed within an emission bench that contains several such units.

As illustrated in Fig. 17.12, most analyzers fit within a standard 19-in. rack and are typically 4 HU high; most are capable of being set for measuring different ranges of their specific chemical.

## Constant-volume sampling systems

In both diesel and gasoline testing we need to dilute the exhaust with ambient air and prevent condensation of water in the collection system. It is necessary to measure or control the total volume of exhaust plus dilution air and collect a continuously proportioned volume of the sample for analysis. The use of a full-flow CVS system is mandatory in some legislation, particularly that produced by the EPA. It may be assumed that this will change over time with developing technology and "mini-dilution systems" will become widely allowed.

CVS systems consist of the following major component parts:

- A tunnel inlet air filter in the case of diesel testing or a filter/mixing tee in the case of gasoline testing, which mixes the exhaust gas and dilution air in a ratio usually about 4:1. There is also a sample point to draw off some of the (ambient) dilution air for later analysis from a sample bag.
- The dilution tunnel made of polished stainless steel and of sufficient size to encourage thorough mixing and to reduce the sample temperature to about 125°F (51.7°C). It is important to prevent condensation of water in the tunnel so in the case of systems designed for use in climatic cells the air will be heated before mixing with gas or the tunnel is outside the cell.

- The bag sampling unit: a proportion of the diluted exhaust is extracted for storage in sample bags.
- The gas sample storage bag array.
- The critical flow venturi, which controls and measures the flow of gas that the turbo-blower draws through the system.
- The blower that induces the airflow through the system, usually mounted well away from the workspaces due to the noise produced.
- The analyzer and control system by which the mass of HC, CO, $NO_x$, $CO_2$, and $CH_4$ is calculated from gas concentration in the bags, the gas density, and total volume, taking into account the composition of the dilution air component.

In the case of diesel testing a part of the flow is taken off to the dilution sampler containing filter papers for determining the mass of particulates over the test cycle.

A typical layout of a CVS system based on a 4 × 4 chassis dynamometer is shown in Fig. 17.12.

US Federal regulations require that the flow and dilution rates can be tested by incorporation of a critical flow orifice check. This requires a subsidiary stainless-steel circuit that precisely injects 99.95% propane as a test gas upstream of the mixing point. The CVS system dilutes this gas according to the flow rate setting, allowing the performance to be checked at a stabilized temperature.

The full-flow particulate tunnel for heavy vehicle engines is a very bulky device. The modules listed previously may be dispersed within a test facility because of the constraints of the building. The tunnel has to be in the cell near the engine and there are some legislative requirements concerning the distance from the tailpipe and the dilution point, ranging from 6.1 m for light-duty to about 10 m for heavy-duty systems.

Similarly, there are limits in some legislation concerning the transit time for HC samples from the sample probe to the analyzer of 4 seconds, which will determine the position within the building spaces of the FID unit.

The analyzers may be in a large and suitable ventilated control room with the venturi and sample bags within another adjoining room. Turbo-blowers are commonly placed on the roof since they often produce high noise levels. Alternatively, for light vehicle engines the analyzers, CVS control, and bag storage may be packaged in one, typically four-bay, instrument cabinet.

The so-called mini-dilution tunnel has been developed to reduce the problems inherent in CVS systems within existing test facilities; however, users must check if its use is authorized by the legislation to which they are working. It is necessary in this case to measure precisely the proportion of the total exhaust flow entering the tunnel. This is achieved by the use of a specially designed sampling probe system that ensures that the ratio between the sampling flow rate and the exhaust gas flow rate is kept constant.

In all CVS systems exhaust gases may be sent to the analyzers for measurement either from the accumulated volume in the sample bags or directly from the vehicle, so-called modal sampling.

In any advanced test facility, such as those involved with SULEV development, there will be three to five or more tapping points created in the vehicle system, from which gas may be drawn for analysis, such as:

1. EGR sample
2. before the vehicle catalytic converter (pre-CAT)
3. after the vehicle catalytic converter (post-CAT)
4. tailpipe sample (modal)
5. diluted sample (sample bags)

From 2 and 3 the efficiency of the CAT(s) may be calculated.

In addition to the tapping points listed previously, there may be several more taking samples from the CVS tunnel for particulate measurement.

### *Health and safety implications and interlocks for constant-volume sampling systems*

The following three fault conditions are drawn from experience and should be included in any operational risk analysis of a new facility:

1. Where the exhaust dilution air is drawn from the test cell and in the fault condition of the blower working and the cell ventilation system intake being out of action, or switched off, a negative cell pressure can be created that will prevent the opening of cell doors. The cell ventilation system should therefore be interlocked with the blower so as to prevent this occurring.
2. In the fault condition of the blower being inoperative or the airflow too low due to too small a flow venturi being used, the hot exhaust gas can back flow into the filter and cause a fire. A temperature transducer should be fitted immediately downstream of the dilution tunnel inlet filter and connected to an alarm channel set at about 45°C.
3. Gas analyzers that have to discharge their calibration gases at atmospheric pressure should have an extraction cowl over them to duct the gases into the ventilation extract; such an extract should be interlocked with the analyzer, preferably by a flow sensor.

## Temperature soak areas for legislative testing

A feature of legislative, chassis dynamometer–based, test procedures is that the vehicle has to soak at a uniform temperature before commencement. The two temperatures often stipulated are between 20°C and 30°C for "ambient" tests and −7°C for cold-start tests, and the soak periods are a minimum of 12 hours long. This requirement needs the efficient use of a climatically controlled building space fitted with temperature recording and windowless so that vehicles are not subjected to direct solar radiation. There are several strategies that may be adopted by vehicle test facilities having to deal with multiple vehicles in order to minimize the footprint of the soak area.

For cold-start testing, vehicles, additional to the one rigged for the first test, may be parked inside the chassis cell; they may then be moved onto the dynamometer by using special wheeled vehicle skids. ISO refrigerated shipping containers that can be parked close to the cell access doors may be used to soak vehicles providing that the transit time, from container to cell, is as short as possible and without engine rotation. In ambient soak areas it may be possible to use vertical stacking frames to store cars two or three high and test them in order of access.

## Implications of choice of site on low-level emissions testing

Test engineers not previously concerned with emissions testing should be aware of a number of special requirements that may not have the same importance elsewhere in other powertrain or vehicle testing facilities; foremost of these is the subject of the test cell site.

The problem with measuring exhaust emission to the levels required to meet the standards set by the latest legislation is that the level of pollution of the ambient air in some locations may be greater than that required to be measured at the vehicle exhaust.

Almost any trace of pollution of the air ingested by the ICE, or air used to dilute the concentration of exhaust gas within the measuring process, will compromise the results of emissions tests. Such pollutants include exhaust gas of adjoining engine cells or fumes from industrial plant, for example, paint or solvents. Cross-contamination via ventilation ducts must be avoided in the design of the facility and it will be necessary to take into account the direction of prevailing winds in siting the cell. Other than siting the facility in an area having, so far as is possible, pristine air, there are three strategies available to help deal with SULEV level measurement:

- Reduction of the exhaust dilution ratio by having a heated CVS system and emissions system that prevents water condensation.
- Refinement or cleaning of the dilution air to reduce the background level of pollutants to $<0.1$ ppm. This supports a CVS system but is expensive and the plant tends to be large and complex.
- Use a partial flow system with "pure" (artificial) air for dilution. These bag mini-dilution systems do not comply with current CVS requirements and require incorporation of an exhaust flow rate sensor.

Time will tell as to which of these methods become preferred and legislative practice, meanwhile investment in emission laboratories sited in areas with high levels of air pollution would not seem to be advisable.

## Heavy-duty test procedures

For an in-depth review and monitoring of international HD diesel emissions and test cycles, readers are recommended to subscribe to DieselNet at https://dieselnet.com/.

Euro VI standards apply to vehicles with a reference mass exceeding 2610 kg and to all M3 and N3 vehicles. The Euro VI standards are also not limited to any particular engine types but apply to all motor vehicles, including those with compression-ignition engines, SI engines (including natural gas/biomethane, petrol, liquefied petroleum gas, and E85), as well as dual fuel engines.

Euro VI emission standards were introduced by Regulation 595/2009. The new emission limits, comparable in stringency to the US 2010 standards, became effective from 2013/2014. The Euro VI standards also introduced particle number (PN) emission limits, stricter OBD requirements and a number of new testing requirements, including off-cycle and in-use PEMS testing. The first test schedule used is the ramped steady-state, World Harmonized Stationary Cycle (WHSC) made up of a sequence of steady-state engine test modes with defined speed and torque criteria at each mode and defined ramps between these modes. It is run from a hot start, following an engine preconditioning stage.

The second test schedule is the World Harmonized Transient Cycle (WHTC) is a transient test of 1800-second duration, with several motoring segments.

The full Euro VI type-approval testing consists of:

- WHSC + WHTC test for diesel engines; WHTC test for positive ignition engines;
- off-cycle emission testing:
  - NTE engine testing over the WNTE cycle and
  - a PEMS vehicle test;
- ISC: in-vehicle PEMS testing (Tables 17.10 and 17.11).

The US EPA has introduced *NTE* emission limits and testing requirements to make sure that heavy-duty engine emissions are controlled over "real-life" conditions. The legislation has required extensive negotiation with manufacturers in order to arrive at appropriate use of the NTE control area by particular truck types.

**TABLE 17.10** EU—EURO VI emission standards for heavy-duty diesel: steady-state testing.

| Test level | Date | Test | CO (g/kWh) | HC (g/kWh) | $NO_x$ (g/kWh) | PM (g/kWh) | PN (no./kWh) |
|---|---|---|---|---|---|---|---|
| Euro VI | 01-2013 | WHSC | 1.5 | 0.13 | 0.4 | 0.01 | $8.0 \times 10^{11}$ |

*WHSC*, World Harmonized Stationary Cycle.

**TABLE 17.11** EU—EURO VI emission standards for heavy-duty diesel: transient testing.

| Test level | Date | Test | CO (g/kWh) | NMHC (g/kWh) | $NO_x$ (g/kWh) | $CH_4$ (g/kWh) | PN (no./kWh) | PM (g/kWh) |
|---|---|---|---|---|---|---|---|---|
| Euro VI | 1-2013 | WHTC | 4.0 | 0.16 | 0.46 | 0.5 | $6.0 \times 10^{11}$ | 0.01 |

*WHTC*, World Harmonized Transient Cycle.

## Abbreviations

| | |
|---|---|
| **AC** | alternating current |
| **CFD** | computational fluid dynamics |
| **CFO** | critical flow orifice |
| **CFV** | critical flow venturi |
| **CLA** | chemiluminescent analyzer |
| **CLD** | chemiluminescent detector |
| **CVS** | constant -volume sampler |
| **DC** | direct current |
| **EAF** | sum of ethanol, acetaldehyde, and formaldehyde |
| **ECD** | electron capture detector |
| **ET** | evaporation tube |
| **Extra High2** | Class 2 WLTC extra-high-speed phase |
| **Extra High3** | Class 3 WLTC extra-high-speed phase |
| **FCHV** | fuel cell hybrid vehicle |
| **FID** | flame ionization detector |
| **FSD** | full-scale deflection |
| **FTIR** | Fourier transform infrared analyzer |
| **GC** | gas chromatograph |
| **HEPA** | high-efficiency particulate air (filter) |
| **HFID** | heated flame ionization detector |
| **High2** | Class 2 WLTC high-speed phase |
| **High3a>** | Class 3a WLTC high-speed phase |
| **High3b** | Class 3b WLTC high-speed phase |
| **ICE** | internal combustion engine |
| **LC** | liquid chromatography |
| **LDS** | laser diode spectrometer |
| **LoD** | limit of detection |
| **LoQ** | limit of quantification |
| **Low1** | Class 1 WLTC low-speed phase |
| **Low2** | Class 2 WLTC low-speed phase |
| **Low3** | Class 3 WLTC low -speed phase |
| **LPG** | liquefied petroleum gas |
| **Medium1** | Class 1 WLTC medium-speed phase |
| **Medium2** | Class 2 WLTC medium-speed phase |
| **Medium3a** | Class 3a WLTC medium-speed phase |

| | |
|---|---|
| **Medium3b** | Class 3b WLTC medium-speed phase |
| **NDIR** | nondispersive infrared (analyzer) |
| **NDUV** | nondispersive ultraviolet |
| **NG/biomethane** | natural gas/biomethane |
| **NMC** | nonmethane cutter |
| **NOVC** | not off-vehicle charging |
| **NOVC-FCHV** | not off-vehicle charging fuel cell hybrid vehicle |
| **NOVC-HEV** | not off-vehicle charging hybrid electric vehicle |
| **OVC-HEV** | off-vehicle charging hybrid electric vehicle |
| **Pa** | particulate mass collected on the background filter |
| **PAO** | poly-alpha-olefin |
| **PCF** | particle preclassifier |
| **PCRF** | particle concentration reduction factor |
| **PDP** | positive displacement pump |
| **Pe** | particulate mass collected on the sample filter |
| **PER** | pure electric range |
| **Per cent FS** | per cent of full scale |
| **PM** | particulate matter emissions |
| **PN** | particle number emissions |
| **PNC** | particle number counter |
| **PND1** | first particle number dilution device |
| **PND2** | second particle number dilution device |
| **PTS** | particle transfer system |
| **PTT** | particle transfer tube |
| **QCL-IR** | infrared quantum cascade laser |
| **RCB** | REESS charge balance |
| **RCDA** | charge-depleting actual range |
| **REESS** | rechargeable electric energy storage |
| **RRC** | rolling resistance coefficient |
| **SSV** | subsonic venturi |
| **USFM** | ultrasonic flow meter |

## References


[1] Worldwide Emission Standards and Related Regulations, Passenger Cars/Light and Medium Duty Vehicles. <www.continental-automotive.com>, 2019.

[2] Directorate General for Internal Policies Policy Department A: Economic and Scientific Policy. Comparative study on the differences between the EU and US legislation on emissions in the automotive sector. P/A/EMIS/2016-02.

[3] DMS500 Mk II Transient Engine Particulate Analyser. Animation at <www.cambustion.com/dms>.

[4] Syft Technologies, Selected Ion Flow Tube Mass Spectrometry (SIFT-MS). <https://www.syft.com/sift-ms/>.

[5] Cambustion. FID 600.

[6] V. Morales, P. Bonnel, 'On-Road Testing With Portable Emissions Measurement Systems' (PEMS), JRC Science Hub, 2017. Available from: <https://wiki.unece.org.20170717_PEMS_Guidance_LDV_2017>.

[7] Cambustion Ltd., Cambridge, CB1 8DH, UK. <http://www.cambustion.com>.

## Further reading

S. S. Amaral, J. A. de Carvalho Jr., M. A. Martins Costa, C. Pinheiro, An Overview of Particulate Matter Measurement Instruments, ISSN 2073-4433. <www.mdpi.com/journal/atmosphere> (open access).

H. Klingenberg, 'Automobile Exhaust Emission Testing', Springer, 1996, ISBN: 978-3-642-80243-0.

F. Rodriguez, Y. Bernard, J. Dornoff, P. Mock, Recommendations for Post-Euro 6 Standards for Light-Duty Vehicles in the European Union, ICCT, 2019. Available from: <https://theicct.org/publications/recommendations-post-euro-6-eu>.

T. Smedley, 'Clearing the Air', Bloomsbury Sigma, 2019, ISBN: 978-1-4729-533-15.

D. Smith, et al., On-line analysis of diesel engine exhaust gases by selected ion flow tube mass spectrometry, 2004, https://doi.org/10.1002/rcm.1702.

Worldwide engine and vehicle test cycles. <https://dieselnet.com/standards/cycles/index.php>.

Chapter 18

# Anechoic (NVH) and electromagnetic compatibility (EMC) testing and test cells

## Chapter outline



Engine Testing. DOI: https://doi.org/10.1016/B978-0-12-821226-4.00018-8

## Introduction

*Note:* The acronym "EMC" is commonly used as a generic term when used to describe test cells that carry out electromagnetic interference (EMI), electromagnetic susceptibility, and EMC testing; this book follows this general practice unless specifically stated otherwise.

The development of the hybrid electric vehicles (HEV) and battery electric vehicles (BEV) has significantly changed the electromagnetic and noise environments studied by the automotive test engineer. In the audible spectrum, dominant frequencies may have increased, while sound pressure levels (SPLs) may have deceased. At the same time EMC and EMI testing has to deal with field generation from ever more powerful direct current (DC) devices and their cables—and a stretching of its test spectrum with the development of high frequency 5G systems and low frequency magnetic fields.

The increase in the number of interconnected digital control units within a modern vehicle together with the increased electrical and magnetic fields generated within the vehicle has made electronic compatibility an even more vital part of powertrain testing; long gone are the days when it was about reducing interference on the long-wave program of the car radio.

Noise and vibration have an important influence on a customer's perception of vehicle quality; therefore the test engineer, while concentrating on the objective nature of testing, has to be continually conscious of the subjective nature of noise, vibration, and harshness (NVH) in their work. In spite of all the advanced modeling and analysis tools and software, there is still a place in the vehicle refinement process for the subjective judgment of human users. All major test facilities will have access to a studio where a "jury," made up of a selection of the target user group, is asked to judge the level of acceptability of recorded and postprocessed sounds.

The objective testing of the noise created by powertrain components and their housings and its attenuation, like that of EMC testing, has to be carried out in an area, so far as possible, devoid of ambient noise and echo, within the studied electromagnetic spectrum—hence the need for anechoic cells.

## Open air test sites versus acoustic anechoic cells

Anechoic means without an echo. To investigate the quantitative and qualitative characteristics of noise, or any other electromagnetic wave energy from the unit under test (UUT), it is important not to have an echo of the source emission corrupting the signal measurement. Such reflected energy waves can have additive or subtractive (canceling) effects on the original transmission.

The everyday concept of an "echo" is the bounce-back of a sound wave, but the same idea holds good for electromagnetic wave energy with frequencies above the audible wavebands. In theory and in limited practice both technologies are able to carry out some of their testing on a flat surface in the wide open air, which is inherently a "free field" over a reflecting plane. An ideal open-area test site (OATS) is an infinite area at a site with no ambient noise and has a perfectly flat reflective surface with a total absence of reflecting obstacles. Since the cost of a semianechoic chamber (SAC) for automotive EMC testing, that attempts to emulate such ideal conditions, is very high, it has encouraged research into the feasibility of finding and using open air sites.

In practice OATS, particularly in urban areas, inevitably suffer from problems of ambient noise, both in the audible and radio frequencies (RFs) fields. For EMI test work to be performed the ambient RF levels at a test site have to be sufficiently low, compared to the levels of measurements. Like all EMC test sites they also have to be kept clear of equipment that are "unintentional radiators" which generate RF energy for use within the device but is not intended to emit electromagnetic energy.

Much of the legislative "drive-by noise test" (also called "pass-by") work is carried out on a specially surfaced and instrumented, open-air track section (certified to ISO 362). However, the theoretically ideal situation of a still day in the flat desert of Arizona is not available to most test engineers so, for EMC and NVH work on a static UUT they require test cells that simulate those anechoic conditions.

In the automotive engine, powertrain, and component test industries, such cells are designed specifically for either electromagnetic compatibly testing working with frequencies in the megahertz and gigahertz ranges, or NVH testing, working in the range of human hearing, around 16 −20,000 Hz.

Anechoic cells fall in the following general classifications:

- *Semi-anechoic chamber (SAC)*: A shielded enclosure whose internal walls and ceiling are lined with material that absorbs electromagnetic energy in the frequency range of interest but whose floor is truly flat and reflective. In the United States, these are commonly and correctly called "hemi-anechoic."
- *Fully anechoic chamber (FAC)*: A shielded enclosure such as the SAC except the ground plane is also formed with sound or RF-absorbing material. Such cells tend to be of much smaller size than SAC and be used for component testing rather than road vehicle or internal combustion engines (ICE) work.

There are some important differences between EMC and NVH anechoic cells, both in the operational and construction features; *they are not dual-purpose facilities*. Moreover, in addition to SAC and FAC cells, EMC testing may make use of gigahertz transverse electromagnetic (GTEM) and reverberation chambers (RVC) described in more detail in Part 2B of the chapter.

## Part 1A. Noise, vibration, and harshness testing and its legislative environment

Vehicle type approval legislation that limits the noise produced by passenger cars has, in the United Kingdom, been controlled since 1929. New cars are now required to meet Europe-wide noise limits (Regulation EU No 540/2014) which, under the test conditions, have been progressively reduced from 82 dB(A) in 1978 to the current limit of 74 dB(A) established in 2016, and by 2026, the limit for most new passenger cars will be 68 dB(A).

In automotive engineering, NVH testing is divided between meeting type approval requirements and the vehicle refinement process; in nonautomotive engine development, it is more concerned with health and safety (H&S) of operators and machinery health monitoring through vibration analysis.

First, some basic definitions of the NVH terms as used in powertrain and automotive engineering:

- Noise is usually considered as unwanted sound. This is distinct from required sound (driver feedback) and desired sound (from vehicle sound system, exhaust note, etc.).
- Vibration is the oscillation emulating from the UUT that is typically, but not exclusively, felt by the operator and passengers rather than being heard.
- Harshness is used to describe the severity and discomfort associated with unwanted sound or vibration, particularly from short-duration events such as antilock braking system (ABS) operation. It is qualitative and based on desired characteristics rather than on quantifiable measurements.

*There is also sound value known as "roughness,"* which is more specific than harshness and while difficult to define it is an imprecise term often used by test engineers. It is a complex effect that quantifies the subjective perception of rapid (15−300 Hz) amplitude modulation of a sound. While applicable to automotive NVH work, the best demonstration of its effect is that of a human being screaming—a scream heard from a person on a roller-coaster can be smooth, while a scream of pure terror has roughness and is perceived with alarm by the hearer (Fig. 18.1).

The development and wider use of electrically powered hybrid vehicles has brought about large reductions in powertrain noise at slow speeds, even to the point that it has caused serious safety concerns for organizations representing visually impaired people and pedestrians. The testing of powertrain sources of noise and particulate pollution, other than the ICE, has given new emphasis to the NVH testing of brakes and from tires running on road surfaces.

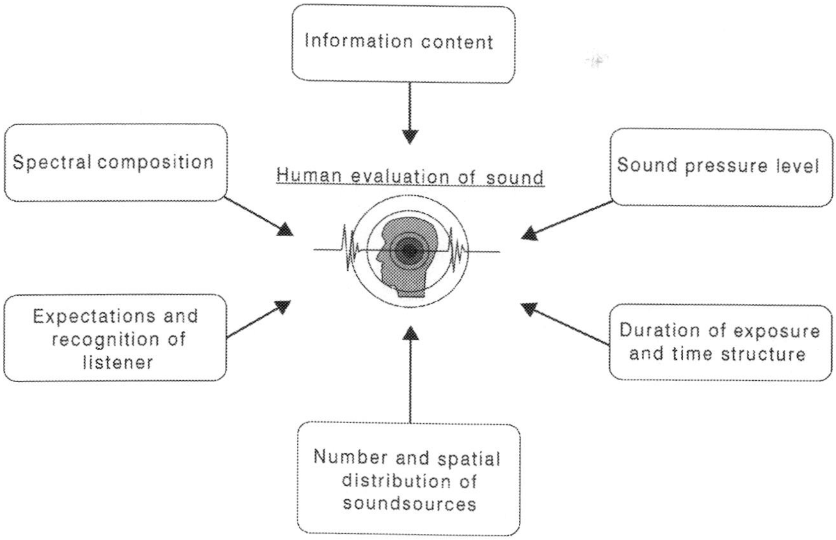


**FIGURE 18.1** A diagrammatic representation of the objective and subjective NVH inputs that make up the human perception of both noise and vibration.

The key legislative standards covering NVH test facilities and their use are:

- ISO 3745:2012 "Acoustics—Determination of sound power levels and sound energy levels of noise sources using sound pressure—Precision methods for anechoic rooms and hemi-anechoic rooms"
- BS EN ISO 3740:2019 entitled "Acoustics. Determination of sound power levels of noise sources. Guidelines for the use of basic standards"

These are concerned with design requirements of acoustic anechoic chambers and the methods of determining sound power levels of noise sources, and they mention certain aspects of anechoic cell design, including:

- desirable shape and volume of the cell,
- desirable absorption coefficient of surfaces,
- specification of absorptive treatment, and
- guidance regarding avoidance of unwanted sound reflections.

## Part 1B. The acoustic anechoic cell structure

A key part of the technical specification of an anechoic cell is known as the "cut-off frequency"; this is the frequency below that the rate of sound level decay in the chamber no longer replicates a "free-field" environment. The lower the cut-off frequency, the better its simulation of a true free field. The maximum cut-off frequency normally required for automotive anechoic cells (both powertrain sub-assembly and vehicle) is around 120 Hz, but most modern facilities are built to have cut-offs below 100 Hz and some down to 60 Hz.

The acceptance test of the anechoic characteristic of a cell is carried out using a broadband generator placed at the geometric center of the cell. The level of sound decay is measured by a microphone physically moved along chords tensioned between the noise source and the high-level corners of the cell. The sound decay measured is compared with the 6 dB distance-doubling characteristic of a free field (inverse square law).

The task of certification of an anechoic cell and the equipment required is a specialist field requiring the services of qualified contractors.

There are certain other points that should be remembered by the nonspecialist if he or she is required to take responsibility for setting up a test facility of this kind, the first of which is the geographical location of the facility. The structure of an anechoic cell should be isolated from any environment in which high levels of airborne noise or ground vibration are generated, such as production plant, nearby public roads, or railways.

Many acoustic anechoic cells built in less than ideal locations have to take the expensive form of a "building within a building" in which the inner structure is isolated by dampers from external vibration.

The walls and ceiling of the chamber have to absorb sound emitted by the UUT and this is usually achieved by completely lining them with pyramidal shapes referred to as "wedges", sometimes "cones" which attenuate sound by both scattering and absorption. The total length (height) from the surface mounting base of the wedges will depend on the lowest frequency range they are designed to absorb. In automotive acoustic anechoic cells, they may be some 700 mm long and behind their base and a mounting frame may be an air gap of some 300 mm to the cell wall. Therefore the effective volume of such spaces is considerably reduced from the full enclosed space; the smaller the cell volume, the more marked the reduction.

Wedges may be individual moldings made of dense foam that are held by simply being held by compression within a steel wire frame having spacings to suit the wedge base shape. Alternative designs are made of perforated steel sheet encasing a foam lining; these are usually formed in rigid rectangular sets consisting of several wedges that are clipped to a rear mounting frame. The metallic surface has the advantage of being usually lighter colors than foam wedges, easier to keep clean and sufficiently rigid to provide a mounting surface for ancillary equipment such as video cameras. Access door frames and furniture can affect the acoustic performance adversely, by creating hard reflective surfaces, so must be carefully incorporated into the overall lining. Windows should not be required for the same reasons and are invariably made unnecessary by the use of color CCTV cameras.

## Special health and safety considerations for anechoic cells

There are H&S risks to all closed anechoic test chambers, the interiors of which can be disorienting and have exit routes that may not be obvious to the inexperienced user or a visitor. Managers and operators should always be aware that first-time visitors may suffer from mild or even severe claustrophobia and be aware of the need to get the sufferer into the open air quickly.

Along with fitting the legally required, illuminated, emergency exit notices above cell exits, the acoustic lining covering exit door latches or operating buttons should be color marked to make their location within the wedges and their opening direction obvious to both operators and visitors.

## Fire detection and suppression in anechoic cells

The detection and quenching systems described in Chapter 3, Test Facility Design and Construction, are relevant to all acoustic anechoic chambers.

In acoustic cells, automatic smoke and flame detectors are vital because the operator has no direct vision of the cell and may be attending to a very localized area of the cell and UUT through a CCTV image. Wide angle color CCTV coverage of the cell is highly recommended.

Since ventilation air flow, and, therefore, air-cooling flows, will be lower in anechoic cells than in other normal powertrain cells, there is a real risk of local overheating, and serious fires in anechoic engine cells have occurred. In EMC cells the danger of localized electrical fires may be higher than normal, due in part to the number of temporary cable runs. In all cases the rule is to reduce the potential and discretionary fire load by restricting the amount of fuel in the cell.

In choosing quenching systems, special consideration must be given to vehicle cells where the volume of the cell is very large compared with the UUT which is the usual source of a fire and which would require very large volumes of any gaseous suppressant. The engine bay cannot be rigged with a rally car extinguishing system because of any non-standard effect on powertrain noise so, as advised in Chapter 10, Chassis Dynamometers, Rolling Roads, and Hub Dynamometers, the fitting of a water mist system, with nozzles below and at the sides of the UUT position, is recommended.

## Treatment and location of services for anechoic cells

Since so much care has to be taken in the creation of a quiet cell environment, it is obvious that any noise created by the fluid and ventilation support systems has to be minimized at source or attenuated, so far as is possible.

### Ventilation services

The ventilation of anechoic cells, particularly when designed for the testing of running ICE, presents particular problems because sufficient flow required to remove heat can create "wind noise." However, since full power running in such cells is normally of the short duration required to take a set of measurements and the precise control of inlet air temperature is not so critical to the sound recording, the full thermodynamic rating of the cell (Chapter 5: Ventilation and Air Conditioning of Automotive Test Facilities and Chapter 6: Cooling Water and Exhaust Systems) is not usually required. The technique adopted for ventilation systems is to have an air flow through the cell of just adequate volume for the test task, entering and leaving at as low a velocity as possible to reduce nozzle and wind noise. This minimization of air velocity into and out of the cell is achieved by making the inlet and discharge ducting as large as possible.

A common design is to make the last 500 mm above floor level of the whole end-wall act as the ventilation air inlet and for the air to exit through ductwork behind the wedges of the upper corner sections of the opposite end of the cell.

### Fluid services

The cooling water piping and flow-control valves can create noise of variable and unpredictable frequencies and at unpredictable times; therefore as much of the system as possible is kept outside the cell and the pipework within the cell is run under floor when possible, an example of such transition tubes is shown in Fig. 18.2. The distance between the temperature control devices and the engine usually means that the thermal inertia of the system is too high to have good transient response but, as with the ventilation air flow, this is rarely critical to the type of recordings being made during NVH testing. However, pumped circulation of the coolant fluid is often needed to reduce the temperature control time lag. Sound transmission through steel pipes needs to be minimized by decoupling with nonmetallic sections.

If an anechoic cell is connected to fluid fuel systems, they have to be fitted with all the isolation valves and interlocks described for other cells in Chapter 7, Energy Storage; in the case of acoustic anechoic cells, the valves are often fitted below the bottom row of the lining wedges. Given that the duration of running in most ICE testing within an anechoic cell is a matter of minutes rather than hours, the use of a small "outboard engine" type of fuel tank may be rigged with the engine, obviating the complication of plumbed in fuel lines.

### Internal combustion engines exhaust piping

Most types of acoustic cell linings can absorb liquids and can become inflammable, so running a hot engine exhaust tube through such linings

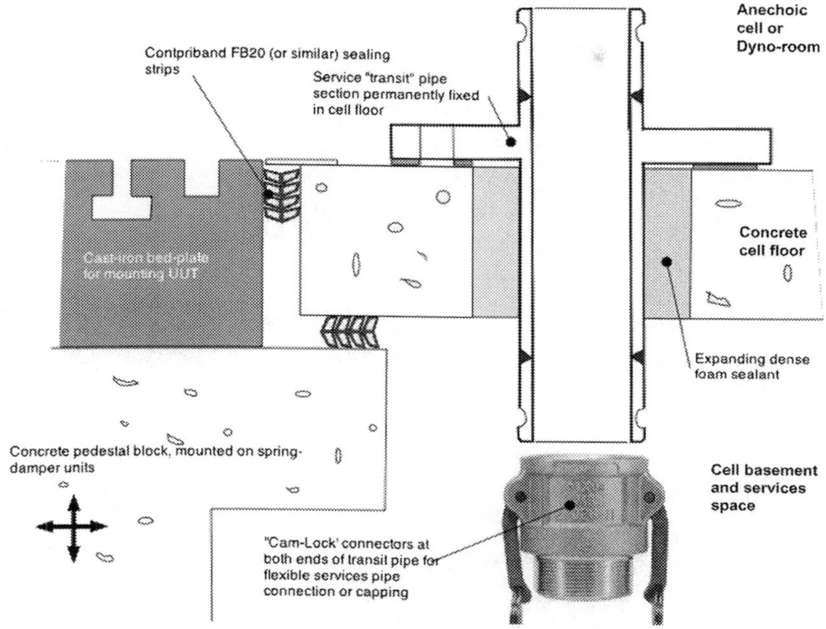


**FIGURE 18.2** A part section through the floor of an anechoic cell running into the cellar space below which houses the support plinth and services. It shows one of several transition tubes for connection to and from the service units. They are plugged or capped when not in use. *Sketch by the Author based on an actual site layout.*

represents an increase in the level of fire hazard. Two solutions are commonly used to run exhausts out of the cell:

1. The engine exhaust is taken out through the floor into a cellar space, which may house the seismic block. The floor-fixed transit tubes need to have a nonrigid decoupling section plus top and bottom caps for installation when not in use, as shown diagrammatically in part section in Fig. 18.2.
2. The exhaust is taken out of the ceiling via a double-skinned tube that is jacketed with circulating cooling water. Such systems must be fitted with a water flow switch in the emergency stop (EM stop) chain in order to prevent the possibility of the cooling water jacket outlet being shut off or blocked, which would lead to a steam explosion. This is not a remote, theoretical hazard but has been experienced by the author.

## Engine mounting

An ICE or any other module of a powertrain, such as an electric motor unit, must be raised in an anechoic cell above floor level more than is usual in other cells, typically to give a gap of 1.0–1.2 m from floor to shaft centerline, to allow microphones to be located at a far-field distance below it. This calls for tall, nonstandard, UUT-mounting arrangements that are frequently prerigged on a skeletonal pallet frame (without any reflective or reverberatory drip tray). If measurements are made too close to the UUT, the SPLs may vary significantly with any small change in the microphone position. This will occur in the "near-field" which is at a distance less than the wavelength of the lowest frequency emitted from the UUT.

## Dynamometer connections in acoustic anechoic cells

In anechoic prime-mover test cells, any dynamometer(s) must be physically separate and isolated acoustically from the cell proper. Since the UUT under test should be ideally in the horizontal center of the anechoic cell space, such an arrangement requires an overlong coupling shaft system, the design of which will present problems discussed in Chapter 11, Mounting and Rigging Internal Combustion Engines for Test.

In many designs, such as that shown in Fig. 18.3, the shaft system is split into two sections, with an intermediate bearing mounted on a support, or pedestal block, inside the anechoic chamber. The shaft section running through the wall, between the dynamometer and the pedestal bearing, will run within a

**FIGURE 18.3** An acoustic semianechoic chamber designed for ICE powertrain or transmission testing having three dynamometer shafts running from bearing pedestals through the cell walls. Note vertical shrouded exhaust pipe in right hand corner. *Reproduced with the permission of IAC UK.*

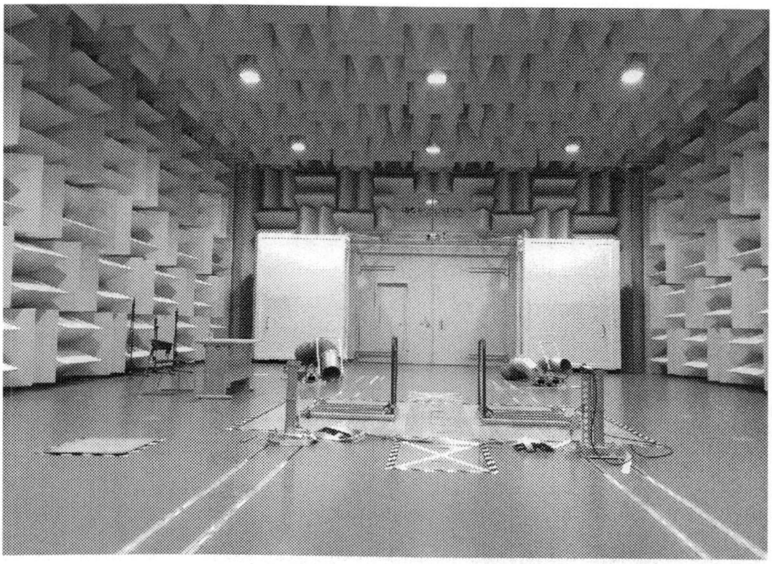

**FIGURE 18.4** A semianechoic chamber fitted with a chassis dynamometer for full vehicle NVH testing. Note lined entry doors are open revealing doors of outer shell building and the two evacuated engine exhaust ducts at floor level. *Reproduced with permission from IAC UK.*

sound-absorbing "silencer tube," designed to minimize transmission of noise from the dynamometer room. The pedestal, which needs to be a massive or nonreverberatory construction, will often be used to house, or support, transducer service connection boxes and EM stop lock-out buttons.

How much sound is reflected, absorbed, or transmitted by the necessary pedestals, microphone stands, or guards in the cell depends on its properties, its size and the wavelength of the sound. In general, the object must be larger than one wavelength in order to significantly disturb the sound.

Fig. 18.4 shows a full vehicle, semianechoic cell having, in accordance with ISO 3745, a cut-off frequency of down to 50 Hz and which lined with 30% open area perforated metal skinned, mineral-wool cored wedges of a type made by IAC UK called Metadyne.

It is fitted with a chassis dynamometer set within a flat reflective floor. In the photograph, the deep inner (wedge lined) doors are open and the outer shell-building doors with a pedestrian access can be seen as can the ceiling mounted lamps set into the wedges.

## External vehicle noise and drive-by noise testing

Testing of drive-by noise is generally carried out on an outdoor section of instrumented track. ISO 362 defines the pass-by test procedures for measuring noise emitted by accelerating road vehicles, and the ECE

EPA (motorcycle), ECE R41 (motorcycle), ECE R51 (automobile), and R138 (tires) regulate the limits of the vehicle exterior noise.

In the ECE R138 standard covering "automobiles without a combustion engine," a minimum sound level is set for vehicles traveling under electrical power. The system required is called the acoustic vehicle alert system, and it dictates that such vehicles must emit at least 56 dB(A) at speeds under 12 mph.

Indoor drive-by testing is carried out indoors but requires a very large NVH test cell with chassis dynamometer and microphone arrays on each side of the vehicle that simulates pass-by driving but the tire noise component will not realistically simulate. ISO 362-3:2016 specifies the procedure for measuring the noise emitted by road vehicles (categories M and N) by using a SAC.

## Instrumentation

Acoustic research makes use of highly sensitive instruments and correspondingly sensitive measurement channels. The design of the signal cabling system must be carefully considered at the design stage to avoid interference, and the hazard formed by a multitude of trailing leads in the cell.

The modules of instrumentation commonly used in NVH test work and mounted on or near the UUT in a cell are:

- single-axis and triaxial accelerometers, filtered and unfiltered, having millivolt outputs;
- charge output accelerometers with similar sensor technology to those used in combustion analysis work mainly for higher temperature applications (e.g. exhaust systems);
- externally polarized precision condenser microphones;
- signal preamplifiers that are mounted close to the transducers and boost their signal to the recording and sound analysis equipment housed in the control room;
- multimicrophone array stands or individual microphone stands or hanging cables; and
- sound level meters conforming to IEC 61672-1.

Accelerometers may be attached to the UUT either by an adhesive or by a fixing stud, drilled, and tapped into the engine, motor or gearbox body.

The collection of raw NVH data in the test cell is only the beginning of a complex test process; NVH postprocessing is a very demanding and time-consuming process carried out by specialist staff using a specially developed commercially available software suite.

## Control and instrumentation room

The advances in digital recording and processing has tended to mean that NVH test cells have short periods of shaft turning and spend most of their time having the UUT prepared for short multistage test sequences, the recorded sounds from which can be recorded, then analyzed, and digitally altered without further running. In anechoic control rooms, UUT performance displays tend to be replaced by banks of multichannel amplifiers and sound analysis equipment. The air conditioning load and heating in these rooms will vary considerably and must be designed accordingly.

While most multidisciplinary test facilities will use the same cell control system, the demands on the engine/dynamometer control hardware and software system is usually much lower in anechoic cells. It is, therefore, possible to use simpler and cheaper control systems as the data acquisition systems used are of a different type to the nonspecialist performance cells. Because of the need to suppress ambient noise and because the typical test sequences tend to consist of warm-up, short stages at different power outputs or inputs, precise control of the cell services is not entirely straightforward. Ventilation in the cell may be turned down to minimum during critical measurements and cooling water kept at a set position to prevent varying valve throughput noise; therefore monitoring UUT temperatures is critical.

## Full vehicle noise: Measurement practice

Semianechoic cells fitted with "quiet" chassis dynamometers (Chapter 10: Chassis Dynamometers, Rolling Roads, and Hub Dynamometers), designed for either acoustic or EMC work, are large and expensive installations required for vehicle refinement, while legislatively required testing requires out-doors or track testing (OATS).

A number of regulations, both national and international stipulate the permitted vehicle, particularly exhaust, noise levels and the methods by which they are to be measured. In Europe it is specified that A-weighted SPL should be measured during vehicle passage at a distance of 7.5 m from the centerline of the vehicle path. The EEC rules (Directive 70/157/EEC) on the permissible sound level and the exhaust system of motor vehicles cover all motor vehicles with a maximum speed of more than 25 km/h. The limits range from 74 dBA for motor cars to 80 dBA for high-powered goods vehicles.

Many *motor sport venues* worldwide have introduced their own noise limits for vehicles using their tracks, normally set at 105 dB but at some sites in the United Kingdom in the range 95–98 dB. With the vehicle stationary this is measured at a three-quarters of rated, maximum, engine speed at 45 degrees and a set distance from the exhaust, relying on the use of hand-held meters operated by scrutineers. Unless care is taken to avoid sound reflection from surroundings, the results of such tests can be highly misleading, nor do they presently measure noise levels coming from engine induction systems. Some racetracks are now fitted with permanent, complex "drive- by" monitoring systems, which allow the authorities to measure peak and average noise levels to keep within the local limits of noise pollution.

## Note concerning testing of tire noise and tire noise classification

As noise from vehicle engines decreases the dominant source of noise for passenger vehicles traveling above 40 km/h becomes tire−pavement interaction noise in the dominant frequency range from 800 to 1200 Hz.

The limits of tire noise are laid down in UN ECE Regulation 117 and EC R661/2009. The testing of tire noise external to the vehicle is not entirely satisfactory but can form the basis for comparison and the noise part of the standard label required, since Nov 2012, to be affixed to all new tires sold within the EU. The EU tire label shows one, two, or three sound waves; one wave is the best performance, three is the worst and is the current limit, while two meets future planed legislation limits and one is a further 3 dB below that.

The other two EU labeled categories are for fuel efficiency and wet grip.

Tire noise testing is done on a vehicle using coast-by methods at 50 mph (80 kph). The measurements are made within a test zone in the form of a square measuring 20 m by 20 m, with microphones placed in the center of the zone at a distance of 7.5 m from the axis of motion of the vehicle. The track's surface should be made of an aggregate with a size of not more than 8 mm. Since this surface does not represent a typical high-speed road surface that might have significantly larger aggregate sizes of around 14−15 mm, the tests give a limited picture of the tires full noise profile. It is also possible that manufacturers could optimize their tires for these test conditions rather than more typical express-way surfaces.

# Part 2A. Introduction: Electromagnetic compatibility testing and its legislative environment

## Electromagnetic compatibility: The evolving task

*Note: In much of the expanding collection of literature concerning EMC testing, the object being tested, whatever its size and type, may be described variously by acronyms such as EUT (equipment under test) or DUT (device under test)—to avoid the confusion of the authors to use the generally understood unit under test (UUT) throughout this chapter and book.*

The spatial and operational separation of a vehicle's power and control subsystems, together with the selection and routing of the interconnecting looms, when all combined into one vehicle system, is referred to as the vehicle's electrical/electronic *architecture*; this creates its own unique electromagnetic *environment*.

There is no legal requirement to test the EMC of each and every automotive module since the manufacturer could obtain type approval for the whole vehicle. However, this is not the general practice of large vehicle manufacturers who will typically require each module and connecting loom of a vehicle's architectural assemblage to be tested to their own specifications, as individual modules, then subsequently as an assembled system, in order to:

- measure, in order to reduce, the EM emissions produced by a module or component during its full range of operation;
- measure, in order to increase, its level of immunity to externally generated electromagnetic radiation;
- measure a module's or component's immunity to damage by electrostatic discharge (ESD) in order to reduce its vulnerability;
- measure the EMC of the complete vehicle system, in its own and its external electromagnetic environment, under all operating conditions.

The electromagnetic environment that must be considered when designing these tests consists not only that generated by the vehicle system itself but also the surroundings in which it may operate. Real examples are when the vehicle is in proximity to beams coming from ground tracking radar at airports or the EM field close to large transformer installations at power stations. Along with the EMC of the vehicle's fixed modules, we also have to consider the function of devices used in or near vehicles that are not part of the vehicle, such as removable GPS systems, mobile phones, CB radios, and biomedical equipment such as cardiac implantable electronic devices, all of which need to be used, without causing or suffering from significant degradation in performance, at random locations within or close to the vehicles.

The new generation of electric and hybrid vehicles is producing a whole new source of RF noise and magnetic fields at higher energy levels than their predecessors. Safety-critical modules such as airbags are multiplying, as are wireless devices, such as forward ranging, collision avoidance radar, and tire pressure transducers, while outside the vehicle wireless battery charging via road embedded circuits are being proposed and tested.

The electronic environment that automotive test chambers need to deal with is made up of wide frequency ranges, which can include:

- FM radio from 70 MHz
- HD radio 535 kHz—108 MHz
- Cellular phones from 700 MHz to 60 GHz (3G, 4G-LTE, and 5G)
- Satellite transmissions: L-band (1–2 GHz) and S-band (2–4 GHz)
- Wi-Fi from 2.4/5.8 GHz
- Dedicated short-range communications 5.9 GHz
- Radio detection and ranging (RADAR), typically within the band of 77 to 79 GHz

Today's automotive EMC test chambers may also be required to test complex and developing vehicle systems, such as light detection and ranging, advanced driver assistance systems (ADAS), and RADAR. Therefore antenna arrays, as well as signal simulators, may have to be installed in a hybrid chamber economically able to use both EMC and antenna measurements to test the performance of these capabilities.

Repeatability of EMC tests can be a problem, particularly for manufacturers carrying out precompliance testing, even when apparently in full compliance with documented test methods. Therefore strict test discipline within a full compliant and unchanging test site using calibrated equipment and the same trained staff is essential.

Iterative test-fail-test cycles are cost and time consuming and are avoided only by close attention to the latest best practices at the circuit design stage.

In the world where some luxury cars have upward of 150 electronic control units installed, the design and implementation of EMC test regimes, at both component and vehicle levels, is becoming ever more demanding and complex and is requiring the employment of ever more specialist facilities and trained staff.

## Electromagnetic compatibility standards and legislation: "Work in progress"

Test facility designers and test engineers need know which of the many National, International, and Proprietary standards covering EMC and their associated test procedures are appropriate to their part of an automotive powertrain system or whole vehicle. The situation is complex and so rapidly developing that it seems inevitable that the legislation will continue to lag the technology it seeks to control.

For BEV and HEV technology to be generally adopted in society there needs standardization of not only safety standards and test procedures but of also physical interfaces. This is nowhere more important than in the vehicle battery charging infrastructure both in its "plug-in" and wireless power transfer forms.

In most of the world, ISO/IEC 17025 is the standard for which most automotive test facilities must hold accreditation in order to be deemed technically competent to issue type approval certification. In the United Kingdom, where approvals are issued through the VCA, there are currently a few exceptions to this requirement (Table 18.1).

The International Commission on Non-Ionizing Radiation Protection (ICNIRP) and directive "2013/35/EU EMF" issued by the European Agency for Safety & Health at Work seek to define the minimum H&S requirements regarding the exposure of workers to the risks arising from electromagnetic fields; there is, however, no specific mention of test procedures or vehicles.

Meanwhile Technical Committees TC106/GEL106 have been working on an automotive standard TS 62764-1 that applies to the assessment of human exposure to low frequency magnetic fields generated by and within automotive

**TABLE 18.1** Some of the major International Standards and some of their SAE equivalents. The key differences between the two standards concern detailed test methods rather than limits.

| International standard | SAE standard | Standard header details |
|---|---|---|
| CISPR 12:2009 | SAE J551-2 | Vehicles, boats, and internal combustion engines. Radio disturbance characteristics. Limits and methods of measurement for the protection of *off-board* receivers. |
| CISPR 25:<br>4th Edition 2016 | SAE J551-4<br>J1113-41 | Vehicles, boats, and internal combustion engines. Radio disturbance characteristics. Limits and methods of measurement for the protection of *on-board* receivers. |
| ISO 7637-2:2011 | SAE J1113-42 | Road vehicles—electrical disturbances from conduction and coupling, methods, and procedures. |
| ISO 11451-x 2015 | | Road vehicles—vehicle test methods for electrical disturbances from narrowband radiated electromagnetic energy. |
| ISO 11452-x 2015 | | Road vehicles—component test methods for electrical disturbances from narrowband radiated electromagnetic energy. |
| ISO 7637-1 2015 | | Road vehicles—electrical disturbances from conduction and coupling. |
| ISO 10605-2008 | | Road vehicles—test methods for electrical disturbances from electrostatic discharge. |

*CISPR*, Comité International Spécial des Perturbations Radioélectriques.

vehicles. This includes the electric vehicle supply equipment and the plugged charging cable sets supplier by the vehicle manufacturer and requires test facilities to use both batteries and battery emulators.

Separately IEC 61980-1:2015 has been drawn up to cover to define the procedures to be applied to "Inductive Charging" and the equipment designed for the wireless transfer of electric power from the grid network electric energy to the batteries (also referred to as rechargeable energy storage system) of electric vehicles. The ultimate goal of this technology is as a system that will allow dynamic electric vehicle charging, this will require test tracks/sites with embedded coil systems.

General automotive standards covering EMC are developed mainly by CISPR (Comité International Spécial des Perturbations Radioélectriques), the ISO and SAE. While CISPR and ISO are organizations that develop and maintain standards for use at the international level, the SAE develops and maintains standards mainly for use in North America.

CISPR/D is responsible for developing and maintaining the standards used to measure the *emissions* produced by vehicles and their components.

ISO/TC22 (Technical Committee 22) is responsible for developing and maintaining the standards used for *immunity* testing of vehicles and their components.

Both the ISO EMC standards and SAE EMC standards are mainly broken into categories, vehicle (SAE J551-xx/ISO 11451-xx) and component (SAE J1113-xx/ISO 11452-xx, ISO 7637-xx, and ISO 10605 for ESD).

In addition to the commercial EMC standards, there are also a series of more stringent military standards such as MIL-STD-461G (2015) that also applies to automotive components, subsystems, and vehicles.

In the United States the SAE spreads the task of defining standards between two committees:

- The *SAE EMI* standards committee covers *immunity* from EMI of automotive systems
- The *SAE EMR* standards committee covers *emissions* radiated from vehicle components that may cause interference to other devices.

In the United States all EMC test facilities able to perform the certification procedure of any product have passed accreditation by the Federal Communications Commission (FCC) and are recognized as a testing laboratory with the appropriate scope of accreditation. In Canada, ISED certification is required, however, in the past reports that show equipment complies with the US FCC standards have been accepted; this situation may be subject to change.

In Europe the situation is different as the CE marking protocols mean that producers may, in effect, self-certify components and modules, while *vehicles must be Type Approved* and "e" marked.

Various test strategies are used; due to the cost of test facilities many component manufacturers in the automotive industry will use (TEM cells, see later) or similar for precertification testing and debugging before then using one of several major international third-party laboratory such as HORIBA MIRA or TÜV Rheinland in Europe, or Eurofins MET Labs and Intertek in the United States, to carry out module or vehicle testing. This allows for the respective tests to be listed upon the ISO 17025 Scope of Accreditation.

An already complicated situation is further fragmented by the major automotive original equipment manufacturers (OEMs), including Ford, GM, and Daimler-Chrysler producing their own global and comprehensive EMC standards and test methods. While these are generally harmonized with, and referenced to, the SAE test methods, all contain detailed variations. The latest Ford document (FMC 1278) is (at the time of writing) 141 pages long and covers every aspect of EMC test methodology required by and of Ford and their Tier 1 suppliers.

## EU type approval "E" and "e" marking

The e-mark is an EU mark and is part of the vehicle homologation requirements for approved vehicles and vehicle components sold into the EU (see Fig. 18.5A). It is the type approval mark required by devices related to all safety-relevant functionality of the vehicle. Such devices will have to be e-mark certified by a certifying authority; manufacturers cannot self-declare and affix an e-mark, unlike most CE marking, where manufacturers' self-declaration is the norm.

Since late 2002, all electronic devices intended for use in vehicles, including aftermarket electronics products, are required to obtain formal type approval for products before placing them on the market. EN 50498:2010 specifies limits and methods of measurement for disturbance emissions and immunity characteristics of *aftermarket equipment.*

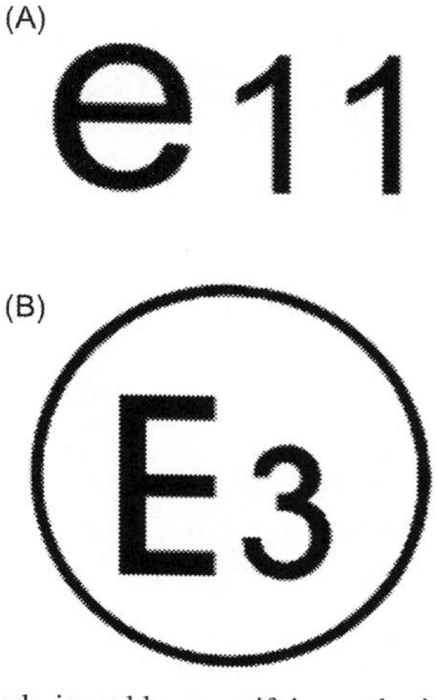


**FIGURE 18.5** (A) The "e" mark, issued by a certifying authority followed by the country code number: E1—Germany, E2—France, etc. 18.5 (B) The "E" mark followed by the country code, in this case for Italy.

The e-mark (Fig. 18.5B) is a United Nations mark for approved vehicles and vehicle components sold into the EU and some other countries under UN-ECE.

Both marks are affixed to approved products with other numbers relating to the legislation under which the approval has been gained.

## Part 2B. Electromagnetic compatibility testing

### Coupling mechanisms between electromagnetic interference source and "victim"

These terms such as "the electromagnetic environment" and "RF noise" can lead one to believe that almost all EMI is due to radiation and that most EMC testing is done in anechoic chambers in which vehicles are "zapped" with powerful beams of energy at RFs.

It is true that in EMC immunity tests vehicles and components are exposed to beams, in an anechoic cell, having frequencies typically between 10 kHz and 18 GHz, at field strengths of 100–200 V/m, and their own emissions may be checked, typically over a frequency range of perhaps 20 Hz up to 18 GHz at set distances of 1, 3, and 10 m.

However, much of the work of EMC test engineers is involved at the level of PCB and wiring loom design, where the coupling between noise source and its victim is frequently by one of the other coupling mechanisms shown in Fig. 18.6. The common results of the various coupling mechanisms give rise, in the vehicle, to phenomena such as voltage drop-outs and fluctuating or continuous disturbance, all of which have to be simulated and any problems found solved, by system redesign or shielding.

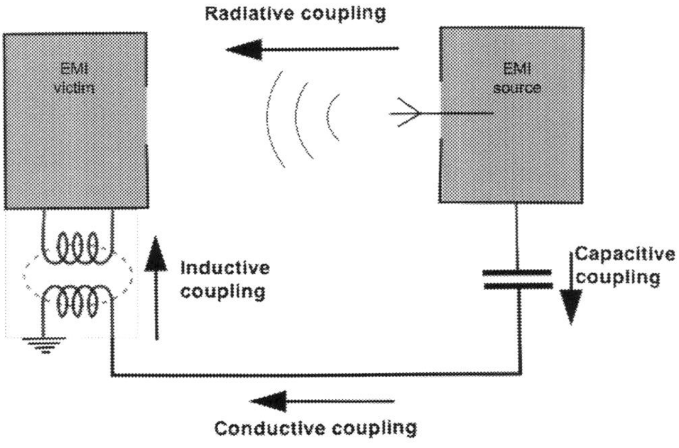


**FIGURE 18.6** Diagram showing the paths of the four different noise coupling mechanisms between an electromagnetic energy source and the victim.

The *permitted performance levels* will depend on the criticality of the subsystem. All vehicle manufacturers grade systems by "class," which range from those that are a convenience up to those that are safety critical. Then the systems are graded for their *function performance status* when subjected to EMI within test limits. Typically, these will be variations on the following grading:

1. The system will function as designed during and after exposure to the disturbance (fully immune).
2. Temporary functional deviation imperceptible to the operator. The system will not suffer permanent damage or loss in I/O parametric values or permanent reduction in functionality.
3. The system may deviate from its design performance within set limits or revert to a fail-safe mode during an exposure to the disturbance, provided it does not affect the safe operation of the vehicle. It should return to normal operation automatically and safely when the disturbance is removed.
4. As in 3 above, but operator action, such as cycling through ignition key functions or replacing a fuse, may be required to return the function to normal after the disturbance is removed.

*Note*: Test engineers need to be aware of ISO 26262 which is a risk-based safety standard that is derived from the International IEC 61508 standard. It applies to electric and/or electronic systems in production vehicles, and the automotive safety integrity level is a key component of that safety standard. Amongst the many strictures of these standards is the statement that none of the systems that are permitted reduction of functionality may have a "knock-on" effect on systems of a higher class that can affect vehicle safety.

## Common or standard tests

The common tests carried out on all products deemed required to meet such legislation as "Electromagnetic Compatibility 2014/30/EU" are

- radiated immunity
- conducted RF immunity
- immunity to ESD
- immunity to electrical fast transient/burst
- surge immunity
- immunity to voltage dips and interruptions
- immunity to magnetic fields

### *Electrostatic discharge or pulse interference*

Before major modules or vehicles are subjected to full EMC testing, it is required practice to test their components and subassemblies for their immunity from damage, or reduction in functionality, caused by ESD.

ESD testing is done at a specifically designed workstation outside an anechoic chamber.

Materials that make up a modern vehicle and its clothed human users vary in their ability to generate and hold an electrical charge through triboelectric charge generation. The antistatic property of a material defines its ability to resist being charged triboelectrically.

During normal vehicle use there can be operations, such as slipping the clutch, when a high triboelectric charge can be generated and actions, such as refueling from a plastic gasoline container that has been sliding on a trunk carpet for hours, might create spark generating, safety-critical situations.

*To deal with such scenarios the EMC test engineer has to anticipate, test, and mitigate.*

Electronic components and modules can be at risk from ESD during installation and maintenance, which is why work areas and tools are chosen to have high "resistivity," the quality that defines the ability to dissipate charge. The human body appears near the top of the + tribological series; therefore workers involved in the construction of PCBs and their testing have to wear earthing wrist straps to prevent a charge building up.

The US-based EOS/ESD Association is an ANSI-recognized standards development organization that coordinates test methods and all matters relating to the protection of equipment from the effects of ESD. For example, ESD Association Standard S5.1 is entitled "Human Body Model (HBM) Electrostatic Discharge Sensitivity Testing" and it covers a procedure for performing ESD susceptibility tests. There is also a European electrostatic standard reference: EN100015.

The most common ESD test procedures are specified in ISO 10605-2008 which applies specifically to road vehicles. The component under test is placed on a wooden table topped with a horizontal coupling plane that is connected to the ground reference plane through a cable that includes bleed resistors of 470K at each end. The component is insulated from the coupling plane by a thin plastic sheet, and the pulses are applied at points judged to be appropriate by an experienced tester.

The test voltages used in most harmonized standards are 4 kV for direct probe contact and 8 kV for air discharge (spark), but many OEM and military tests use voltages of 15 or 25 kV for modules that are directly accessible from outside the vehicle, such as keyless entry devices or diagnostic system plug sockets.

## *Assessment of human exposure to low frequency magnetic fields*

IEC/TS 62764-1:2019 covers the measurement procedures of low frequency magnetic field levels generated by electrical equipment in the vehicle's environment with respect to human exposure. The frequency range covered goes down to 1 Hz and the traction current fluctuations experienced in the vehicle are typically in the 1 Hz to 3 kHz band. Cable routing and layouts involving twisting will feature in testing of human exposure to such fields since shielding only becomes effective at the upper end of the frequency range.

### *The implications of a 5G future*

As the technology advances to cover the future, vehicle to vehicle and 5G infrastructure legislation will continue to chase the technology. Currently, the relevant legislation UNECE Reg. 10 is limited to 1 GHz emissions and 2 GHz immunity; however, the planned 5th edition IEC CISPR 25 has plans to raise this to 5925 MHz.

There are currently (Jan 2020) no test methods to deal with the 26 GHz mm waveband or the 40 GHz frequency proposed for 5G by such organizations as the International Amateur Radio Union.

## Unit under test to antenna separation distances, near field, and far field

*Near field* measurements of electric or magnetic fields can be defined as being done at a separation distance of within one wavelength of the studied frequency. It is extremely difficult to accurately measure the absolute field strength in the near field because of the complex and spatially variable relationship between the E (electric) and H (magnetic) fields. Therefore much legislative testing is carried out at distances of 3, or preferably, 10 m which is in the *far field.* Measuring at 10 m means that measurements of a stable wave of the electric field strength are taken at least one wavelength away at a frequency of 30 MHz.

EN 55011/CISPR 11 states that for radiated emission testing, a 3 m separation test site can be used for "small equipment" defined within the standard.

## Electromagnetic compatibility anechoic radio frequency test facilities

As with acoustic anechoic cells, the cost of EMC anechoic cells increases steeply with the increase in the usable volume; therefore the designers at project specification stage (Chapter 1: Test Facility Specification, System Integration, and Project Organization) must carefully consider the maximum size of the UUT to be installed and the range frequencies being used.

The UUT will range from components, which may be tested in small gigahertz transverse electromagnetic (GTEM) "boxes" (as given later), to full vehicles ranging from bikes to trucks that will require full sized chambers. Further capital cost rises are incurred when there is the need for running vehicles under load, requiring a chassis dynamometer, and/or rotation of the vehicle, requiring a turntable (see "installed plant" as given later).

If the UUT is a prime mover, either a motor or ICE, then a dynamometer, mounted externally to the cell, will be required. If electric motor modules are being tested, the dynamometer will need to be a high-speed four-quadrant machine and will require an intermediate bearing block, usually within the cell, as discussed earlier with acoustic cells.

The test frequencies to be generated and/or studied will have an effect on cell size and in some designs the shape; this is a technical environment undergoing rapid development.

The EMC testing of vehicles incorporating the developments of advanced driver-assistance systems (ADAS), whose antennas are highly integrated with the automotive body will require a hybrid chamber capable of measuring the signal path of wireless over-the-air (OTA), and antenna pattern measurements.

All these developments demand a well-audited functional specification and supporting business plan for any project producing a new or modified EMC test facility.

Since EVs and communication systems are rapidly developing, consideration has to be made, so far as is possible, for future proofing the proposed cell and its equipment.

It is recommended that EMC chamber expertise is sought at the concept stage and that the project design is organized using an expert EMC design system within a building information modeling (BIM) software suite so that the core and special structural requirements are correctly incorporated within the whole project.

## Other electromagnetic compatibility test spaces

The testing of small components and printed circuit boards EMC testing can be carried out in one or more types of *transverse electromagnetic mode (TEM)* containment "box" that can be used to establish a standardized EM field within a small shielded environment.

In emission and immunity testing within a cell or OATS, in order to ensure far field conditions, a minimum distance of 3 m between antenna and UUT is required, but since far field conditions, having an impedance of 377 W, also describe a TEM wave, the alternative smaller test environments that provide TEM wave propagation are accepted for EMC tests.

While the great advantage of this family of test devices is that they are very much cheaper than full SACs and faster to set up than most OATS testing, their limitation is usually that the upper usable frequency is restricted by the physical dimensions.

A *GTEM* cell is a high frequency version of the TEM cell and is used as an alternative facility for EMC testing. According to IEC61000-4-3 the standard environment for radiated immunity tests is a screened enclosure, "large enough to accommodate the UUT whilst allowing adequate control over the field strength." The GTEM cell takes the form of a tapered box section with a wide end termination consisting of a 50 Ω resistor board for low frequencies and pyramidal absorbers for high frequencies. Within the enclosure an internal conductor surface "the septum" runs along the upper third of the cell, and the test volume is beneath it. Commercially available GTEM cells have a septum height between 250 and 2000 mm and have outer dimensions of between 1.4 and 8 m in length, 0.7 and 4.5 m in width, and 0.5 and 3.2 m in height.

Research [1] indicates that GTEM cells can be used up to 2.3 GHz for immunity tests and at least up to 4.2 GHz for emission tests.

## Reverberation chambers (RVC) for electromagnetic compatibility testing

A RVC or mode-stirred chamber can be used for a more limited range of automotive EMC testing than a full anechoic chamber, but they are often capable of working at much higher RF field energy levels; their use is covered by IEC 61000-4-21.

They normally take the form of a rectangular enclosure, with perfectly conducting walls, inside which the electromagnetic field distribution is characterized by a standing-wave pattern created by the reflection from the walls. However, it is possible to change the standing-wave pattern within the cavity with "stirrers," which take the form of zigzagged reflectors mounted on a rotating shaft; these effectively change the boundary of the chamber inside a RVC. There are cost advantages of using a RVC:

- The high conductivity allows for high field power levels to be generated from moderate levels of input power.
- The chamber construction costs are lower than an anechoic chamber because no expensive and heavy absorbing lining is needed.
- The RVC provides a realistic simulation for electronics functioning within a cavity such as an electrical cabinet. Currently (2020), a new standard ISO 11451-5 is being developed for vehicle reverberation testing.

The largest and highest powered chambers are used for military work, while automotive versions tend to be used only in subassembly testing and have internal volumes of around 60 $m^3$ with a door that is a size capable of allowing a 2.2-m-tall electrical cabinet to be installed.

Once again, it is recommended that any building project incorporating an EMC chamber is organized using an expert EMC design system within a BIM software suite so that the core and special structural requirements are correctly incorporated within the whole project.

## Electromagnetic compatibility facility location, structure, and lining

As discussed earlier in this chapter, acoustic anechoic cells have to be isolated from external noise and vibration; for the same reasons of test integrity, anechoic cells working in the RF waveband have to be insulated and isolated from electromagnetic noise coming from outside the test field. Also similar to acoustic, for anechoic testing the reference, is the ideal OATS with a conductive ground plane which for EMC is defined in IEC CISPR 12. The test cell structure is built to form a Faraday cage, the shielding skin being formed from solid metal sheeting with absolutely no gaps in the construction.

## Electromagnetic compatibility anechoic cell linings and support structures

The lining materials of EMC anechoic chambers are designed to absorb radio wave and microwave energies and are referred to as "radar absorbing material" (RAM).

Since all the performance of all types of anechoic lining declines as the angle of incidence of the waves deviates from normal (90 degrees), "the bigger the better" for chambers of a particular CISPR separation designation.

RAM can be made from two distinctly different types of material, although both types are commonly met in different areas of the same chamber:

1. Closed-cell polyurethane or polystyrene foam molded into a steep pyramidal shape and "dosed" (impregnated) with carbon, then painted with a fire-resistant paint. The principle of this type of absorber is to mimic free space by "progressive impedance"; that is, absorbing the radio or microwave by internal reflection through a resistive structure that has a progressively larger shape, during which the wave energy is being converted into heat. The absorbing performance is directly related to the wavelength of the energy, and the figure quoted (in dB) relates to waves with a normal angle of incidence.
2. Ferrite tiles, which are usually made as 100 mm × 100 mm × 6 mm thick pieces with a central fixing hole. They are made from sintered iron/nickel material, which is ground to a precise shape so that large wall sections can be precisely assembled without intertile gaps. Since they are of about the same density as steel, the chamber support frame and roof structure need to be suitably substantial and they have to be attached to a backing layer. In some forms the tiles are supplied in panels ready assembled with a tuned dielectric backing layer. Ferrite linings work by "magnetic permeability" and operate well at lower frequencies, where the wavelength is too long for pyramidal absorbers to work effectively.
3. Hybrid linings and mixed-cell linings. The attenuating performance of ferrite tiles is modified by placing wedge- or pyramid-shaped absorbers in front of them, and hybrid linings are often used where the chamber needs to operate at frequencies above 1000 MHz. Mixed-cell linings of both ferrite tiles and pyramids are used in many large automotive cells with the patches of pyramid or hybrid lining in large patches at a normal incidence to the vehicle under test and the ferrite tiles over the remainder of the surface (see Fig. 18.7).

**FIGURE 18.7** A vehicle mounted on a chassis dynamometer within a large EMC test chamber lined with pyramidal absorption material. The log-periodic antenna is fitted to a wheeled support frame. Photograph courtesy of Horiba-Mira Ltd.

All support structures and fittings used inside EMC cells should be made of nonconductive material with low dielectric constant factors such as some modern polymers, although not ideal wood has been widely used because of its low cost.

## Auxiliary plant installed within an electromagnetic compatibility anechoic cell

Large automotive EMC cells may contain a $+/-$ 190° turntable that allows the test vehicle, or any other mounted UUT, to be turned in a directional energy beam or present a different aspect to a receiver. In the most complex facilities a chassis dynamometer is mounted within a turntable; this allows a vehicle, running under load, to be presented at 360° to the energy beam or receiver.

Whatever the configuration of its mounting, most EMC chassis dynamometer installations, up to the recent past, had DC drives rather than alternating current (AC) because the latter can produce high levels of RF noise; however, recent design practice means that all drive cabinets may be shielded below ground by concrete. The most modern designs use completely shielded AC motor designs. Whatever the technology used, such dynamometer installations should have no detectable RF-radiation in the frequency range between 30 kHz and 3 GHz. Such large, complex, and expensive chambers may have dimensions, internal to the lining, of 20 m × 13 m × 8 m high.

For radiated emission measurement, antennas are used as transducers. The standard that defines the requirements for antennas to be used in EMC measurement is currently CISPR 16-1-1: 2019 which specifies the characteristics and performance of equipment for the measurement of radio disturbance in the frequency range 9 kHz to 18 GHz.

CISPR 16 lays down the requirements that have to be met by fully and semi-anechoic chambers designed for EMC testing and often have a subdesignation relating to the distance of separation between the UUT and the antenna. Thus chambers may be designated CISPR 16 − 10 m or CISPR 16 − 3 m.

## Health and safety in radio frequency cells

As mentioned in the first part of this chapter, there are H&S risks posed by all closed anechoic test chambers, the interiors of which can be disorienting and have exit routes that may not be obvious to the inexperienced user. Managers and operators should always be aware that first-time visitors may suffer from mild or even severe claustrophobia and be aware of the need to get the sufferer into the open air quickly.

The health effects of exposure to high levels of RF energy have been the subject of much discussion for years; health scares led to rules covering the leakage from microwave ovens and still lead to concerns about 5G mobile telephone mast locations today. One of the major sets of guidelines is that issued by the International Council on Non-Ionizing Radiation Protection (ICNIRP).

It is not disputed that there are two health effects of exposure to high-energy RF fields, which are heating of the human body and RF shocks and burns (electrostimulation) resulting from direct contact with an RF radiator such as an antenna.

Although some scientists believe that exposure to low-level RF fields for long periods of time can result in other harmful effects, none of these non-thermal effects are currently incorporated in major standards or guidelines.

The key H&S requirement of EMC cells is the control of access and ensuring that unsupervised, solo work is not carried out in any chamber, be it in an open or closed state.

Each test facility has to develop and strictly enforce their own rules, based upon published guidelines and according to the type of work being carried out, which might range between checking car radio reception during engine cranking, to testing military equipment for immunity against the energy pulse of a nuclear weapon.

## Fire suppression in electromagnetic compatibility anechoic cells

The fire load in any closed test cell depends to a large degree on the nature of the UUT. The reader is referred to the section in this chapter entitled "Fire detection and suppression in anechoic cells" and sections in Chapter 3,

Test Facility Design and Construction, which discuss the requirements for the ventilation of closed cells and the various types of fire suppression strategies recommended for all automotive test facilities.

Since EMC testing involves work, not only in large sealed cells but also in small "boxes" and open tabletops, all operatives must be trained in the correct operation of hand-held fire extinguishers; see section entitled "Handheld Fire Extinguishers" in Chapter 3, Test Facility Design and Construction.

## References

[1] A. Nothofe, et al., The Use of GTEM Cells for EMC Measurements, National Physical Laboratory, January 2003. ISSN 1368-6550.

## Further reading

M.J. Crocker (Ed.), Handbook of Noise and Vibration Control Hardcover, John Wiley & Sons, Inc., 2007. 23 Oct, ISBN-13: 978-0471395997.

Sandberg, Ejsmont, Tyre/Road Reference Book, Informex, 2002. ISBN-13: 978-9163126109.

T. Williams, EMC for Product Designers, fifth revised ed., Newnes, 2016. 28 Sept. ISBN-13: 978-0081010167.

Chapter 19

# The pursuit and definition of accuracy

## Chapter outline



## Introduction

It is a repeated statement of the obvious that the purpose of engine testing is to produce data, and that the value of that data, particularly when transformed into "information," depends critically on its accuracy. Yet modern instrumentation, data processing, simulation, and modeling have tended to mask or even obscure questions of accuracy and to give an illusion of precision to experimental results that is often totally unjustified. In some people's minds, "repeatability" of test results is used as a proxy for the accuracy of those results, which is a false concept since, as can be seen in Fig. 19.1, it is possible to be repeatably and precisely wrong. Taking a wider view of repeatability of test results, it should be remembered that much of the latest legislation and testing of the automotive powertrain has been aimed at correlation with real life. It will not have escaped the reader's attention that vehicular journeys in real life, at any scale, are almost by definition unrepeatable; so like the detailed data within the Real Driving Emissions test results have to be subjected to a simple "did not exceed" test or to some degree of statistical analysis.

In addition to test sequence variability, the measurement accuracy of individual instruments is another part of the total combined measurement uncertainties,

Engine Testing. DOI: https://doi.org/10.1016/B978-0-12-821226-4.00019-X

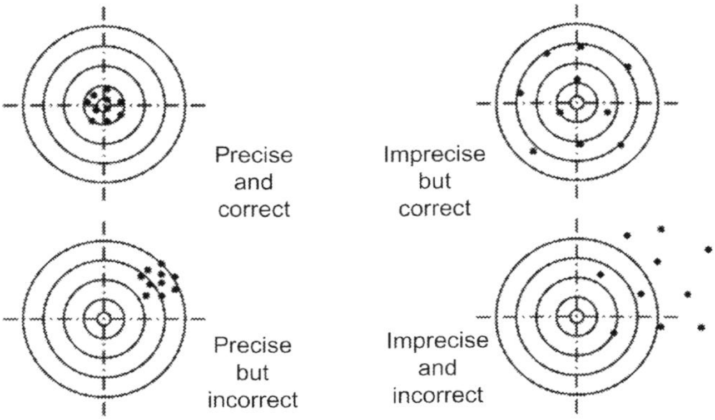


**FIGURE 19.1** A "target" diagram that seeks to disambiguate the meaning of the terms precise, imprecise, correct, and incorrect. *Redrawn from the guide to the Expression of Uncertainty in Measurement.*

a subject that is addressed in the Guide to the Expression of Uncertainty in Measurement (GUM), JCGM 100:2008 and ISO 3534-1:2006 [1].

Along with accuracy and precision, instruments will have an inherent measurement resolution that is the smallest change in the physical quantity that produces a change in the measurement. Appropriate accuracy in the test process presents the biggest challenge facing the test engineer, who, along with a complete understanding of the unit under test, must master the following skills:

- experience in the correct application of instruments;
- knowledge of methods of calibration and an awareness of the different kinds of error to which instruments are subject, both individually and collectively (as part of a complex system);
- a critical understanding of the relative merits and limitations of different methods of measurement and their applicability to different experimental situations;
- an understanding of the differences between true and observed values of experimental quantities;
- an appreciation that repeatability and accuracy are separate and distinctly different attributes of data;
- an appreciation of the effects on raw data of the signal processing chain;
- an appreciation of the changes caused by variations in climatic conditions and in the support services of the test laboratory; and
- an appreciation of the importance of data channel identification, data storage and retrieval, and postprocessing systems.

This is a very wide range of skills, only to be acquired by experience.

*The first essential is to acquire, as a habit of mind, a sceptical attitude to all experimental observations: all instruments tend to be liars.*

For those test engineers engaged in tests where the data being collected is within the "noise band" of the instrumentation, such as some ultralow emission work, it is necessary that they develop a mindset that seeks answers to help verify the work rather than problems that negates it.

## The display of data during testing: analog versus digital

Display screens in the control room provide near instantaneous but nonpermanent communication of information between the test processes and a human observers so, as the display of data is the final link, before storage, between the measuring process and the test engineer, it needs to be in a form that is appropriate to the task expected of that observer.

Data can be presented to the tester, through display screens or by discrete instruments, in either an analog or a digital form, and it is vital to design any instrument display in such a way that the advantages of each are exploited to the best effect of the testing being carried out.

Digital displays show the measured value as digits, the apparent accuracy of which depends on the number of decimal points below unity displayed, but since digital displays are able to display very small values in an easily discernible form, they can be said to be more accurate than analog displays. When dealing with data channels that display quasistatic data digital displays, they are easier to use because they display a numeric value.

However, the digital display of a value gives an illusion of accuracy that may be totally unjustified. The temptation to believe that a reading of, say, 97.12°C is to be relied upon to the second decimal place is very strong, but in the absence of convincing proof, it must be resisted. An analog indicator connected to a thermocouple cannot, in most cases, be read to closer than 1°C and the act of reading such an indicator is likely to bring to mind the many sources of inaccuracy in the whole temperature measuring system. Replace the analog indicator by a digital instrument reading to 0.1°C, and it is easy to forget that all the sources of error are still present: the fact that the readout can now discriminate to 0.05°C does not mean that the overall accuracy of the measuring system is producing a reading within comparable limits.

Analog display of data has a different message to give to the observer to that of a digital display, because it shows how far the measured point is from other values above and below it, along with the rate and direction of change. This difference between the ways that the human brain processes digital and analog displays of numerical information is far from subtle and should be considered carefully by control equipment designers; when rate of change or proximity to a critical value needs to be observed, an appropriately configured analog display is easier for most staff to use intuitively.

Consider the classic examples of the display of the "artificial horizon" and rate of change of altitude in a traditional aircraft cockpit; in both cases, while the values, angle from horizontal and altitude, have a specific value, it is the rate and direction of their change that is important to the pilot. From the very first ICE test cells until the present day, the values of speed and torque have been shown in analog form, more recently in both analog and digital forms to serve two different functions largely driven by the way in which the human brain deals with those forms.

## Some general principles

1. While cumulative measurements are generally more accurate than rate measurements, when taken over a period of fluctuating engine powers, they do not indicate where high and low values are recorded. Examples are the measurement of speed by counting revolutions over a period of time and the measurement of fuel consumption by recording the time taken to consume a given volume or mass.
2. Be wary of instantaneous readings. Few processes are perfectly steady. This is particularly true of flow phenomena, as will be clear if an attempt is made to observe pressure or velocity head in a nominally steady gas flow. A water manometer fluctuates continually and at a range of frequencies depending on the scale of the various irregularities in the flow. No two successive power cycles in a spark-ignition engine are identical. If it is necessary to take a reading instantaneously, it is better to take a number in quick succession and statistically analyze them.
3. Be wary of signal damping. Often, it is entirely sensible but sometimes leads to serious misrepresentation of the true nature of a fluctuating value.
4. A related problem: "time slope effects." If one is making a number of different observations over a period of nominally steady-state operation, make sure that there is no drift in performance over this period.
5. Individual measurement events must be recorded in the correct order for the correct causal deductions to be made; the more complex the data streams and number of sources, the more difficult this becomes.
6. The closer one can come to an "absolute" method of measurement, for example, pressure by water or mercury manometer or deadweight tester, force, and mass by deadweights (plus knowledge of the local value of $g$; see Chapter 9: Dynamometers: The Measurement and Control of Torque, Speed, and Power), the less the likelihood of error.
7. In instrumentation, as in engineering generally, simplicity is a virtue. Each elaboration is a potential source of error.

## Definition of terms relating to accuracy

This is a particularly "gray" area [2,3]. The statement "this instrument is accurate to $\pm 1\%$" is in fact entirely meaningless in the absence of further definition. Table 19.1 lists the terms commonly used in any discussion of accuracy.

**TABLE 19.1 Definitions of various terms that are used in any discussion of accuracy.**

| | |
|---|---|
| Error | The difference between the value of a measurement as indicated by an instrument and the absolute or true value. |
| Sensing error | An error arising because of the failure of an instrument to sense the true value of the quantity being measured. |
| Systematic error | An error to which all the readings made by a given instrument is subject; examples of systematic errors are zero errors, calibration errors, and nonlinearity. |
| Random error | Errors of an unpredictable kind. Random errors are due to such causes as electrical supply "spikes" or friction and backlash in mechanisms. |
| Observer error | Errors due to the failure of the observer to read the instrument correctly or to record what he or she has observed correctly. |
| Repeatability | A measure of the scatter of successive readings of the same quantity. |
| Sensitivity | The smallest change in the quantity being measured that can be detected by an instrument. |
| Precision | The smallest difference in instrument reading that it is possible to observe. |
| Average error | Take a large number of readings of a particular quantity, average these readings to give a mean value, and calculate the difference between each reading and the mean. The average of these differences is the average error; roughly half the readings will differ from the mean by more than the average error and roughly half by less than the average error. |
| Correlation error (as defined in this book and by common practice in the test industry) | The difference between the values of a measurement made of the same parameter (under the same claimed test conditions) in two different cells or at two different sites. |

## An examination of statements regarding accuracy

We must distinguish between the accuracy of an observation and the claimed accuracy of the instrument that is used to make the observation. Our starting point must be the true value of the quantity to be measured (Fig. 19.2A).

The following are a few examples of such quantities associated with engine testing:

- fuel flow rate
- air flow rate
- output torque
- output speed
- pressures in the engine cylinder
- pressures in the inlet and exhaust tracts
- gas temperatures
- motor phase voltages and currents
- DC currents and voltages including ripple

By their nature, none of these quantities are perfectly steady, and decisions as to the sampling period are critical.

## Sensing errors

These are particularly difficult to evaluate and can really only be assessed by a systematic comparison of the results of different methods of measuring the same quantity.

Sensing errors can be very substantial. In the case discussed earlier regarding the measurement of exhaust temperature, the author has observed errors of over 70°C while balancing fuel pumps on large diesel engines (Fig. 19.2B).

Typical sources of sensing error include the following:

1. mismatch between temperature of a gas or liquid stream and the temperature of the sensor,
2. mismatch between true variation of pressure and that sensed by a transducer at the end of a connecting passage,
3. inappropriate damping or filtration applied within the instrument's measuring chain,
4. failure of a transducer to give a true average value of a fluctuating quantity,
5. air and vapor present in fuel systems,
6. time lag of sensors under transient conditions,
7. errors arising from the necessity of sensing and amplifying very small source signal values, along with phase delay errors associated with signal filtering, and
8. errors due to positioning, droop/offset of non-permanent, clamp-on sensing devices for electrical signals.

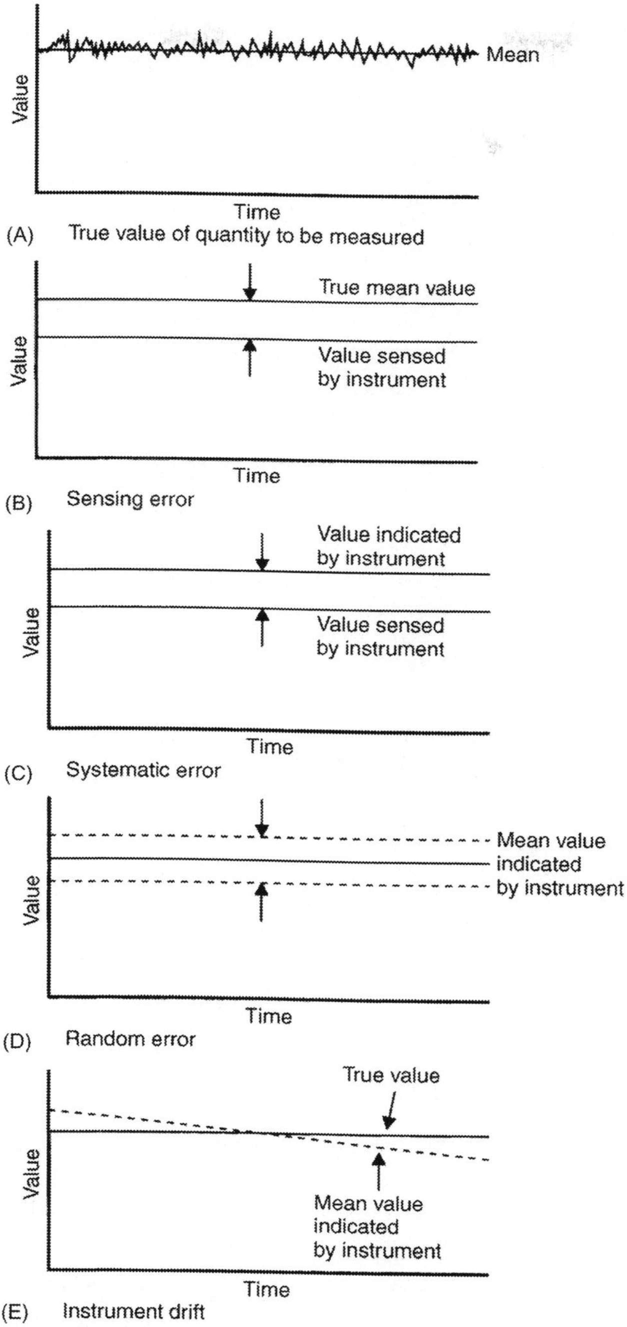


**FIGURE 19.2** Various types of instrumentation errors. (A) true value to be measured plotted against time—on the same time scale: (B) sensing error, (C) systematic error, (D) random error, and (E) instrument drift.

Note that the accuracy claimed by its manufacturer does not usually include an allowance for sensing errors. These are the responsibility of the user.

### Systematic instrument errors

Typical systematic errors (Fig. 19.2C) include the following:

1. Zero errors—the instrument does not read zero when the value of the quantity observed is zero.
2. Scaling errors—the instrument reads systematically high or low.
3. Nonlinearity—the relation between the true value of the quantity and the indicated value is not exactly in proportion; if the proportion of error is plotted against each measurement over full scale, the graph is nonlinear.
4. Dimensional errors—for example, the effective length of a dynamometer torque arm may not be precisely correct.

### Random instrument errors

Random errors (Fig. 19.2D) include:

1. effects of stiction in mechanical linkages,
2. effects of friction, for example in dynamometer trunnion bearings,
3. effects of vibration or its absence, and
4. effects of electromagnetic interference (see Chapter 4: Electrical Design Requirements of Test Facilities and Chapter 18: Anechoic and Electromagnetic Compatibility Testing and Test Cells).

### Instrument drift

Instrument drift (Fig. 19.2E) is a slow change in the calibration of the instrument as the result of:

1. changes in instrument temperature or a difference in temperature across it;
2. effects of vibration and fatigue;
3. fouling of the sensor, blocking of passages, etc.; and
4. inherent long-term lack of stability.

From all of the above, it will be clear that it is entirely straightforward to give a realistic figure for the accuracy of an observation.

## Instrumental accuracy: manufacturers' claims

One of the significant variables in all power train testing is ambient (atmospheric) conditions, as sensed in the test cell; therefore any instrument specification should qualify the conditions under which the measurements were made. Manufacturers often state instruments are corrected for "standard conditions" without specifying them, leading to confusion and errors.

In Chemistry the International Union of Pure and Applied Chemistry (IUPAC) the *standard condition* for temperature and pressure (STP) since 1982 is defined as a temperature of 273.15K (0°C, 32°F) and an absolute pressure of exactly $10^5$ Pa (100 kPa, 1 bar).

The National Institute of Standards and Technology (NIST) of the United States uses a temperature of 20°C (293.15K, 68°F) and an absolute pressure of 1 atmosphere (14.696 psi, 101.325 kPa). This standard is also called *normal temperature and pressure* (abbreviated as NTP)

Looking critically at the statement "the instrument is accurate to within ± 1% of full-scale reading," at least two different interpretations are possible:

1. No reading will differ from the true value by more than 1% of full scale. This implies that the sum of the systematic errors of the instrument, plus the largest random error to be expected, will not exceed this limit.
2. The average error is not more than 1% of full scale. This implies that about half of the readings of the instrument will differ from the average by less than 1% full scale and half by more than this amount. However, the definition implies that systematic errors are negligible: we are only looking at random errors.

Neither of these definitions is entirely satisfactory. In fact, the question can only be dealt with satisfactorily on the basis of the mathematical theory of errors.

## Uncertainty

While it is seldom mentioned in statements of the accuracy of a particular instrument or measurement, the concept of uncertainty is central to any meaningful discussion of accuracy. Uncertainty is a property of a measurement, not of an instrument. The uncertainty of a measurement is defined as the range within which the true value is likely to lie, at a stated level of probability.

It is noteworthy that in most scientific disciplines, data, when presented graphically, has each data point represented by a cruciform indicating the degree of uncertainty or error band; this practice is rarely seen in engine test data and helps to confirm an unjustified illusion of exactness.

The level of probability, also known as the confidence level, most often used in industry is 95%. If the confidence level is 95%, there is a 19:1 chance that a single measurement differs from the true value by less than the uncertainty and one chance in 20 that it lies outside these limits.

If we make a very large number of measurements of the same quantity and plot the number of measurements lying within successive intervals, we will probably obtain a distribution of the form similar to theoretical curve that is known as a normal or Gaussian distribution (Fig. 19.3). This curve is derived from first principles on the assumption that the value of any "event" or measurement is the result of a large number of independent causes (random sources of error).

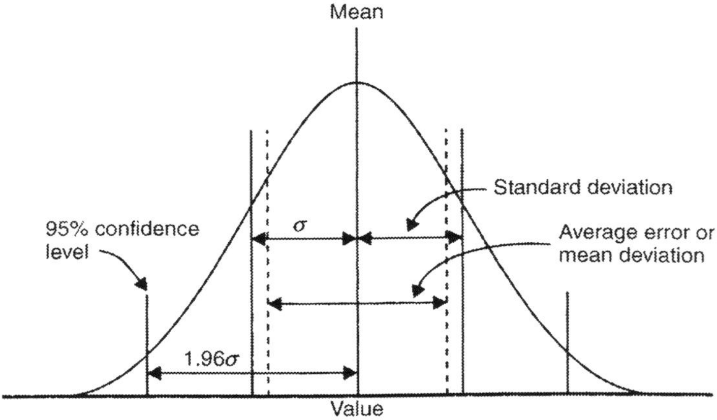


**FIGURE 19.3** Normal or Gaussian distribution.

The normal distribution has a number of properties shown in Fig. 19.3.

- The *mean value* is simply the average value of all the measurements.
- The *deviation* of any given measurement is the difference between that measurement and the mean value.
- The $\sigma^2$ (sigma$^2$) is equal to the sum of the squares of all the individual deviations divided by the number of observations.
- The *standard deviation* $\sigma$ is the square root of the variance.

The standard deviation characterizes the degree of "scatter" in the measurements and has a number of important properties. In particular, the 95% confidence level corresponds to a value sigma = 1.96. Ninety-five percent of the measurements will lie within these limits, and the remaining 5% in the "tails" at each end of the distribution.

In many cases the "accuracy" of an instrument as quoted merely describes the average value of the deviation, that is, if a large number of measurements are made, about half will differ from the true or mean value by more than this amount and about half by less. Mean deviation = $0.8\sigma$ approximately.

However, this treatment only deals with the random errors; the systematic errors still remain. To give a simple example, consider the usual procedure for checking the calibration of a dynamometer torque transducer. A calibration arm, length 1.00 m, carries a knife-edge assembly to which a deadweight of 10.00 kg is applied. The load is applied and removed 20 times and the amplifier output recorded. This is found to range from 4.935 to 4.982 V, with a mean value 4.9602 V. At first sight, we could feel a considerable degree of confidence in this mean value and derive a calibration constant:

$$k = \frac{4.9602}{10 \times 9.81 \times 1.0} = 0.050563 \text{ V/N m}$$

The 95% confidence limit for a single torque reading may be derived from the 20 amplifier output readings and, for the limiting values assumed, would probably be about $\pm 0.024$ V, or $\pm 0.48\%$, an acceptable value.

There are, however, four possible sources of systematic error:

- The local value of $g$ may not be exactly $9.81 \text{ m/s}^2$ (see Chapter 9: Dynamometers: The Measurement and Control of Torque, Speed, and Power).
- The mass of the deadweight may not be exactly 10 kg.
- The length of the calibration arm may not be exactly 1.00 m.
- The voltmeter used may have its own error.

In fact, none of these conditions can ever be fulfilled with absolute exactness. We must widen our 95% confidence band to take into account these probably unknown errors.

This leads straight on to our next topic.

## Traceability

The term traceability tends to mean different things in practice to hardware and software engineers.

So far as the test facility's calibration engineer is concerned, calibration traceability refers to an unbroken chain of comparisons relating an instrument's measurements to a known and acknowledged standard. The calibration of that instrument to a traceable standard determines that instrument's level of precision and accuracy.

Traceability in software testing, which is not a subject of this book, is the ability to record and trace tests forward and backward through the development lifecycle. Tests are traced forward to test runs and backward to requirements. Traceability in software testing is often done using a traceability matrix.

Many countries of the world have their own national standards for weights and measures that are maintained by a National Metrological Institutes that provide the highest level of standards for the measurement traceability in that country. In the United Kingdom the National Physical Laboratory (NPL) is the national measurement standards laboratory while in the United States the NIST fulfills that and a wider industrial support role.

In a well-run test department, QA certified to ISO 9001:2015 (see Chapter 2: Quality and Health and Safety Management) traceability of calibration will be a well-defined part of the procedures. The standard of work produced by any test facility is directly proportional to the care taken in keeping their instruments, from torque-spanners and rev-counters to dynamometers and fuel weighers, correctly calibrated.

The full "traceability ladder" of, for example, the deadweight referred to in the example of the dynamometer calibration might look like this:

- the international kilogram (Paris)
- the British copy (NPL)

- national secondary standards (NPL)
- local standard weight (portable)
- our own standard weight

*Note*: The kilogram (kg) is the SI unit of mass and its definition, as of May 20, 2019, (to which the authors find it difficult to relate) is:

"The kilogram is defined by taking the fixed numerical value of the Planck constant h to be $6.62607015 \times 10^{-34}$ when expressed in the unit J·s, which is equal to $kg \cdot m^2 \cdot s^{-1}$, where the meter and the second are defined in terms of c and $\Delta\nu_{Cs}$." (It is also very close to the mass of 1 L of water!)

## Combination of errors

Most derived quantities of interest to the test engineer are the product of several measurements, each of them subject to error. Consider, for example, the measurement of specific fuel consumption of an engine. The various factors involved and typical 95% confidence limits could be as follows:

- torque $\pm 0.5\%$
- number of revolutions to consume measured mass of fuel $\pm 0.25\%$
- actual mass of fuel $\pm 0.3\%$

The theory of errors indicates that in such a case, in which each factor is involved to the power 1, the confidence limit of the result is equal to the square root of the sum of the squares of the various factors, that is, in the present case the 95% confidence limit for the calculated specific fuel consumption is:

$$\sqrt{0.5^2 + 0.25^2 + 0.3^2} = \pm 0.58\%$$

## The number of significant figures to be quoted

One of the commonest errors in reporting the results of experimental work is to quote an inappropriate number of significant figures, some of which are totally meaningless, thus giving the illusion of accuracy once more, often engendered by the misuse of a spreadsheet formatting.

Consider a measurement of specific fuel consumption, readings as follows:

| | |
|---|---|
| Engine torque | 110.3 N m |
| Number of revolutions | 6753 |
| Mass of fuel | 0.25 kg |

Specific consumption equals:

$$\frac{3.6 \times 10^6 \times 0.25}{2 \times \pi \times 6753 \times 110.3} = 0.1923 \text{ kg/kW h}$$

If we assume that each of the three readings have a 95% confidence limit, then the specific fuel consumption has a 0.58% limit that is between 0.1912 and 0.1934 kg/kWh; therefore it is meaningless to quote the figure closer than 10192 kg/kWh.

## Absolute and relative accuracy

In a great deal of power train and vehicle test work—perhaps the majority—we are interested in measuring relative changes and the effect of modifications. The absolute values of a parameter before and after the modification may be of less importance so it may not be necessary to concern ourselves with the absolute accuracy and traceability of the instrumentation. Sensitivity and precision may be of much greater importance. Some test engineers have unhealthy respect for the "repeatability" of results in a given test cell, while not considering that this can be almost completely disconnected from accuracy of results.

## The cost of accuracy: a final consideration

The cost of accuracy is a subject that frequently arises during discussions about the design of test facilities and test procedures. An engineering manager with responsibility for the choice of instrumentation should be aware of the danger of mismatch between what he thinks he requires and what is actually necessary for an adequate job to be done. The cost of an instrument to perform a particular task can vary by an order of magnitude; in most cases the main variable is the level of accuracy offered, and this should be compared with the accuracy that is really necessary for the job in hand.

## References

[1] JCGM 100:2008, GUM 1995 with minor corrections. Evaluation of measurement data—guide to the expression of uncertainty in measurement. <https://www.bipm.org/utils/common/documents/jcgm/JCGM_100_2008_E.pdf>.

[2] A.T.J. Hayward, Repeatability and Accuracy. ERA Technology Ltd; New edition 1992 ISBN-13: 978-0700804917

[3] P.J. Campion, J.E. Burns, A. Williams, A code of Practice for the Detailed Statement of Accuracy, HMSO, 1973. National Archives ref. DSIR 31/22.

# Index

*Note*: Page numbers followed by "*f*" and "*t*" refer to figures and tables, respectively.

## D

## E

## G

## N

## Q



## R

## S

## U

## V



## W

## Z

Printed in the United States
By Bookmasters